River Basin Management in the Twenty-first Century

Understanding People and Place

River Basin Management in the Twenty-first Century

Understanding People and Place

Editors

Victor R. Squires
Visiting Professor, College of Grassland Science
Gansu Agricultural University, Lanzhou
China and International Consultant
Adelaide, Australia

Hugh M. Milner
Water Management Consultant
Northbridge, NSW
Australia

Katherine A. Daniell
Research School of Social Sciences and
Crawford School of Public Policy
The Australian National University, Canberra
ACT, Australia

CRC Press
Taylor & Francis Group
Boca Raton London New York

CRC Press is an imprint of the
Taylor & Francis Group, an **informa** business
A SCIENCE PUBLISHERS BOOK

CRC Press
Taylor & Francis Group
6000 Broken Sound Parkway NW, Suite 300
Boca Raton, FL 33487-2742

Cover illustrations

- The photograph of the reservoir at the top left hand corner and the photograph at the top right hand corner of the montage are reproduced by courtesy of Water Resources Bureau of Gansu, NW China.
- Other illustrations were provided by various authors in this book and are reproduced with their kind permission.

Printed on acid-free paper

International Standard Book Number: 978-1-4665-7962-0 (Hardback)

Library of Congress Cataloging-in-Publication Data

River basin management in the twenty-first century : understanding people and place / editors: Victor Squires, Hugh M. Milner, Katherine Daniell.
pages cm
Includes bibliographical references and index.
ISBN 978-1-4665-7962-0 (hardcover : alk. paper) 1. Watershed management. 2. Watersheds. I. Squires, V. R. (Victor R.), 1937- II. Milner, Hugh M. III. Daniell, Katherine A., 1981-

TC413.R58 2014
333.91'62--dc23

2014018144

Visit the Taylor & Francis Web site at
http://www.taylorandfrancis.com

CRC Press Web site at
http://www.crcpress.com

Science Publishers Web site at
http://www.scipub.net

Scope and Purpose

Scope and Purpose

The book is written by leading authorities on the current and latent issues relating to river basin management (RBM) and brings to the reader an up to date analysis and provides a window into this important subject. A key aspect of the work is that of achieving *balance*. Technical approaches can address some issues such as hydropower generation, flood mitigation, provision of water for irrigated agriculture to feed burgeoning populations, recreation and navigation, etc. but balancing these is the negotiated outcome of social processes. Ultimately it is these social processes that are the hard part, and often the stumbling block, for improving RBM.

This book is divided into four Parts.

PART 1 IWRM—Principles and Practices

Integrated Water Resources Management (IWRM) is seen by many as the way forward towards food, water, and energy security. It ideally offers a way and means to better manage water across a territory; managing demand and using limited water more efficiently; adopting new policies in order to cope with climate change and variability, including flood and drought events; increasing water supply through the use of non conventional water resources, especially waste water; adopting approaches of stakeholder participation and information exchange; and raising public awareness of the value of water.

This Part comprises 6 chapters detailing experience with IWRM formulation, transfer, adaptation and implementation in both developed and developing countries. It draws on the rich experience of water management practitioners and puts into sharp focus the strengths and limitations of the IWRM approach. ***Mukhtarov*** **and** ***Cherp*** take a global perspective. ***Maurel et al.,*** provide insights into how IWRM can be merged with territorial development to better account for people and place in river basin management. ***Mitchell*** focuses on experiences from Canada while ***Ffolliott*** **and** ***Brooks*** give a synoptic overview of experience from USA. The little known situation in China's arid north west, where inland rivers predominate, is outlined by ***Li* and *Squires*** in their study on the Shule River basin in north-west China, and ***Marr* and *Raut*** examine issues and experiences in working with local farmers in India.

PART 2 Transboundary River Management and Politics

No problem is so vexed as how to manage rivers that pass through several jurisdictions, especially international transboundary rivers. It has been predicted that access to water will create conflict between countries, even if initial conflict eventually leads to heightened cooperation. In Africa, central Asia, west Asia and the Americas, some countries are already arguing fiercely over access to rivers and inland seas, and confrontations could arise as water shortages grow. Countries currently or potentially involved in international disputes over access to river water and aquifers include: Turkey, Syria and Iraq (the Tigris and Euphrates Rivers); Israel, Jordan, Syria and Palestine (the Jordan River and the aquifers of the Golan Heights); India and Pakistan (the Punjab Rivers); India and Bangladesh (the Ganges and Brahmaputra Rivers); China and South-East Asian countries (the Mekong River); Tajikistan, Kyrgyzstan and Uzbekistan (the Amu Darya and Syr Darya Rivers); Ethiopia, Sudan and East African riparian countries, including Kenya, Tanzania, Rwanda, Burundi, Uganda and Egypt (the Nile River) and Iran and Turkmenistan over the Atrek River and Caspian sea.

The 4 chapters here draw on examples from many countries. ***Hassenforder* and *Noury*** examine 8 case studies on transboundary water management projects drawn from their work in 4 continents, ***Kibaroglu* and *Ahmetova*** address the real life issues in the Tigris-Euphrates river basin, ***Sullivan*** deals with the largest river in Southern Africa that rises in Lesotho, flows across south Africa and enters the Atlantic ocean via Namibia. Water management policy and practice in the Nile River system receives scrutiny from ***Thuo* and *Riddell*.**

PART 3 Water Management Policy, Politics and Economics

Water, especially freshwater, is such a vital resource. Policies and projects focused on freshwater ecosystem alterations have been carried out through much of modern history, with the intensity of modifications increasing in the early to mid-1900s. Common waterway modifications, such as the construction of dams and irrigation channels, inter-basin connections and water transfers, can impact on the hydrology of freshwater systems, disconnect rivers from floodplains and wetlands, and decrease water velocity in riverine systems. This, in turn, can affect the seasonal flow and sediment transport of rivers downstream, impacting on fish migrations and changing the composition of riparian ecosystems. All of these issues require a balanced approach to their resolution. Legislation, policy formulation and the role of socio-economic forces are all part of the complex matrix that represents modern day responses to increasing demand for water and its dwindling supply relative to global population.

The five chapters examine aspects of the responses of societies concerned about ensuring a continuing supply of freshwater to service the needs of agriculture, industry, domestic use and the environment. ***Du et al.,*** use the Yellow River Commission in China as a case study of how a large but mainly arid country supporting the world's largest human population has legislated to manage and allocate water from one of the world's longest rivers, ***Loch et al.,*** elaborate on the issues and conflicts involved in managing Australia's largest river system that services water users in five separate jurisdictions. ***Xu et al.,*** present an analysis of how China has tackled the management of a large inland river basin in an arid part of north-west China, and ***Krutov et al.,*** summarize the present situation in the Aral Sea basin and examine the role of the Republic of Tajikistan in the Inter-state Aral sea commission. ***Kingsford et al.,*** analyze the issues in the Lake Eyre basin in central Australia.

PART 4 People and Place

The successful implementation of river basin management, integrated or not, depends on how the local stakeholders (urban, rural, industrial, environmental, etc.) behave in relation to water. It depends on their perceptions of the role of the water. Place is paramount because inevitably those upstream will have different priorities from those downstream in a river basin. Agriculture, collectively the world's biggest user of freshwater, lays claim to vast quantities of water to produce food for the worlds' burgeoning population. Increasing awareness of impending water shortages (at crisis point in some countries or regions) and the concern about ecosystems invariably lead to a clash of opinions over water allocation priorities. Some of these issues are dealt with in this Part of 5 chapters. ***Squires*** looks at the pivotal role of people and the importance of place. The national water policies in Nepal are outlined and assessed by ***Pradhan et al.,*** while ***Wenger*** looks at the consequences of devastating floods, often exacerbated by human interference with rivers, and the lessons drawn from experiences on four continents. ***Plant et al.,*** then investigate the importance of information and communication arrangements for people working together across the Thau water territory in France. Finally ***Daniell, Milner* and *Squires*** provide an overview of a number of key issues raised in this book.

There are no quick or easy solutions to the complex land and water problems faced by many countries. If this book can help in the process of advancing better RBM, we will feel truly rewarded.

The Editors are grateful to Raju Primlani and the team in the Editorial Department for their help and support in the preparation of the manuscript for publication and to our colleagues who provided feedback and advice on the earlier drafts. Thanks are due to those who prepared or upgraded the graphics or supplied photos.

Victor R. Squires, Adelaide
Hugh M. Milner, Sydney
Katherine A. Daniell, Canberra

Contents

PART 2: Transboundary River Management and Politics

PART 3: Water Management Policy, Politics and Economics

List of Contributors

Adamson, David
Senior Research Fellow, Risk and Sustainability Management Group, School of Economics, Colin Clark Building, The University of Queensland, St Lucia QLD 4072.
Email: d.adamson@uq.edu.au

Ahmetova, Radkha
The Boston Consulting Group, Istanbul, Turkey.
Email: aysegul.kibaroglu@okan.edu.tr

Barreteau, Olivier
Irstea, G-Eau Lab Montpellier, 361, rue J.F. Breton B.P. 5095 34196 MONTPELLIER Cedex 5 France.

Bertacchini, Yann
I3M—Information, Milieux, Médias, Médiations Lab, University of the South Toulon-Var, University Avenue B.P. 20132 83957 LA GARDE Cedex France.

Bhawani Shankar Dongol
Senior Program Officer, KRBM-SHL, WWF Nepal, PO Box 7660, Kathmandu Nepal.
Email: bhawani.dongol@wwfnepal.org

Brooks, Kenneth N.
Professor Emeritus, Department of Forest Resources, University of Minnesota, 621 Wildwood Lane, Stillwater, Minnesota, 55082.
Email: brook007@umn.edu

Chao, Zhuang
Research Institute of Environmental Law, Law School, Wuhan University, Hubei, 430072 P.R. China.

Cherp, Aleh
Professor at the Department of Environmental Sciences and Policy, Central European University, Hungary and Associate Professor at the International Institute for Industrial Environmental Economics at Lund University.

Costelloe, J.F.
University of Melbourne, Victoria, Australia.

Daniell, Katherine A.
Research Fellow, HC Coombs Policy Forum, ANIPP, Crawford School of Public Policy & Centre for Policy Innovation, Research School of Social Sciences, The Australian National University, Canberra ACT 0200, Australia.
Email: katherine.daniell@anu.edu.au

Du Qun
Professor, Research Institute of Environmental Law, Law School, Wuhan University, Hubei, 4340072 P.R. China.
Email: qdu@wuhan.edu.cn

Ffolliott, Peter F.
Professor Emeritus School of Natural Resources and the Environment, University of Arizona, 2107 Calle de la Cienega Tucson, Arizona, 85721 USA.
Email: ffolliott@ag.arizona.edu

Hassenforder, Emeline
AgroParisTech, IRSTEA G-EAU and Australian National University, Canberra, Australia.
Email: emeline.hassenforder@irstea.fr

Kamolidinov, Anvar
Senior Research Fellow of the Tajikistan Branch of the Scientific Information Centre of Interstate Commission for Water Coordination (SIC ICWC), 16, Behzod street, Dushanbe, 734013 Tajikistan.
Email: anvarkamol@gmail.com

Kibaroglu, Aysegul
International Relations Department, Okan University, Istanbul.
Email: aysegul.kibaroglu@okan.edu.tr

Kingsford, Richard T.
Centre for Ecosystem Science, University of NSW, SYDNEY 2052, Australia.
Email: richard.kingsford@unsw.edu.au

Krutov, Anatoly
Principal Research Fellow, State Oceanographic Institute, Moscow, Russian Federation.
Email: ankrutov@yahoo.com

Li Yuhong
Department of Water Resources Gansu Province, Lanzhou, P.R. China.
Email: yhli2000@aliyun.com

Loch, Adam
Early Career Development Fellow, Centre for Regulation and Market Analysis, School of Commerce, The University of South Australia, GPO Box 2471, Adelaide, SA 5001 Australia.
Email: adam.loch@unisa.edu.au

Marr, Andrew
SMEC International Pty Ltd, PO Box 1052, North Sydney NSW 2059 Australia.
Email: andyjmarr@gmail.com

Maurel, Pierre
Irstea, G-Eau lab Montpellier, France.
Email: pierre.maurel@teledetection.fr

Milner, Hugh
Freelance International Water Resources Management Consultant, Northbridge NSW Australia.
Email: hugh.milner@gmail.com

Mitchell, Bruce
Department of Geography and Environmental Management, University of Waterloo, Ontario, Canada N2L 3G1.
Email: mitchell@uwaterloo.ca

Mukhtarov, Farhad
School of Public and International Affairs, Azerbaijan Diplomatic Academy, Azerbaijan; Researcher at the Department of Biotechnology, Delft University of Technology, The Netherlands.
Email: fmukhtarov@gmail.com

Neera Shrestha Pradhan
Water and Adaptation Specialist, International Centre for Integrated Mountain Development (ICIMOD), GPO Box 3226, Kathmandu, Nepal.
Email: nspradhan@icimod.org

Noury, Benjamin
Consultant, Transboundary water issues, Oxyo-water, France.
Email: benjamin@oxyo-water.com

Plant, Roel
Irstea, TETIS Lab, Montpellier, France/ISF, and University of Technology of Sydney, Australia.
Email: Roelof.Plant@uts.edu.au

Rahimov, Sulton
First Deputy Minister of Land Reclamation and Water Resources of the Republic of Tajikistan, Dushanbe, 734054 Tajikistan.
Email: r.sulton@gmail.com

Raut, Ashok
SMEC International Pty Ltd, PO Box 1052, North Sydney NSW 2059 Australia.

Riddell, Phil
Freelance International Water Policy and Management Specialist, Pays de Gex, France; currently Water Resources Management Adviser at the Ministry of Electricity, Dams, Irrigation and Water Resources, South Sudan.
Email: philriddell@hotmail.com

Sheldon, Fran
Australian Rivers Institute, Griffith University, Brisbane, Australia.
Email: f.sheldon@griffith.edu.au

Spooner, Simon
Senior Consultant, Atkins Global, UK.
Email: simon.spooner@atkinsglobal.com

Squires, V.R.
College of Grassland Science, Gansu Agricultural University, Lanzhou, P.R. China and International consultant, Adelaide, Australia.
Email: dryland1812@internode.on.net

Sui Lichun
School of Economics, Lanzhou University, Lanzhou, 730000, P.R. China.

Sullivan, Caroline A.
School of Environment, Science and Engineering, Southern Cross University, NSW, Australia.
Email: caroline.sullivan@scu.edu.au

Suman Sijapati
President, International Network on Participatory Irrigation Management, P.O. Box 8975 EPC 1115, Kathmandu, Nepal.
Email: sijapati@wlink.com.np

Thuo, Simon
Policy & Strategy advisor on water resources management; Institutional and social inclusion specialist for water development, Kenya, Rwanda and Uganda; Member, Global Water Partnership Reference Group-technical support for water, climate & development.
Email: simonthuo@gmail.com

Umesh Parajuli
Freelance Consultant, Water Resource Management Engineer, 132 Ichchu Nadi Marga, Baluwatar, Kathmandu, Nepal.
Email: unparajuli@gmail.com

Wenger, Caroline
Australian National University, Canberra, Australia.
Email: caroline.wenger@anu.edu.au

Wheeler, Sarah
Senior Research Fellow, Centre for Regulation and Market Analysis, School of Commerce, The University of South Australia, GPO Box 2471, Adelaide, SA 5001.

Xu Hui
School of Economics, Lanzhou University, Lanzhou, 730000, P.R. China.

Zhang Dawei
State Key Laboratory of Grassland and Agro-Ecosystems, School of Life Science, Lanzhou University, Lanzhou, 730000, P.R. China.

List of Figures

List of Tables

List of Boxes

Preamble

An Introduction to People and Place in River Basin Management

Victor R. Squires,[1,*] Katherine A. Daniell[2] and Hugh M. Milner[3]

Integrated Water Resources Management (IWRM) is seen by many as the way forward towards food, water, and energy security. It ideally offers ways and means to better manage water across a territory; managing demand and using limited water more efficiently; adopting new policies in order to cope with climate change and variability including flood and drought events; increasing water supply through the use of non conventional water resources and managing water quality, especially of waste water; adopting approaches of stakeholder participation and information exchange; and raising public awareness of the value of water.

River Basin Management (RBM), a subset of IWRM, can be characterized in a number of ways, although it often entails working through sets of trade-off decisions, where the potential (or very real) benefits from one choice are relinquished in favour of another choice that is perceived as more desirable or beneficial (see Loch et al., this volume; Mitchell, this volume). The development of RBM decision-making processes have been given further weight in recent decades by public policy requirements to satisfy triple-bottom-line (i.e., economic, social and environmental) objectives. Sustainable management is an often-stated but difficult to define objective of management of any river basin. It entails judicious

[1] Visiting Professor, College of Grassland Science, Gansu Agricultural University, Lanzhou, China and International Consultant, Adelaide, Australia.

[2] Fellow, HC Coombs Policy Forum, ANIPP, Crawford School of Public Policy & Centre for Policy Innovation, Research School of Social Sciences, The Australian National University, Canberra ACT 0200, Australia.

[3] Freelance International Water Resources Management Consultant, Northbridge, NSW Australia.

* Corresponding author: dryland1812@internode.on.net

governance arrangements to manage alternative choices between economic, environmental, social, cultural, and other outcomes that will support and allow future generations to gain similar or greater value from the basin, its water and environment, as current generations.

People and place are major determinants of success of any measures taken. There will be continuing struggle between the needs of people (in terms of economic and social dimensions) and the needs of place (in terms of the environment). Such a struggle is made even more complicated by the fact that the needs of place overlap with economic and social dimensions as well (see Squires, this volume). We characterize *people* as the economic, social and community aspects of RMB, while *place* is comprised of environmental attributes that vary from one river basin to another, but also within river basins, with important aspects including upstream users vs. downstream or the reconciling the water needs of urban vs. rural (mainly irrigated agriculture). Water allocation is another dilemma. Retaining a balance between people (i.e., economic focus on water market and rural social welfare improvements via infra-structure projects) and place (i.e. environmental flow provision) issues puts pressure on RBM authorities.

Over-allocation presents an issue when the full volume of water that is able to be extracted at a point in time exceeds an environmentally sustainable level of extraction. As the various case studies in this volume show, various countries have adopted market-based instruments as a means of reallocating water between various users. Effective markets manage rising demand for water among competing users and ensure that it is used efficiently for desired ends, while still promoting environmental, economic and social sustainability. Managing a river basin as a single entity, rather than its management by various local administrations, has the potential to facilitate recognition of social, economic and environmental factors. However, it should be remembered that management of the basin remains an inherently political process, and it would be naive to assume that technical information alone will be sufficient to arbitrate disputes between key stakeholders; notably those related to the economic, social, cultural and environmental aspects within the basin.

At a time when populations are rapidly rising, demand for food is on the increase and climate change is setting in, it is inevitable that water for agriculture must have high priority. There needs to be more market-based instrument policy development to increase farmers' adaptive capacity to climate change. The efficiency and effectiveness of on and off-farm irrigation infrastructure investment need careful reconsideration (e.g., because of return-flow issues, increased future energy costs, increased future water charges). The increasing over-extraction of groundwater and its use as a substitute for surface water allocations will also gain increasing prominence over the next few decades. There will be an increasing need for markets to offer new products that optimize flexibility in water use for both irrigators, other community members (including urban residents) and the environment. Such developments include trade in allocations for environmental flows, counter-cyclical trade between irrigation and environmental water holders, option contracts in both rural and urban markets.

Trading requires well defined and administered water rights, and comprehensive public information on water resources, water environments and water use. In many river basins these conditions are not yet met and introducing water trading systems only threatens sustainability by encouraging exploitation by powerful and elite groups. Until low level water rights and information systems are established, control of water use should be exercised by authorities that have access to information and capability to understand the changes which will be induced in natural water systems. Checks should be encouraged through regular public status reports of the river system by these authorities so that judgment can be made on how the 'triple bottom line' is being met.

Climate change greatly exacerbates the effects of mismanagement while introducing major additional challenges of its own. Sea level rise for instance could result in salinization and, in extreme cases, permanent inundation of major food producing areas, especially in South and South East Asia but also in the Nile delta, regions which are characterized by vast areas of cropland in the coastal lowlands. Even where permanent inundation is not a risk, climate change induced storm surges could result in catastrophic flooding, sometimes with saline water, of the same areas. Sea level rise could also compromise the sustainability of essential ecosystems in the coastal shallows, ecosystems on which, as we have already seen, major food chains depend. Increasing temperature (and the other factors mentioned above) will also affect these ecosystems—hence the claim regarding the need to find alternative sources of food for the 1 billion people that currently depend more or less directly on them for their food security.

The severity of floods and droughts is forecast to increase under the influence of more powerful heat-driven climate systems. Management of extreme conditions will be put to the test.

Climate change is also causing glaciers both to retreat and to thaw early (Li & Squires, this volume). Glacial retreat, which results from annual melt rates that are greater than annual precipitation rates, compromises the long term ability of glaciers to supply water using sectors downstream. These crucially include large areas of irrigated agriculture that have hitherto depended on such glacial melt such as those in Central Asia (see Krutov et al., this volume). That this is a problem will be obvious, but there is something else. Early thaws mean that instead of being used for productive purposes, significant amounts of water leave the system before ambient temperatures downstream are high enough for crops to be planted. Thus not only is the overall glacial resource diminishing, the usefulness of that which remains is also diminishing—a double whammy as it were. A similar picture is also emerging with respect to rainfall both in terms of its total quantum but also in its seasonal distribution and intensity. Water supplies depend on runoff from mountains, much of which originates as snowmelt in forested watersheds. Water is one of the critical ecosystem services provided by forests and grasslands. Vegetation has a large effect on the water budget through both transpiration and interception, with an inverse relationship between forest cover and streamflow demonstrated for many forested landscapes.

In snow-dominated, forested catchments, water yields are affected by the energy budget of the forest, which determines the accumulation and melt characteristics of the snowpack, and by the magnitude of evapotranspiration, that is, amount of vegetation; both can be manipulated by forest management. The impact of any reduction of vegetation on snow accumulation is significant and is a result of the combined effect of interception loss and alteration of the depositional pattern. Clearing of forests may lead to accelerated soil erosion and increased sedimentation.

Because changing climate is expected to dramatically affect the amount and seasonal distribution of rainfall and snowpack, land managers have acknowledged the need for new strategies and effective approaches to address these changes.

Although experts are not as yet fully confident in the convergence of their models, especially with respect to inland continental areas, the degree of consistency that is emerging shows that many important food producing areas (existing or potential) will become hotter and drier. Thus the overall water resource represented by precipitation that results in usable runoff will trend downwards. Additionally, in typical situations, precipitation events are expected to become more intense rendering the water less manageable, in that less of the water that does fall is retained in the root zone, the natural drainage systems (including any aquifers) and artificial storage dams—the double whammy once again. Conversely, where rainfall is expected to increase—and this is expected in some important food producing areas—there is the associated risk of flooding, thereby increasing either the risks or costs associated with sustainable agriculture and having implications for urban and industrial areas downstream.

Sustainable water management and allocation is a complex issue involving legal, governance, institutional, policy and economic factors, as well as wise use practices. But this does not obviate the need to get it right. The effects of poor water management and allocation on food security can be summarized as follows.

Where irrigation is concerned, over-abstraction can lead to water logging and hence long-term or permanent soil deterioration and reduced productivity. It furthermore reduces environmental stream flows, which clearly reduces access to water for productive (as well as other) uses by downstream stakeholders. These uses include not only more irrigation, but also capture fisheries which in some locations have a vital food security role. Capture fisheries are also severely compromised as a result of gene pool degradation when water bodies (including rivers) become fragmented due to badly planned storage, over-abstraction or wastage of water.

Excessive sedimentation has a detrimental effect on marine fisheries. This should not be taken to mean that sedimentation is bad per se. In fact in most large river systems the opposite is true. The vast, complex and usually economically significant food chains living within these systems have generally evolved on the basis on an annual flood and turbidity cycle. Disruption of these cycles by means of badly planned dams, excessive abstractions and unseasonal sediment loads (which can be less than required as well as more) can have a catastrophic effect on fisheries on both the rivers themselves and the marine environments into which they discharge.

Finally on sediment, is the fact that natural sediment loads carried during normal flooding usually increase fertility when left in the soil when the flood recedes. It was this very benefit that sustained Egypt as a superpower for thousands of years. Construction of the Aswan High Dam means that this sediment no longer reaches the farms along the Nile Valley upstream (Thuo and Riddell, this volume). Farmers now depend on expensive artificial fertilizers to do the job. When these are unaffordable, food security suffers; and when they are financially affordable, use tends to be excessive, with predictable environmental cost increases. To compound the difficulty of devising and implementing better RBM is the need to provide for trans-seasonal storage in many cases due to the seasonality of river flow, and take account of increasing potential for competition at the point of use, and sensitive transboundary issues.

As river basins have been developed with greater levels of infrastructure and intensity of agriculture, urbanization and industry, the need to move beyond traditional local-level and individual governance of water and land access and rights has increased. Large infrastructure systems such as dams, canals, pressurized water distribution systems and large-scale water treatment plants rely upon high levels of technical expertise for their effective management. Likewise complex water allocation arrangements, including environmental flow releases, the use of markets and legislation, can also not effectively function without high levels of monitoring and evaluation, and the associated capacity to enforce compliance. This has driven the development of large water bureaucracies in many places. However, especially for the latter issues of allocation, it has necessitated the need for increasing engagement with stakeholders, including agricultural, community, public sector and industry water and land managers across river basins and at different levels of government. The organizational challenges associated with the development of a range of participatory approaches that can be used to allow these stakeholders to negotiate decisions and work more effectively together is one of the most significant global water governance issues (Daniell 2012). To what extent such challenges can be overcome and how in each river basin is very much a matter for people, the specificities of the places they inhabit and the governance systems supported by the governments of the countries in which they lie.

References

Daniell, K.A. 2012. Co-engineering and Participatory Water Management: Organisational Challenges for Water Governance, Cambridge University Press, Cambridge, UK, 346 p.

Krutov, A., S. Rahimov and A. Kamolidinov. 2014. Republic of Tajikistan: Its Role in theManagement of Water Resources inthe Aral Sea Basin. pp. 325–345. *In*: V.R. Squires, H.M. Milner and K.A. Daniell (eds.). River Basin Management in the Twenty-first Century: Understanding People and Place. CRC Press Boca Raton.

Li, Y and V.R. Squires. 2014. Integrated Water Resources Management of Inland River Basins in the Hexi Corridor, Gansu, China with Special Reference to the Shule River Basin. pp. 82–100. *In*: V.R. Squires, H.M. Milner and K.A. Daniell (eds.). River Basin Management in the Twenty-first Century: Understanding People and Place. CRC Press Boca Raton.

Loch, A.S. Wheeler and D. Adamson. 2014. People versus Place in Australia's Murray-Darling Basin:Balancing Economic, Social Ecosystem and Community Outcomes. pp. 275–303. *In*: V.R. Squires, H.M. Milner and K.A. Daniell (eds.). River Basin Management in the Twenty-first Century: Understanding People and Place. CRC Press Boca Raton.

Mitchell, B. 2014. Addressing Implementation Deficits Related to IWRM in Canada. pp. 42–59. *In*: V.R. Squires, H.M. Milner and K.A. Daniell (eds.). River Basin Management in the Twenty-first Century: Understanding People and Place. CRC Press Boca Raton.

Squires, V.R. 2014. People and Place as Determinants of Success or Failure to Meet Management Objectives in River Basins. pp. 371–396. *In*: V.R. Squires, H.M. Milner and K.A. Daniell (eds.). River Basin Management in the Twenty-first Century: Understanding People and Place. CRC Press Boca Raton.

Thuo, S. and P. Riddell. 2014. Political Economy versus ComparativeAdvantagein the Nile Basin: Short Term Advantages or Long TermGain? pp. 216–243. *In*: V.R. Squires H.M. Milner and K.A. Daniell (eds.). River Basin Management in the Twenty-first Century: Understanding People and Place. CRC Press Boca Raton.

PART 1

IWRM—Principles and Practices

Integrated Water Resources Management (IWRM) is seen by many as the way forward towards food, water, and energy security. It ideally offers a ways and means to better manage water across a territory; managing demand and using limited water more efficiently; adopting new policies in order to cope with climate change and variability including flood and drought events; increasing water supply through the use of non conventional water resources, especially waste water; adopting approaches of stakeholder participation and information exchange; and raising public awareness of the value of water.

This Part comprises 6 chapters detailing experience with IWRM formulation, transfer, adaptation and implementation in both developed and developing countries. It draws on the rich experience of water management practitioners and puts into to sharp focus the strengths and limitations of the IWRM approach. ***Mukhtarov* and *Cherp*** take a global perspective, ***Ffolliott* and *Brooks*** give a synoptic overview of experience from USA while ***Mitchell*** focuses on experiences from Canada. The little known situation in China's arid north west, where inland rivers predominate, is outlined by ***Li* and *Squires*** in their study on the Shule River Basin in north-west China. ***Marr* and Raut** examine issues and experiences in working with local farmers in India and **Maurel et al.,** provide insights into how IWRM can be merged with territorial development to better account for people and place in river basin management.

1

The Hegemony of Integrated Water Resources Management as a Global Water Discourse

Farhad Mukhtarov[1,*] and Aleh Cherp[2]

SYNOPSIS

The early form of Integrated Water Resources Management (IWRM) emerged in the USA in the 1900s in order to manage interactions between water, land, eco- and social systems. By the end of the last century, IWRM has become a globally prominent policy concept. We concern ourselves with three questions, namely, a) "why did IWRM become a globally popular concept"?; b) "how did IWRM become a globally popular concept"?; and c) "what are the effects of IWRM being a globally popular concept"? We argue that this popularity can be explained in term of a neo-Gramscian concept of hegemony and the three-dimensional model of power. The hegemony of IWRM relies on: a) providing material incentives to engage with IWRM; b) directing normative persuasion in order to create and diffuse the norms; and c) building up organizational hierarchies to support IWRM

[1] Assistant Professor at the School of Public and International Affairs, Azerbaijan Diplomatic Academy, Azerbaijan; Researcher at the Department of Biotechnology, Delft University of Technology, The Netherlands.
Email: fmukhtarov@gmail.com

[2] Professor at the Department of Environmental Sciences and Policy, Central European University, Hungary and Associate Professor at the International Institute for Industrial Environmental Economics at Lund University.

[*] Corresponding author

planning. Using water management in Kazakhstan as a case study, we demonstrate some of the risks associated with an uncritical embrace of IWRM which may stem from its global hegemony.

Keywords: fragmentation, global water initiatives, holistic management, IWRM, Kazakhstan, neo-Gramscian, neoliberalism, technocratic elites, transnational actors, USA

1 Introduction

It is widely held that current practices of water resources management on the global scale are inadequate (Watkins 2006). The roots of this inadequacy lie not so much in poor financing or technology or natural scarcity of water, as in "poverty, inequality and unequal power relationships, as well as flawed water management policies that exacerbate scarcity" (Watkins 2006: 1). Many agree that the *fragmentation* of water management by sectors, resources and users is the main cause of the problem (e.g., UNEP 1994). In response to this consensus, the need for a holistic management approach has been advocated in the form of Integrated Water Resources Management (IWRM) presented as a tool for efficient, equitable and sustainable development and management of the world's limited water resources and for coping with conflicting demands for water (UN-Water 2008: 4).

Since its emergence in the early 20th century in the conservation movement of the American president Theodore Roosevelt, Integrated Water Resources Management (IWRM) has taken various forms and is currently embraced by most international policy actors. IWRM can be defined as the process which "promotes coordinated development and management of water, land and related resources, in order to maximize economic and social welfare in an equitable manner without compromising the sustainability of vital systems" (GWP-TAC 2000: 22).

Today, IWRM boasts global popularity and is currently being implemented in over 100 countries (UN-Water 2008; 2012). The UN-Water, United Nations coordinating mechanism for global water initiatives, monitors the progress towards IWRM planning and has produced two status reports (UN-Water 2008; 2012). In addition, the official mandate of the United Nations Commission for Sustainable Development (UN CSD) includes the facilitation and monitoring of the IWRM efforts globally (Baumgartner and Pahl-Wostl 2013). Promoted by international organizations, transnational actors and the Internet, IWRM ideas travel from the international to the national policy arena and are widespread on a global scale (Mukhtarov 2008).

Despite its sweeping popularity with academia, policy-makers and general public, there is little agreement on what IWRM actually constitutes

(Biswas 2004a; 2004b). There is an on-going debate on the basic meaning, utility, scope and nature of IWRM and over thirty IWRM definitions can be found in the literature (Saravanan et al. 2009; Mukhtarov and Gerlak forthcoming).

There are many fronts on which the IWRM debate is taking place. The supporters of IWRM, such as Mitchell (1990; 2005) and White (1998), argued for the need of reform towards more integration and coordination in water management, however challenging this task may be. At the same time, donors and think-tanks continue to promote IWRM planning and monitoring procedures (Baumgartner and Pahl-Wostl 2013: 3). Those opposing IWRM, in turn, argued that the beliefs in integration are 'idealistic' (Walther 1987), that integration is inevitably impeded by the politics (Saravanan et al. 2009), and that IWRM has degenerated into a policy 'buzzword' which served technocratic elites who continue business as usual under the new banner (Biswas 2004b; 2008).

One of the most contested is the relationship of IWRM with the neoliberal approaches to water governance, which can be defined as "a politically guided intensification of market rule and commodification" (Brenner et al. 2010: 184). The privatization of water supply and sanitation services and the greater role of the private sector in irrigation and hydropower have often been mentioned under the banner of IWRM (ICWE 1992; Conca 2006). Another heated debate is about the importance of public participation in IWRM and whether it is a façade promoted by IWRM for what is mostly technocratic measured. While the ethos of participation is strong in the discourse of IWRM supported by the Dublin Principles (ICWE 1992) and various Global Water Partnership (GWP) guidelines, there is no literature known to us which would examine the extent to which the views of stakeholders have been accommodated. The recent literature on IWRM has argued that the debate has been polarized "with theoreticians and donors on the one side promoting and requesting IWRM definitions, plans and monitoring procedures, and practitioners on the other side who are torn between living up to the expectations of donors and simply trying to "get on with their job"" (Baumgartner and Pahl-Wostl 2013). Despite this polarization, new attempts to define and frame IWRM still emerge; for example Groenfeldt (2013: 14) suggested that a new concept of "*water ethic*" must be built upon IWRM as it "incorporates a holistic view of water which gives particular recognitions to environmental sustainability social welfare, and governance arrangements".

The diversity of definitions of IWRM and the heat of the debate around this concept are not surprising *per se*. What is striking is that despite being vaguely defined and lacking any solid proof of effectiveness on the ground, IWRM has become popular on the international water policy arena (Biswas 2004b; 2008; Mollinga et al. 2006). The ubiquitous scope of IWRM's implementation under the conditions when it is not well-defined and agreed upon, and the policy expectations attached to IWRM nevertheless make it important to understand what is behind the dominant position of this policy discourse in the water policy arena. There are numerous studies on IWRM implementation and effectiveness. However, very few studies attempt to explain the popularity of IWRM worldwide with a systematic and theoretically grounded framework (e.g., Biswas 2004a).

This paper examines the reasons for the strong global stature of IWRM by looking at the academic and policy literature which discusses global water governance. We concern ourselves with three questions, namely, a) "why did IWRM become a globally popular concept"?; b) "how did IWRM become a globally popular concept"?, and c) "what are the effects of IWRM being a globally popular concept"?

In Section 2, we discuss the theoretical framework for understanding the domination of IWRM, namely the neo-Gramscian ideas on hegemony. Section 3 makes the case for the hegemonic status of IWRM by discussing its prominence globally, whereas Section 4 addresses the process and mechanism through which IWRM has come to dominate the global water discourse. In Section 5, we illustrate the hegemonic status of IWRM and the impacts it produces at the national level with an example of Kazakhstan national water policy. Finally, Section 6 discusses the implications of the hegemonic status of IWRM for Kazakhstan and broadly and concludes the paper with suggestions for the future research.

2 A neo-Gramscian Approach to Global Water Discourse

The phenomenon of integrated water management is certainly not new in water governance, various authors have argued for integration since the 1900s conservation movement led by Theodore Roosevelt (e.g., Hays 1959). What *is* new, however, is the centrality of integrated water resources management as a discourse in global governance of water resources. In other words, the multiple initiatives, events and efforts pursued by a wide array of global actors in the area of water management have coalesced around one single policy concept of IWRM, which gave it an additional symbolic and discursive dimension. A discourse theory is a promising approach to analyse and explain how certain policy concepts become prominent, or hegemonic in neo-Gramscian conception, to what effect, and what conditions and mechanisms are involved in this process.

2.1 Discourse theory and hegemony

We approach IWRM as a policy discourse, which in a broad sense can be defined as "an ensemble of ideas, concepts and categories through which meaning is given to social and physical phenomena, and which is reproduced through an identifiable set of practices" (Hajer 1995: 44). Discourse analysis accentuates the role of structure, institutions, symbols, identities, and language in constituting agents and shaping their preferences and behaviour.

Despite the rich literature on discourse theory and its application to the environment, no unified approach exists for examination of the path through which discourses become dominant (Mukhtarov and Gerlak 2013). Diverse approaches such as critical discourse analysis (Fairclough 1992), discursive policy analysis (Hajer 1995), and interpretative policy analysis (e.g., Yanow 2009) exist to approach policies as discourses. A prominent discursive approach to the study of politics is an adaptation of post-Marxist thought of an Italian politician and social theorist *Antonio Gramsci*. This approach focuses on explaining the dominance of some discourses over others and has been elaborated by Laclau and Mouffe (1985) and more recently by Howarth (2010), Newell and Levy (2005), and Newell (2008b).

The starting premise of this work is Gramsci's analysis of the dominance of one social class over another through the means of a mixed use of persuasion and coercion. *A neo-Gramscian perspective* on discourse analysis emphasizes the notion of *hegemony* and focuses on the description and explanation of how some discourses acquire *hegemony* over others. The notion of hegemony "refers to the alignment of material, institutional and discursive power" in such a way which favours certain coalitions of actors sustaining the status-quo (Newell 2009: 38). The persistence and the strength of hegemony as a form of power stems precisely from the submission of the dominated to the ruler, or rather, in the normalization in their eyes of the relationship of domination. The neo-Gramscian thinkers emphasize the importance of co-opting, or enrolling others rather than dominating them in an exercise of durable form of power (e.g., Cox 1993). According to Newell and Levy (2005: 50), hegemony is rooted in the institutions of civil society, such as the church, academia, and the media, which play a central role in ideological reproduction, providing legitimacy through the assertion of moral and intellectual leadership and the projection of a particular set of interests as the general interest.

This school of thought is also explicit about the mechanisms through which hegemony is achieved. Having developed the original thoughts of Gramsci to the area of discourse analysis, Laclau and Mouffe (1985) posit that hegemony is inseparable from the hegemonic coalitions of actors who coalesce in order to promote a certain system of power relations. For Gramsci, hegemony brings about "not only a unison of economic and political aims, but also intellectual and moral unity" (Gramsci 1971: 182). Thus, hegemony can be seen as the most sophisticated and durable form of power (Zeitoun 2008). Furthermore, as a lenses to study policy issues, the neo-Gramscian perspective draws attention to a combination of ideological means and material concessions by certain policy actors in order to build, sustain or block alliances and promote certain 'ways of knowing' policy issues (Fischer and Forester 1993).

Newell and Levy (2005) and Newell (2008b) have suggested that actors engage across three pillars of power in their struggle. On the *material level,* there are various rewards and punishments for compliance to the order of hegemony. On a *discursive level,* the frames of seeing reality are provided and sustained so that to maintain the relations of domination. And on an *organizational level,* coalitions between actors are built in order to support the ideational and material pillars of a certain discourse.

The three pillars approach to hegemony has been applied to explain the role of business in international environmental politics (Newell and Levy 2005), the dominance of agricultural biotechnology in India and China (Newell 2008a), as well as in Argentina (2008a), the spread of forest certification (Bloomfield 2012). Although in a slightly different context, the concept of hegemony has also been applied quite broadly to the study of transboundary water governance (Zeitoun and Warner 2006; Wegerich 2008), and discourses in water although under the title of 'sanctioned discourses' (Allan 2003). The next section uses the three pillar model of hegemony in order to explain the lasting prominence of IWRM as a discourse at the global water arena.

3 The Current Hegemony of IWRM as Global Water Discourse

As mentioned earlier, IWRM has received much attention during the last 15 years, since the Dublin International Conference on Water and the Environment in January 1992 and the Rio de Janeiro United Nations Conference on the Environment and Development in June 1992 (GWP 2005; Watkins 2006). A number of high-profile organizations have embraced IWRM throughout this period. Examples include the United Nations Development Programme (Watkins 2006), the United Nations Environmental Programme (UCC-IWRM), World Bank, the Asian Development Bank (2006), the World Water Council and the EU (EU Water Framework Directive). The concept was mentioned in the UNEP's Agenda 21 (Article 18) (UNEP 1994), as well as the UN World Summit on Sustainable Development (WSSD) in 2002.

The Johannesburg Plan of Implementation accepted at the Johannesburg World Summit on Sustainable Development in 2002 (WSSD) required that countries-signatories produce national IWRM and Water Efficiency plans by 2005. Thus, IWRM became an institutionalized international obligation,[1]

[1] Although formally binding, no enforcement mechanism have been discussed, and the Plan of Implementation remained legally toothless.

while UN-Water now conducts regular assessments of the planning progress world-wide. The latest UN-Water assessments of IWRM implementation have been carried out in 2008 and 2012 and show an increasing trend in countries adopting the concept (UN-Water 2008; 2012). As a summary of the dominating status of IWRM in the global water arena, Conca (2006: 127) wrote the following:

> IWRM has become *the* discursive framework of international water policy—the reference point to which all other arguments end up appealing. Much like a thoroughly picked-over concept of sustainability, IWRM combines intuitive reasonableness, an appeal to technical authority, and an all-encompassing character of such great flexibility that it approaches vagueness. … Vague or not, actors in each of the other institution-building venues analysed in this book routinely appeal to IWRM arguments, concepts, and rhetoric to bolster their respective positions.

One striking conclusion of scholars interested in IWRM is its panacea-like features (Ingram 2008; Pahl-Wostl et al. 2012). In many of policy reports, IWRM has been promoted as relevant and beneficial to implement regardless of the context and problems at hand. Another observation is that IWRM has acquired a life of its own as a symbolic concept (Molle 2006; Mollinga et al. 2006), and IWRM ideas may travel with very little regard to the problems on the ground which they are supposed to tackle. The hegemony of IWRM best presents itself in the promotion of the taken-for-grantedness of its universal relevance and policy value. At the same time, the meaning and interpretation of IWRM have remained very broad:

> What this, by no means exhaustive, list of examples of 'buy-ins' to the IWRM agenda suggests is that different categories of people appropriate the different meanings of 'integration' in different ways and for different purposes. This is only to be expected: the same is true for the participation and privatization notions, or any other policy concept (Mollinga et al. 2006: 30).

The hegemonic status of the IWRM discourse is also supported by the concept of 'the universal' in the thought of Laclau, which is to a certain extent similar to an idea of 'floating signifiers' of Levi-Strauss. In Laclau's words "(t)he universal is an empty place, a void which can be filled only by the particular, but which, through its very emptiness, produces a series of crucial effects in the structuration/restructuration of social relations" (Laclau 2000: 58). In that sense, IWRM is the 'universal' panacea-like concept which produces the effects of structuration of social relations favourable to technocratic and engineering elites. The IWRM then is a 'floating signifier', or a concept which is "lacking any sense (and) serve(s) to take on any meaning that is given them" (Callinicos 2007: 270).

Thus, it occurs that the ill-defined nature of IWRM and its malleability to take various meanings, in fact only contributes to its popularity at the global scale. Its appeal lies in the fact that it can act as a 'boundary object' (Jeffrey and Gearey 2006; Mollinga et al. 2006) to allow for discussing water resources among various actors with different backgrounds. At the same time, in addition to conceptual vagueness, it has the warm normative appeal of what Molle (2008: 132) called 'nirvana concepts': "(a)lthough, just as with nirvana, the likelihood that we may reach them is admittedly low, the mere possibility of achieving them and the sense of 'progress' attached to any shift in their direction suffice to make them an attractive and useful focal point".

The emptiness of IWRM in terms of the content has not only attracted a universal agreement on its acceptance as a 'lingua-franca' of global water governance, it has also played a role in smoothing up a number of sensitive conflicts in the area of water resources, such as the debate on neo-liberalization of water governance, the debate over the scale at which water resources are managed best, and about the roles and responsibilities of various policy actors. The contentious politics of water needed at least a semblance of a consensus at a global scale, which has finally precipitated around the notion of IWRM (Conca 2006; Baumgartner and Pahl-Wostl 2013).

However, understanding the hegemony of IWRM without discussing the actors and strategies involved in this process is impossible, and therefore the next section will deal with the process through which IWRM has achieved its hegemonic status.

4 The Pathway to IWRM Hegemony: A Three Dimensional Model of Power

Among the actors who actively promote IWRM are the International Network of Basin Organizations (INBO), GWP, World Water Council (WWC), and United National Development Programme's (UNDP) initiative of capacity building (CAP-Net), who legitimize and institutionalize the norms of integration as part of good water governance. Global actors may pursue various strategies in order to promote a particular discourse, or frame a discourse to reflect changing values of civil society, leverage incentives for stakeholders to buy in certain discourses, construct and market best practice examples, and link to other discourses and concepts (Mukhtarov and Gerlak 2013). Using the three-pillar model of hegemony developed by Newell (2008b), we can illustrate below how IWRM has been advanced on three grounds: the material, the ideational and the organizational.

On the *material* side, IWRM provides better access of actors who adopt it to international funding: preparation of IWRM national plans is often funded by international organizations or research councils. This could be observed in dozens of IWRM projects implemented with donor funding in Central Asia and Africa (Mukhtarov 2009; UN-Water 2012). At the same time, the framing of IWRM as a cheaper solution also contributes to material incentives for its adoption. For example, the UK Department for Environment, Food and Rural Affairs (UK DEFRA) sent a clear message that the 'Making Space for Water' programme for integrated flood risk management is cheaper than conventional engineering flood defence (UK DEFRA 2005). As Mukhtarov (2009) illustrated by Kazakhstan's national IWRM planning and the narrative of sustainable human development adopted in the South-eastern Anatolia Project in Turkey, material incentives also exist for individuals championing IWRM in national contexts. IWRM advocates gain access to international expert and policy networks, and boost their prestige at home. Subsequently, these champions are imitated and change rhetoric and practice in their national and local networks of influence.

Apart from the material incentives, the strong *ideational* appeal makes it hard to ignore IWRM for nation states and individuals. Often IWRM represents entrance of a country into a community of progressive (western) nations concerned with the current state of water resources management and eager to improve it (Tarlock 2008), just as the dams in the 20th century symbolized the modernity era and the imagined victory of humanity over nature. On the ideational side, linking IWRM to sustainability and other values, has acquired a strong normative power and the "taken-for-grantedness" that presumes that IWRM is good under all conditions (Ingram 2008). An important part of the ideational hegemony of IWRM is the constant presentation of normative guidelines on implementation and the success stories. The IWRM Tool-Box released in 2003 and regularly refreshed since then represents the epitome of normative guidance on 'how-to-do' approaches to the concept. The Tool-Box provides 'best practice' examples and models of IWRM in various contexts and sectors. The recent INBO handbook on IWRM and the Leibniz Institute for Regional Development and Spatial Planning (IRS) guidelines on implementing IWRM (Beveridge et al. 2012) provide further examples of support to promoting the image of IWRM as a desirable, implementable and a proved manner of dealing with water resources sustainably.

The *organizational* pillar of the IWRM hegemony is represented by international organizations and formal and informal networks which facilitate and develop this discourse. The hegemony of IWRM thus comes from the proliferation of professional membership organizations, specialized publications, professional journals, international congresses, technical meetings and issue-oriented global summits, bringing to both ideational domination and the organizational manifestation of the hegemony (Conca 2006: 132; Zeitoun 2008). The UNESCO's International Hydrological Programme prepared a draft report of IWRM in river basins, sub-basins and aquifers. It stated the following:

> The new organizations still do not have real influence globally to assist co-ordination or actions on a global scale. … As a result, there is no entity in the world that stands out as the leader in co-ordinating knowledge of IWRM actions. As a result, there are many dispersed efforts that are not strong or effective. Even those of INBO (International Network of Basin Organizations) fall mostly in the category of "event publicity" and have no real basis for co-ordination (UNESCO-IHP 2007: 29).

Thus, the organizational pillar of the IWRM hegemony seems to be the weakest, incomplete and would require further strengthening if the IWRM is to retain its dominant status. A recent study of the global water governance by Baumgartner and Pahl-Wostl (2013) further corroborated on our view by arguing that "UN-Water has just started to scratch on the surface of the issue" referring to the rise of the IWRM discourse. They have argued that while the involvement of UN-Water is welcome and the promotion of IWRM falls within its mandate, coordination with GWP and other global actors is necessary to avoid duplication of tasks and rivalry.

Not only the three pillars of power have contributed to the rise of IWRM, the historical conditions in the 1990s and 2000s have been favourable as well. First of all, there was a clear *institutional vacuum* at the global level in terms of policies, legal regimes or frameworks involving in-land water management amidst the greater recognition of water as a global issue of extreme importance (Varady and Iles-Shih 2005; Conca 2006). By the 1990s, the UN-designated periods, events and other initiatives had not resulted in any consistent strategy to deal with diverse water problems (Varady and Iles-Shih 2005; Conca 2006). Thus, there was an acute need to accommodate deep conflicts over fundamental issues regarding water. In other words, there was a need for a *"consensus"* on global water governance which required an agreed upon policy concept. Secondly, expert networking and "conferencing" had built-up by the 1990s and resulted in the increasing professionalization of the water policy field. This resulted in the positioning of IWRM on the political agenda as a distinct subject where management and economics played as important role as engineering. Thirdly, and most importantly, the sustainability discourse created a window of opportunity for IWRM to become popular, as it is still often conceived as a mere extension of sustainability thinking in the water sector. Thus, the convergence of the three pillars of hegemony and favourable historical conditions have catapulted IWRM into the global prominence by the early 2000s. At the same time, the global hegemony of IWRM has direct bearing on water planning at the national level, especially in cases with involvement of donors and international consultants. In the next section, we discuss some of the adverse impacts the uncritical embrace of IWRM may have at the national level.

5 The IWRM Hegemony in the Context of Kazakhstan

The water resources of Kazakhstan, a country in Central Asia, are poorly managed (UNDP 2003; 2005). Among the most notable problems are industrial pollution of rivers and lakes, the shrinking of the Balkhash lake, the competition for water between hydroelectricity production and irrigation, inefficient water use and transportation, especially in agriculture, and the absence of water demand management (UNECE 2008). Comparatively little attention has been paid to water quality with the traditional and inherited from the Soviet epoch focus on water quantity, especially due to the transboundary character of water resources management in Central Asia at large as discussed in more detail in another chapter in this volume (see Krutov et al. 2014). Water supply and irrigation infrastructure is dilapidated, water efficiency in irrigation is as low as 50–60%, and water lost in the pipe system causes water-logging and the salination of land. The inefficient use of irrigation water results in estimated 200 million USD lost for Kazakhstan in crop value, whereas at the Central Asian scale this figure was 1.7 billion USD or 3% of the GDP of the region in 2007 (Borishpolets and Babadjanov 2007). One of the biggest problems in Kazakhstan is the poor access to drinking water sources for the population, mostly in rural areas. According to UNECE (2008), over 39% of the population did not have permanent access to safe drinking water in 2006. This is currently a priority area for the government as it is implementing the State Programme on "Drinking Waters: 2002–2010" (Genina 2007).

Generally, the dare state of water management in Kazakhstan, the transboundary character of many of its problems, the Soviet legacy of inter-dependencies between Kazakhstan and its neighbours, and the crucial importance of water for survival in the region have elevated the issue on the political agenda both regionally and globally. As a result, Central Asia has been an area of much attention from donors and international consultants alike since the dissolution of the Soviet Union (Mukhtarov 2013).

Overall, the main cause of the crisis is in the poor water management system and such problems as centralized administration, poor and over-bureaucratised communication between government agencies, and weak cross-sectoral co-ordination amidst common fragmentation of responsibilities (Zimina 2003). A large number of stakeholders are engaged in water management, and there is no single agency well-placed to co-ordinate policies and their delivery. For example, according to the Water Code of Kazakhstan, articles 37–40, the Committee for Water Resources is the main state agency charged with water-use planning and authorization (Parliament of the Republic of Kazakhstan 2003). However, many tasks of the committee went beyond its capacity with the low staff levels and weak organization (UNECE 2008). With such conditions in Kazakhstan and the international popularity of IWRM as a preferred framework for water management, many have called for a project on Integrated Water Resources Management in Kazakhstan.

Among many internationally funded water projects, the UNDP Project "Preparation of the National IWRM and Water Efficiency Plan for Kazakhstan" has introduced the idea of national IWRM planning to Kazakhstan in 2005. The project was initiated jointly by the Norwegian and the Kazakh governments, supported by the UNDP, GWP and the UK Department for International Development (UNDP 2008). It has resulted in the preparation and eventual approval of the plan by the State Budget Programme of Kazakhstan for 2009–2011 (Nikolaenko, personal communication, August 16, 2008; Nee, personal communication, February 02, 2009). The plan recommended restructuring several governmental agencies, such as the Committees for Water Resources, the introduction of river basin councils, the national information system for monitoring of the water use and quality, cost-recovery, improvements in water efficiency, capacity building and education programs.

The hegemony of IWRM meant that there was virtually no resistance to the introduction of this concept to the policy arena of Kazakhstan within the government actors. On the other hand, outside of the government there has been little awareness and support to IWRM.

> However, it became apparent. … that very few people knew what IWRM is. Many had heard of it and even used the term quite freely but did not actually understand its concept. Some dismissed it as a 'western concept' that has no applicability to Kazakhstan. Others were concerned that the introduction of IWRM and the integration that is its main point would weaken their organisations by removing or reducing their functions and budget allocations. The first forum was therefore very difficult as the assumption of a general understanding was incorrect and there was little support for IWRM outside of those organisations directly involved in water resources management. Subsequent forums included presentations to educate participants and to reduce their concerns (Hannan 2006: 6).

Even those actors who welcomed IWRM disagreed about its meaning. For example, Dukhovny and Sokolov (2003), the representatives of the Central Asian branch of the GWP, saw IWRM as a systemic approach to water management reminiscent of a rationalistic tradition of comprehensive rational planning (see their comments on the draft Kazakh IWRM plan consultation document (UNDP-IWRM 2007)). Others believed that IWRM was similar to the *river basin management plans* practiced in the USSR since the 1970s and still existing in Kazakhstan (Kazgiprovodkhoz official, personal communication, August 15, 2008). Called the "complex schemes of use and protection of water resources," those basin level plans include the inventory of all water and related land objects and socio-economic trends. In addition, since 1986 there have been eight government-based river basin organizations (or Basin Water Authorities) in the country, which means that river basin management had already been introduced in Kazakhstan. This led the government of Kazakhstan and some independent experts to claim that Kazakhstan has been complying with IWRM for the last twenty years (Kazgiprovodkhoz official, personal communication, August 15, 2008). In short, IWRM is viewed by policy actors in Kazakhstan in three different ways:

1) as a *process of management* that is new to Kazakhstan and needs to be established from scratch through a comprehensive legal and institutional reform;
2) as a *managerial addition* to the old system of "river basin plans" (e.g., the former manager of the UNDP IWRM project, Alexander Nikolaenko, has defined the 'new version of IWRM' as adding the environmental and participatory elements to the previously practiced 'schemes'); and
3) as completely *identical to the* already practiced *"river basin plans"*.

Such disagreement on the essence of a policy is typical to hegemonic concepts as they represent 'floating signifiers' designed to enrol support through illusionary agreements. The context of Kazakhstan, where river basin planning and comprehensive rational planning approach to water had been practiced before, contributed to the confusion over its meaning.

As mentioned earlier, prior to the commencement of the UNDP-IWRM project which formally introduced IWRM planning, there were some fifteen international IWRM-related projects in Kazakhstan. However, no single dominant interpretation of IWRM had emerged and, therefore, no normative notion of what was good and therefore should have been implemented. As Tirtishniy (2005) put it, the Global Water Partnership promoted IWRM in Kazakhstan, but it did not suggest *how* to make it work (Tirtishniy 2005). There is an unresolved tension between the notion of IWRM identical to the Soviet schemes of river basin management in the 1980s and the notion of IWRM that embodies public participation, equal consideration of environment and economic interests (ICWE 1992). This is an important testimony to the discursive richness of water policy. Such diversity needs to be made explicit through the creation of various venues where policy deliberation can take place.

When the hegemony of IWRM makes it a 'floating signifier', it is important that there is an opportunity for actors to engage in a discussion of what it means and how to implement it. Such deliberative process is necessary to make discursive diversity explicit and reach the normative consensus on its practical implementation. In the case of Kazakhstan such deliberation has been attempted by the UNDP-IWRM project with an extensive consultation of the concept notes and draft plans. However, this has not resulted in a consensus over its meaning and the path of implementation.

6 Reflections on the IWRM Hegemony and Conclusions

There are three major risks or negative aspects of hegemony of a policy concept. First of all, the hegemony is based of ideational power which in turn is promoted by the knowledge elites and experts. Therefore, the hegemony promotes the expert knowledge that is often based on codified and model-based epistemology which Mukhtarov and Gerlak (forthcoming) called the '*prescriptive way of knowing*'. By privileging the model-based expert knowledge, the IWRM hegemony pushes aside other ways of knowing, such as indigenous knowledge, or knowledge based on values and ethics as opposed to the science. This expert-oriented inclination of IWRM has been well noted by Conca, who argued that 'the central forum of IWRM in global water politics is the global expert conference, not the diplomatic arena; its currency is the task force report, not the treaty' (Conca 2006: 127). This elitist character of the discourse suggests that local knowledge is often overlooked in national discourses.

The second negative aspect of the hegemony of IWRM, is the inclination of the 'universal' to take prevalence over the particular in policy context. The IWRM rhetoric obscures the importance of the context in policy issues. The recent work on contextual relevance in water resources management and the absence of panaceas emphasizes this malady of IWRM (Brugnach and Ingram 2012; Pahl-Wostl et al. 2012). This can be observed in the case study of Kazakhstan presented above where IWRM has been adapted and the managerial and quality aspects of water management were prioritised under the conditions where the priorities of water management have lied with the quantity aspects of water availability for irrigation and the aspects of infrastructure. Indeed, the case of Kazakhstan asserts that international discourses, actors and funding have been instrumental in putting the

IWRM plan on the policy agenda of Kazakhstan. The policy context, the pre-existent discourses and institutions in Kazakhstan and the interests of policy actors have been as important as the external drivers of the policy reform of Kazakhstan.

The third negative impact of hegemonic concepts is that of actors buying into the language without committing to the principles of a certain policy or approach. In other words, the 'floating signifiers' in Levi-Strauss's words produce 'fake agreements' between actors, on the one hand allowing for more boundary space and opportunity to build a shared understanding, and on the other hand, building the dialogue on completely false assumptions and misunderstandings. That could also be observed in Kazakhstan where the adoption of the language of IWRM has effectively meant reinforcement of the technocratic elite and the agenda of the comprehensive river basin planning practiced in the 1980s in the Soviet Union times. Studying the hegemonic discourses and revealing the mechanisms through which they acquire power and pose danger to water resources sustainability in order to ameliorate those, therefore, remains an important goal of both academics and practitioners.

In summary, we would like to recall the three questions which we set forth for ourselves in the introduction. The first question concerned the reasons for the global domination of IWRM, and we argued in this chapter that the convergence of historically favourable period in the 1990s and 2000s and the interests of global policy actors have been key in the rise of IWRM. The second question concerned the mechanisms through which IWRM rose to power, and we argued that the neo-Gramscian approach to global water governance with the three-pillar model of power helps explaining how IWRM has reached its hegemony. Finally, the third question asked about the global and national impacts of the domination of IWRM, and we illustrated the risks of the hegemony of IWRM by the example of national water planning in Kazakhstan. Our analysis showed that hegemonic concepts are risky in three ways:

1) they privilege expert knowledge and abstract-scientific way of knowing (as opposed to value-based knowledge or knowledge which emerges from practice);
2) they denigrate the importance of the specific and particular in policy contexts; and
3) they incentivize actors to adopt the vocabulary of integrated management and participation without real commitment to the latter, which creates difficulty in meaningful monitoring of the progress towards IWRM on the ground.

References

Allan, T. 2003. IWRM/IWRAM: a new sanctioned discourse? SOAS Occasional Paper number 50. http://www.soas.ac.uk/water/publications/papers/file38393.pdf (accessed March 12, 2013).

Baumgartner, T. and C. Pahl-Wostl. 2013. UN-Water and its role in global water governance. Ecology and Society 18(3): 3.

Beveridge, R., J. Monsees and T. Moss. 2012. The IRS Handbook: Analysing Institutional and Political Contexts of Water Resources Management Projects. The Leibniz Institute for Regional Development and Spatial Planning. http://www.fona.de/mediathek/pdf/IRS_IWRM_en.pdf (accessed May 15, 2013).

Biswas, A. 2004a. From Mar del Plata to Kyoto: an analysis of global water policy dialogue. Global Environmental Change 14: 81–88.

Biswas, A. 2004b. Integrated water resources management: a reassessment. Water International 29(2): 248–256.

Biswas, A. 2008. Integrated water resources management: is it working? Water Resources Development 24(1): 5–22.

Bloomfield, M.J. 2012. Is forest certification a hegemonic force? The FSC and its Challengers. The Journal of Environment & Development 21(4): 391–413.

Borishpolets, K. and A. Babadjanov. 2007. Water resources of Central Asia in the context of regional co-operation. Moscow: Moscow State Institute of International Relations: Analytical Remarks. In Russian.

Brenner, N., J. Peck and N. Theodore. 2010. Variegated neoliberalization: geographies, modalities, pathways. Global Networks 10(2): 1–41.

Brugnach, M. and H. Ingram. 2012. Ambiguity: the challenge of knowing and deciding together. Environmental Science and Policy 15: 60–71.

Callinicos, A. 2007. Social Theory: A Historical Introduction. Polity Press, Cambridge.

Conca, K. 2006. Governing Water: Contentious Transnational Politics and Global Institution Building. MIT Press, Cambridge, Massachusetts.

Cox, R. 1993. Gramsci, hegemony and international relations: an essay in method. pp. 49–67. *In*: S. Gill (ed.). Historical Materialism and International Relations. Cambridge University Press, Cambridge.

Dukhovny, V. and V. Sokolov. 2003. Lessons on co-operation building to manage water conflicts in the Aral sea basin. UNESCO IHP report.

Fairclough, N. 1992. Discourse and Social Change. Polity Press, Cambridge.

Fischer, F. and J. Forester. 1993. The Argumentative Turn in Policy Analysis and Planning. UCL Press, London.

Genina, M. 2007. The development of a new water code in the Republic of Kazakhstan. Politikon 13(1): 21–31.

Global Water Partnership. 2003. IWRM Toolbox Version 2. http://www.gwpforum.org/gwp/library/ToolBox%20Version%202%20.pdf (accessed 15 May, 2013).

Global Water Partnership. 2005. Catalyzing change: a handbook for developing IWRM and water efficiency strategies. GWP report.

Global Water Partnership Technical Advisory Committee (GWP-TAC). 2000. Integrated water resources management: Background Paper No. 4. Global Water Partnership Technical Report number 4.

Gramsci, A. 1971. Selections from the Prison Notebooks. New York: International Publishers.

Groenfeldt, D. 2013. Towards a new water ethic. Global water news: the Bonn Declaration on global water security, No. 13, August 2013.

Hajer, M. 1995. The Politics of Environmental Discourse: Ecological Modernization and the Policy Process. Oxford University Press, Oxford.

Hannan, T. 2006. The process of preparing a national IWRM and water efficiency plan for Kazakhstan. The UNDP Water Governance Facility Report.

Hays, S. 1959. Conservation and the Gospel of Efficiency: The Progressive Conservation Movement, 1890–1920. Harvard University Press, Cambridge.

Howarth, D. 2010. Power, discourse, and policy: articulating a hegemony approach to critical policy studies. Critical Policy Studies 3(3-4): 309–335.

Ingram, H. 2008. Beyond universal remedies for good water governance: a political and contextual approach. The Sixth Biennial Rosenberg Water Policy Forum on Water For Food: Quantity and Quality in a Changing World, 24–27 June 2008, Zaragoza, Spain. http://www.zaragoza.es/contenidos/medioambiente/cajaAzul/IngramACC.pdf (accessed May 15, 2012).

International Conference on Water and the Environment (ICWE). 1992. The Dublin Statement on Water and Sustainable Development. http://www.wmo.int/pages/prog/hwrp/documents/english/icwedece.html (accessed May 15, 2013).

Jeffrey, P. and M. Gearey. 2006. Integrated water resources management: lost on the road from ambition to realization? Water Science and Technology 53(1): 1–8.

Kazgiprovodkhoz official. Consultant for the IWRM-UNDP project (Personal communication, 15 August, 2008).

Krutov, A., S. Rahimov and R. Kamolidinov. 2014. The Republic of Tajikistan: its role in the management of water resources in the Aral Sea Basin. pp. 325–345. *In*: V. Squires, H. Milner and K. Daniell (eds.). River Basin Management in the Twenty-first Century: Understanding People and Place. CRC Press (this book).

Laclau, E. 2000. Identity and hegemony: the role of universality in the constitution of political logics. pp. 44–89. *In*: J. Butler, E. Laclau and S. Zizek (eds.). Contingency, Hegemony, Universality. Verso, London.

Laclau, E. and C. Mouffe. 1985. Hegemony and Socialist Strategy. Verso, London.

Mitchell, B. 2005. Integrated water resources management, institutional arrangements, and land-use planning. Environment and Planning A 37: 1335–1352.

Mitchell, B. 1990. Integrated Water Management: International Experiences and Perspectives. Belhaven Press, London and New York.

Molle, F. 2006. Planning and managing water resources at the river-basin level: emergence and evolution of a concept. The International Water Management Institute report.

Molle, F. 2008. Nirvana concepts, narratives and policy models: insights from the water sector. Water Alternatives 1(1): 23–40.

Mollinga, P., A. Dixit and K. Athukorala. 2006. Integrated Water Resources Management: Global Theory, Emerging Practice, and Local Needs. Sage Publications, New Delhi.

Mukhtarov, F. 2008. Intellectual history and current status of integrated water resources management: a global perspective. pp. 167–185. *In*: C. Pahl-Wostl, P. Kabat and J. Möltgen (eds.). Adaptive and Integrated Water Management: Coping with Complexity and Uncertainty. Springer Verlag and Heidelberg GmbH and Co. KG, Heidelberg.

Mukhtarov, F. 2009. The hegemony of integrated water resources management: a study of policy translation in England, Turkey and Kazakhstan. Ph.D. diss., Central European University.

Mukhtarov, F. 2013. Translating water policy innovations into Kazakhstan: the importance of context. pp. 113–127. *In*: Vinke de Kruijf, Cheryl De Boer, J. Vinke-de Kruijf, G. Ozerol and J. Bressers (eds.). Water Governance, Policy and Knowledge Transfer: International Studies on Contextual Water Management. Earthscan, London.

Mukhtarov, F. and A. Gerlak (forthcoming). Ways of knowing integrated water resources management: towards versatility of knowledge. Policy Sciences.

Mukhtarov, F. and A. Gerlak. 2013. The rise of the river basin organizations: discourse and political economy perspectives. Global Governance 19(2): 307–326.

Nee, Viktor. Consultant for the IWRM-UNDP project (Personal communication, 02 February, 2009).

Newell, P. 2008a. Lost in translation? Domesticating global policy on genetically modified organisms: comparing India and China. Global Society 22(1): 115–136.
Newell, P. 2008b. The political economy of global environmental governance. Review of International Studies 34(3): 507–529.
Newell, P. 2009. Bio-hegemony: the political economy of agricultural biotechnology in Argentina. Journal of Latin American Studies 41(1): 27–57.
Newell, P. and David Levy. 2005. The Business of Global Environmental Governance. MIT Press, Cambridge, Massachusetts.
Nikolaenko, A. Former IWRM-UNDP project manager (personal communication, 16 August, 2008).
Pahl-Wostl, C., L. Lebel and E. Nikitina. 2012. From applying panaceas to mastering complexity: toward adaptive water governance in river basins. Environmental Science & Policy 23: 24–34.
Parliament of the Republic of Kazakhstan. 2003. Online (in Russian), online.zakon.kz/Documents/?doc_id=1042116 (accessed 23 February 2014).
Saravanan, V., T. McDonald and P. Mollinga. 2009. Critical review of integrated water resources management: moving beyond polarised discourse. Natural Resources Forum 33: 76–86.
Tarlock, D. 2008. Water security, fear mitigation and international water law. Hamline Law Review 31: 703–728.
Tirtishniy, E. 2005. Comments and suggestions to the concept note for the national IWRM and water efficiency plan. Atasu NGO report.
UK Department for Environment, Food and Rural Affairs (DEFRA). 2005. Making space for water: taking forward a new government strategy for flood and coastal erosion risk management in England: first government response to the autumn 2004 Making Space for Water consultation exercise. London, Defra.
United Nations Development Programme (UNDP). 2003. Human development report: water as a key human development factor. UNDP, Almaty.
United Nations Development Programme (UNDP). 2005. Kazakhstan national integrated water resources management & water efficiency plan. UNDP IWRM project office, Almaty.
United Nations Development Programme (UNDP). 2008. National integrated water resources management (IWRM): planning and transboundary dialogue in Central Asia. http://europeandcis.undp.org/public/show/ (accessed 02 February, 2013).
United Nations Development Programme Project on Integrated Water Resources Management (UNDP-IWRM). 2007. Draft programme on integrated water resources management and water efficiency in Kazakhstan. UNDP-IWRM, Almaty.
United Nations Environmental Programme (UNEP). 1994. Agenda 21. Online, URL: http://sustainabledevelopment.un.org/content/documents/Agenda21.pdf, accessed, May 15, 2013.
United Nations Education, Scientific and Cultural Organization, International Hydrological Programme (UNESCO, IHP). 2007. State of the art review on IWRM. UNESCO-IHP, Paris.
United National Economic Commission for Europe (UNECE). 2008. Environmental Performance Review: Kazakhstan. UNECE, Geneva.
UN-Water. 2008. Status report on integrated water resources management and water efficiency plans for CSD 16. United Nations, New York, USA. Online, URL: http://www.unwater.org/downloads/UNW_Status_Report_IWRM.pdf.
UN-Water. 2012. The UN-water status report on the application of integrated approaches to water resources management. http://www.unwater.org/downloads/UN-Water_Report_for_Rio_Flyer.pdf (accessed 15 May, 2013).
Varady, R. and M. Iles-Shih. 2005. Global water initiatives: what do the experts think? Report on a survey of leading figures in the World of Water. Presented at workshop on impacts of mega-conferences on global water development and Management, Bangkok, Thailand.

http://udallcenter.arizona.edu/gwi/pdfs/GWI%20Bangkok%20Presentation.pdf (accessed 15 May, 2013).
Walther, P. 1987. Against idealistic beliefs in the problem-solving capacities of integrated resource management. Environmental Management 11(4): 430–446.
Watkins, K. 2006. Human development report. Beyond scarcity: power, poverty and the global water crisis. United Nations Development Programme.
Wegerich, K. 2008. Hydro-hegemony in the Amu Darya Basin. Water Policy 10(5): 71–88.
White, G. 1998. Reflections on the 50-year international search for integrated water management. Water Policy 1: 21–27.
Yanow, D. 2009. Ways of knowing: passionate humility and reflexive practice in research and management. The American Review of Public Administration 39: 579–601.
Zeitoun, M. 2008. Power and Water in the Middle East: The Hidden Politics of the Palestinian-Israeli Water Conflict. I.B. Tauris & Co. Ltd., London & New York.
Zeitoun, M. and J. Warner. 2006. Hydro-hegemony—a framework for analysis of trans-boundary water conflicts. Water Policy 8: 435–460.
Zimina, L. 2003. Developing water management in the south Kazakhstan. pp. 87–105. *In*: S. O'Hara (ed.). Drop By Drop: Water Management in the Caucasus and Central Asia. OSI, Local Governance Initiative, Budapest.

2

Beyond IWRM: Developing Territorial Intelligence at the Local Scales

Pierre Maurel,[1,*] Roel Plant,[2] Olivier Barreteau[3] and Yann Bertacchini[4]

SYNOPSIS

Even if IWRM has been a dominant paradigm for more than 70 years and is now promoted worldwide, it is still facing a major challenge of practical implementation. To cope with this issue, we investigate an approach which combines IWRM and territorial development to better incorporate people and place in river basin management. The territory is used as an integrative concept between the bio-physical processes (including water), the knowledge and engineering capacities of the territory (including for water management) and a shared sense of place, of territoriality (including a water component). Developing a territory with an explicit water dimension requires knowledge brokers between top-down technocratic public policies (including water policies) and bottom-up development project supported by local population and politicians. Specific STICA (socio-technical information and communication arrangements) can be imagined by these knowledge brokers to bridge the gap between engineers, experts, local people and elected officials.

Keywords: France, Information and Communication, IWRM, knowledge brokers, People, Place, Spatial Representation, Territorial Intelligence, Territory

[1] Irstea, UMR TETIS Lab, Montpellier, France.
Email: pierre.maurel@teledetection.fr
[2] Irstea, UMR Montpellier, France/ISF, University of Technology, Sydney, Australia.
[3] Irstea, G-Eau Lab Montpellier, France.
[4] I3M Lab, University of the South Toulon-Var, France.
* Corresponding author

1 Introduction

The water sector, along with transport and energy, is considered as one of the key infrastructural requirements of any modern society for industrial, agricultural, economic, social and cultural development (Jeffrey and Gearey 2006).

A holistic approach is frequently advocated for water resources management since it is not contingent upon a sole domain or sphere (Mitchell 2006). Among these approaches, Integrated Water Resources Management—IWRM (and Adaptive Management—AM) has emerged as a dominant paradigm in the discursive framing of international water policy over the past twenty years (Conca 2006) even if the concept has been around for 70 years among international institutions (Biswas 2004). Linked implicitly with the rational planning paradigm (see Mukhtarov and Cherp, this volume), IWRM is now worldwide promoted (Global Water Partnership 2000). Even if an agreed definition of IWRM has never been established, the most quoted one is probably the one provided by the Global Water Partnership (2000): "a process which promotes the coordinated development and management of water, land and related resources, in order to maximize the resultant economic and social welfare in a equitable manner without compromising the sustainability of vital ecosystems".

IWRM draws attention to a diverse range of values and perceptions associated with water between interdependent players, connections between biophysical dimensions of water and governance, institutional innovations (Jaspers 2003), relations between knowledge production modes and water management outcomes, aspects of uncertainty and complexity, linkage between action and social learning (Pahl-Wostl et al. 2008a) and between adaptive management and public participation (Pahl-Wostl et al. 2008b). Key-features of IWRM are triple: adaptation, institutions and governance (see also Mitchell 2006; Ffolliott and Brooks, this volume).

The implementation of IWRM as understood by the Global Water Partnership has been described in one of its technical chapter which constitutes a statement of necessary governance conditions and offers a toolbox of good practices based on a standard decision cycle adapted to IWRM (Global Water Partnership 2004).

However, this holistic approach of water management has been strongly criticized by several researchers and practitioners during the last decade (see Mukhtarov and Cherp, this volume), as illustrated by the following fundamental question: "Why it has not been possible to properly implement a concept that has been around for some two generations in the real work for macro and meso-level water project and programs?" (Biswas 2005; Mukhtarov and Cherp, this volume).

The main critique is that IWRM remains strongly conceptual, its definition quite elusive and fuzzy, with a profound gap towards practical implementation and operational results (Biswas 2004; Jeffrey and Gearey 2006). During the 2006 World Water Forum, water management professionals pointed out that there can't be a single blueprint for moving towards IWRM because it remains deeply dependent on local contexts. This point makes it very difficult to develop a generic and overall description, casting further doubt on the practical implementation of IWRM.

Other detractors point to its over-reliance on a regulatory regime, and its developed world policy prescription (Lankford and Hepworth 2010) as well as its expert-driven agenda (Conca 2006). For Jeffrey and Gearey (2006, p. 4), it reveals the schizophrenic character of IWRM paradigm, as part modernist and part post-modernist: "However, whilst IWRM reflects this post-modernist inspired agenda through its emphasis on contextual relevance, wider participation in planning and decision making, and responsive and reflexive practice, it remains rooted, by and large, in a "predict and prepare" paradigm. It is, therefore, more akin in practice to the contingency planning approaches of the 1960s and 1970s than to the adaptive management frameworks promoted during the 1990s".

Among the IWRM principles which have been identified (at least in part) as characteristic of many national, regional and basin-scale strategies, several deal with information availability, capacities building and collective learning within heterogeneous stakeholders (IWA UNEP 2002). This raises a first crucial issue of information design, storage and retrieval as well as analytical tools to determine the effects of policies or action plans on water systems (McDonnell 2008). But information availability and rational analytical capabilities remain insufficient as long as water managers and local communities do not dispose of the capacities to make it meaningful and to transform it into actionable knowledge to anticipate developments (Johnson 2013). A post-modern perspective both considers knowledge as conditional and particular (i.e., knowledge as a function of experience and thereby neither absolute nor general), while recognizes the complexity of the natural world, and of the relationships between nature and society. Exploring "the knowledge gap between IWRM policy and practice" remains an important issue for a successful implementation of IWRM and required to abandon the positivist approach still widely dominant within the water engineers and planning communities (Jeffrey and Gearey 2006). To bridge this gap, effective communication is all the more essential due to the increasing number of interactions among heterogeneous stakeholders and the difficulties to combine expert and non-expert knowledge, even if this process is fruitful (Pahl-Wostl et al. 2008a). Effective communication can be considered as a means to reflect and reinforce social relations or "communities" (Squires, this volume; Ffolliott and Brooks, this volume).

New communication patterns can help to build up new communities. Within these communities, new representations and new "meanings" of reality can develop (Wenger 2000).

Another critique deals with one of the major supposed benefits of using IWRM as a framework, the mixing of viewpoints from different players (users, planners, developers, sciences, policy makers). This mixing of viewpoints is suggested to lead to a holistic understanding of the situation and solutions to be implemented. However, IWRM can be considered as not holistic since it considers water as the most important resource and excludes other resources (land, food, air, biodiversity, ...) and other human activities and vital needs (land planning, housing, transport, economic development, recreation, culture, ...) (Medema et al. 2008). Problems that can occur due to a lack of integration between water and other policy sectors are now widely recognized (Slater et al. 2004; Carter et al. 2005; Bouleau et al. 2010). To avoid an unmanageable integration of all resources and activities in a single framework, Biswas (2004) suggests the development of collaboration and coordination between the existing institutions rather than trying to create institutions in charge of multiple resources.

Finally, in the daily practice of water management, others point out that there is little consideration of the social and pluri-institutional dimensions (Timmerman and Pahl-Wostl 2008). The same analysis has been made in a recent systematic review of water vulnerability assessment tools (Plummer et al. 2012).

These implementation capacity issues of IWRM echo Gilbert White's statement that "the problems of accurate analysis of inter-sectoral linkages and of achievement of institutional reforms in the planning process are formidable. It would be sanguine to expect early or easy solutions" (White 1998).

In this chapter, our contribution to this complex issue of practical implementation of IWRM thus remains modest and tries to bring an additional approach centered on the territorial dimension of development at local scale which includes the water component. As explained further in this chapter, the term "territorial" is used from a francophone geographer's perspective and is much closer in the Anglo-Saxon world to the concept of "place" than to the concept of "territory".

The key question that we want to develop here is: What are the concepts and the lessons of territorial development that can be integrated in IWRM to better incorporate people and place in river basin management?

In the first part of this chapter, we will investigate from a theoretical perspective the francophone concept of "territory" and will show how the territory, described as a complex system, can be used as an integrative concept between the bio-physical processes (including water), the knowledge and engineering capacities of the territory (including for water management) and a shared sense of place, of territoriality (including a water component). We will then focus on the informational and communicational processes, explored in the territorial intelligence literature that sustains the emergence and consolidation of a territory. Developing a territory requires knowledge brokers between top-down technocratic public policies (including water policies) and bottom up development project supported by local population and politics. Specific STICA (socio-technical information and communication arrangements) can be imagined by these knowledge brokers to bridge the gap between engineers or experts, various stakeholders and lay people engaged in the design and the implementation of a territorial project partially based on water issues. A specific focus is given to spatial representation of the phenomena and their functionalities within STICA.

The final discussion tries to highlight which lessons derived from this theoretical approach can be transposed to other places and contexts than France.

2 Theory

2.1 Territory as an Integrative Concept

2.1.1 Concept clarification

It is first necessary to clarify the meaning that we give to the term "territory" as it varies significantly between, and even within, French and English literature despite the same Latin etymology (Debarbieux 1999). While the English approach has remained close to the original definition of the territory based on an ethological understanding, i.e., an area with boundaries and controlled by a political power (Elden 2010), French-speaking social sciences have gradually expanded the concept by adding the social, symbolic and cultural dimensions and transposing it to other scales than the nation-state. In France, this is probably due to the decentralization process that led to the creation of new local authorities which challenged the legitimacy of the state to ensure the development in an exclusive way.

A single definition of the word "territory" does not exist in French literature and there are even several types of "territories", intertwined with each other in the real world, the main ones being:

- Political and administrative territories: which corresponds to the original meaning of the word (see above), that is still in force in Anglo-Saxon countries. In France, this definition initially applied to the nation-state now extends to all political and administrative divisions with a political representation and capacity for action: region, departments, communes, and more recently, inter-communal bodies.

- Biophysical territories: beyond the strict ethological territories of the animal world, the most significant example in France is that of 'water territories'. The water laws of 1964 and 1992 have carved up the national territory into large hydrographic basins and small basin catchments, each of them with their own governing body (Ghiotti 2006). In the United States, several movements also tried to create institutional divisions based on biogeographical limits (McGinnis 1999).
- Territories with a sense of belonging: these territories have been highlighted by researchers in human geography from the links between territories, individual identities and collective identities: "The territory reflects a sense of ownership of the geographical space, both economic, ideological and political, by human groups that develop a particular representation of themselves, their history, their singularity" (Di Méo 1998).

More recent territories have also been analyzed, such as the territories of mobility related to the movement of individuals or social groups (Bonnet and Desjeux 2000), or the digital and virtual territories as they appear in the digital space (Le Groupe Going and Bertacchini 2004).

In Anglo-Saxon countries, other concepts than the "territory" have been developed to overcome the emphasis on the political dimension given to the term territory (Debarbieux 1999; Elden 2010): the so-called quantitative geographers brought the concept of "space", humanist geographers those of "place", while the new geography developed those of "landscape". It is finally the concept of "place" that has been particularly explored and debated by scholars in U.S., Canada and UK: these authors invest the word "place" with a social, cultural and political dimension that contains a critique of political territory, its rigid delimitation, and the state control that is coextensive to it" (Raffestin 2012, p. 126).

The English concept of 'place' is probably the closest of the French concept of "territory" used in this chapter because we will deal mainly with local areas at communal and inter-communal levels. It is therefore with the meaning of "place" that the non-French reader should try to understand the meaning of the word "territory" as used here.

2.1.2 Territory represented as a complex system

To introduce this section, we adopt the definition of the term 'territory' given by Claude Raffestin, a famous human geographer in the French speaking academic community: "Territory refers to human labor exercised on a complex combination of mechanical, physical, chemical, organic, and other forces and actions. Territory is a reordering of spaces and their contents, whose order is to be found in the informational systems which humans dispose of by virtue of belonging to a culture" (Raffestin 2012, p. 129).

When considering a territory as a complex, open and dynamic system, it can be described through a systemic approach (von Bertalanffy 1968), that is to say "as a set of interacting dynamic components organized towards a goal" (de Rosnay 1975, p. 93). To represent the territory as a system, several scholars have adopted the meta-model of Schwarz (Schwarz 1992) which is based on three nested and interrelated levels:

- The physical level of materiality and energy: it includes the materiality of the physical world and its objects, whether real (natural or anthropogenic) or symbolic (e.g., a place of memory). Since the Schwarz meta-model is a dynamic model, it also includes the change of state of objects due to natural or anthropogenic phenomena such as floods, fire, but also changes of activities or economic conditions.
- The logical level of information and representations: it is a matter of interpretation and conceptualization by the local actors of physical reality and its various representations (academic, expert or lay). This level also includes planning tools and programs of actions developed by public bodies, NGO or professionals involved in the physical level.
- The symbolic level of identity and self- reference: this level includes the symbolic elements to which local players refer to define the territory, to distinguish it from other territories, while marking their sense of belonging to it.

This static description of reality is completed by a dynamic dimension, inside which four phases are distinguished: drift, hazard, metamorphosis and stability. The drift sequence corresponds to the progressive change of a system towards a state of greater disorder, causing tensions and instabilities. The hazard period is the time of the exacerbation of tensions, preventing the system to remain stable. The system can then regress and disappear, or continue to evolve while keeping the same structure, or change its structure. The metamorphosis corresponds to the period of reorganization of the system to reach a new stable state which necessarily leads to new developments.

Transposed to the territory, these changes correspond to the cyclical process of territorialization, deterritorialization and reterritorialization (TDR) described by Claude Raffestin from the previous works of Deleuze and Guattari (1987). The temporal intervals between these various phases of the process can be quite short or, on the contrary, quite long. This obviously depends on the actors and a range of ecological and symbolic conditions (Raffestin 2012).

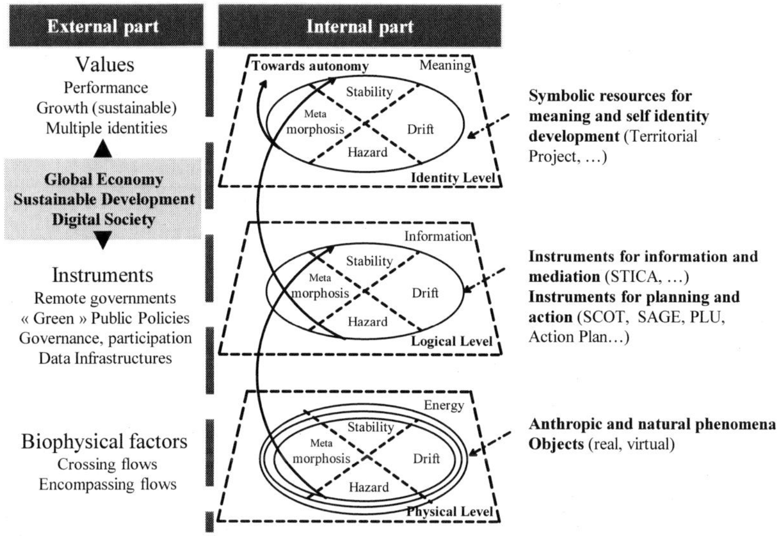

Figure 2.1. Territory as a complex and socially constructed system.

In a territorialization phase, a territory, considered simultaneously at the three intertwined levels of the meta-model, becomes self-organizing and autonomous, acquires capacity of self-analysis and monitoring, and develops a shared awareness of its own image and a self-identity in reference to other territories. Such a process cannot just be announced, it is always rooted in a specific territorial culture in terms of governance. It requires: i) a deliberate collective will to develop a share and meaningful vision of the future which can be transcribed in a political project and a program of actions, ii) a networking process between stakeholders, and iii) specific resources (energy, information, mediators), in order to spread out.

2.1.3 Water management, territorial development and their respective instruments

In the field of water management, the development of IWRM approach has led to the creation of 'water territories' (Ghiotti 2006) based on watershed, catchment or river basin limits, on ad hoc institutions and governance systems, with specific planning instruments and program of actions (Ffolliott and Brooks, this volume).

In France, there are 5 hydrographic basin districts which mostly correspond to the boundaries of existing water agencies created by the water act of 1964. The water act of 1992 gave new competencies to water agencies. They were given charge of planning water uses, defining priorities in water allocation and setting general objectives for river restoration in basin master plans called *Schémas Directeurs d'Aménagement et de Gestion des Eaux (SDAGE)* and acknowledged later as the official French planning instrument under the WFD. The competent authority is the river basin coordinator (*Préfet coordonnateur de bassin*) supported by State employees of the *délégation de bassin* and employees of the water agency.

In some areas, SDAGEs designate threatened water bodies where additional regulations should be negotiated locally. For this purpose a local water commission is enacted (*Commission Locale de l'Eau, CLE*) by the *Préfet*. Members are State representatives (at most 25%), water users (at least 25%) and local elected representatives (at least 50%). The CLE has the duty to set up a local plan of water management (*Schéma d'Aménagement et de Gestion des Eaux*, SAGE) setting local objectives and appropriate regulations to meet the objectives. A SAGE has to be approved by the *Préfet de Département* and should comply with SDAGE.

However, neither water agencies nor basin-coordinator prefects have the power to implement measures themselves. They rely on "maîtres d'ouvrages", i.e., local authorities such as the municipalities competent in water projects and private owners.

In the field of land planning and local development, a new decentralization effort was carried out in France at the end of the 1990s, characterized by the empowerment of institutional and planning inter-municipal authorities (so-called *communauté de communes, communauté d'agglomération* and *communauté urbaine*), with an indirect political legitimacy,[1] mandated to collect their own taxes. To compensate the political weakness of these new territories, their elected representatives have worked on their symbolic construction in order to develop a sense of place among the population. They create narratives, they promote cultural heritage and create new cultural or social events at this new inter-municipal scale.

The SRU[2] act endowed these inter-municipal authorities with a new planning instrument, the Territorial Cohesion Blueprint (*Schéma de Cohérence Territoriale—SCoT*). A SCoT is a 10-year strategic planning tool to align sectorial policies relating to urban planning, housing, social diversity, transport and commercial facilities. Although a SCoT remains essentially a highly technocratic tool with a strong urban and environmental focus (to cope with climate change and biodiversity loss issues), it shall rely on a long term (10 to 15 years) **territorial project**, a form of political vision of the future of the territory covering the area of a single or several inter-

[1] Institutional inter-municipal authorities have a political organ, the community council, composed of elected officials, especially mayors, appointed by the municipal councils of the member municipalities.

[2] Solidarity Urban Renewal Act of December 2000.

municipal authorities. Indeed, planning documents in France and more generally in Europe are increasingly designed as instruments of social mobilization, to support dialogue, rather than as tools to regulate land use and allocate building rights (Pinson 2006, p. 635). A SCoT is set up by the local inter-municipal authorities (the elected representatives, such as the mayors, and their technical staff) under the control of the State (the *Préfet* and its technical departments) and with the participation of the stakeholders and inhabitants of the territory.

At the lower level, each municipality belonging to a SCoT must develop its own local land-planning document (*Plan Local d'Urbanisme*, PLU), which shall also reflect a long-term political vision, that elected representatives or candidates have to defend in front of their electors.

In terms of hierarchy between the different planning instruments, the Act of April 21, 2004 which transposed the WFD into French law, encouraged the alignment of land planning and water policies, giving priority to the last ones. Article L122-1 of the Urban Code related to the SCoT was amended to stipulate that the SCoT must now be compatible with the orientations of the SDAGE and with the protection objectives of the SAGE. Since municipal land-planning documents (PLU) have to be compatible with the ones at the inter-municipal (SCoT), PLU consequently have also to be compatible with the SDAGE and the SAGE.

This means in concrete terms that by law the water component must now be clearly taken into account in the territorial planning documents and, as we have seen, in political territorial projects at the local scale. When applying the meta-model of territory, water appears at all three levels: it is one of the material resources of the physical level, it is also a conceptual object of analysis and management at the logical level, and there is sometimes a symbolic element contributing to a sense of territorial belonging at the identity level. The challenge, therefore, is how to articulate water territories with politico-administrative territories and to develop a sense of ownership among people, that is to say territories that generate feelings of belonging (3rd level of the meta-model).

However in France, the professional routines of spatial planners, the democratic legitimacy of local elected representatives and the major expectations of the population towards their politicians are still dominated by an urban and economic view of local development. Water considerations are still limited and require huge efforts of mediation for raising awareness among the decision makers and the population.

In addition, if the complexity and multi-dimensionality of the territory by adopting a systemic approach is recognized, it becomes pointless to seek a rational territorial division that would be optimal and consistent between the human and biophysical phenomena, including the water, and the politico-administrative boundaries, to deploy the instruments of public territorial action (Ortiz 1994). Scales and levels being irreducibly multiple, the real issue is the relationship between these territories, both in a functional sense but also in terms of multi-level governance and shared strategic vision.

2.2 Territorial Intelligence: An Informational and Communicational Approach of the Territory

2.2.1 Conceptual clarification

In the definition of the territory given by Raffestin as well as in the conceptual model of territory presented above, one can point out the importance given to information and communication. As suggested by Lotman (1985), territory can be considered as space informed by the semiosphere—that is, the signifying system of signs from which the actor draws the informational resources for action. Raffestin (2012) considers a territory as an outcome from the projection of two kinds of labor, as constitutive categories of territoriality, by a community in a given space. The first type of labor, based on energy, generates material production in the physical space of the meta-model while the second one, based on information, is important to the production of representations in the logical space, both to understand the complexity of the territory and to act on it.

Based on the French understanding of the concept of territory, different groups of researchers from the community of Information and Communication Sciences began from the 1980s to investigate various dimensions of territorial development. Among these fields of research, works on 'Territorial Intelligence'[3] first began in the late 1990s (Bertacchini 2000), providing an original focus of Information and Communication Sciences on the process of collective action for endogenous territorial development.

Territorial intelligence is defined as *"an information and anthropological process, regular and continuous, initiated by local actors, physically present and/or distant, who intend to appropriate land resources by mobilizing and transforming the energy of the territorial system into a project capacity"* (Bertacchini 2004a, p. 3).

[3] The term "territorial intelligence" is also used by other scholars working on strategic approaches (business intelligence) for the economic development of territories.

Information and communication: shifting towards an interactionist and ecological model

Territorial intelligence relies on a different understanding of information and communication than the one's based on the ''code model''. The "code model" corresponds to Shannon & Weaver's mathematical model of communication (encoding and decoding), which was developed to optimize the transmission of data and information through electronic means, and now, through digital networks (Shannon and Weaver 1949). This model has been wrongly used for decades to deal with human communication and is still culturally dominant in our occidental societies, especially within the rational communities of engineers, planners, experts and scientists (Le Coadic 2004).

Territorial intelligence is based on the constructivist and interactionist paradigms of information developed in the 60s and the 70s by the group of Palo Alto that defined information as *"a difference which makes a difference"* (for a given knowledge in someone's mind), resulting from intersubjective communication between individuals (Bateson 1972, p. 448–466). This interactionist approach gives the priority to a user perspective of information and integrates the issue of meaning-making while the code model described above reflects a positivist information producer approach.

In the field of land planning and natural resources management, we also advocate adopting an ecological and perceptive approach of information as developed by James Gibson (1979). This approach assumes that our natural environment contains information and meanings immediately available, directly perceptible and autonomous, which do not depend on acts of communication and cognitive interpretation by humans. These statements are based on invariance within a more or less complex system, i.e., constant and regular relationships between facts, events or situations, because of physical or ecological laws. By example, a weathervane informs of the wind direction, even if no human observed it. This information present in the environment, however, can be captured by sensor technology, but also by actors accustomed to the particular environment with its constraints, its invariants and its dynamics. Thus, actors such as farmers, fishermen, hunters, naturalists, hikers, use forms of knowledge that are required for action and are built over long periods of time through intensive practice in the natural environment in which they operate.

In an epistemic and scientific approach of territorial intelligence focusing on the environmental dimension of the territory, this naturalistic concept of information appears as particularly relevant. Indeed, it ties up with a wider trend related to the progress of sustainable development as a major and global paradigm and should foster the recognition of local ecological knowledge as underlined by a recent bibliometric analysis (Brooks and

McLachlan 2008). The environmental sociologist Carole Barthélémy also highlights the social issues of such a change (Barthélémy 2005). According to her, emerging practices of co-management of environmental resources leads to the recognition that local stakeholders are not only resources users, but also potential experts. The use of this kind of knowledge thus becomes a new challenge for the social recognition and legitimacy of these territorial stakeholders. However, in reality local knowledge is rarely formalized in a manner that is at par with explicit knowledge from the technical and scientific spheres. Local knowledge often falls within the category of tacit knowledge (Polanyi 1983). The challenge of successfully using tacit knowledge in territorial intelligence processes requires transformation into explicit knowledge (Nonaka and Takeuchi 1995) by using different methods of data capture and processing, then subjecting the new knowledge to various tests of legitimacy in order to be recognized as "confirmed facts".

The issue of meaning-making

In a territorialization phase as described above, a central issue is the existence among people of a collective meaning attached to the territory. Meaning emerges from informational and communicational processes at work between local stakeholders which helps to develop a system of shared representations, and a certain culture. From a territorial intelligence perspective, the question of meaning can be assimilated to that of territoriality. Territoriality is a true informational and communicational phenomenon that transforms a given space into a territory for a local society (Bertacchini 2002).

Meaning emerges in each of the three levels of the meta-model:

(i) the physical space, when actors appropriate resources of this shared area for their activities. This appropriation is a more or less regulated, recognized and negotiated process within the network of stakeholders concerned by these resources,

(ii) the logical space, when actors produce, exchange and adhere to shared representations of the territorial reality (made of physical or conceptual interrelated objects), as well as to negotiated agreements for the use of land resources in the physical space,

(iii) the identity space, when actors refer to shared symbols and social norms, and to a common vision of the future expressed for a specific political project which constitutes the main symbolic resource to consolidate the identity.

When a territory chooses an endogenous development based on territorial intelligence, it will seek to develop in the logical level of the meta-model a substrate, called "formal capital" (Bertacchini 2004b), necessary for this development: it is a set of values, codes, rules, forms of interaction and coordination, data, information, explicit knowledge contained in digital documents, identified and shared by local actors to formulate and implement a collective local development project.

A shared territoriality, as described above, is the main vector for strengthening this type of capital. It can be fostered by new top-down land planning policies and their instruments such as SCoT or SAGE (see above): These policies can enrich the formal capital of a territory by bringing new habits of public participation, regulation of social relationships, use of shared territorial resources, new data, information and knowledge for a better understanding of the territorial situation, and new symbols strengthening the territorial identity. Indeed, these new public policies force local territories to organize themselves to formalize local realities in diagnosis documents and to express political choices in development projects. These top-down statutory requirements mechanically increase the formal capital of the territory. However, the formal capital will be more or less significant and its use will depend on whether planning is conducted as a technocratic mandatory exercise, or as a broad, collective and meaningful process to forge a common destiny.

2.2.2 Engineering team and facilitators as human resources for territorial mediation

An endogenous approach as outlined above assumes that the territory itself is potentially holding all the knowledge and skills necessary to its own development. Mobilizing them in a territorial intelligence process needs to inform and to communicate. Local mediation arrangements can facilitate communication among stakeholders, both to establish first contacts, to make the most of existing skills or to develop new ones required for the design and/or the implementation of the project.

Once again, top-down regulatory procedures of territorial planning can be the trigger for networking local actors by imposing a form of "mandatory communication" (Thoenig and Duran 1996) without which non-cooperative stakeholders might be marginalized. Problem formulation and solving require collaborative learning, both cognitive and relational, a kind of "joint conceptualization" (Thoenig and Duran, op.cit.). Indeed, these procedures go hand in hand with governance arrangements, information and communication processes, steps more or less imposed by a logic of project-based management, each of which provide opportunities for exchanges and interactions. The statutory requirement to build up a "territorial project", despite all the ambiguities around this concept (Maurel 2012), also helps local actors to develop and express a strategic vision, strengthening their identity at the third level of the meta-model (see above).

In practice, these procedures require the availability of a "territorial engineering team" (Landel 2007). This team must not only master regulatory planning procedures, but also be able to fulfill the aspirations of a local society symbolized by a singular political project and to translate aspirations in the normative frameworks imposed by regional, national and European territorial bureaucracies. In this case, territorial engineering concerns all local stakeholders, including civil society. It covers not only the understanding of territorial phenomena and the development of technical projects, but also all organizational, institutional, social and individual changes which allow the territory to gain in reflexivity, autonomy, adaptability and affirmation of its own identity.

In the field of IWRM in France, the importance of "*animateurs*" in charge of sub-basin has been outlined (Richard-Ferroudji 2008). More than a job, it is sometimes perceived as a vocation. These professional staff act as mediators, collecting people's opinions and reformulating them in adequate arenas. But they also have to deal with personalities and fragile trade-offs. They need to express empathy towards members of the water community they intend to build and towards the environment itself. *Animateurs* communicate to build a common ground and promote a basin public interest vis-à-vis more local claims.

These territorial engineering resources embrace a broad range of skills and tools because they address various actors with heterogeneous representations and codes. Informational, communicational and mediation issues occupy a central place, either in micro-events which mark the life of a "learning territory" (Herbaux 2007), or in more permanent macro-arrangements for sharing information, signs and their interpretation, or for building and perpetuating the symbols of the territorial identity.

2.2.3 Informational and communicational arrangements as additional means for territorial mediation

To formalize the informational and communicational dimensions at work in a territorial intelligence process, we use the concept of Distic[4] (in English: socio-technical information and communication arrangements) developed by the I3M Laboratory. I3M defines the 'STICA' as "a place of mediation made up of multiple semiotic, esthetic and technical factors in interaction which link up social actors through sensory and mediated means."[5]

[4] Distic: *Dispositif Socio-Technique d'Information et de Communication.*

[5] http://i3m.univ-tln.fr/Seminaires-DISTIC.html?var_recherche=distic.

A '*STICA*' consists of three inseparable entities in a relation of reciprocal co-determination: (i) The "media product(s)" (text, speech, film, graphics, tables, maps, hypermedia, etc.), which require the mastery of specific languages (e.g., a mapping language). Media products are disseminated by means of agents of mediation, either technical (channel, display device, etc.) or human (expert, facilitator, etc.), (ii) the "area of social cooperation for production" characterized by the intentionality of the designers of the STICA and the media products, (iii) the "area of social cooperation for reception" in which the participants are not just receivers, not simple message decoders, but autonomous and reflexive social subjects, with multiple resources, who can divert the interpretations and uses of the media products originally expected by the designers (Berten 1999).

Different forms of mediation can take place within a STICA: technological mediation between the individual and the technique, social mediation between actors, and semio-cognitive mediation between the thought and media products (Meunier and Peraya 2010).

Preferring the concept of STICA to those more classical of information systems, IC-Tools (Maurel et al. 2007) or communication plans is consistent with the shift of information paradigm required by territorial intelligence as advocated above. It allows us to highlight the intersubjective dimension of communication, the intentionality of the designers of the STICA as well as the autonomy of the users.

Several kind of STICA exists and they can play various functionalities according to their place in the territorial decision making process and the intentionally of their designer (Bertacchini et al. 2013). It is a cyclical process, organized in several iterative phases, in response to a problem, either identified locally or imposed externally (e.g., a SCoT imposed by the State to coastal municipalities): (i) the organization of actors affected by the problem (ii) a phase of intelligence (inventory, diagnosis, prioritized issues) (iii) a prospective phase to imagine the future to address these issues (iv) a modeling and choice phase (comparison of scenarios) (v) a phase of development and implementation of an action plan (vi) a phase of monitoring and evaluation to measure the effects of actions and possibly redirect the remaining ones. See Maurel et al., this volume, for a fuller discussion of these matters and a concrete example on a case study.

3 Discussion

To face the issue of IWRM practical implementation, we proposed (Plant et al., this volume) in this chapter a complementary approach centered on the territorial dimension of development at local scale which includes the water component. The key question that we wanted to develop here was: What are the concepts and lessons of territorial development that can be integrated in IWRM to better incorporate people and place in river basin management?

We first defined the concept of "territory" in the French context and have shown that in the English-speaking community, the closest concept, when applied at local scale, was not the one of "territory" but instead the concept of "place". This concept of territory can be considered as relevant because of its potential for mainstreaming the water dimension in a wider territorial field that includes other aspects that mobilize as much, or even more, the public and elected officials: housing, transport, economic and cultural activities. By representing the territory as a complex system using a meta-model in three levels (physical, logical and identity), we have added the symbolic and meaningful dimension of a shared territorial project for the future, which can generate feelings of a collective destiny and help the territory to emerge and become autonomous in relation to other territories.

Building such a territory is based on a collective process that we called "territorial intelligence", in which the informational and communicational aspects are central. Top down public policies, which still remain largely sectoral, at least in Europe and France, may constitute triggers for the engagement of local actors when these sectoral issues can be articulated within a wider local political project for the development of the territory. In this approach, the main integration effort therefore is carried out locally, both at technical and political level, and requires specific conditions and resources that are not all transferable to other contexts.

However several general limitations can also be outlined in this type of approach. The absence of direct universal suffrage in election for inter-municipal institutions weakens the engagement of the elected officials and those of citizens in building a shared and legitimate vision for the future at the scale of inter-municipal territories where the implementation IWRM is relevant. The technical instruments of planning and management which are mobilized remain sectoral. Their coordinated articulation leads to governance systems and costs of transaction that are prohibitory for the whole of the actors. The first concerned are the engineers who have to cope with participation fatigue. They must also find a balanced relationship with their elected officials to avoid their disengagement and to help them understanding this new territorial complexity.

Thus, it seems that this attempt of integrated management through a territorial approach by coordinating various scales and sectoral policies with their own instruments is also reaching some limits. It is likely that new paradigms and new instruments, fewer, simpler but intrinsically more integrative will have to be found in the future.

References

Barthélémy, C. 2005. Les savoirs locaux: entre connaissances et reconnaissance. VertigO—la revue électronique en sciences de l'environnement [En ligne] 6(1).
Bateson, G. 1972. Steps to an Ecology of Mind: Collected Essays in Anthropology, Psychiatry, Evolution, and Epistemology. Chandler Press.
Bertacchini, Y. 2000. Information et veille territoriales: représentation du complexe local et émergence d'un projet d'intelligence territoriale. Thèse de doctorat en Sciences de l'information et de la Communication, Université d'Aix-Marseille III, Laboratoire CRRM, 256 p.
Bertacchini, Y. 2002. Intelligence territoriale. Volet 1. Territoire et territorialités. Toulon, septembre 2002, Presses Technologiques, Collection Les ETIC.
Bertacchini, Y. 2004a. Entre information & processus de communication: l'intelligence territoriale. Les Cahiers du Centre d'études et de Recherche, Revue Humanisme et Entreprise, 267 p.
Bertacchini, Y. 2004b. Le territoire, une enterprise d'intelligence collective à organizer vers la formation du capital formel. Communication & Organisation 25: 35–51.
Bertacchini, Y., P. Maurel, P. Deprez and R. Plant. 2013. Spatial information & communication arrangements: a contribution to territorial intelligence. International Journal of Computer and Information Technology 4(1): 19–28.
Berten, A. 1999. Dispositif, médiation, créativité: petite généalogie. Hermès 25: 33–47.
Biswas, A.K. 2004. Integrated water resources management: a reassessment—a water forum contribution. Water International 29: 248–256.
Biswas, A.K. 2005. Integrated water resources management: a reassessment. pp. 319-336. *In*: A.K. Biswas, O. Varis and C. Tortajada (eds.). Integrated Water Resources Management in South and South-East Asia. Oxford University Press, Oxford.
Bonnet, M. and D. Desjeux (dir). 2000. Les territoires de la mobilité. Sciences sociales & sociétés, Puf., 224 p.
Bouleau, G., A. Richard-Ferroudji, E. Mostert, S. Junnier, I. Borowski Masser, E. Interwies, D. Ridder, P. Bots, S. Barone, P. Maurel, G. Abrami, F. Cernesson and S. Richard. 2010. I-Five European Project—Research Report No. 3 & 4: innovative instruments and institutions in implementing the Water Framework Directive: Lessons learnt for the second implementation cycle of the WFD. Case Study Cross Comparison & the QuickScan method. Montpellier, France, 77 p.
Brooks, R.K. and S.M. McLachlan. 2008. Trends and prospects for local knowledge in ecological and conservation research and monitoring. Biodiversity and Conservation 17(14): 3501–3512.
Carter, N., R.D. Kreutzwiser and R.C. de Loe. 2005. Closing the circle: linking land use planning and water management at the local level. Land Use Policy 22(2): 115–127.
Conca, K. 2006. Expert networks: the elusive quest for integrated water management. pp. 123–165. *In*: K. Conca (ed.). Governing Water: Contentious Transnational Politics and Global Institution Building. MIT Press, Cambridge, MA.
de Rosnay, J. 1975. Le macroscope. Paris.
Debarbieux, B. 1999. Le territoire: Histoires en deux langues. A bilingual (hist-)story of territory. pp. 33–46. *In*: C. Chivallon, P. Ragouet and M. Samers (eds.). Discours scientifique et contextes culturels. Géographies françaises à l'épreuve postmoderne. Bordeaux, Maison des Sciences de l'homme d'Aquitaine.
Deleuze, G. and F. Guattari. 1987. Capitalism and Schizophrenia: Volume 2: A Thousand Plateaus. Minneapolis, MN [1980] translated by B Massumi, University of Minnesota Press.
Di Méo, G. 1998. Géographie sociale et territoire. Nathan université, Paris.
Elden, S. 2010. Land, terrain, territory. Progress in Human Geography 34(6): 799–817.

Ffolliott, P. and K. Brooks. 2014. Integrated management of water and other natural resources in river basins in the United States of America. pp. 60–81. *In*: V.R. Squires, H.M. Milner and K.A. Daniell (eds.). River Basin Management in the Twenty-first Century: Understanding People and Place. CRC Press (this book).
Ghiotti, S. 2006. Les Territoires de l'eau et la décentralisation. La gouvernance de bassin versant ou les limites d'une évidence. Développement durable et territoires [En ligne] (Dossier 6: Les territoires de l'eau).
Gibson, J.J. 1979. The Ecological Approach to Visual Perception. Houghton-Mifflin, Boston.
Global Water Partnership. 2000. Integrated water resources management. Technical Advisory Committee (GWP-TAC)—TAC Background Paper, 4 p. Stockholm, Sweden.
Global Water Partnership. 2004. Integrated water resources management (IWRM) and water efficiency plans by 2005. Why, what and how? Technical Advisory Committee (GWP-TAC)—TAC Background Paper, 10 p. Stockholm, Sweden.
Herbaux, P. 2007. Intelligence territoriale. Repères théoriques. Paris.
IWA UNEP. 2002. Industry as a Partner for Sustainable Development: Water Management. IWA/UNEP, London.
Jaspers, F.G.W. 2003. Institutional arrangements for integrated river basin management. Water Policy 5: 77–90.
Jeffrey, P. and M. Gearey. 2006. Integrated water resources management: lost on the road from ambition to realization? Water Science & Technology 53(1): 1–8.
Johnson, D. 2013. Governing frameworks for sharing actionable knowledge. The Electronic Journal of Knowledge Management 11(2): 127–138.
Landel, P.A. 2007. Entre politique publique et action publique: l'ingénierie territoriale. pp. 117–122. *In*: A. Faure and E. Négrier (eds.). Les politiques publiques à l'épreuve de l'action locale. Critiques de la territorialisation. L'Harmattan.
Lankford, B. and N. Hepworth. 2010. The cathedral and the bazaar: monocentric and polycentric river basin management. Water Alternatives 3: 82–101.
Le Coadic, Y.F. 2004. Usages et usagers de l'information, Armand Colin/ADBS Editions.
Le Groupe Going and Y. Bertacchini. 2004. Territoire physique et territoire virtuel. Quelle cohabitation? Revue ISDM 6: 77–89.
Lotman, J.M. 1985. La Semiosfera. Marsilio Editor, Venice, Italia.
Maurel, P. 2012. Signes, données, representations spatiales: des éléments de sens dans l'élaboration d'un projet de territoire intercommunal. Application au territoire de Thau. Thèse de doctorat en Sciences de l'Information et de la Communication (sous la dir. Yann Bertacchini), Université du Sud Toulon/Var, 583 p.
Maurel, P., M. Craps, F. Cernesson, R. Raymond, P. Valkering and N. Ferrand. 2007. Concepts and methods for analysing the role of IC-tools in social learning processes for river basin management. Environmental Modelling and Software 22: 630–639.
McDonnell, R. 2008. Challenges for integrated water resources management: how do we provide the knowledge to support truly integrated thinking? Water Resources Development 24(1): 131–143.
McGinnis, M. 1999. Bioregionalism. Routledge, London.
Medema, W., B.S. McIntosh and P.J. Jeffrey. 2008. From premise to practice: a critical assessment of integrated water resources management and adaptive management approaches in the water sector. Ecology and Society 13(2): 29–29.
Meunier, J.P. and D. Peraya. 2010. Introduction aux théories de la communication. 3ème édition. De Boeck, Bruxelles.
Mitchell, B. 2006. IWRM in practice: lessons from Canadian experiences. Contemporary Water Research & Education pp. 51–55.
Mukhtarov, F. and A. Cherp. 2014. Hegemony and contextual governance: integrated water resources management and global water discourse. pp. 3–21. *In*: V.R. Squires, H.M. Milner and K.A. Daniell (eds.). River Basin Management in the Twenty-first Century: Understanding People and Place. CRC Press (this book).

Nonaka, I. and H. Takeuchi. 1995. The Knowledge-Creating Company. Oxford University Press, New York.

Pahl-Wostl, C., E. Mostert and J.D. Tàbara (eds.). 2008a. Social learning in water resources management. Ecology and Society 13(1): 493–586.

Pahl-Wostl, C., J. Newig and D. Ridder. 2008b. Linking public participation to adaptive management. pp. 150–173. *In*: P. Quevauviller (ed.). Groundwater Science and Policy. An International Overview. S., Cambridge.

Pinson, G. 2006. Projets de ville et gouvernance urbaine. Revue Française de Science Politique 56(4): 619–651.

Plant, R., P. Maurel, O. Barreteau and Y. Bertacchini. 2014. Developing territorial intelligence in the Thau, France. pp. 446–466. *In*: V.R. Squires, H.M. Milner and K.A. Daniell (eds.). River Basin Management in the Twenty-first Century: Understanding People and Place. CRC Press (this book).

Plummer, R., R. de Loë and D. Armitage. 2012. A systematic review of water vulnerability assessment tools. Water Resources Management 26: 4327–4346.

Polanyi, M. 1983. Tacit Dimension. Peter Smith Publisher Inc., London.

Raffestin, C. 2012. Space, territory, and territoriality. Environment and Planning D: Society and Space 30: 121–141.

Richard-Ferroudji, A. 2008. L'animateur de basin versant: Insuffler vie à une communauté de l'eau. *In*: C. Gramaglia (ed.). L'eau: un bien commun à composer, Cosmopolitiques 17. Editions Apogée.

Schwarz, E. 1992. A generic model describing the complexification and autonomization of natural systems and its epistemological consequences. 6th International Conference on Systems Research Informatics and Cybernetics, Baden-Baden.

Shannon, C.E. and W. Weaver. 1949. The Mathematical Theory of Communication. University of Illinois Press.

Slater, S., S. Marvin and M. Newsom. 2004. Land use planning and the water sector. Town and Country Planning Review 65: 375–397.

Squires, V. 2014. People and place as determinants of success or failure to meet management objectives in river basin management. pp. 371–396. *In*: V.R. Squires, H.M. Milner and K.A. Daniell (eds.). River Basin Management in the Twenty-first Century: Understanding People and Place. CRC Press (this book).

Thoenig, J.C. and P. Duran. 1996. L'État et la gestion publique territoriale. Revue française de science politique 46(4): 580–623.

Timmerman, J. and C. Pahl-Wostl. 2008. The adaptiveness of IWRM. Analysing European IWRM Research. London, Moltgen, J. (eds.) IWA Publishing.

Von Bertalanffy, L. 1968. General System Theory: Foundations, Development, Applications. George Braziller, New York.

Wenger, E. 2000. Communities of practice and social learning systems. Organization 7(2): 225–246.

White, G. 1998. Reflections on the 50-year international search for integrated water management. Water Policy 1(1): 21–27.

3

Addressing Implementation Deficits Related to IWRM in Canada

Bruce Mitchell

SYNOPSIS

Three experiences from Canada highlight that an ecosystem or holistic approach to water management has been implemented effectively at various spatial scales. The purpose here is to highlight those experiences, and then identify 'lessons learned' related to facilitating action related to the concept of Integrated Water Resource Management (IWRM).

Keywords: British Columbia, Fraser River Basin, indigenous, Lake Simcoe, lessons learned, multi jurisdictional cooperation, Ontario, USA

1 Effective Implementation of Integrated Water Resource Management: Examples

1.1 Ontario Conservation Authorities

Thirty-six catchment-based Conservation Authorities operate in Ontario, based on legislation passed in 1946. Conservation Ontario (2012), the umbrella organization for the authorities, refers to their basic approach as 'integrated watershed management' (IWM), which, in its words, "... focuses on water and related resources, including aquatic and terrestrial

Geography and Environmental Management, University of Waterloo, 200 University Avenue West Waterloo, Ontario N2L 3G1, Canada.
Email: mitchell@uwaterloo.ca

biodiversity and ecosystems, addressing the inter-relationships between them and the broader socio-economic systems which they support," an approach analogous to IWRM.

The Grand River Conservation Authority (GRCA) and the Lake Simcoe Region Conservation Authority (LSRCA) each received the international Thiess River Prize, in 2000 and 2009, respectively, awarded in Australia. The prize recognizes "excellence in river, waterway or catchment management." A key consideration is "measurable outcomes".

The GRCA was recognized for long-term restoration in its 6,800 km^2 catchment in southwestern Ontario, especially for water quality improvement. The LSRCA received the award because over 20 years the phosphorus in its 3,303 km^2 catchment had been reduced by 35%, and dissolved oxygen in deep waters of Lake Simcoe had increased by over 200% (Lake Simcoe Region Conservation Authority 2008; 2010; 2011; Ontario MOE 2009). Thus, the two CAs were recognized for documented positive outcomes based on their integrated approaches to water management (Mitchell, in press).

1.2 Areas of Concern and Remedial Action Programs in the Great Lakes ecosystem

During 1987, the United States and Canada amended the Canada-US Great Lakes Water Quality Agreement (GLWQA). The GLWQA committed both countries to restore and maintain the chemical, physical and biological integrity of the Great Lakes ecosystem. The 1987 Protocol provides for Remedial Action Plans (RAPs), for Areas of Concern (AOC). AOCs were identified as 'hot spots', usually a harbour, estuary or bay, with serious environmental degradation.

In 1987, 43 AOCs were identified, 15 entirely or partially in Canada, with five shared with the United States. Three of the AOCs in Ontario have been delisted by the Canadian federal and Ontario provincial governments, and the International Joint Commission: Collingwood Harbour (November 1994), Severn Sound (January 2003) and Wheatley Harbour (April 2010). Two other AOCs have been delisted in the United States (Oswego River, New York, January 2006; Presque Isle Bay, Pennsylvania, February 2013). To be delisted, a RAP had to have been developed and implemented through three stages. The third stage involves implementing recommended actions and monitoring progress. The five AOCs were delisted because changes relative to explicit goals in the RAPs had been verified.

1.3 Fraser River Basin, British Columbia

The Fraser River basin occupies about 234,000 km^2 in British Columbia, and enters the Strait of Georgia in the Greater Vancouver area. In 1985, the Fraser River Estuary Management Program (FREMP) was created to facilitate a coordinated approach by federal, provincial, municipal and other organizations to inter-related water issues in the estuary. In 1992, a basin-wide approach was established through the Fraser Basin Management Program (FBMP). Guided by a multi-stakeholder board, the FBMP focused on the entire basin and aimed to achieve sustainability through coordinating and collaborating initiatives, enhancing institutional arrangements, identifying priorities, and modifying public views. Audits were conducted related to various management issues, mechanisms and processes for improved coordination were introduced, demonstration projects were started, State of the Basin Reports were initiated, and a 'Charter of Sustainability' was prepared. Progress was achieved, but challenges arose due to modest funding, lack of accountability, and resistance from some government agencies (Watson 2005, 420).

The Charter of Sustainability became the launching pad for a subsequent organization, the Fraser Basin Council (FBC), established in 1997. Its goal is to achieve sustainable development for the entire basin through collaboration among governments at federal, provincial and municipal levels, Aboriginal peoples, the private sector and civil society organizations.

Watson (2005, 437–438) evaluated the FBC, and concluded significant progress has occurred, with key contributors being a common vision, adaptive capacity, adequate resources, independence, balanced representation and power, and tangible outputs and outcomes. The FBC is not an operational organization such as the CAs in Ontario, nor as focused as the RAPs in the Great Lakes system, but represents another model demonstrating tangible success in implementing an integrated approach to basin-wide management (Dorcey 1997; Hanna 1999).

2 What is the Implementation Challenge?

Managers must address diverse issues to translate strategies and plans into actions through IWRM (Montgomery et al. 1995; Hooper et al. 1999; Mitchell 2005; 2006; 2009a; 2009b; 2011; Joseph et al. 2006; White and Boswell 2007; Biswas 2008a; 2008b; Morrison and Brown 2010). Challenges include at least the following.

(1) Water problems are usually connected to other resources, such as land, forests and fish, and therefore inter-relationships need attention. A dilemma is how much to include in an integrated approach.

(2) Water has many dimensions (e.g., quantity/quality, upstream/downstream, groundwater/surface water), all requiring attention.
(3) The boundaries of catchments or river basins rarely align with political or administrative boundaries, creating governance issues.
(4) Water-related problems often have developed and evolved over decades, with numerous causes and inter-connections. Thus, scientists and managers deal with environments characterized by complexity, uncertainty and change, and, as a result, have incomplete knowledge.
(5) Economic and social systems need attention along with the hydrological system. As a result, scientific or technical knowledge is necessary but not sufficient, and processes to engage stakeholders are needed.
(6) Various stakeholders often focus on their own needs and interests, and do not always consider implications for others. Such a 'siloed' perspective is frequently reinforced by public agencies which, due to mandates, structure and legal accountability, focus on one part or subset of a river basin system.
(7) Needs within a river basin or catchment compete with other societal needs, meaning that sufficient financial and human resources are not usually allocated.

Given the above characteristics, the need is to have capacity to address significant and complex problems, rarely with complete knowledge, in a collaborative and coordinated manner, if scarce resources are to be allocated effectively, efficiently and equitably for IWRM to be successfully implemented.

3 Overcoming Implementation Deficits

In this section, 10 factors or variables relevant to effective IWRM implementation in the three experiences in Canada are reviewed. Observations are based on research literature, agency reports, comments from respondents to a set of questions related to implementation, and over 40 years of research focused on and experience with IWRM. While formidable implementation challenges exist related to IWRM, much has been learned related to implementation in various management contexts (Pressman and Wildavsky 1971; Gunn 1978; Sabatier and Mazmanian 1983; Mazmanian and Sabatier 1989; Goggin et al. 1990; Weale 1992; MacKenzie 1993; Hill and Hupe 2002; O'toole 2004; Brynard 2009).

3.1 Recognize Context, Develop Custom-designed Solutions and Ensure Appropriate Scope

Successful implementation is more likely if IWRM is designed with reference to the reality of a specific time and place. In contrast, a generic IWRM template is less likely to lead to success, other than for a basic data system. This view is highlighted by Heathcote (1998, 391), who remarked that "... a single framework for effective integrated water resources planning is not possible for all countries, or even for all regions within a single country...". Her rationale was each county or region has distinctive environments and natural resources, demographic patterns, economic conditions, and political, governance and legal structures. As a result, she concluded that "Planning programs may therefore fail if they are not successful in matching projects to watershed conditions,...". Her view is reinforced by Gurtner-Zimmermann (1995, 241–242), after a review of RAP programs, who observed that, "... the approach of an advanced RAP... has evolved from all-encompassing in plan development to pragmatic and focused on high-priority issues to enhance chances of accomplishing key goals in the implementation phase."

McLaughlin and Krantzberg (2011, 391–392) identify another contextual feature deserving attention. They differentiate between acute and chronic deficits in policy implementation. They (2011, 392) argue that acute deficits occur in situations in which "... ingrained behaviors of organizations and conflicting political agendas create political uncertainty that undermines the continuation and efficacy of management interventions." They suggest acute deficits are likely to be caused by "... competing problem definitions and political priorities that potentially inhibit policy processes and the achievement of policy outcomes... When these factors dominate, there is little that those carrying out policy can do to overcome them." They also suggest that chronic deficits related to implementation may become the most challenging because, "... they become an accepted fact of life and less likely to attract appropriate action ..." in contrast to acute deficits which usually "... represent "crises" that tend to attract disproportionate political or public attention." Chronic and acute deficits thus need to be understood in their local context to be overcome.

In terms of ensuring appropriate scope, Mitchell (1990, 4–5) has cautioned that while there is an intuitive appeal to examine an array of variables and their relationship, casting the net too wide may make it difficult to complete a watershed plan in a timely manner, and to establish priorities. As a result, he argued that at a strategic level, it is sensible to take a comprehensive perspective. However, at an operational scale, an integrated approach is more appropriate. What is the difference? By definition, a comprehensive approach indicates attention to all variables and relationships. In contrast, an integrated approach focuses on those variables and relationships that cause significant variance in a system and are amenable to management interventions. Thus, both perspectives reflect a systems or holistic approach, but an integrated perspective is more focused and pragmatic.

3.2 Maintain a Long-term Perspective

A long-term perspective is necessary, given that many problems addressed by IWRM normally have taken years if not decades to emerge, and are often complex. Thus, immediate solutions are unlikely. A sustained and long-term commitment can be difficult to obtain, however, especially since the time frame for many elected decision makers usually is until the next election. Such individuals often want short-term tangible results, which can be pointed to as endorsements of their decisions.

Over 20 years ago, Hartig and Zarull (1992, 263) mentioned the importance of a long-term perspective. In their view:

> It took decades to manifest the degree and extent of toxic substance contamination in Areas of Concern, and it would be naïve to think that such problems could be resolved in a short period of time. Therefore, it must be recognized that RAPs are a long-term process in which numerous obstacles must be overcome to sustain progress.

It is worth emphasizing that the comment immediately above, as well as many of those in the preceding and following sections, was provided several decades ago. Thus, we have known for decades about such challenges in implementing IWRM, and thus have no excuse not to anticipate and address them.

3.3 Identify a Vision

If degraded conditions are to be reversed, or existing conditions enhanced, a vision is needed regarding a desirable future, because IWRM is a 'means' not an 'end'. Thus, the desired future should be identified before IWRM is applied.

Various researchers have emphasized the need for a clear vision. For example, Hartig and Law (1994, 862), commented that explicit partnership agreements assist "... stakeholders to embrace a common vision and new institutional structures... A partnership agreement is typically a succinct vision statement and a set of principles that are used to resolve a problem. ...". They highlighted that a key aspect of such agreements is recognizing common ground among partners. Such recognition, in their view, "... increases trust among stakeholders."

Krantzberg (2003, 641) concurs, stating that a "clearly articulated and shared vision" is essential for implementation to succeed. A similar perspective was offered by McLaughlin and Krantzberg (2011, 393). In their view, an explicit need is for "... complete understanding of, and agreement upon, the objectives to be achieved, and that these conditions persist throughout the implementation process."

It is sometimes stated that if you do not know where you want to go, any road will do. This truism is a reminder of the importance of having an articulate vision before using IWRM to achieve it.

3.4 Create Legitimacy and Credibility for the Vision, Goals, Objectives, Strategies, Plans and Projects

Several elements provide legitimacy or credibility for IWRM: a legal foundation; political commitment; appropriate policies, administrative support, and human resources; sufficient funding; and, suitable governance arrangements, including partnerships. The more of these present, the more probable will be successful implementation.

Regarding a *legal foundation*, Hartig and Zarull (1992, 267) concluded that, for successful implementation, "... RAPs should be incorporated into law, either by amalgamating RAPs/Areas of Concern with existing statutes or by developing new statutes ... Such statutes must include the direction, authority, and funding to develop and implement RAPs." Gurtner-Zimmermann (1995, 235) strongly agrees, stating that RAPs have required reliance on local-level legislation to achieve 'formal legitimacy'.

To achieve successful implementation with an integrated approach, Calbick et al. (2004, 48, 49) called for a "... an empowering statute that contains a clear description of authority and responsibilities, ...", because, in their view, "The agencies with a mandate are provided with clear and consistent directives from their empowering legislation. Moreover, the statutes for these agencies more than adequately identify the causal mechanisms, as well as provide sufficient jurisdiction over enough factors to enhance the likely attainment of the desired goals, ...". At the same time, they realized that a statutory foundation does not guarantee implementation success because of difficulties that may arise from "... interlaced, overlain, and sometimes contradictory legislation, which engenders conflict."

A staff member from the GRCA stated that enabling legislation is one of the most important factors for successful implementation. Given that stability of a watershed organization is an important factor in the long-term success of watershed management, then in the view of that person, "Enabling legislation provides the framework for administrative, funding and regulatory procedures, partnership building and legitimacy of actions." Staff of the LSRCA concur. As they observed, "In 1946, the *Conservation*

Authorities Act was legislated enabling the creation of local, watershed-based organizations called Conservation Authorities. Inherent in the legislation was the need for municipalities to partner and request the formation of a conservation authority. Since 1951, this municipal watershed partnership has been the foundation of success for the Lake Simcoe Region Conservation Authority. It is the basis of our governance model and is critical to achieving alignment of broad, collective efforts, including resources, within the Lake Simcoe watershed." Furthermore, the staff noted that subsequent legislation provided additional legitimacy. Specifically, they observed that "In 2008 and 2009, the Province of Ontario introduced the *Lake Simcoe Protection Act* and the *Lake Simcoe Protection Plan*. These important initiatives have strengthened alignment of effort through both policy and expanded collaborations and partnerships."

Political commitment is also helpful. According to Gurtner-Zimmermann (1995, 240), "Political support and funding have been particularly important for the advancement of RAP." Furthermore, in the view of his respondents, "political support, which includes the categories "funding", "time", "administrative staff", and "program coordination" is most difficult to get in the RAP process."

Hartig and Law (1994, 856) concur, observing, "… getting support from the entire watershed may be an onerous task, as not all political jurisdictions may be willing to get involved."

Hartig and Zarull (1992, 266) highlighted the importance of human and financial resources. In their words, "The availability of adequate human and financial resources is a common concern… Available resources with state, provincial and federal government environmental programs are stretched to the limit, and citizens have limited time and resources to bring to the planning process." One option to address such inadequacies was identified by Hartig and Law (1994, 860) who observed, "Resource limitations and increasing responsibilities within government call for cooperative initiatives that engage a wider segment of stakeholders in remediation and prevention."

McLaughlin and Krantzberg (2011, 392) agreed, noting that lack of funding often was a key limitation. Furthermore, in their view, "… too much can be expected too soon, especially when changes in attitudes or behaviors are required. This is common where funds are available only within an unrealistically short time, shorter than the program can effectively absorb them." They concluded that when essential funding is not available then 'acute bottlenecks' commonly emerge.

In the view of a staff member at the GRCA, "Insufficient funding leads to a piecemeal, issue-oriented approach to dealing with resources issues, with disproportional time and effort spent on soliciting funds for projects and programs, rather than implementation. A system of apportioning costs and benefits equitably across a watershed can help secure consistent municipal participation and funding." The latter comment is insightful, and highlights that a well-constructed funding model can also contribute to engagement by and funding contributions from municipal levels of government.

Governance is critical (Grigg 2010; Lockwood et al. 2010; Wick and Larson 2012). Hartig and Law (1994, 860) stated, "Once a clear road map… has been determined, RAP institutional frameworks need to compare the recommended actions with the existing authorities and regulatory and nonregulatory mechanisms to identify any gaps. If gaps exist, RAP institutional frameworks should be empowered to recommend or help create alternative means to implement RAP recommendations, monitor progress, and ensure responsible parties are held responsible." Calbick et al. (2004, 51) concur, arguing that multi jurisdictional cooperation, "…. is critical for actually making the implementation efforts successful." Specifically, they commented that,

> Finding a path through the maze of special-purpose government agencies, …, all possessing different jurisdictional boundaries and mandates is an extraordinarily difficult task. Typically, these agencies jealously guard their decision-making authority and are reluctant to relinquish any power pursuing the possibility of achieving a greater public good. Moreover, different jurisdictions have different, sometimes competing, or conflicting, objectives and interests.

Given the above challenges, McLaughlin and Krantzberg (2011, 391) conclude that, "The creation of conditions for ordered, collective action… is a matter ultimately of governance…". Thus, governance deserves careful attention, along with legal, political, financial and human resource matters.

3.5 Ensure One or more Leaders or Champions

Strong consensus exists that leaders or champions are important for facilitating implementation of IWRM. The explanation is basic. Given many needs, interests and values of diverse stakeholders, as highlighted in preceding and following sections, sustained leadership is needed to help create a shared vision, convince stakeholders that the 'common good' must be a priority, and facilitate understanding and collaboration among many participants.

Such a need has been recognized for some time. To illustrate, MacKenzie (1993, 142–143), after reviewing experience with three RAPs, recommended a need to "Identify key individuals who can guide the RAP process through all its permutations. There are many pressure points at which the RAP process can break down: the data are complex, individuals may feel threatened, decisions may become conflict-laden and time consuming among others. At these times, key individuals may demonstrate leadership and use the strength of their personalities and professional positions to maintain forward momentum."

Based on her experience with the RAP in Collingwood, Ontario, Krantzberg (2003, 648) noted that a key element for successful implementation of IWRM was to "Find a strong leader and stay focused on the task at hand." She also commented that it helps to "Engage local leaders who are committed to their community and can affect change" (645).

The significance of 'leadership' has been recognized for decades. For example, almost 25 years ago, Mazmanian and Sabatier (1989) highlighted the need for strong leaders or champions, and emphasized they need to have both managerial skills and political acumen. The many complications and challenges encountered in achieving collaboration and coordination in IWRM can lead to discouragement and inclination to disengage. Especially at such vulnerable stages, the value of an effective leader or champion can be critically important.

3.6 Use a Multi-stakeholder Approach

Many interests and needs have to be considered during IWRM, extending over environmental, economic and social dimensions. As a result, IWRM cannot be based only on technical or scientific expertise. Capacity has to be created to engage with stakeholders, and, if they believe they have ownership related to processes and outcomes, the more likely they are to be supportive and become engaged partners. Such results are the ideal, but often challenging to obtain, because competing and/or conflicting interests and aspirations exist. However, much has been learned about applying a multi-stakeholder approach.

Views based on experience in implementing IWRM reflect the importance of multi-stakeholder engagement. For example, Hartig and Law (1994, 858–859) observed,

> Broad-based participation and shared decision-making power are viewed as key success factors... institutional frameworks provide a forum for mutual understanding, help establish trust, and help orient decision makers to an ecosystem approach. Further, sharing decision-making power within institutional frameworks can also be a good mechanism of getting organizations to participate through enlightened self-interest....

To achieve meaningful stakeholder engagement, Hartig and Law (1994, 860) recommended striving for: (1) agreement on goals, objectives, and benchmarks to confirm progress; (2) identification of key actions, responsibilities, and sequencing of actions; (3) demonstration of commitments from responsible participants; and, (4) identification of activities to resolve issues which could delay implementation.

The above views are reinforced by Krantzberg (2003, 646–647), who suggests three factors for implementation success:

> *Ownership*: Ownership means that when agreeing to the plan, each member overtly recognizes and takes responsibility for the resource implications for its stakeholder group. Ownership results in pride in delivery, which sustains the process;
> *Respect*: Derived from a common purpose and reliability, respect strengthens the group's credibility and ability to solve challenges as a team; and,
> *Incentives*: Incentives for achieving shared goals can differ among participants and must be respected. It is acceptable if different stakeholders extract different benefits by meeting shared goals.

A planner from the GRCA explained its approach to stakeholder engagement. "Successful plans—ones that do not collect dust—focus on a manageable scale, remain adaptable and flexible, link issues to local concerns, and build partnerships to motivate action. For this reason, the people and organizations … responsible for implementing actions need to be actively engaged in the IWM process. Collaboration builds trust and consensus, leverages effort and stretches capacity, and ensures on-going implementation and change will occur at the local level." A similar view was expressed by staff at the LSRCA. In their view, "Engaging multiple stakeholders in IWM is critical. Programs and projects must be locally based and relevant—whether science, protection, education or communication. This cannot be achieved without the meaningful participation and engagement of all community stakeholders." Such experiences and perspectives corroborate the views expressed by others.

Finally, two key lessons have been learned by the GRCA. As expressed by a staff member, one has been to develop "… processes that are collaborative, yet streamlined to match stakeholder capacity and sustain interest and enthusiasm. This requires conscious and deliberate effort and a commitment and willingness to pool resources and work collectively to resolve issues of mutual concern." The other is that "… we have to approach participation in a very strategic, yet meaningful way. To cast a wide net and to have 'all interests' at the table is sometimes an exercise in frustration… you need to have some key individuals/organizations at the table—those that can effect change and those that can influence or confound change if not onboard. Inclusivity should be invited always, but in reality, only those people who have an interest will be involved." Such practical insight reminds us that achieving collaboration and engagement is not easy, and demands conscious attention and effort.

3.7 Be Adaptable and Flexible

It was earlier noted that IWRM managers and decision makers function in a changing world characterized by complexity, uncertainty, and conflict. As a result, flexibility and adaptability are needed. Indeed, an approach characterized as 'adaptive management' is well developed, and needs to be incorporated into IWRM.

The importance of flexibility and adaptability has been highlighted by many. Hartig and Zarull (1992, 270) argued that, "plans must be updated periodically to reflect new data, information or new technologies…" In addition, Hartig and Law (1994, 860) noted "… the need for more flexible and adaptable institutional frameworks to coordinate interrelated programs and initiatives to achieve user restoration." A similar view has been presented by MacKenzie (1993, 143) when she observed that it is important to "Recognize that the RAP is unlikely to proceed in a predetermined and orderly fashion. The ecosystem approach needs ample time and opportunity to evolve."

Jones and Taylor (2009, 251–253) believe that one implementation challenge is due to previous successes which often resulted from solving relatively easy problems, creating a more valuable ecosystem but with more difficult problems remaining. In their view,

> … the problems we seek to solve today and into the future are more challenging and complex, primarily because they are more subtle, and involve many interacting ecosystem components. By reducing the overwhelming effects of a few, large problems … we have exposed a more complex array of more subtle problems, for which solutions are less easy to determine. …
>
> … questions are far more difficult, and since we cannot presume to know the answers, they call for us to manage adaptively, that is, to use management decisions themselves as a tool to reduce uncertainty and seek acceptable solutions.

Finally, McLaughlin and Krantzberg (2011, 393) provide a refreshingly candid observation, when remarking that, "In the presence of inherent uncertainty and confounding factors (and therefore, conversely, in the absence of unequivocal cause and effect relationships), we face [implementation] deficit conditions that cannot be resolved entirely." No approach can totally overcome uncertainty, complexity and change, but applying an adaptive management approach offers opportunity for enhancing implementation.

3.8 Identify, Monitor and Assess Outputs and Outcomes

While a clear vision is needed to indicate a desirable future, capacity to monitor and assess outputs and outcomes relative to goals and objectives also is needed. Too often, sufficient human and financial resources are not allocated to this task for IWRM projects.

The staff at the LSRCA argues that it is critical to ensure that "…, public confidence must be built on a foundation of sound science." They highlighted that "Watersheds are complex and dynamic systems intimately connected to human activities, locally and globally. LSRCA's strength and credibility lays in its commitment to understand and translate the science of the watershed in ways that are meaningful and relevant to its full range of municipal, provincial and federal governments, and community partners. Through its science programs, LSRCA collaborates and partners with many stakeholders to: (1) monitor, report and forecast watershed conditions, improvements and trends; (2) prepare sub watershed plans to respond to local and regional issues and opportunities; (3) identify priorities for ecosystem restoration and align implementation efforts; (4) identify and evaluate emerging threats to watershed health, such as climate change; (5) explore, test and evaluate new technologies and best management practices for effectiveness in the Lake Simcoe watershed; and, (6) transfer knowledge and technology regionally, provincially, nationally and internationally."

Krantzberg (2003, 645, 646) emphasized the importance of including what she labelled 'quantifiable endpoints' which "… signify success and the achievement of the goals" as well as "allow the group to recognize progress, prioritize actions, and reach consensus…". In addition, she stressed that absence of targets and lack of knowing what progress is being made towards them can "… represent a real challenge to progress. In times of scarce resources, it is imperative that RAP practitioners affirm that investments in particular active interventions are appropriate and productive." Finally, she observed that when stakeholders in a watershed or Area of Concern see incremental progress is occurring they become more likely to volunteer to become participants (Krantzberg (2003, 648)).

Pre-conditions for credible monitoring and assessment can be identified. In the view of McLaughlin and Krantzberg (2011, 393), pre-conditions should be objectives that are "… clearly defined, specific and preferably quantified, comprehensively agreed to and understood, mutually compatible and supportive, and provide a blueprint against which policy implementation can be monitored." If such characteristics are to be realized, objectives cannot be couched "… in vague and evasive terms,…". Furthermore, attention has to be given to the reality that some 'official' objectives may be incompatible with each other, and that other 'unofficial' goals can exist and be promoted by stakeholders pursuing their self-interests. If such reality is not acknowledged and addressed, it becomes challenging to create a credible monitoring and assessment program.

3.9 Facilitate Information Sharing and Develop Effective Communication to Stakeholders

Information and education have to be provided during preparation, initiation and rollout of an IWRM program, so that stakeholders can learn about accomplishments and challenges, and enhance understanding regarding aspirations and progress. Furthermore, information and education is not an end, but should be a means to achieve deeper engagement (LSRCA staff).

Two comments support the above statement. First, Hartig and Law (1994, 861) concluded that stakeholder education should help to clarify roles and responsibilities, and also meet the education or training needs of all groups. Second, Gurtner-Zimmermann (1995, 244) noted that a key goal of education should be to help participants develop new perspectives about issues, and also to broaden their outlook through gaining understanding and respect for other views.

A fine line exists between information and/or education, and propaganda, to promote alleged merits of a program. To be credible, information and education should reflect the characteristics highlighted above by Hartig and Law, and by Gurtner-Zimmermann.

Communication can be a double-edged sword. When communication is done well, it can be a key element to cultivate trust and respect among stakeholders. However, communication can consume significant amounts of time and resources, leaving managers little time for substantive issues.

Hartig and Zarull (1992, 265, 266) highlighted several positive outcomes of effective communication. In their view, both effective communication and cooperation are necessary if different regulatory, management and public groups are to be able to understand and work with each other. In their view, the obvious positive outcome will be the creation and/or strengthening of effective coalitions. "Coalitions are essential to elevate the priority given to specific remedial actions, to obtain funding for implementation, and to develop the "strength in numbers" or substantial agreement and trust between diverse interests needed to take action."

However, they also (266) argued that because few citizens are knowledgeable about or comfortable with technical aspects of problems addressed by IWRM initiatives, scientists have to learn to communicate with non-technical individuals. Unfortunately, few scientists develop such skills, often making communication about technical or scientific issues a challenge.

Quality communication, or lack thereof, has implications for other attributes needed for effective implementation of IWRM. For example,

McLaughlin and Krantzberg (2011, 394) observed that "Lack of adequate communication and coordination by the lead agencies of governments has been a chronic problem of the Great Lakes governance regime." Their comment reminds us that, by being effective with one element (communication), it is often possible to enhance other elements (e.g., governance) needed for successful implementation.

3.10 Utilize Demonstration Projects to highlight Tangible Progress, and Profile and Celebrate Achievements

A prevalent view is that producing tangible outcomes, and celebrating them, is a powerful way to keep stakeholders engaged in IWRM. In contrast, with no or little evidence of progress, the probability increases that stakeholders withdraw, or continue with less commitment. This view is reflected by Krantzberg (2003, 649) who remarked about the "… benefit from a greater emphasis on measuring, celebrating, and marketing successes, and building the local capacity to sustain progress."

In that context, Hartig and Zarull (1992, 272) argued that "A record of success must be built into the RAP process to keep the momentum going." They suggested that one way to celebrate such progress is to publish annual progress reports about a specific IWRM program. Furthermore, if IWRM programs are being implemented in several river basins in a jurisdiction, they recommend preparing annual 'state-of-the-IWRM' events" regarding all river basin programs (Hartig and Zarull 1992, 273).

4 Implications

Attention to the above 10 characteristics or elements should help managers to improve implementation. None are 'silver bullets', and challenges always should be expected in moving IWRM from concept to action. McLaughlin and Krantzberg (2011, 394) highlight why difficulties are always likely. As they comment,

> "Our collective response to threats to ecological function and the revitalization of ecosystem resilience … requires an institutional arrangement capable of better reconciling the complex of competing political authorities, pressures, priorities, and points of view. But in remodeling governance, consider that such deficits are to be expected, in that they are so often due to social-ecological behaviors that cannot fully be predicted. A "wise" policy-maker therefore expects that policies will only partially achieve intended outcomes and, at the same time, implementation of those policies will produce unanticipated consequences that would be avoided preferably."

Given the above comment, managers need to be realistic. IWRM initiatives are unlikely to be rolled out without surprises or some unintended negative outcomes. Furthermore, such a perspective emphasizes the need to monitor; to acknowledge surprises, incomplete progress, and failures; and, to 'recalibrate' based on learning. And, evaluators of IWRM experiences are unanimous that variables causing significant implementation problems include inadequately defined or understood environmental problems; lack of sufficient funding and qualified staff; inadequate inter-program coordination; and, poor governance arrangements.

Another consideration has been highlighted by staff of the LSRCA. In responding to a question about relative importance of variables, they stated it is very difficult to identify the most significant. The reason, in their words, is that, "Perceptions, priorities and 'critical' implementation aspects change over time as progress is achieved and in response to IWM system feedback. In addition, these ... aspects—and their weighted importance—constantly respond to the program and activity at hand; for example, some aspects are more critical for new strategic direction and policy, while others are more critical for on-the-ground stewardship. In other words, these ... aspects can be somewhat cyclical in nature..."

In the context of the two sensible observations above, over decades we have learned how to improve implementation. The following ideas from Krantzberg (2003, 649) thus are appropriate to conclude this chapter. Implementation success will be enhanced when governments serve as facilitators and partnership builders, as well as provide resources and technical assistance which leverage local resources, mutually agreed decision-making processes, establishment of common objectives as well as methods for public participation and dispute resolution, and strong local leadership.

5 Acknowledgements

I am grateful to Barbara Veale of the Grand River Conservation Authority who reviewed a draft of this chapter and provided constructive suggestions which have been incorporated.

6 References

Biswas, A.K. 2008a. Integrated water resources management: is it working? Water Resources Development 24: 5–22.

Biswas, A.K. 2008b. Current directions: integrated water resources management—a second look. Water International 33: 274–278.

Brynard, P.A. 2009. Mapping the factors that influence policy implementation. Journal of Public Administration 44: 557–577.

Calbick, K.S., T.I. Gunton and J.C. Day. 2004. Watershed resources planning and management: lessons learned from comparative case studies. pp. 33–55. *In*: D. Shrubsole (ed.). Canadian Perspectives on Integrated Water Resources Management. Canadian Water Resources Association, Cambridge, ON.

Conservation Ontario. 2012. Protecting the Great Lakes, Integrated Watershed Management Approach. News Byte, April 23rd, Conservation Ontario, Newmarket, ON, http://www.conservationontario.ca.

Dorcey, A.H.J. 1997. Collaborating towards sustainability together: the Fraser River Basin Management Board and Program. pp. 167–199. *In*: D. Shrubsole and B. Mitchell (eds.). Practising Sustainable Water Management: Canadian and International Experiences. Canadian Water Resources Association, Cambridge, ON.

Goggin, M.L., A. O'M. Bowman, J.P. Lester and L.J. O'Toole. 1990. Implementation Theory and Practice: Toward a Third Generation. Scott, Foresman/Little, Brown Higher Education, Glenview, Il.

Grigg, N.S. 2010. Governance and Management for Sustainable Water Systems. IWA Publishing, London.

Gunn, L. 1978. Why is implementing so difficult? Management Services in Government 33(4): 169–176.

Gurtner-Zimmermann, A. 1995. A mid-term review of Remedial Action Plans: difficulties with translating comprehensive planning into comprehensive actions. Journal of Great Lakes Research 21: 234–247.

Hanna, K.S. 1999. Integrated resources management in the Fraser River estuary; stakeholder's perceptions of the state of the river and program influence. Journal of Soil and Water Conservation 54: 490–498.

Hartig, H.H. and M.A. Zarull. 1992. Keystones for success. pp. 263–279. *In*: H.H. Hartig and M.A. Zarull (eds.). Under RAPs: Toward Grassroots Ecological Democracy in the Great Lakes Basin. University of Michigan Press, Ann Arbor.

Hartig, H.H. and N.L. Law. 1994. Institutional frameworks to direct the development and implementation of great lakes remedial action plans. Environmental Management 18: 855–864.

Heathcote, I.W. 1998. Integrated Watershed Management: Principles and Practice. John Wiley and Sons, New York.

Hill, M. and P. Hupe. 2002. Implementing Public Policy: Governance in Theory and Practice. Sage Publications, London.

Hooper, B.P., G.T. MacDonald and B. Mitchell. 1999. Facilitating integrated resource and environmental management: Australian and Canadian perspectives. Journal of Environmental Planning and Management 42: 747–766.

Jones, M.L. and W.W. Taylor. 2009. Challenges to the implementation of the ecosystem approach in the Great Lakes basin. Aquatic Ecosystem Health and Management 2: 249–254.

Joseph, C., T.I. Gunton and J.C. Day. 2006. Implementation of resource management plans: identifying keys to success. Journal of Environmental Management 88: 594–606.

Krantzberg, G. 2003. Keeping remedial action plans on target: lessons learned from Collingwood Harbour. Journal of Great Lakes Research 29: 641–651.

Lake Simcoe Region Conservation Authority. 2008. Integrated Watershed Management Plan. Lake Simcoe Region Conservation Authority, Newmarket, ON.

Lake Simcoe Region Conservation Authority. 2010. Watershed Report Card: 2009. Update. Lake Simcoe Region Conservation Authority, Newmarket, ON.

Lake Simcoe Region Conservation Authority. 2011. Strategic Plan, Focused Future 2014. Lake Simcoe Region Conservation Authority, Newmarket, ON.

Lockwood, M., J. Davidson, A. Curtis, Z. Stratford and R. Griffith. 2010. Governance principles for natural resource management. Society and Natural Resources 23: 986–1001.

MacKenzie, S.H. 1993. Ecosystem management in the Great Lakes: some observations from Three RAP sites. Journal of Great Lakes Research 19: 136–144.

Mazmanian, D.A. and P.A. Sabatier. 1989. Implementation and Public Policy. University Press of America, Lantham, MD.
McLaughlin, C. and G. Krantzberg. 2011. An appraisal of policy implementation deficits in the Great Lakes. Journal of Great Lakes Research 37: 390–396.
Mitchell, B. 1990. Integrated water management. pp. 1–21. *In*: B. Mitchell (ed.). Integrated Water Management: International Perspectives and Experiences. Belhaven Press, London and New York.
Mitchell, B. 2005. Integrated water resource management, institutional arrangements, and land-use planning. Environment and Planning A 37: 1335–1352.
Mitchell, B. 2006. IWRM in practice: lessons from Canadian experiences. Journal of Contemporary Water Research and Education 135: 51–55.
Mitchell, B. 2009a. Implementation gap. IWRA Update 22(3): 7–12.
Mitchell, B. 2009b. Integrated water resources management. pp. II.3-1 to II.3-15. *In*: Guide to Hydrological Practices: Management of Water Resources and Application of Hydrological Practices, 6th Edition, Volume II. World Meteorological Organization, Geneva.
Mitchell, B. 2011. The implementation gap: from policy to action. pp. 56–59. *In*: Water Innovation Forum: A Competitive and Innovative Agricultural Sector. Lawrence National Centre for Policy and Management, University of Western Ontario, London.
Mitchell, B. (in press). Water risk management, governance, IWRM and implementation. *In*: U. Fra (ed.). Risk Governance: The Articulation of Hazards, Politics and Ecology. Springer.
Montgomery, D.R., G.E. Grant and K. Sullivan. 1995. Watershed analysis as a framework for implementing ecosystem management. Water Resources Bulletin 31: 369–386.
Morrison, P.J. and R.R. Brown. 2010. Avoiding the presumptive policy errors of intergovernmental environmental planning programmes: a case analysis of urban stormwater management planning. Journal of Environmental Planning and Management 53: 197–217.
Ontario Ministry of the Environment, Ministry of Natural Resources, Lake Simcoe Region Conservation Authority. 2009. Lake Simcoe Protection Plan. Queen's Printer, Toronto.
O'toole, L.J. 2004. The theory-practice issue in policy implementation research. Public Administration 82: 309–329.
Pressman, J.L. and A. Wildavsky. 1971. Implementation: How Great Expectations in Washington are Dashed in Oakland. University of California Press, Los Angeles.
Sabatier, P.A. and D.A. Mazmanian. 1983. Can Regulation Work? The Implementation of the 1972 California Coastal Commission. Plenum, New York.
Watson, N. 2005. Partnerships for integrated land and water management: lessons from the Fraser basin. pp. 409–446. *In*: D. Shrubsole and N. Watson (eds.). Sustaining Our Futures: Reflections on Environment, Economy and Society. University of Waterloo, Department of Geography Publication Series No. 60, Waterloo, ON.
Weale, A. 1992. Implementation failure: a suitable case for review? pp. 141–160. *In*: E. Lykke (ed.). Achieving Environmental Goals: The Concept and Practice of Environmental Performance Review. Belhaven, London.
White, S.S. and M.R. Boswell. 2007. Planning for water quality: implementation of the NPDES Phase II Stormwater Program in California and Kansas. Journal of Environmental Planning and Management 49: 141–160.
Wick, A. and K.L. Larson. 2012. Water, people and sustainability—a systems framework for analyzing and assessing water governance regimes. Water Resources Management 26: 3153–3171.

4

Integrated Watershed Management of Water and Other Natural Resources in River Basins of the United States

Peter F. Ffolliott[1,*] and Kenneth N. Brooks[2]

SYNOPSIS

The challenge of managing water and other natural resources in a river basin has become increasingly more difficult as populations of people increase, demands on the natural resources increase, and new management technologies become available. One approach to meeting this challenge is through integrated watershed management. Application of this approach in the United States is the topic of this chapter. Integrated watershed management is aimed primarily at minimizing adverse impacts to soil and water resources, sustaining flows of high-quality water, rehabilitating degraded watersheds in poor condition, or varying combinations of these practices while maintaining a flow of watershed-based commodities and amenities and sustaining environmental quality. People must take into account natural processes and institutional processes in planning for the integrated management of natural resources in a river basin. Institutional mechanisms for implementing integrated watershed management practices include a set of effective policies. Importantly, developing appropriate institutional

[1] Professor Emeritus, School of Natural Resources and the Environment, University of Arizona, Tucson, Arizona, USA.
Email: ffolliott@ag.arizona.edu
[2] Professor Emeritus, Department of Forest Resources, University of Minnesota, St. Paul, Minnesota, USA.
Email: brook007@umn.edu
* Corresponding author

mechanisms to encourage integrated watershed management also require responsible management agencies, land owners or administrators, profit-oriented private enterprises, and other interested organizations and people to be actively involved in the decision-making process. Interactions of economic, social, and political forces with the technical aspects of integrated watershed management are often fostered by locally-led partnerships, councils, or cooperatives where the involved organizations and people have equal voice in the decision-making process. It is emphasized that a participatory planning process is paramount to selecting integrated watershed management practices that are feasible to implement on a river basin that is characterized by a diversity of stakeholders with multiple goals and objectives.

Keywords: Arizona, benefits obtained, informal social guidelines, integrated watershed management, legal and regulatory requirements, locally-led initiatives, management practices, Minnesota, participatory planning, policies, water rights

1 Introduction

The challenge of managing water and other natural resources in a river basin has become increasingly more difficult as populations of people increase, demands on the natural resources on watershed landscapes increase, and new technologies become available. As a result, people either willingly or unwillingly degrade or destroy their environments at a more rapid rate than in the past. At the same time that the challenges for watershed managers increase, there are also opportunities to slow down or halt degradation and destruction of these often fragile landscapes and, by doing so, create sustainable environments in which people live and attempt to satisfy their basic needs. Several approaches to land stewardship have been put forth by people to meet these challenges. One approach is called *integrated watershed management*, another approach is called *integrated river basin management*, while still another approach is called *integrated natural resources management* or *integrated water resource management* when the focus is placed directly on the water resources involved (Gregersen et al. 2007). All of these approaches embrace similar ideas when it comes to fundamental principles of integrated landscape management. The authors of this chapter use the term *integrated watershed management* because a watershed is frequently considered the basic hydrologic unit for planning and implementing watershed management practices within a river basin.

The fundamental objective of integrated watershed management is developing, managing, and sustaining production systems that are suited to the environmental conditions confronted and diversity of natural resources within these environments that can be sustained for future generations of people (Gregersen et al. 2007; Brooks et al. 2013). What distinguishes the integrated watershed management approach from other watershed management practices is its holistic consideration of the linkages among this objective and the management practices aimed at meeting it. Accepting the fact that a river basin and its contributing watersheds are made up of land units of varying biological and physical characteristics with each often defined by different ownership or administrative control, the interactions among these units and the managerial activities undertaken on them such as livestock, wood, or agricultural production, infrastructure development, and increased urbanization are all considered within the framework of integrated watershed management approach.

The premise of this chapter is that effective combinations of technical information and institutional support are essential in providing lasting benefits to people. Achieving this purpose is possible by implementing integrated watershed management practices.

2 Integrated Watershed Management

To appreciate the contributions of integrated watershed management requires recognition of what is meant by a watershed and its relationship to a larger river basin. A *watershed*—also called a *catchment* by many people—is a topographically delineated area of land that is drained by a stream system (Quinn et al. 1995; Ffolliott et al. 2002; Brooks et al. 2013). It is represented by the total land area above a selected point in a stream system that drains past that point. A watershed is a hydrologic unit that can be used by managers as a physical-biological unit and a social-political-economic unit in planning and managing natural resources as a sustainable flow of benefits. A *river basin* is defined similarly to a watershed but it is larger in scale. As mentioned above, a river basin it includes all of the contributing watersheds within the basin. For example, the Mississippi River Basin, the Congo River Basin, and the Amazon River Basin encompass all of the watersheds that drain within and through their basins and tributaries.

2.1 Tenets of integrated watershed management

Integrated watershed management of water and other natural resources is based on the recognition of the varying interrelationships among land-use activities, soil and water resources, and the linkages of hydrologic processes between upland watersheds and downstream areas (Quinn et al. 1995; Gregersen et al. 2007; Brooks et al. 2013). More specifically, integrated watershed management incorporates soil and water conservation and land-use planning into a broader framework by focusing on the following points.

The interactions of soil and water with other natural resources on a river basin and its contributing watersheds impact people both beneficially and negatively. In turn, people affect the nature and severity of these interactions by the ways in which they use these natural resources and the rates and quantities of natural resources that they use. The intermingling effects of these interactions follow watershed boundaries not political boundaries.

Water flows downhill regardless of how people define their social or political institutions. Therefore, what people living in one political or administrative unit do on the upland watersheds in a river basin can significantly affect the political or administrative units situated further downstream. Moreover, these interactions often cross political or administrative boundaries. As a consequence, what seems appropriate management and use of natural resources on one political or administrative unit upstream might not be the best management or use of the natural resources for another political or administrative unit because of the possible undesirable downstream effects such as increased flooding events or further pollution of water supplies. Integrated watershed management brings these off-site effects—called *externalities* by economists—into the planning process (see *Participatory Planning* below). These off-site effects include both the beneficial and negative effects of a watershed management practice at one location on other locations within a river basin and the beneficial and negative effects of these activities at a specified time in the future.

An integrated watershed management approach also provides a more effective and efficient framework for attaining or maintaining the development and sustainable use of natural resources than would otherwise be achieved, while the watershed management practices implemented represent the tools for making this framework operational (Gregersen et al. 2007; Brooks et al. 2013). People must recognize that implementing watershed management does not depend only on the physical and biological characteristics of the landscapes. Institutional factors such as government rules and regulations and the cultural background of local people need to be fully integrated into solutions that meet the environmental, economic, and social objectives of the people.

2.2 Management practices

Integrated watershed management practices can be grouped into three general categories or varying combinations of these categories. The categories are minimizing adverse impacts to soil and water resources to maintain a watershed in good condition; sustaining flows of high-quality water originating on a watershed; and rehabilitating a degraded watershed in poor condition to restore or increase ecosystem productivity, or varying combinations of these practices. While many forms of these three categories of watershed management practices are likely to occur within a river basin and its contributing watersheds at a point in time (Ice and Stednick 2004; Zou et al. 2010; Brooks et al. 2013), people should remember that both the natural-resource production and environmental protection obtained with these practices are equally important in their implementation.

2.2.1 Minimizing adverse impacts

Precipitation infiltrates into the soil horizon but does not contribute to excessive soil erosion on a watershed in good condition. Streamflow response to a precipitation event is slower on a watershed in good condition than on a degraded watershed in poor condition and the baseflow of perennial streams is sustained between storms. Watershed within a river basin in good condition can degrade to a poor condition because of overgrazing by livestock; excessive cutting of trees or removal of other vegetation; high-severity wildfires; extremes in weather conditions such as excessive amounts of rainfall leading to flooding, prolonged dry periods; or severe wind-storms uprooting trees, or improperly aligned roads (Gregersen et al. 2007; Brooks et al. 2013). It can be necessary to protect a watershed that is moving toward poor conditions from further deterioration by preventing an increase in soil loss caused by these occurrences while controlling the rapid streamflow response to precipitation that often occurs because of a loss of water storage (Anderson et al. 1976; Swank and Crossley 1988; Ice and Stednick 2004).

Among the actions proposed to mitigate the degradation of a watershed are allowing livestock grazing only within the capacity of the land to sustain livestock grazing; curtailing logging activities and clearing of other types of vegetation on sites sensitive to increased soil erosion; establishing riparian (stream-side) buffer strips of vegetation to maintain the hydrologic integrity of stream channels; and preventing construction of roads near stream systems, maintaining effective drainage of other roads, and seeding unused secondary roads with perennial herbaceous plants to minimize soil erosion (Gregersen et al. 2007; Brooks et al. 2013). With respect to the latter, some secondary roads might have to be closed to use by people to prevent an increase in soil erosion. These actions are designed to improve watershed conditions while sustaining ecosystem-based, multiple-use benefits and accommodating people's concerns about protecting rare, threatened, and endangered plant and animal species. Concurrence by the people to be affected by these interventions is necessary and, in many instances, can be obtained with these people participating in the planning effort prior to their implementation (see *Participatory Planning* below).

2.2.2 Sustaining flows of high-quality water

Sustaining flows of high-quality water is a major focus of integrated watershed management practices. Planning and implementing environmentally, ecologically, and economically sound management practices that are acceptable to local people and the general society are paramount in achieving this fundamental goal (Brooks et al. 2013). However, streamflow volumes, stormflow peaks, and low flows of water are often affected by livestock grazing practices, logging activities, vegetative conversions, wildfire, and road construction, and, therefore, these activities must be carefully implemented to prevent unwanted hydrologic impacts.

Streamflow volumes can sometimes increase after thinning, harvesting, or otherwise eliminating the forest overstory on a watershed; replacing deep-rooted tree species with shallow-rooted tree species or shrubs; replacing tree species with high-interception capacities with species of low-interception capacities; or replacing forest overstories comprised of comparatively high water-consuming trees with lower water-consuming herbaceous plants (Bosch and Hewlett 1982; Whitehead and Robinson 1993; Ice and Stednick 2004). The observed increases in streamflows are not generally sustainable in the long-term, however, and return to their original volumes once the watershed has recovered from these manipulations.

Control of non-point pollution of streamflows has focused on implementation of what are called *Best Management Practices* since passage of the Clean Water Act by the Congress of the United States. These practices are designed to control the delivery of sediment, nutrient, and other pollutants into receiving waters by non-structural (vegetative) or structural (engineering) measures (Brown et al. 1993; Ice and Stednick 2004; National Research Council 2008). Best Management Practices evolve through negotiations among people with interests in maintaining the integrity of stream systems by minimizing the movement of pollutants into bodies of surface water (lakes, reservoirs, etc.) and groundwater aquifers; and control soil erosion on upstream watersheds through the re-vegetation of disturbed sites. Best Management Practices to mitigate the erosion-sedimentation-nutrient processes are currently known for many silvicultural treatments, livestock-grazing practices, road-related disturbances, and agricultural activities (Moore et al. 1979; Lynch et al. 1985; Chaney et al. 1990). Monitoring efforts have indicated that the implementation of best management practices can result in compliance with established water-quality standards in many areas (Edwards and Willard 2010).

2.2.3 Watershed rehabilitation

A concern for declining health of the watersheds in a river basin has often led to the implementation of management practices to restore the proper hydrologic functioning of the affected watersheds. These declines in watershed health are caused mostly by excessive increases in soil loss (Hudson 1981; Morgan 1995). Situations that are susceptible to increasing soil erosion are sloping landscapes with shallow soils; sites with soils of low hydraulic conductivity; and where the removal or denudation of a vegetative cover is likely to occur. Establishing a vegetative cover on a degrading site is the best means of preventing increased soil erosion in the long-term (Ice and Stednick 2004; Brooks et al. 2013). Mechanical methods of controlling soil erosion including creating contour trenches, fallow strips, or pitting on sloping landscapes to reduce overland flows of water and the entrained soil particles are more temporary in nature (Brooks et al. 2013). Establishing buffer strip of vegetation can help to protect riparian corridors and edges of other water bodies from upland perturbations such as excessive livestock grazing or tree cutting and wildfires (Comerford et al. 1992; Corner and Bassman 1993; Neary et al. 2010) and, in doing so, restore lost hydrologic functioning.

Roads can be indispensable in implementing sustainable integrated watershed management practices. However, much of the increased in soil loss on a watershed in poor condition is attributed to improper road design, construction, and location. Fortunately, many of these problems can be eliminated before construction activities begin (Megahan 1977; Burroughs and King 1989). For example, minimizing the area in roads by reducing mileage and disturbance; establishing and maintaining a vegetative cover to protect cutbanks, fill slopes, and other sites of exposed soil; avoiding steep gradients that are less stable; and properly sizing, spacing, and maintaining culverts to avoid washouts should be considered in the planning stage.

2.3 Benefits obtained

Integrated watershed management practices takes place within the reality that river basins and their contributing watersheds function in response to climatic regimes, natural-resource capabilities, and land-use patterns that are independent of their respective boundaries. It is not surprising, therefore, that the benefits of integrated watershed management practices vary widely. However, the benefits that are obtained depend upon physical and biological conditions of the managed landscapes involved and on appropriate regulations, controls, and incentives. Two examples of the benefits of integrated watershed management implemented in diverse regions of the United States are presented below.

2.3.1 Arizona

Experimental watersheds in the ponderosa pine (*Pinus ponderosa*) forests of the Salt-Verde River Basin in northern Arizona were established to study and quantify the benefits of integrated watershed management practices in this region. One of the practices evaluated was a combined strip-cutting and thinning treatment designed to increase streamflow volumes, maintain timber production, and enhance other ecosystem services (Baker and Ffolliott 1999). One-third of the watershed studied was cleared of trees in strips varying from 15 to 25 m in width and oriented in the direction of the land slope (uphill-downhill). The intervening leave strips were thinned to improve growth of the residual trees. A streamflow increase of almost 25% was observed annually until re-establishment of vegetation in the strip-cuts. While the practice eliminated trees in the strip-cuts, even-aged stands of trees were retained in the alternating leave strips to maintain the integrity of the forests on the watershed. Livestock forage increased in response to the removal of trees in the strip-cut and reduced stocking of trees in the leave strips. Habitats for indigenous wildlife species improved because of the increase in forage plants, retention of protective cover in the leave strips, and creation of the effects (ecotones) between the strip-cuts and leave strips. Increases in soil losses were minimal, and, therefore, sustainability of the benefits was achieved.

2.3.2 Minnesota

Wetlands within the agricultural-production region of the Minnesota River Basin were drained in the past to expand production of the annual crops. However, the drainage network and ditch system increasing the flows of water off the land and into stream channels. These activities contributed to excessive sedimentation, stream-channel instability, and impaired water quality. An array of integrated watershed management practices has been implemented to rectify these environmental problems (Brooks et al. 2013). Trees and perennial grasses were planted on hillslopes and along riparian corridors to reduce the overland flows of water and, as a result, mitigate the delivery of eroded soil particles and nutrients into the stream channels. Many of the wetlands have been restored, stream channels that were degraded earlier have been rehabilitated, and the increased carbon sequestration might reduce atmospheric carbon. It is anticipated that the perennial crop-cover and wetland restoration will provide greater ecosystem resilience than the earlier annual cropping systems into the future.

3 Institutional Considerations

People must take into account two sets of processes in planning for the management and use of high-quality water, preferred livestock forage, wood fiber for lumber and other products, desirable wildlife habitats, and scenic recreational settings on a landscape. There are the *natural processes* that are expressed in terms of biological and physical relationships and there are the *institutional processes* that are equally important as natural processes in achieving effective and responsible stewardship of a river basin (Quinn et al. 1995; Gregersen et al. 2007; Brooks et al. 2013). It is the intersection of these two sets of processes that largely identifies the viable options available to people in the decision-making process.

Most of the natural limitations occur because of the status of natural processes on a watershed although some of these limitations are created by people's manipulations of the natural resources and environmental impacts of these manipulations. However, all of the institutional limitations are created by people. One of the complexities involved on the institutional side of the "equation" is that there are generally a large number of people and organizations with diverse interests involved in the use and management of natural resources. While people are able to modify some of the natural conditions of watershed ecosystems through the applications of biological and physical measures, they can also improve the institutional arrangements within which the use and management of the natural resources take place. To accomplish the latter requires that the involved people possess an understanding of the legal and regulatory requirements, informal social guidelines, established policies, and social, economic, and environmental values that are relevant to the situation confronted.

3.1 Legal and regulatory requirements

The most important of the institutional considerations are those that have been sanctioned by laws and regulations; that is, the formalized rules that guide and enforce the conduct of people in their use and stewardship of natural resources (Cortner and Moote 1999; Gregersen et al. 2007). Included are the laws and regulations that deal with the ownership and rights to use the land and natural resources on a watershed. The requirements that are related to integrated watershed management are established largely by the need to sustain the flows of high-quality water and use of other natural resources within varying ownership patterns and appropriate rights to use categories. Water rights differ among the states issuing the rights but must always comply with other water-related and environmental legislation and requirements.

3.1.1 Surface water rights

People need to understand and appreciate surface water rights when considering how to effectively and efficiently use water. Nobody actually owns the water that flows through the stream channels of a river basin. Rather, water is usually recognized as public property that is subject to established rights of use as the water flows from a watershed and through a river basin. What people own or control, therefore, are the rights to use this water.

Three laws govern the right to use surface water in the United States (Ffolliott and Gregersen 2004; National Research Council 2008). In 29 states, the owners of land that is adjacent to a stream have the right to use of water from the stream if this use does not jeopardize the water rights of other landowners along the stream (Ffolliott and Gregersen 2004). This is called *riparian law*. The doctrine of *prior appropriation* applies in nine states. Under this doctrine, the first person to claim a limited water supply, regardless of where that person lives, has the first right to use this water with the claim restricted to water that can be used beneficially. Any other person is entitled to claim water in excess of the first claimant's use with still further claims made by people for the successive amounts of unclaimed water until the supply of water is exhausted. A hybrid of riparian law and the prior appropriation doctrine is followed in 10 states while two states have unique rules of water use.

Surface water rights become more complicated when water flows out of multiple watersheds in a large river basin consisting of multiple watershed boundaries or from one country to another (Wolf 2006; Gregersen et al. 2007). Issues of conflict confronted and governance protocols are often resolved through negotiated treaties, agreements, or compacts in this case.

3.1.2 Groundwater rights

Laws and regulations related to the use of groundwater resources must also be recognized for effective integrated watershed management to occur. For example, the *American Rule* that prevails throughout much of the western United States allows overlying land-owners to capture and use groundwater for on-site uses in a reasonable (non-malicious) manner (Ffolliott and Gregersen 2004). However, this law prohibits the transportation of groundwater resources for off-site uses. Of concern to many managers of water resources is the integration of surface water and groundwater resources into a comprehensive management framework. While many people have often considered these two water resources as separate entities in the past, more recently people have come to recognize that there are hydrologic linkages between surface water and groundwater resources that should be considered (Blomquist et al. 2004). Recognition and appreciation of the respective rights of people to these two supplies is necessary to effectively manage the water resources in a river basin and its contributing watersheds in a more comprehensive manner.

3.1.3 Other regulations

Implementation of integrated watershed management practices in the United States must comply with the requirements of federal, state, and local statutes and ordinances adopted to regulate the effects of land and water management of a river basin. Underpinning federal regulation are the requirements to comply with the National Environmental Policy Act that specifies that federal management agencies must carefully weigh the environmental impacts of proposed management practices prior to taking major actions. Another federal regulation of relevance to integrated watershed management is contained in the Clean Water Act that address the need to protect surface water bodies and groundwater aquifers from an array of physical, chemical, and biological pollutants. Among the other federal statutes that can impact on the implementation of integrated watershed management practices include the Safe Drinking Water, the Endangered Species, the Clear Air, and the Federal Insecticide, Fungicide, and Rodenticide Acts.[1] States in the United States also have laws and regulations that apply to integrated watershed management practices. Most of these regulations are administered by a state agency with responsibilities to protect public interests specified by the state. Some states have also been delegated the authority to administer some of the federal regulations by the responsible federal agency. Regulatory requirements and laws applicable to management of land and water resources in the private sector include the need to prepare land-use plans, zoning ordinances, permits, and federal and state specifications in some instances.

3.2 Informal social guidelines

Less explicit than laws and regulations are informal social guidelines on how water and other natural resources are used and managed that evolved through time from the cultural background and inherent heritage of local

[1] The reader is referred to Gregersen et al. (2007), Brooks et al. (2013), the appendix of the report on the "Hydrologic Effects of a Changing Forest Landscape" published by the National Research Council (2008) and the Internet for detailed information on these and other federal regulations impacting on integrated watershed management in the United States.

people. Many of these guidelines have been accepted into local legal systems without formal legislative enactment or enforcement. Nevertheless, it is necessary that these social values are properly reflected in the planning and decision-making processes for a river basin to agree on an appropriate strategy for the use and management of water and other natural resources (Cortner and Moote 1999; Gregersen et al. 2007). It is imperative, therefore, that managers are aware of these social guidelines. If not already known, this awareness is often obtained through the participatory planning process (see below). Failure to recognize these values can cause inappropriate or adverse reactions by people to integrated watershed management practices and, as a consequence, result in more formalized rules and regulations that might limit the options that people have had available to them on how natural resources are used and managed.

3.3 Policies

Institutional mechanisms necessary for implementing integrated watershed management practices generally include a set of effective policies (Brooks et al. 1994). Policies are considered to be the operational rules for the management of a river basin and its contributing watersheds. Policies are often established through an iterative process of successive approximations of resulting actions with the consensus-building efforts moving toward a blending of people's short-term needs and long-term goals (Quinn et al. 1995; Gregersen et al. 2007; Brooks et al. 2013). However, policy-makers in the United States can face a number of challenges in dealing with the problems and issues confronted in achieving an acceptable combination of people's needs and goals.

As mentioned earlier, while the responsibilities of social and political institutions rarely coincide with watershed boundaries, watershed boundaries often become the scale that governs most of the processes that ultimately determine the productivity and sustainability of natural-resource use. Diverse groups of stakeholders concerned with the acceptable resolution of conflicting issues are usually confronted by policy-makers (Gregersen et al. 1994). There are also public sectors that impact on the use and sustainability of the natural resources within a river basin that represent a variety of economic enterprises that are dependent on watershed resources such as those found in more heavily-populated downstream urban settings. These sectors can carry more "political clout" than only considerations of upland conservation.

3.3.1 Policy components

A policy consists of the formal establishment, implementation, and enforcement of the appropriate institutional arrangements to guide a land-use choice that is made by people (Quinn et al. 1995; Cortner and Moote 1999). An effective policy consists of three essential and largely related components regardless of the purpose of the policy (Bromley 1992). These components are *intentions*, *rules*, and *compliance*. Policy intentions reveal what people hope to accomplish within the framework of the institutional arrangements confronted. Policy rules are the institutional arrangements that can constrain some people and liberate others. Compliance is the essential component that converts promises and proclamations into meaningful results (compliance is does not necessarily mean that the policy rules are restrictive in a negative sense but rather outlines the procedure to properly apply the rules). All three of these components are necessary in formulating and then enforcing a policy.

3.3.2 Policy effectiveness

Formulating and enforcing policies to achieve integrated watershed management can be a significant challenge. Such policies can already exist in some instances but they might not be effective. Therefore, an initial step in evaluating the effectiveness of established policies is assessing the policies to identify weakness in their formulation or enforcement (Gregersen et al. 1994). Viable options for overcoming the policy weaknesses must then be considered. Existing policies might be modified to improve their effectiveness while the formulation of a new policy can become necessary at other times (Quinn et al. 1995; Gregersen et al. 2007; Brooks et al. 2013). In either instance, policy-makers, watershed managers, and other stakeholders should build on local expertise and, importantly, recognize that what might work for one agency, organization, or other group of people might not be feasible elsewhere. The involved people also need to know how the condition of a river basin might be changing so the policies that are implemented conform to the changing conditions.

Three groups of issues are often addressed in determining the effectiveness of established policies (Gregersen et al. 1994; Brooks et al. 2013). One group of issues concerns who gains and who losses with implementation of the policies in question. Included in these issues are impacts of the policies on people grouped, for example, by their income, location on a watershed or river basin, and gender, age, and occupation. A second group of issues relates to whether the gains in people's welfare that result by enforcing the policies can be sustained. Included here is whether there is an adequate combination of natural resources to meet people's needs; the security of people's access to these resources; and maintenance of the resource capabilities on a long-term basis. Recognizing the concerns of people about possible changes in benefits and costs because of a watershed practice must be addressed by the policies represent the third group of issues confronted.

The concerns embedded within these three groups of issues are frequently manifested by people's willingness to pay or not to pay for an integrated watershed management practice and their knowledge of the consequent relationships between the incremental benefits and costs of the management practice expressed in terms of *economic efficiency*. Economic efficiency relates to the relationships between total benefits and costs that are valued by market prices or expressed in economic values when market prices are not appropriate. Therefore, an analysis of economic efficiency encompasses the direct inputs determined by a financial analysis of inputs and outputs for which payments are made; and the indirect effects that are not included in a financial analysis because they are not bought or sold in a marketplace but affect society's welfare as a consequence of the management practice. These indirect effects can be increased flood prevention and sustainability of healthy watershed systems.

While market prices are used in valuing inputs and outputs in a financial analysis, people's willingness to pay is the primary measure of value in an analysis of economic efficiency. Market prices are used where these prices adequately reflect the willingness to pay. Otherwise, shadow prices can provide a measure of people's willingness to pay. For example, surrogate market prices are often used when the benefits and costs of a watershed management practice are not themselves valued in the market but where clear substitutes are found in the marketplace. The market prices of these substitutes are then used to estimate of the willingness of people to pay.

3.4 Other Institutional considerations

Institutional arrangements to facilitate appropriate market prices and non-market incentives and encourage investments in research, education, and effective extension activities are also needed to achieve the goals and objectives of integrated watershed management (Quinn et al. 1995; Gregersen et al. 2007; Brooks et al. 2013). Additionally, institutional mechanisms that allow for the accounting of environmental benefits and costs associated with an integrated watershed management practice must be considered (National Research Council 2005). However, there are several reasons why achieving integrated watershed management through these mechanisms can be difficult.

Watershed boundaries and political boundaries rarely coincide as mentioned repeatedly in this chapter. As a consequence, biological and physical processes do not always occur on the same scale or within the same boundaries as social, political, and economic activities. Another difficulty in formulating effective institutional arrangements for the implementation of integrated watershed management practices is that administrative and management responsibilities, sustained economic development opportunities, and extension activities are frequently shared by several organizations. It is unlikely therefore, that only one organization will possess the required coordination authority for a planned watershed management practice.

That the full array of integrated watershed management goals and benefits are not totally understood or might be assigned a low priority by some people represents still another difficulty in formulating institutional mechanisms to effectively deal with the management and use of the natural resources considered. In fact, existing institutional mechanisms might hinder the effectiveness of integrated watershed management although not necessarily intentionally (Gregersen et al. 2007). For example, institutional arrangements that encourage a specific land-use activity are frequently formulated with a primary purpose of increasing the availability of commodities with little consideration for sustainability of the land-use activity; the desired level of production of these commodities; and the off-site effects of the land-use activity that can result.

Developing institutional mechanisms to encourage integrated watershed management also require land owners, responsible management agencies, profit-oriented private enterprises, and other interested stakeholders to become actively involved in the decision-making process (Gregersen et al. 2007; Brooks et al. 2013). Furthermore, society must be aware of the increasing importance for the sustainable use of land, water, and other natural resources and the associated linkages upon which integrated watershed management practices are built. All groups of people involved in the use of water and other natural resources and their perceptions and attitudes relative to these resources need to be adequately identified in developing relevant policies and other institutional mechanisms. Importantly, the impacts of these policies and other institutional mechanisms must be continuously assessed as they evolve through time.

4 Locally-Led Initiatives

Interactions of economic, social, and political forces with the technical aspects of integrated watershed management are often fostered by the creation of locally-led initiatives. These initiatives frequently evolve from past planning activities that proved less than equitable to people. Locally-led initiatives in the United States generally manifest themselves in partnerships, councils, or cooperatives of varying organizational structures and responsibilities among the interested parties. The basic idea behind these initiatives is that all of the involved people should have equal voice in the decision-making process (Toupal and Johnson 1998; Griffin 1999). Successful locally-led initiatives usually begin with a dialogue among local stakeholders and representatives of the responsible management and regulatory organizations. The participation of people in these initiatives is "recruited" from a variety of stakeholders with a diversity of interests and concerns relative to the management of a watershed and larger river basin.

Locally-led initiatives can overcome a loss of trust that people might have in the effectiveness of integrated watershed management practices and the consequent enforcement of these practices by regulatory agencies. Successful initiatives also improve the access that lay people have to technical information. For example, every person with Internet access is able to receive information on almost any subject relating to the management and use of natural resources on a river-basin landscape. Partners in these initiatives can then address issues and concerns about integrated watershed management before a planning process based on incorrect assumptions proceeds and, with this information, provide immediate feedback to management and regulatory agencies.

5 Participatory Planning

A key part of successful integrated watershed management is a participatory planning process to select management practices that are considered feasible. This process involves a planning expert or team of planning experts; a watershed manager or group of watershed managers with varying skills and responsibilities; a decision-maker or group of decision-makers representing different public sectors; and an assemblage of interested stakeholders. The organization of human, capital, and natural resources that is required to carry out relatively straightforward management tasks on a relatively small watershed scale can require little formal planning when simple goals and objectives can be satisfied with only few constraints in a relatively short period of time. However, a more detailed participatory planning process is necessary to select the most appropriate integrated watershed management practices to implement on a river basin or its contributing watersheds that are characterized by a diversity of stakeholders with multiple goals and objectives to be sustained within a framework of numerous constraints in the long-term. This second planning process is the focus of the following discussion.

5.1 Overview

There are three general components in participatory planning for this latter case (Ffolliott et al. 1999; Gregersen et al. 2007; Brooks et al. 2013). The *goals* and *objectives* of a proposed integrated watershed management practice are commonly established on the basis of a comprehensive problem analysis of on-going and potential watershed management alternatives. Concurrently, the biological-physical, social, cultural, financial, and political *constraints* that are associated with the proposed alternatives must also be considered. Appropriate *management techniques* for implementing the management alternatives are necessary to satisfy the goals and objectives within the framework of the identified constrains. This form of participatory planning requires organizing, analyzing, and integrating goals and objectives, constraints, and management techniques in a way that decision-making is achieved more efficiently and effectively than when an unplanned approach is used.

In only a few instances is it likely that a planner or planning team will have sufficient information to make risk-free decisions about the actions and reactions to a proposed integrated watershed management practice. As a consequence, participatory planning requires both judgment and flexibility in its execution (Gregersen et al. 2007; Brooks et al. 2013). If people have learned one thing from their past experiences it is that a watershed management practice seldom unfolds as originally planned. Those people who are able to adjust to the changes in technical and institutional conditions are further ahead than those who rigidly follow an original plan despite changes in the conditions confronted. Therefore, planners need to anticipate the adjustments that are identified in the planning process and, in doing so, work cooperatively in addressing the inevitable changes in opportunities and constraints that will take place as planning and eventual implementation moves forward.

5.2 Participatory planning context

Integrated watershed management provides an array of benefits to people in an environmentally-sound manner while sustaining the hydrologic functioning of a river basin and its contributing watersheds as discussed throughout this chapter. To accomplish this purpose, planners, managers, and other interested people must recognize and appreciate the flows of both on-site and off-site effects and the benefits that can result from implementation of the management practices (Gregersen et al. 2007; Brooks et al. 2013). Three interrelated points should be considered within the context of producing these benefits. Management practices (inputs) and costs associated with the practices must be known. The biological-physical effects and environmental changes associated with the management practices should also be identified and the socio-economic changes including both benefits and costs associated with the biological-physical effects and environmental changes must also be known.

The effects and benefits obtained from a proposed integrated management practice should be defined as changes with and without implementation of the practice (Gregersen et al. 1987; Ffolliott et al. 1999; Brooks et al. 2013). For example, by increasing plant productivity that has been declining on a watershed, one should consider the differences in plant productivity between the conditions with and without the management practice. Plant productivity might still be declining after the management practice has been implemented but at a slower rate than observed before the practice was imposed. Therefore, there is an increase in plant productivity in comparison to what it would have been without the practice and this increase represents a benefit that should be valued.

5.3 Steps in participatory planning

Participatory planning has no real beginning or end. However, the process must start somewhere. A logical starting point is before a problem or opportunity in relation to an integrated watershed management approach has been identified (Ffolliott et al. 1999; Gregersen et al. 2007; Brooks et al. 2013). In this case, the planning process begins with monitoring and evaluating past activities and then identifying problems and opportunities with this information. Sequential steps in the process include determining the main features of these problems and opportunities; setting the goals and objectives necessary to overcome the problems or take advantage of the opportunities; developing strategies for action; and identifying the constraints to these strategies for action. Identifying viable management alternatives to implement the strategies for action within the defined constraints follow.

Appraising and evaluating the environmental, economic, and social impacts of proposed integrated watershed management practices and assessing the uncertainties associated with these appraisals and evaluations is the next step in the process. This step involves an appraisal of the economic alternatives of the proposed watershed management practices (Gregersen et al. 1987; 2007). While watershed managers often deal mostly with biological-physical processes, they should also have an understanding of the economic implications of what they are doing. It is important, therefore, to know the economic values of each of the management alternatives considered to be feasible; how much each of these alternatives will cost to implement; and cash flow required to sustain each of the alternatives through time (Gregersen et al. 2007; Brooks et al.

2013). This information generally becomes available in comprehensive economic appraisals of the alternative watershed management practices.[2] The final step in the participatory planning process is ranking the alternative watershed management practices evaluated in terms of their feasibility for implementation and presenting the implications of risk and uncertainty for the each of these practices (Gregersen et al. 1987). One can estimate probabilities to the various outcomes in identifying risk while quantitative measures of the probabilities cannot be generated in the case of uncertainty. However, one might develop "subjective probability estimates" for the aspects of a management practices that are of highest interest in considering uncertainty. But, these estimates might do more harm than good because subjectivity in the planning process should not be hidden. Therefore, a straightforward analysis of how measures of the values (i.e., net present values, internal rates of return, benefit-cost ratios, etc.) of the alternative management practices change with different assumptions concerning the values of key parameters is generally suggested.

The responsibilities of the planning personnel can stop at this point in the process unless recommendations on which of the alternatives evaluated should be selected for implementation and the timing and approach to its implementation are requested. If this request is made, a ranked set of the alternatives is often presented to the decision-making party. The criteria followed in arriving at this ranking should also be made available to the decision-maker. Importantly, only the responsible decision-maker should ultimately determine which of the alternatives considered should be selected for implementation.

5.4 Continuous process of participatory planning

What has been presented in this discussion is a participatory planning process of moving from problem or opportunity identification through an evaluation phase to recommending an integrated watershed practice for implementation. This process is continuous in most instances (Stuth et al. 1991; Ffolliott et al. 1999; Gregersen et al. 2007). Participatory planning can also be an iterative process with information concerning outcomes of the management actions taken into account and any emerging problems constantly fed back into the process. This information is useful in making changes in the selected plan of action when necessary. More formally, the process of collecting and evaluating information related to the performance of a watershed management practice is part of the monitoring and

[2] The reader is referred to Dixon and Hufschmidt (1986), Gregersen et al. (1987), Winpenny (1991), Dixon et al. (1994), and other references on economic valuation techniques for details of the economic appraisal process.

evaluation tasks mentioned earlier. There should be continuous feedback of information relative to the performance of the management practice that can then be synthesized, analyzed, and incorporated into strategies and suggestions or actions that are passed back to the responsible manager and decision-maker. These continuous exchanges of information lead to useful interactions among planners, managers, and the involved public in ensuring the benefits of integrated watershed management practices are achieved.

6 Summary

Challenges of managing watershed landscapes within a river basin do not necessarily arise solely from physical limitation nor a lack of technical knowledge. The limitations of production and abundance found in social and political structures and the economic relationships among the involved people are equally important. Needed natural resources might be available but their sustainable use depends largely on the collective efforts of the people. A watershed manager can help these efforts by implementing solutions for enhancing the productivity of the natural resources to provide the greatest long-term array of benefits to people into the future. This is the purpose of integrated watershed management. However, decision-makers might have relatively short-term goals or objectives in mind because they frequently focus mainly on the immediate and often more popular concerns of people. It is necessary, therefore, for the decision-makers to appreciate that the conservation and proper use of the natural resources in a river basin and its contributing watersheds depend mostly on long-term, continuous, and often expensive commitments.

References

Anderson, H.W., H.D. Hoover and K.G. Reinhart. 1976. Forests and water: effects of forest management on floods, sedimentation, and water supply. USDA Forest Service, Pacific Southwest Forest and Range Experiment Station, General Technical Report PSW-18, Berkeley, California.

Baker, M.B., Jr. and P.F. Ffolliott. 1999. Interdisciplinary land use along the Mogollon Rim. pp. 27–34. *In*: M.B. Baker, Jr. (ed.). History of Watershed Research in the Central Arizona Highlands. USDA Forest Service, Rocky Mountain Research Station, Research Paper RMRS-29, Fort Collins, Colorado.

Blomquist, W., E. Schlager and T. Heikkila. 2004. Common Watershed, Diverging Streams: Linking Institutions and Watershed Management in Arizona, California, and Colorado. Resources for the Future, Washington, DC.

Bosch, J.M. and J.D. Hewlett. 1982. A review of catchment experiments to determine the effect of vegetation changes on water yield and evapotranspiration. Journal of Hydrology 55: 3–23.

Bromley, D.W. 1992. Institutions and institutional arrangements in environmental policy. pp. 1–20. *In*: Environmental policy analysis in developing countries. U.S. Agency for

International Development, Environment and Natural Resources Policy and Training Project, Rosslyn, Virginia.
Brooks, K.N., P.F. Ffolliott, H.M. Gregersen and K.W. Easter. 1994. Policies for sustainable development: the role of watershed management. U.S. Agency for International Development, Environment and Natural Resources Policy and Training Project, Policy Brief 6, Washington, DC.
Brooks, K.N., P.F. Ffolliott and J.A. Magner. 2013. Hydrology and the Management of Watersheds. Wiley-Blackwell, Hoboken, New Jersey.
Brown, T.C., D. Brown and D. Brinkley. 1993. Laws and programs for controlling nonpoint source pollution in forest areas. Water Resources Bulletin 22: 1–13.
Burroughs, E.R. and J.G. King. 1989. Reduction of soil erosion on forest roads. USDA Forest Service, Intermountain Forest and Range Experiment Station, General Technical Report INT-264, Ogden, Utah.
Chaney, E., W. Elmore and W.S. Platts. 1990. Livestock grazing on western riparian areas. U.S. Environmental Agency, Washington, DC.
Comerford, N.B., D.G. Neary and R.S. Mansell. 1992. The effectiveness of buffer strips for ameliorating offsite transport of sediment, nutrients, and pesticides from silvicultural operations. National Council of the Paper Industry for Air and Stream Improvement, Technical Bulletin 631, New York, NY.
Corner, R.A. and J.H. Bassman. 1993. Contribution of legal buffer zones to non-point source pollution abatement following timber harvesting in Northeast Washington. p. 245. *In*: B.H. Tellman, H.J. Cortner, M.G. Wallace, L.F. DeBano and R.H. Hamre (tech. coords.). Riparian Management: Common Treads and Shared Interest. USDA Forest Service, Rocky Mountain Research Station, General Technical Report RM-226, Fort Collins, Colorado.
Cortner, H.J. and M.A. Moote. 1999. The Politics of Ecosystem Management. Island Press, Covelo, California.
Dixon, J. and M. Hufschmidt (eds.). 1986. Economic Valuation Techniques for the Environment: A Case Study Workbook. Johns Hopkins University Press, Baltimore, Maryland.
Dixon, J.A., L. Fallon Scura, R.A. Carpenter and P.B. Sherman. 1994. Economic Analysis of Environmental Impacts. Earthscan Publications, London.
Edwards, K.A. and K.W.J. Willard. 2010. Effectiveness of forestry best management practices for reducing sediment and nutrient losses in the eastern United States. Journal of Forestry 108: 245–249.
Ffolliott, P.F. and H.M. Gregersen. 2004. Institutional limitation to management and use of riparian resources. pp. 359–374. *In*: M.B. Baker, Jr., P.F. Ffolliott, L.F. DeBano and D.G. Neary (eds.). Riparian Areas of the Southwestern United States: Hydrology, Ecology, and Management. Lewis Publishers, Boca Raton, Florida.
Ffolliott, P.F., I. Moshe and T.W. Sammis. 1999. A planning process. pp. 171–178. *In*: T.W. Hoekstra and M. Shachak (eds.). Arid Land Management: Toward Ecological Sustainability. University of Illinois Press, Urbana, Illinois.
Ffolliott, P.F., M.B. Baker, Jr., C.B. Edminster, M.C. Dillon and K.L. Mora. 2002. Land Stewardship through Watershed Management: Perspectives for the 21st Century. Kluwer Academic Publishers, New York, NY.
Gregersen, H.M., K.N. Brooks, J.A. Dixon and L.S. Hamilton. 1987. Guidelines for economic appraisal of watershed management projects. Food and Agriculture Organization (FAO) of the United Nations, FAO Conservation Guide 16, Rome, Italy.
Gregersen, H.M., K.N. Brooks, P.F. Ffolliott, A.L. Lundgren, K.W. Easter, B. Belcher, K. Eckman, R. Quinn, D. Ward, T.A. White, S. Josiah, Z. Xu and D. Robinson. 1994. Assessing natural resources policy issues: a framework for developing options. U.S. Agency for International Development, Environment and Natural Resources Policy and Training Project, Draft Policy Paper, University of Minnesota, St. Paul, Minnesota.
Gregersen, H.M., P.F. Ffolliott and K.N. Brooks. 2007. Integrated Watershed Management: Connecting People to their Land and Water. CAB International, Oxfordshire.

Griffin, C.B. 1999. Watershed councils: an emerging form of public participation in natural resource management. Journal of the American Water Resources Association 35: 505–518.

Hudson, N.W. 1981. Soil Conservation. Cornell University Press, New York, NY.

Ice, G.G. and J.D. Stednick (eds.). 2004. A Century of Forest and Wildland Watershed Lessons. Society of American Foresters, Bethesda, Maryland.

Lynch, J.A., E.S. Corbett and K. Mussallem. 1985. Best management practices for controlling nonpoint-source pollution on forested watersheds. Journal of Soil and Water Conservation 40: 164–167.

Megahan, W.F. 1977. Reducing erosional impacts of roads. pp. 237–261. *In*: S.H. Kunkle and J.L. Thames (eds.). Guidelines for Watershed Management. Food and Agriculture Organization (FAO) of the United Nations, FAO Conservation Guide 1, Rome, Italy.

Moore, E., E. James, F. Kinsinger, K. Pitney and J. Sainbury. 1979. Livestock grazing management and water quality protection. U.S. Environmental Protection Agency, EPA 910/9-79-67, Washington, DC.

Morgan, R.P.C. 1995. Soil Erosion and Conservation. Longman, Essex.

National Research Council. 2005. Valuing Ecosystem Services: Toward Better Environmental Decision-making. The National Academies Press, Washington, DC.

National Research Council. 2008. Hydrologic Effects of a Changing Forest Landscape. The National Academies Press, Washington, DC.

Neary, D.G., S. Buschmann and P.F. Ffolliott. 2010. Function of riparian vegetation in retaining sediment in watersheds. Hydrology and Water Resources in Arizona and the Southwest 40: 31–44.

Quinn, R.M., K.N. Brooks, P.F. Ffolliott, H.M. Gregersen and A.L. Lundgren. 1995. Reducing degradation: designing policy for effective watershed management. U.S. Agency for International Development, Environment and Natural Resources Policy and Training Project, Working Paper 22, Washington, DC.

Stuth, J.W., J.R. Conner and R.K. Heitschmidt. 1991. The decision-making environment and planning paradigm. pp. 201–223. *In*: R.K. Heitschmidt and J.W. Stuth (eds.). Grazing Management: An Ecological Perspective. Timber Press, Portland, Oregon.

Swank, W.T. and D.A. Crossley (eds.). 1988. Forest Hydrology and Ecology at Coweeta. Springer-Verlag, New York, NY.

Toupal, R.S. and M. Johnson. 1998. Conservation partnerships: indicators of success. USDA Natural Resources Conservation Service, Technical Bulletin 7, Washington, DC.

Whitehead, P.G. and M. Robinson. 1993. Experimental basin studies—an international and historical perspective of forest impacts. Journal of Hydrology 145: 217–230.

Winpenny, J.T. 1991. Values for the Environment: A Guide to Economic Appraisal. Her Majesty's Stationery Office, London, England.

Wolf, A. 2006. Transboundary fresh water dispute database. Oregon State University, Corvallis, Oregon. Available at http://www.transboundarywaters.orst.edu.

Zou, C.B., P.F. Ffolliott and M. Wine. 2010. Streamflow response to vegetation manipulations along a gradient of precipitation in the Colorado River Basin. Forest Ecology and Management 259: 1268–1276.

5

Integrated Water Resources Management of Inland River Basins in the Hexi Corridor, Gansu, China with Special Reference to the Shule River Basin

Yuhong Li[1,*] and Victor R. Squires[2]

SYNOPSIS

The three inland river basins in the Hexi Corridor in Northwest China are the major production base for agriculture in Gansu Province. To achieve this status required extensive use of water and large-scale land conversion to create irrigated artificial oases in this arid environment. The impacts of over-use of water were far reaching. This chapter details the history of development and reviews the current status of natural resource utilization. Particular emphasis is given to a case study of the Shule Basin, one of the key inland river systems in NW China. High priority to the sustainable management of water resources in the inland basins has been directed by the Chinese Central Government in the "12th-Five Year Plan". For the sustainable management of the water and related resources of inland river basins Chinese water agencies are trying to develop integrated river basin management approaches.

[1] Department of Water Resources Gansu Province, Lanzhou, China.
Email: yhli2000@aliyun.com
[2] Visiting Professor, Gansu Agricultural University, Lanzhou International Consultant, Adelaide, Australia.
Email: dryland1812@internode.on.net
* Corresponding author

Keywords: desert encroachment, desertification, ecosystem, government policy, oasis development, over abstraction of groundwater, water allocation, water resources, water trading, water user associations

1 Background

Inland or internal drainage systems (also called Endoheic) are closed hydrologic systems. Their surface waters drain to inland terminal locations (Fig. 5.1) where the water evaporates or seeps into the ground, having no access to discharge into the sea (see Kingsford et al., this volume).

There are 18 arid inland basins in China and many of these are in the Hexi Corridor in Gansu Province (Li and Squires 2008; Xu et al., this volume). Lately governments at all levels have placed high priority on their sustainable management. In the decades beginning 1990 several central government-funded river basin rehabilitation projects were launched in Gansu Province, north-western China to make a quick response to the serious water shortage problems currently experienced in inland river basins. There are many dimensions to the underlying causes of problems in dealing with water shortages including: lack of knowledge and technology, inadequate policies and regulations, little policy coordination, inconsistencies between government policy and actions within a river basin, and lack of awareness of measures that could most effectively redress the problems. If not addressed, the problems in river basins are exacerbated by economic development, migration, changing land use and climate change.

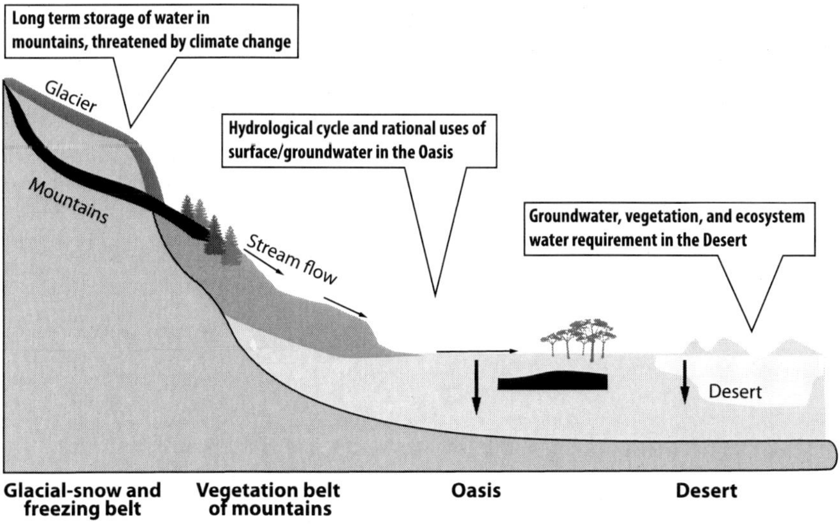

Figure 5.1. Schematic showing the relationship of the elements in typical inland drainage basin in the Hexi Corridor, China.

Some river basins, such as the Shiyang River Basin, are failing to meet sustainable water resource use targets and there is a socially, environmentally and economically expensive program of remediation underway with an uncertain likelihood of success and unclear end points. Some other inland river basins such as Shule River Basin (the focus of this chapter) are at an earlier stage of development so that well planned and executed approaches to integrated river basin management offer the best option for minimizing future degradation of natural resources, social dislocation, and costly programs of repair. Experiences gained in China's inland river basins (including the Tarim basin in neighboring Xinjiang (Jiang et al. 2005) offer approaches that could be used in other inland basins and possibly scaled up for larger river basins in arid regions of the world.

In recent years the Central Government of China required Gansu Province to give utmost priority to the sustainable use of water resources in the inland river basins. The river basin management agencies have made great efforts to seek domestic and international advice and support in regard to developing a system, principally for developing water sector reform policies, for integrated river basin management suitable for their local conditions.

2 Change of Water and Land Use in the Shule River Basin

The Shule River Basin is located at the western end of the Hexi Corridor in Gansu Province. The total basin area is around 113,600 km^2. The basin is in a hyper-arid desert climate zone with annual precipitation less than 50 mm, and potential evaporation is greater than 3,000 mm, annual sunshine time is from 3033 hours to 3246 hours. The elevation of the basin is in the range of 1100 m to 1900 m. The main river is 670 km long, average annual runoff 10.31 million m^3. The Shule River Basin has a long history of irrigated agriculture. Livestock and gravity irrigation farming were developed in the Han Dynasty (BC 206) and Tang Dynasty (AD 618). There are three medium-sized reservoirs built since the 1960s, including Changma, Shuangta and Chijinxia reservoirs, with a total storage capacity 468 million m^3. There are three large-scale irrigation districts namely Changma, Shuangta and Huahai irrigation districts (IDs), with command areas totaling 94,000 ha, which predated completion of Changma Reservoir, and 17 hydropower stations with installed capacity of 87,000 kW.

2.1 Water Resource Balance Analysis on the Shule River Basin

For the purposes of this study the Shule River Basin comprises only the Shule River and its tributaries and the Shiyou River which diverts a part of

Shule River water for irrigation in the Huahai Irrigation District and receives a part of the drainage water from the Changma Irrigation District in the Shule River Basin. The total annual surface and groundwater resources in the Shule River Basin are 1082 million m^3 (or 1082 gigaliters, Gl, of which 1031 million m^3 is from the Shule River and 51 million m^3 is from the Shiyou River), as shown in Table 5.2. Before the implementation of the Shule River Agriculture Development and Resettlement (SRADR) Project that was funded by the World Bank and Chinese Government (see Section 4.3), the total water diverted for irrigation and industrial uses from the Changma headwork and from the Shuangta and Chijin reservoirs as well as extracted from groundwater aquifers was 709 million m^3, the coefficient of utilization of water resources was below 65.5%.

After the implementation of the SRADR Project, the Shule River runoff was regulated by Changma Reservoir, the available water supply for the year of p = 50% and the year of p = 75 % respectively are as shown in Tables 5.3 and 5.4.

The planned groundwater supply to different sectors is as shown in Table 5.4.

Table 5.1 Water resources of the Shule River Basin (million m^3).

Items		Resources	Remarks
Surface water	Shule River	1031	
	Shiyou River	51	
	Total	1082	
Groundwater	Changma ID	784	Non-double counted part of s.w and g.w resources. 22.0
	Shuangta ID	205	Non-double counted part of s.w and g.w resources. 30.0
	Huahai ID		
Total water resources		1134	Only non-double counted g.w resources are included

Notes: *s.w* is surface water and *g.w* is groundwater

Table 5.2 Available water supply for the year p = 50%.

Water supply probability	Irrigation district	Available water supply (million m^3)			
		Total	River water	Spring water	Groundwater
P = 50%	Project total	1039.40	878.84	82.60	77.96
	Changma ID	641.11	590.87		50.24
	Shuangta ID	292.93	188.30	82.60	22.03
	Huahai ID	105.36	99.67		5.69

The water demand of different sectors for the year of p = 50% is as shown in Table 5.5. From this table, it can be seen that after implementation of the SRADR Project the total water demand 1039.40 million m^3 is less than the total amount of water resources of the river basin 1134 million m^3, the demand can be met, but the coefficient of utilization of water resources in the basin is 1039.40/1134 = 91.7%.

The exploitable groundwater resources are mainly calculated on the basis of groundwater balance for each irrigation district as a whole, without taking into consideration of the location and ways of groundwater utilization and the different options of surface water supply. The major recharge sources of groundwater resources include seepage losses from river channels, irrigation canals, deep percolation of irrigation water in the field, and the major discharge items are evaporation from shallow water table and extraction of groundwater. Before construction of Changma Reservoir the recharge and discharge items were in balance. After the construction of Changma Reservoir, new intake structures and canals, expansion and

Table 5.3 Available water supply for the year p = 75%.

Water supply probability	Irrigation district	Available water supply (million m^3)			
		Total	River water	Spring water	Groundwater
P = 75%	Project total	1024.32	804.11	70.20	150.01
	Changma ID	633.77	548.52	-	85.25
	Shuangta ID	285.99	160.52	70.20	22.03
	Huahai ID	104.56	95.07		5.69

Table 5.4 Planned groundwater supply to different sectors post SARDR project (million m^3).

Irrigation District (ID)	Groundwater extraction					Exploitable groundwater resources
	Irrigated agriculture	Irrigation of tree belts	Urban and rural domestic use	Industrial water use	Subtotal	
Changma ID	69.85	4.10	6.50	4.80	85.28	112.00
Shuangta ID	46.67	3.50	3.00	2.10	55.27	58.10
Huahai ID	6.89		1.50	1.10	9.49	10.00
Total	123.41	7.60	11.00	8.00	150.01	180.10

Table 5.5 Water demand (in 2007) for different sectors (million m^3).

Water demand	Total (all IDs)	Changma ID	Shuangta ID	Huahai ID
Total demand	1039.40	641.11	292.93	105.36
Irrigation use	930.07	542.96	284.33	102.76
Industrial use	82.75	82.75		
Other uses	26.60	15.40	8.60	2.60

improvement of existing canals through lining, the distribution of recharge to groundwater and available groundwater resources in different areas changed accordingly. The major discharge item, groundwater evaporation, also changed in response to fluctuation of water table depth.

The water balance analysis shows that whilst water demand for economic development can be met, the coefficient of water utilization of 91% in the basin is very high. Since the ecological environment in the basin is fragile, more attention should still be paid to the conservation of the ecological environment. Some measures for the improvement of ecological environment, such as planting of artificial protection forests, growing forests for fuels, restoration and preservation of natural vegetation by stabilizing Gobis and dunes, are included in the project development plan. With rapid development of social economy and urbanization there will be an increased water demand for industrial and domestic uses. It is necessary to take water saving measures, such as adjustment of areas for agriculture, forestry and grassland for animal husbandry, adjustment of areas for cereal and economic crops, conjunctive use of surface water and groundwater, adoption of improved field irrigation techniques, as well as reuse of drainage water for irrigation of salt tolerant crops, grass and trees.

2.2 Groundwater development and management

Many tube-wells were constructed in the basin before the SRADR project and the total number of tube-wells in the project area was 958, of which 419 were in Changma ID, 330 in Shuangta ID, and 209 in Huahai ID. The total area solely irrigated by groundwater in the Changma ID alone was around 4,000 ha. Later, most of the groundwater recharge comes from the seepage losses from irrigation canals and deep percolation of irrigation water in the fields, which is more or less uniformly distributed over the whole irrigated areas. In areas solely dependent on groundwater, without surface irrigation recharge, the exploitable groundwater resources are very limited, and continuous extraction causes overdraft and mining of the aquifer. The development of groundwater cannot be sustainable.

The water table depth is the key index for the sustainable development of water resources and protection of ecological environment in the basin. If the water table depth exceeds a maximum limit the natural vegetation such as trees, shrubs and grass will die and land degradation will develop. If the water table rises, soil salinization will occur, hence keeping the water table at the required depth is a key issue for the development of water resources. In the design of groundwater development the water table depth should be predicted by groundwater modeling, the total water demand for preservation of ecological environment can be determined by the groundwater evaporation from the water table in the non-irrigated areas and this should be included in the overall water balance.

2.2.1 Groundwater management in Shule River Basin

At present in fact there is no effective groundwater management in the Shule River Basin. The principal water administrative institution at county level, namely county water affairs bureau is responsible for issuing water permits for extraction of groundwater and construction of tube-wells, but the quantity of extraction is not properly controlled. Since the surface water and groundwater are interconnected, the groundwater recharge comes mostly from surface water irrigation. Groundwater is used in a period of water shortage from May to June to supplement the gap of water supply. The uncontrolled exploitation of groundwater in some localities can cause overdraft and mining of groundwater aquifers, while in other areas with excessive use of surface water may induce water table rise and soil secondary salinization. Uncontrolled groundwater utilization not only causes environmental problems within the IDs, but also has ecological impact on the uncultivated areas and the whole oasis. Therefore it is necessary to have integrated surface and ground water management in the basin, the extraction of groundwater should be rigorously controlled according to a master plan of water allocation and distribution, and the water table depth should be controlled in accordance to the requirements of both crop growth in the IDs and the ecological balance in non-irrigated area and the whole basin. For the conjunctive use of surface and groundwater, the tube-wells should be managed by Shule River Basin Authority (see below), and unified water charges, determined as the weighted average of both surface and ground water prices should be collected.

2.3 The impact of SRADR project on the ecology in Shule Basin

The construction of Changma dam in the upstream and lining of main irrigation canals had an impact on spring flow at the front edge of the deluvial and alluvial fan in the basin. Before dam construction and lining the main canals, the seepage losses from the old river channels and canals, as well as the recharge of river water to groundwater due to overflow in the flood period, was estimated to be more than 60 million m^3, and the groundwater emerged at the edge of the deluvial and alluvial fan as springs. After the implementation of the SRADR project the recharge from the river channels and canals is considerably reduced, consequently the discharge from springs was also reduced. Since the river flow is regulated by the Changma reservoir, and the canal efficiency is increased, much more river water can be diverted to areas where spring water is relied upon for irrigation. Overall the negative impact of project implementation on the spring flow can be well compensated by the positive impact of increased water supplied by canals.

2.3.1 Land reclamation and salinity control

The salinity of irrigation water can be as high as 1.992 g/l, which is about the allowable upper limit for irrigation (2 g/l), especially in arid areas such as Shule River Basin with very fragile ecological conditions. In downstream areas irrigated by Shuangta Reservoir water, it is necessary to take measures to reduce the salinity of irrigation water to prevent secondary soil salinization.

2.3.2 Artificial oasis expansion and ecological environment protection

Before the implementation of the project, 65% of the water resources in the basin were used in the different economic sectors. After the project, the coefficient of water resources utilization went to 91%, an increase of 40%, while the irrigated area (or the artificial oasis) increased by 140%. The increase of the area of artificial oasis and its perceived impact on the ecology is of public concern. At the present time, there are no proven criteria to protect the ecological environment in the arid zone. Based on the study of the relationship between the water resources and the evolution of the eco-system in arid region, the structural features and the stability of the ecosphere are characterized by the degree of desertification, the proper scale of the transition zone and the oasis, as well as the intensity of development of the artificial oasis.

- The driving factor for the degree of desertification is precipitation, which determines the basic pattern of desert and non-desert zones in arid regions.
- The driving factor for the proper scale of the transition zone and the oasis is the scope and the intensity of the runoff activity, the degree of the vegetative coverage and the source of moisture determine the performance of the transition zone for the protection of oasis.
- The driving factor for the intensity of the development of the artificial oasis is the development and utilization of the water and land resources.

The development and construction of an artificial oasis induces the readjustment of the structure of the ecosystem, and the limits for the development of an artificial oasis in different areas are quite different. The utilization of water and land resources resulting from large scale development of artificial oasis, changes oasis structure and leads to shifts in the relative importance of surface runoff and groundwater on the most sensitive transition zone. Precipitation alone, in the absence of runoff, is not sufficient to support the continued existence of the vegetative cover in the transition zone, and the reduction of vegetative cover will cause the degradation of the transition zone and the expansion of the desert area.

The ecology of the northwestern inland arid areas can be classified into three types: (i) The area of basically stable ecology; (ii) The area of unstable ecology; and (iii) The area of most unstable ecology. The Shule Basin is classed as type III, the area of the most unstable ecology. Since the basin is located in the extreme arid region, with rainfall only from 47.4 mm to 67.4 mm, desertified land in the Shule Basin accounts for 86% of the total area of the basin. The area of the transition zone from desert to oasis accounts for 79% of the non-desert area, is very fragile and strongly dependent on the water resources in the form of runoff, and hence it is very sensitive to the expansion of the oasis. The area of the oasis should be restricted in order to preserve the predominant role of the transition zone in protecting the oasis. The ratio of the areas of the transition zone and the oasis should be kept at the range from 2:1 to 3:1. The total area in the Shule River Basin is 41,000 km^2, in which the Gobi and sandy desert areas account for more than 90% of the plain area. Before the implementation of the SRADR project the gross oasis area is 60,000 ha the area of the transition zone is about 213,333 ha, the total non-desert areas is about 273,333 ha: the ratio of the areas of the transition zone and the oasis was 3.4. After the implementation of the SRADR project, when the total irrigated area reaches 98,000 ha, the ratio of the areas of the transition zone and the oasis is about 1.72, which is close to the criteria required for the protection of the ecological environment. A series of measures should be taken for the protection of ecological environment both in the irrigated area and in the transition zone. These include planting of ecological protection forests, growing woodlots for fuels, preservation and restoration of natural vegetation, as well as revegetation to stabilize Gobi and sand dunes. To create better ecological conditions, it is recommended that the area of land converted to cropland could be reduced and the transition zone around the oasis could be increased and protected, and in these areas the water table should be kept at a depth less than the maximum depth required for the growth of the natural vegetative cover.

3 Approaches to IRBM in Shule River Basin

3.1 Current problems in Shule River Basin

The encroaching Kumutage desert is threatening the Dunhuang oasis and the Mogao grottoes in Dunhuang City. The 1,600-year-old Mogao grottoes, listed as world cultural heritage by UNESCO, are famous for their statues, wall paintings and Buddhist art. The natural barrier of wetlands which protected the advance of the desert is shrinking and desertification of the oasis is worsening because construction of dams in the upstream of the Shule River to divert water for irrigation has reduced groundwater recharge and flow to downstream wetlands.

3.2 Management strategy to cope with water crisis

Conflicts and competing water use between human and nature can no longer continue in the Shule basin if its important cultural and biological heritage is to be conserved. Immediate action is required to prevent disastrous consequences of desertification of wetlands and oasis. Balanced allocation of scarce water resources between socio-economic development and healthy ecosystems is the only way to achieve harmonious co-existence of human society and nature. Thus in the "12th-Five-Plan" period, Chinese Central Government decided to fund a program entitled "Dunhuang Water Resources Rational Development & Ecosystem Protection Project", the goal of the project is to develop an optimized plan for conjunctive use of surface and groundwater resources to meet water demand for sustainable socio-economic development, particularly to protect Dunhuang oasis and Mogao grottoes, while keeping environmental flow for healthy ecosystems in the downstream of Shule River Basin.

In the past a strong interactive surface and groundwater system existed in the Shule River Basin. Construction of dams and diversion of river flow for agricultural development in the upstream has depleted the water resources originating from the Qilian Mountains leading to dry riverbeds and cessation of river flows to downstream wetlands. The combination of decreased groundwater recharge and increased groundwater abstraction resulted in continuous decline of groundwater level, which has caused damage to groundwater dependent ecosystems and to accelerated land degradation.

3.3 Drivers for IRBM in Shule River Basin

(i) Rationale for establishment of a basin management agency

Gansu Province Government made a significant decision in 1995 to launch a program entitled "Shule River Agriculture Development and Resettlement

Project" (SRADR) that was co-financed by the World Bank and Chinese Government, with primary tasks of developing irrigation, constructing water infrastructures, converting of 54,600 ha of rangeland to irrigated cropland and resettlement of 0.2 million people. A key issue facing the project was the limited water supply within the Shule River Basin. A special study of the basin's ecology to determine the "ecological demand" for water comparing available water supply and projected requirements for irrigation and other uses was completed in 2002 and showed that although the project had not as yet produced negative impacts, the overall water balance was extremely tight or negative when ecological water requirements of the three downstream nature reserves are considered, even with the reduced resettlement target of 0.1 million people. The study also indicated that groundwater over-extraction was severe in some areas even before the SRADR project was completed. This was due to intensified human activities in the past decades. This meant that appropriation of project land and water by local farmers would mean less water for those people who planned to resettle in the project area; and a large-scale water saving program would be needed to maintain the basin water balance. The study recommended that strong and effective integrated river basin management was essential to protect basin water resources and ecology and to ensure the long-term sustainability of the SRADR project. The project legal covenant had included provision for establishment of a river basin agency by the end of 1998.

Because there are many agencies and water users at various levels involved in the Shule River Basin it is a complicated issue to set up a new agency with the provincial level powers and authority needed for integrated river basin management, including integrated management of both surface and ground water, efficient water use by all users, application of water conserving technology and management, and effective salinity control and stakeholder participation in environmental management. In consideration of the wide range of water management issues and various levels of government involved, Gansu Province established the Gansu Shule River Basin Water Resources Management Commission at the provincial level in October 2001. The new commission was expected to greatly assist comprehensive development and management of the Shule River Basin. However, no significant progress was made in making the new commission operational until much later.

Eventually, by the end of 2004, Gansu Province Government established the Shule River Basin Water Resources Management Bureau (SRBWRMB) with quota of 667 employees in total to be responsible for integrated river basin water resources management. SRBWRMB is affiliated to the Department of Water Resources Gansu Province (DWRGP). After considerable consultation, reconciliation and compromise SRBWRMB as a public entity has set up 9 offices and divisions. SRBWRMB is also directly in charge of the 3 large irrigation districts and 17 hydropower stations. The primary responsibilities of SRBWRMB include following: to implement China's Water Law and relevant decrees; be responsible for comprehensive rehabilitation of Shule River Basin, to develop river basin water resources planning; to carry out integrated water resources management in the basin; to formulate overall water allocation plan and annual water distribution plan; be responsible for protection, monitoring and evaluation of water resources in the basin; be responsible for integrated management and rehabilitation of major rivers and river channels in the basin, to develop flood control and drought combating plans, and to carry out physical works related to flood control and drought combating; be responsible for managing groundwater and surface water utilization, collecting water charges, coordinating and resolving water use disputes in the basin.

The SRBWRMB carried out a large amount of infrastructural work. The majority of main and branch canals in the irrigation districts within the basin were lined with concrete slabs, and an irrigation district control information system was developed. A series of hydrology monitoring stations were set up at main basin sections, and more than 50 groundwater observation wells were constructed. The basin water resources plan was formulated in 2006, and river basin surface water and groundwater allocation model was developed. More than 120 farmer water user associations (WUAs) were established to cover the three irrigation districts. All these made a good foundation for integrated basin water resources management (see also Xu et al., this volume, for experience in neighboring Hei He basin).

(ii) Challenges and problems to tackle

The new basin management agency was faced with challenges and issues as follows: the previous project development and construction entity was set up in 1995, its main mandates were to reclaim so-called wasteland and develop water resources, to relocate 0.1 million poor farmers to the basin, to improve livelihood of both native residents and re-settlers and promote their peaceful and harmonious co-existence. In 2004 the new basin management agency was established to carry out integrated water resources management. Although they had no experience and knowledge in integrated basin water resources management, the new basin management agency had to be transformed from engineering construction management oriented entity to an integrated basin resource management agency in a short time to achieve sustainability of socio-economic development, resources utilization and healthy eco-environment in the basin.

The SRBWRMB has a significant number of staff and some technical capacity following the SRADR project. However, there is no effective overall river basin management committee coordinating the policies and activities of governments, primary and secondary stakeholders in the river basin.

The SRBWRMB needs assistance to support the further development of the system for river basin governance, in particular to provide advice on the development of the necessary regulations and implementing water law and decrees, and in assessing the water resources and planning for its sustainable management.

The SRBWRMB desperately needs to improve its capacity relevant to river basin governance, policy and science based management, river basin planning and community participation. It is necessary to develop and use hydrological and biophysical modeling to support river basin management; to study science based approaches (such as sustainable yield studies, scenario based modeling) to deal with risks and uncertainties in the changing river basin environment such as climate change, population expansion and economic development, etc.; to run preliminary case studies with recently changed policy settings and scenarios of threats to sustainable water resources management; to further investigate policy measures including water allocation, water permitting and market based approaches which assist water use and agricultural adjustment; to include water abstraction caps, ecosystem targets and water markets in river basin planning; to develop appropriate policies to be adopted by local water management bureaus and governmental agencies at different levels. Eventually a research institute was set up in 2012 to provide science support to the basin authority.

4 Water Allocation and Water Entitlements in Shule River Basin

The application of water trading and water markets to the inland river basins has been increasingly becoming a hot topic for discussion. The purpose would be to help implement the new national water policy aimed at improving river basin management, to deal with the threats to sustainable water use in the basins. It is noted that a Sino-Australian research team has carried out research on feasibility of water entitlements trading in Shule River Basin in November 2011. The team noticed that some Chinese water managers were excited about starting water markets. After site visiting and discussions the team was cautious about how quickly this could happen, however, taking the following factors into account: (i) plots of household farmland are very small; (ii) it may be too expensive to meter the water for each farmer household, and water metering may only be feasible for groups of households or water user associations (WUAs); (iii) according to China's law individual entitlement to land and the trading of land is limited in China, making it unwise to expect water rights and water trading to develop any time soon; (iv) in general the cultural leaning towards "command and control" and the lack of statutory rights and rules would be challenges to work through. These factors have proved to be issues, but now much of what the river basin authority is doing in water resources management, both organizationally and technically, is very significant and impressive.

4.1 Water distribution within a WUA

People might be puzzled about how water allocation for households worked for the Shule Basin, where entitlements are proposed but not yet in place. Taking a typical WUA as an example, having five user groups, each with 20 households, with 300,000 m^3 of water altogether (i.e., each household has about 3,000 m^3). This WUA might have about 70 ha of land (i.e., each household has about $^2/_3$ ha).

It was found that in Shule Basin the 300,000 m^3 of water entitlement is not written down anywhere. Rather each year a new irrigation plan is worked out, which says how much water each WUA will get. Each WUA will first specify how much land is to be sown to wheat—and according to the irrigation norms this will get 8400 to 9000 m^3 per ha, depending on the soil. Hops will get 10500 m^3 per ha, malt 9000 m^3, cotton 7500 to 12000 m^3. Cabbages, tomatoes, leeks, etc. are mostly grown in greenhouses, using groundwater applied under pressure via drip or micro-jet systems, and the allocation of groundwater is arranged separately.

Water management stations in IDs compile the bids from WUAs into the irrigation plan, and submit it to the branch canal station, that delivers water to tertiary canals (normally one of these canals for each WUA). Out of this process each WUA will receive a specified volume of water for that season, and will make a distribution plan covering each household, plus the losses—one household might get water for two hours, another for three hours, etc. Each household will end up knowing how much water is to be delivered to its land that season, and the consequent tariff. A question is raised, if farmers want to switch from wheat to cotton, which takes more water, do they get more? Yes, if it's in the plan. If all the WUAs want to start growing cotton, they can do this, but if there's not enough water, all the WUAs will have a deficit—i.e., their crops will be under-irrigated, and not as good.

4.2 Water use and new settlers in Shule River Basin

This section compares the experiences in two inland basins in the Hexi Corridor. Development in the Shiyang basin began earlier and many mistakes were made. The experience gained by the Department of Water Resources Gansu in the Shiyang Basin were incorporated into the planning for the Shule Basin. The Shule River Basin is about the same size as the Shiyang River Basin, but it has a slightly smaller population of about 1.5 million, with a lower proportion engaged in agriculture (more are employed in steel-making, power-generation, service, tourism, etc.). Irrigated farmland amounts to 100,000 ha. It is even more arid in the Shule Basin than in the Shiyang Basin. A lot of the water emanating from the mountains in the Shule Basin goes underground at the top of the alluvial fan, or else in canals and on farms, but subsequently re-emerges as surface water again.

The resources in Shule River Basin have not been over-exploited to the same extent as in the Shiyang River Basin. But the natural river flows have recently been diverted and there is a danger of over exploitation. As many as 100,000 people, mostly from ethnic minority groups, have been moved here from mountainous areas further southeast in Gansu, where it was overcrowded and people farmed terraces on steep slopes, causing severe erosion. With funds from the World Bank, over the course of about 10 years since 1996 Changma Reservoir has been constructed, new canals have been built opening up some 50,000 ha to irrigated farming, more than 6 new townships and 46 villages have been created, and support has been available with setting up of 6 new high schools, 50 primary schools, 46 clinics, 6 seed supply stations, 10 livestock service stations, etc.

It was initially planned to accept 200,000 new settlers, who were living in dire poverty. Water resources were proposed to be 91% utilized. But concerns about sustainability came to receive more attention, an ecological study was carried out (as discussed in section 4.3 in this chapter), and the assumed utilization rate was revised down to 65%, to protect the lower reaches. In practice the utilization rate remains at about 84%: now shortage of water and competition for the available water are certainly ongoing issues.

4.3 Specifying water rights to suit the hydrology

It is noted that reservoir capacity is low compared with usage. Changma reservoir holds less than 200 million m^3, considerably less if dead storage is subtracted; with the two other main reservoirs, namely Shuangta and Chijin reservoirs, the combined active capacity in the Shule Basin is less than 500 million m^3 in total. This compares with total usage of more than twice that amount. The situation is comparable in the Shiyang River Basin, where there are eight medium-sized reservoirs, and total active capacity is just 370 million m^3.

The SRBWRMB has given out two rights to use water to businesses (a power-generation company, 8 million m^3 and a steel plant, 4 million m^3) in return for investments in water savings. These rights were on-going entitlements. No distinction had been made in the Shule Basin between permanent and temporary water rights. The feasibility of seasonal allocations (temporary rights) on regulated systems in the inland basins has been questioned because of uncertainty in the hydrology and water distribution. The volumes that can be actually be used are not certain and hence the volume of water available to meet entitlements is not secure. Each river basin has its own drought management plan to cope with water crisis and in a dry year all water rights are reduced proportionately.

Each year there are two deliveries of water down to the lower storages, the three irrigation districts and beyond to terminal wetlands. The first water delivery starts from mid March to the end of June, and the second from late August to mid November. In between, in mid July the sluice gate is opened for two weeks to try to flush out sediment near the dam wall, but only about 10% of the sediment can be dealt with, which is why the reservoir life is limited.

The rainy season in Shule Basin is very brief and intense, in the warm months from mid July to early September. Preceding this, there is a high demand for irrigation in spring from April to June. In the inland basin in late November, farmers are just finishing off their "winter irrigation". The water freezes in the soil, helping to prevent wind erosion over the dry winter months, having benefits for soil quality, and providing moisture for crops when the ice melts in the spring. It can be seen that framing a system of tradable water rights, particularly rights to water in the current year or season, requires good understanding of the specific nature of the hydrology and of the water demands in the basin.

5 Summary of Lessons and Experience on Enhancing IRBM in Northwest China

IRBM approaches are being taken to manage water-stressed inland river basins in China. The long term goal is the sustainable management of the water and related resources of inland river basins in China by adapting world leading approaches to inland river conditions. The Shule, Heihe, Shiyang River Basins are chosen as representatives of the inland basins in Northwest China which are at different stages of development, environmental degradation and water resources governance. The Heihe River Basin was a previously prioritized basin, the Shiyang River Basin is currently a national level priority of the Chinese Central Government until 2020, and the Shule River Basin has been elevated to the same level in the new "12th-Five Year" Plan. The results of various rehabilitation programs funded by the Chinese Government will potentially be extended to the other inland river basins. Some of the conditions and measures in northwest China will also have future relevance for other river basin agencies worldwide.

The management experience and practice in Shule River Basin is focused on future directions for the management of stressed river basin, with a special emphasis on technical and policy aspects. The overall outcomes and direction include:

(i) To set up integrated river basin management systems for the inland river basins including: the scope of operational policies, science based investigations to support policy development, and approaches that are responsive to changing conditions such as climate change, increased environmental requirements, development pressures, etc., development and application of computer modeling to support river basin management.

(ii) To improve understanding and capacity of institutions and water managers to adapt policies for the integrated management of inland river basins in Northwest China. Water Law and river basin regulations can provide effective and powerful management tools with river basin agency only if IWRM is prioritized. Reviewing the lessons and experience in river basin management in past decades, it is considered that government policy has had the major effect on the state of the water resource. The adverse effects of poor government policy may have lessened in recent years as policy (at least at the national level) has adjusted to reflect a more holistic view of how to manage water resources and ecosystems.

(iii) To take a participatory approach to manage water resources at grass-root level. Local community based participatory approach can generate more economic and social benefits as well as political influence. As an innovative form of the participatory approach, water users associations (WUAs) have rapidly developed in Shule River Basin. The basin authority had plans for establishment of WUAs in the new resettlement areas at the time the resettlers relocated and the local on-farm irrigation facilities were built. However, at the beginning some of the WUAs established were led by village chiefs or other local leaders, instead of being led by farmers, which is one of the basic principles for independent, democratic, sustainable WUAs. Eventually the WUAs all were converted to farmer leadership. Establishment of WUAs as a useful bottom-up approach should be advocated to encourage more and more local farmers to participate in water management (see Chapter 13 by Xu et al., on the role of water users associations in Zhangye City, Heihe River Basin, China, this volume).

Many of these developments in water use and land reclamation have often been in response to political and economic development factors that sometimes conflict with goals of maintaining basin ecosystem stability. The absence of an integrated river basin management agency was one of the primary causes of ecosystem degradation. Establishment of effective basin agency that is supported by science and technology advancement as well as effective legal framework is the key to solve the problem of competing water use.

Under the centrally planned economy system in past decades, development of the large-scale artificial oases had a major impact on the eco-environment, including the important riparian areas. Most surface water resources in the upper part of the catchment have been regulated by dams and reservoirs, giving the ability to control the distribution of water resources and operate water markets. Water allocation plans should give secured allocation of water to the environment.

Acknowledgments

We are grateful to Peter Waldheim for preparing Figure 5.1. and to the Gansu Water Resources Bureau and the World Bank Project Management Office for access to project files and documents

References and further Readings

DWRGP. 2010. Gansu Province Water Resources Bulletin 2009, Department of Water Resources Gansu Province (DWRGP), Lanzhou, China.

GPSRBWRMB. 2007. Project Completion Report on the Gansu Hexi Corridor Shule River Project, compiled by Gansu Province Shule River Basin Water Resources Management Bureau (GPSRBWRMB), Yumen Gansu, China, February 2007.

GPWRHDI. 1995. Feasibility Study Report of Hexi Corridor (Shule River) Agriculture Irrigation and Resettlement Development Project, compiled by Gansu Province Water Resources and Hydropower Design Institute (GPWRHDI), Lanzhou, China, December 1995.

GPWRHDI. 1996. Preliminary design report of the Hexi corridor (Shule River) Agriculture Irrigation and Resettlement Development Project, Compiled by Gansu Province Water Resources and Hydropower Design Institute (GPWRHDI), Lanzhou, China, December 1996.

Jiang, Leiwen, Tong Yufen, Zhao Zhijie, Li Tianhong and Liao Jianhua. 2005. Water resources, land exploration and population dynamics in arid areas—the case of the Tarim River Basin in Xinjiang of China. Population and Environment 26(6): 471–503.

Kang, E., G. Cheng, Y. Lan and H. Jin. 1999. A model for simulating the response of runoff from the mountainous watersheds of inland river basins in the arid area of northwest China to climatic changes. Science in China Series D: Earth Sciences 42(1): 52–63.

Kingsford, Richard, Justin Costello and Fran Sheldon. 2014. Lake Eyre Basin- Challenges for managing the World's Most Variable River System. pp. 371–396 *In*: Victor R. Squires, Hugh M. Milner and Katherine A. Daniell (eds.). River Basin Management in the Twenty-first Century: Understanding People and Place. CRC Press, Boca Raton.

Li, S. and G. Chen. The Study on the Rational Development and Utilization of Water Resources and the Protection of the Ecological Environment in North Western Areas. Yellow River Publishing House, December 2002.

Li Yuhong and Victor Squires. 2008. Case Study 5: Hexi Corridor, Gansu. pp. 151–170. *In*: Victor R. Squires, Xinshi Lu, Qi Lu, Tao Wang and Youlin Yang (eds.). Degradation and Recovery in China's Pastoral Lands. CABI, Wallingford.

Wang, Y., Q. Feng, L. Chen and T. Yu. 2013. Significance and effect of ecological rehabilitation project in inland river basins in northwest China. Environmental Management 52(1): 209–220. doi: 10.1007/s00267-013-20130077-x. Epub 2013 May 29.
World Bank. 2007. Implementation Completion and Results Report on the Gansu Hexi Corridor Project, June 2007.
Xu, H., L. Sui, Y. Li and D. Zhang. 2014. The Role of Water User Associations in Integrated Water Resource Management in HeiHe River Basin, China. pp. 304–324. *In*: V.R. Squires, H.M. Milner and K.A. Daniell (eds.). River Basin Management in the Twenty-first Century: Understanding People and Place. CRC Press (this book).
Zhou, L. and G. Yang. 2006. Ecological economic problems and development patterns of the arid inland river basin in Northwest China. Ambio 35(6): 316–8.

Websites

Victorian Water Register website: www.waterregister.vic.gov.au
The Department for International Development website: http://www.dfid.gov.uk
The Australian Agency for International Development website: http://www.ausaid.gov.au
The Worldbank website: http://www.worldbank.org
The Water Resources Demand Management Assistance Project website: www.wrdmap.com

6

The Critical Role of Local Farmers in IWRM in Ghaghra-Gomti Basin, India

Andrew Marr* and Ashok Raut

SYNOPSIS

This chapter summarizes the outcomes of one component of a World Bank funded Water Sector Restructuring Project in the Indian state of Uttar Pradesh. This component investigated how agricultural production might be increased through management of water-logging in a 540,000 ha study area of Jaunpur Branch Canal of the Sarda Sahayak Canal System.

The goal shared by the Government of Uttar Pradesh (GoUP) and the farmers, some of whom are landless laborers, is to increase the value of agricultural production. Water-logging which leads to land degradation in the agricultural areas, and under certain conditions to soil salinity and sodicity, both threatens this goal and offers an opportunity for achieving it.

A water balance model and economic analysis showed that operating the State-run canal system conjunctively, using the groundwater system to provide needed storage, allows greater supply reliability; that small shallow tubewells, owned and operated by farmers, can increase timeliness in water delivery; and options then emerge for crop diversification and hence greater value in agricultural production.

SMEC International Pty Ltd, PO Box 1052, North Sydney NSW 2059 Australia.
*Corresponding author: andyjmarr@gmail.com

Keywords: aquifer management, crop yields, decision support system, economic analysis, farmer's income, groundwater, irrigation, land degradation, legislation, population, salinity, sodicity, tube wells, water balance model, water logging, World Bank

1 Introduction

In November 2004, a Concept Paper was issued by a sub-committee of government specialists of the Government of Uttar Pradesh (GoUP) in response to the growing problem of water logging and land degradation in the agricultural areas of the state. One of the key findings of the sub-committee was that groundwater usage for irrigation should be increased in areas where surface and sub-surface water logging were occurring. Since sub-surface water logging is most prevalent in areas with good access to canal water, the finding reinforced the findings of previous studies that conjunctive use of surface and ground water for irrigation is necessary to address water logging. The sub-committee recognized that adoption of conjunctive use of surface and groundwater sources of irrigation will require the active participation of farmers with support from government.

In the same year that the sub-committee issued its findings, the GoUP, as part of the World Bank-funded Uttar Pradesh Water Sector Restructuring Project (UPWSRP), commissioned the consultant SMEC International to develop two Decision Support Systems (DSS) and to apply these DSS to the preparation of basin plans for two areas of the state. The first DSS was for a pilot area, the Jaunpur Branch Canal command area. The second DSS was for a larger area that included the pilot study area, the Ghaghra-Gomti basin. Only the DSS for the pilot area is discussed here although many findings also apply to the larger area. The management units adopted for the DSS are not drainage basins, the normal management unit for IWRM, but are actually the command areas of canal systems. It was considered that the IWRM approach could be applied to these management units because the most significant and actively managed water movement in these units is from the canals to the irrigated areas, with drainage from the catchments to the drains and rivers being of secondary importance. These management units are also adopted by the Irrigation Department that has primary responsibility for the distribution of surface water for irrigation.

The findings of the GoUP sub-committee provided the basis for the development of the basin plans under the DSS project. The DSS project mapped the spatial features of the land and water resources within the study areas, assessed the works and management changes required for effective conjunctive use of surface and canal water for irrigation, and provided the socio-economic rationale for the recommended development and management plans for the basins.

In October 2013, a new credit agreement was signed between the Government of India (GoI) and the World Bank to help build the institutional capacity needed to increase agricultural productivity in Uttar Pradesh. The activities will support a more holistic strategy for aquifer management, improving governance through water user associations, and focusing on rehabilitating and modernizing existing irrigation systems in critical areas.

2 The Study Area

2.1 Location and physical features

The State of Uttar Pradesh is located on the Indo-Gangetic Plain in the north of India, extending from Delhi in the west to Varanasi (Benares) in the east, and from the Nepal border along the Himalayan foothills in the north to the Ganges River in the south (Fig. 6.1). The state has an area of 24.3 million hectares and a population in 2001 of 165 million (199.6 million in 2011), making it the most populous state in India. Lucknow, the state capital, is located close to the geographic center of the state.

The pilot DSS covers the Jaunpur Branch command area located, at its closest point, about 125 km south-east of Lucknow. This area was selected for pilot studies and works in the first phase of UPWSRP. This study area has a gross area of 542,600 ha, a Culturable Command Area of 275,000 ha (as defined by GoUP) and a population density of 697 persons/km^2. The Census data for 2001 and 2011 show that total population increased by 17.5% over the decade in the five administrative Districts covering the study area, and that 92.5% were classified as rural residents. The elevation of the study area ranges from about 100 m asl to 77 m asl. Approximately 78% of the study area has a slope of less than 1%, with the remainder having a slope from 1–3%. Within the study area there is little relief, although the drains and rivers are incised into the surrounding plain. From satellite imagery, the many paleo-channels that are visible indicate the numerous changes in channel locations over the millennia.

2.2 Climate

The climate of the study area is typical of the Gangetic Plain of north India with a Monsoon season from mid-June to October, a Winter season from November to March and a Pre-monsoon season from April to mid-June.

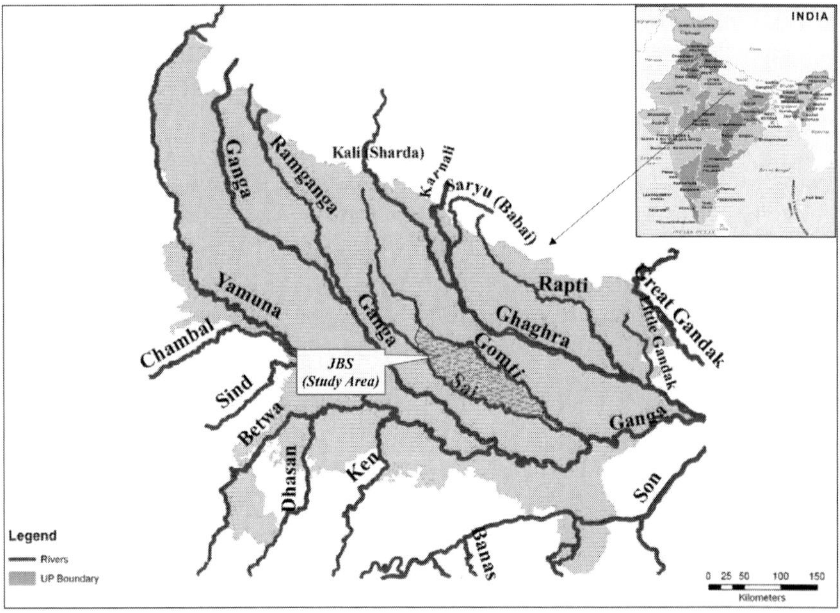

Figure 6.1. Uttar Pradesh is a province in northern India. The study area, south of Lucknow, is indicated by the square.

Annual rainfall is about 900 mm, with nearly 90% falling in the Monsoon season. Maximum temperatures of up to 45°C occur in May–June while over-night minimums as low as 4–5°C occur in December–January.

2.3 Soils

The soils are sedimentary. The surface soil is loamy while sub-surface soils are mostly fine-loamy, with some areas of coarse-loamy soil and one area of fine-silty soil.

2.4 Ground water

Ground water is contained in the sediments that underlie the Study Area. These sediments extend hundreds to thousands of meters below ground level to bedrock. In the study area there is an unconfined (phreatic) aquifer that extends 40 to 70 m below ground level. This aquifer is composed of inter-bedded layers of sand, silt and clay with some gravels and "kankar" (calcium carbonate nodules). Groundwater levels within the study area range from about 2 m to 10–12 m below the surface, over most of the area, based on the post-monsoon 2002 data (GoUP, 1996; RSAC, 2002).

2.5 Agriculture

Yield of rice and wheat is assessed by the Department of Agriculture each year based on sampling from a scattering of small areas. Statistics for the five administrative districts covering the study area are shown in Table 6.1. These show that there is a wide range of yields for different blocks (groups of villages) within each district, and that there is quite a range in the averages for the various districts that cover the study area.

Table 6.2 compares the crop yield in the study area with yield statistics for other regions in India and beyond. This shows that the yields achieved in the study area are low by comparison with the larger regions that contain the study area, as well as by comparison with other countries.

Crop data for 2002–03 for the five administrative districts covering the study area show that Rice (Monsoon season) covers 32% of the area cultivated during the year and Wheat (Winter season) covers 38%. The only other significant crops were Maize (Monsoon season—2.4% of cultivated area), Lentil (Winter season—6% of cultivated area) and Sugarcane (perennial—2.2% of cultivated area). There was very little cultivation during the Pre-monsoon season. About 10% of the cultivated area was uncropped. Table 6.3 shows the various classifications of uncropped land, while Table 6.4 shows the seasonal cropped areas.

Table 6.1 Yield of Rice and Wheat in Districts covering study area (2003–04).

District	Rice Yield from Nyaya Panchayats (quintal/ha)			Wheat Yield from Nyaya Panchayats (quintal/ha)		
	Minimum	Maximum	Average	Minimum	Maximum	Average
Pratapgarh	7.11	21.43	16.37	15.92	28.25	20.73
Jaunpur	2.60	22.10	10.96	13.42	35.76	20.88
Raebareli	12.86	30.67	20.46	11.90	30.47	21.50
Sultanpur	9.90	37.42	18.19	14.93	34.40	25.23
Barabanki	16.60	29.63	24.75	20.20	34.28	26.32

Source: MOA 2004, Note: 1 quintal[1] = 100 kg

[1] 1 Quintal = 100 kilograms

Table 6.2 Comparison of Crop Yield in UP with other locations (2003–2004) (quintal/ha).

Crop	Study Area	GGB	U.P.	Punjab	Haryana	India	China	Egypt	U.S.A.	Avge of World
Rice	17.10	19.68	21.87	36.94	27.49	30.00	60.74	94.31	74.48	38.37
Wheat	22.70	23.13	27.94	42.07	39.66	26.17	39.07	61.50	29.74	26.65
Maize	11.60	11.82	13.92	29.81	25.33	21.14	48.54	77.11	89.24	44.72

Source: Statistical Diary, Agricultural Dept., U.P. & Fertilizer Patrika

Table 6.3 Land use classification for Uncropped Land (2003–04 as per Board of Revenue, UP).

Land Use Classification	Area (ha)	% of Total
Total waste and fallow land	101,920	19%
Barren and culturable waste area	11,305	2.1%
Barren and unculturable land	13,040	2.2%
Other fallow land	23,775	4.4%
Current fallow land	53,800	10.3%
Pasture land	3,012	1%
Land under miscellaneous trees and grove	18,922	3%
Forest area	1,390	0.2%
Land put to non-agricultural use	64,150	12%

Table 6.4 Agricultural statistics of study area (2005–06).

Description	Area (ha)	Crop Intensity
Net area sown (single, double or triple cropped)	353,228	
Seasonal cropped areas:		
Monsoon season (Kharif)	229,314	64.9%
Winter season (Rabi)	271,098	76.8%
Pre-monsoon season (Zaid)	17,025	4.8%
Total cropped area (in any season)	517,437	146.5%

2.6 Irrigation

The Sarda irrigation canal system was commissioned in the study area in 1928 as part of a system to provide "protective irrigation" which aimed to alleviate famine in the event of drought. The study area was at the tail of the canal system and supply was unreliable. In the early 1980s it was connected instead to the newly constructed Sarda Sahayak Canal to improve the reliability of irrigation supply. The canal system is designed to reach approximately 50% of the study area (UPID 1985), however, a census of irrigators shows only 24% use canal irrigation while 53% use private shallow tubewells (MID 2001). These statistics may not reflect the actual situation because studies have shown that many irrigators use canal water when available but rely on shallow tube wells to irrigate their crops when there is no canal water. Also, there may be under-reporting of the area benefitting from canal supply because an irrigation service fee should be paid for such usage. Table 6.5 shows the sources of irrigation supply by area irrigated.

Table 6.5 Irrigation sources in study area (2005–06).

Source of Irrigation	Area Irrigated (ha)	% of Net Sown Area
Total irrigated area	273,057	77.3%
Canal irrigation	84,632	24.0%
State tube well irrigation	608	0.2%
Private tube well irrigation	187,657	53.1%
Other source of irrigation	160	0.1%

Note that the statistics do **not** allow reporting of areas irrigated by both canal and tube well

It is generally acknowledged that the provision of surface irrigation from the canal system has enhanced production of cereals, particularly rice and wheat. However, water-logging and associated soil salinity/sodicity in large areas near the heads of canal command areas has reduced cropped areas, crop yields and crop diversity. In particular, legume crops that are particularly sensitive to salinity and sodicity have been adversely affected, while tree crops have also been adversely affected by shallow water tables as well as salinity.

Although irrigation from ground water using Persian wheels and dug wells dates back over a thousand years in UP, in the past few decades approximately 3.5 million shallow tube wells with electric or diesel-powered pumps have been installed in the state to tap the shallow ground water.

In the study area the unconfined aquifer is the main source of ground water, with about 130,000 shallow tube wells in place (MID 2001). The installation of shallow tube wells has benefitted from the free boring and pump subsidy schemes funded by GoUP. These schemes are directed particularly at small-holders and socially disadvantaged farmers. Available data indicates around 37% of shallow tube wells in the study area were subsidized. The 2001 census also recorded 566 state-owned deep tubewells that extended down to a second confined fresh aquifer. Between the two fresh water aquifers is a saline aquifer confined between two clay layers.

In UP, over 99% of shallow tube wells are privately owned and this is likely to be true for the study area also. About 85% of shallow tube wells in the study area are fitted with diesel pumps (typically 4 HP to 10 HP) with the remainder fitted with electric pumps (typically 3–5 kW) or other devices (Fig. 6.2).

From field observations, it appears that many irrigators now use flexible hoses of about 100 mm diameter to deliver the ground water to the fields.

While some subsidies are available for boring shallow tube wells and pump purchase, operators of shallow tube wells pay full market price where diesel fuel is used. The state has endeavored to provide free electric power for shallow tube wells, but the supply and distribution is extremely unreliable in rural areas, so this option is rarely pursued. There is some evidence of a market for water from shallow tube wells, with the market price reflecting the cost of production. This market may be imperfect because studies suggest that many irrigators are captives of the existing shallow tube well owners. This occurs because the state will not subsidize borings or pumps close to an existing shallow tube well. As a result, irrigators have a limited choice of ground water irrigation supplier (Pant 2004; Shah 2001). Recently, the use of flexible hoses for distances up to 200 m potentially allow irrigators a greater choice of ground water supplier because each irrigator typically has access to several shallow tube wells within 200 m of their landholding.

Figure 6.2. A typical shallow tubewell.

3 Socio-economic Issues

Population of the study area was 3.8 million persons (Census of India 2001) and the density was 697/km^2. The majority of the cropped area is held in small landholdings. Farm size distribution is given in Table 6.6. Over the whole study area, the average farm size is only 0.54 ha, and 65.4% of landholders have 0.5 ha or less.

In eastern UP about 79% of the population relies directly on agriculture for sustenance (Shah 2001), and this figure is likely to be higher in the study area. Most agricultural produce is likely to be consumed locally, either within the village or neighboring villages, because transport and marketing infrastructure is poor and access to markets is very limited.

Table 6.6 Land Holding classification of Study Area (2004–05).

Farmers Category	Size of Land Holding		Percentage		Avg. Land Holding (ha)
	No.	Area (ha)	No.	Area (ha)	
Marginal Farmers (0.1 to 1 ha)	648,968	215,041	86.3%	52.6%	0.33
Less than 0.5 ha	491,670	112,907	65.4%	27.6%	0.23
0.5 to 1 ha	157,298	102,134	20.9%	25.0%	0.65
Small Farmers (1 to 2 ha)	72,057	95,857	9.6%	23.4%	1.33
General Farmers (more than 2 ha)	30,995	97,974	4.1%	24.0%	3.16
2 to 4 ha	25,458	66,733	3.4%	16.3%	2.62
4 to 10 ha	5,356	28,664	0.7%	7.0%	5.35
More than 10 ha	181	2,577	0.0%	0.6%	14.2
Total	**752,020**	**408,872**	**100%**	**100%**	**0.54**

3.1 Institutional setting

An analysis of institutional issues affecting water resources management in Ghagha-Gomti Basin was provided by Danton and Marr (2005). Within India, the states are responsible for management of land and water resources. The Central government is involved in monitoring and provision of technical support, but also influences state activities through funding. Policies providing a framework for the development and management of water resources were developed from the late 1990s onwards. A National Water Policy was adopted in 2002 and updated in 2012. A State Water Policy for UP was released in 1999 (amended 2004), and a wide-ranging Water Management and Regulatory Commission Act for UP was legislated in 2008. In 2010, a Bill was also drafted to regulate groundwater within the State (the Uttar Pradesh (UP) Ground Water Conservation (Protection & Development) Bill) based on the GoI's groundwater model bill (1970/2005). An updated model bill was released by GoI's Planning Commission in 2011.

For effective implementation of the State Water Policy, an adequate organization is essential. The Government has set up a State Water Board under the chairmanship of the Chief Secretary, with the following subsidiary units:

- State Water Planning Office: To plan the water resources of the State including ground water on a basin/sub-basin unit considering also drainage and flood control; to provide a technical secretariat for the State Water Board and to be a regulatory body for water resources.
- State Water Resource Data Centre: To collect and store water supply and water resources data.
- Management Information System (M.I.S).

Other responsibilities of the State Water Board include enactment of legislation for regulation and control of surface and ground water resources and their conjunctive use; and administrative and legislative reforms for ensuring users participation in management and decentralization of authority.

Within UP, there are numerous government agencies involved in the development and management of water resources. The two key agencies of the Central government are the Central Water Commission and the Central Groundwater Board. These agencies are primarily involved in monitoring the resources and providing technical support, but they also influence the development and management of water resources by providing advice on the technical and economic aspects of major projects or policies initiated by the states.

Within the GoUP there are several key agencies involved in water resources. The Irrigation Department is responsible for development and operation of canal systems. Their responsibility ends at the heads of the Minor canals where management by local water users takes over. The Minor Irrigation Department is responsible for development and operation of small scale irrigation systems, using either surface or ground water sources, as well has having the capability to construct shallow tube wells. There are two agencies responsible for drinking water, Rural Water Supply Department for rural water supply and Jal Nigam for urban. The State Groundwater Department is mostly involved in monitoring and assessment of groundwater, although it also has a role in providing subsidies for tubewell construction by individual irrigators. Some of the responsibility for groundwater development and management has also been devolved to the lowest tier of government, the Panchayats. There is some overlap with state level agencies in this role, and there are issues with technical capacity and resources because this level of government also has responsibility for a wide range of social and development activities. There are two key state agencies responsible for environmental issues, the Ministry of Environment and the Pollution Control Board.

The major use of water within the study area is irrigated agriculture. As there is no industrial usage, the only other use is for animal and human consumption. This accounts for less than 1% of total usage. As almost all crops with the study area rely on some form of irrigation, there is a strong link between water resources management and agricultural production. In recognition of the importance of agriculture to the state, the GoUP has an Agriculture Production Commissioner (APC) who is second only to the Chief Secretary in the state bureaucracy. Several agencies that are vital to agriculture are overseen by the APC including the Agriculture Department, the Minor Irrigation Department and the State Groundwater Board. One anomaly in this arrangement is that the Irrigation Department is not overseen by the APC. It is unclear why this is so, although this may reflect the historically powerful position of the Irrigation Department. It has been an important agency for over 125 years. In the decades following independence in 1947, the Department administered huge budgets as national government money was directed to the construction within UP of one of the largest irrigation systems in the world, with around 74,000 km of canals (Fig. 6.3).

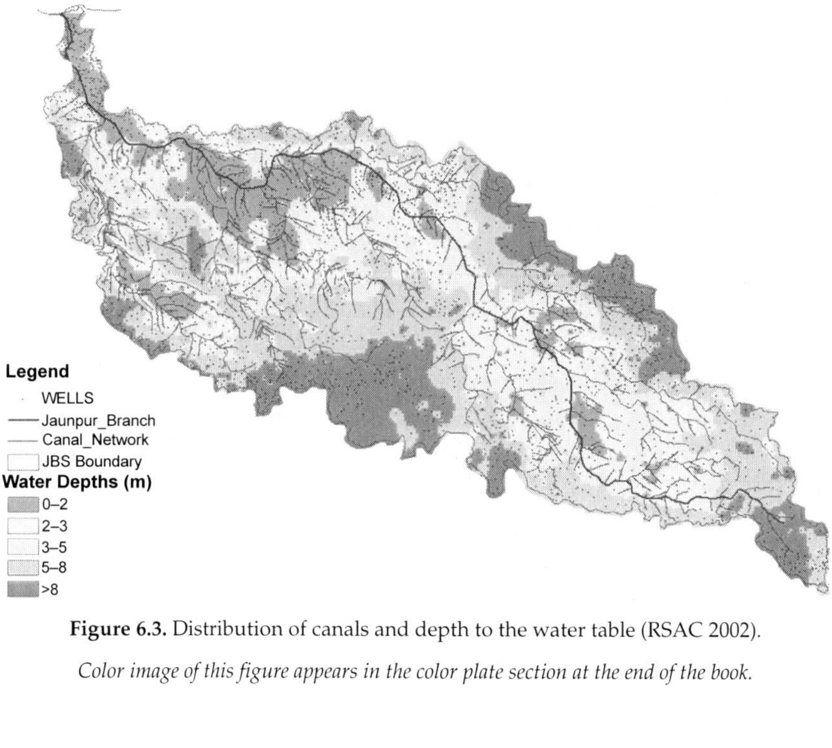

Figure 6.3. Distribution of canals and depth to the water table (RSAC 2002).

Color image of this figure appears in the color plate section at the end of the book.

From the early 1990s, the construction role of the Department has decreased while the operation and maintenance role has increased in importance. With this changing role, it is logical that the Irrigation Department should be brought within the oversight of the APC so that coordination with other agencies involved in agricultural production is improved.

4 History of Design and Operation of the Canal System

Planning of new or upgraded irrigation systems within the state has been based on relatively simple technical procedures (UPID 1982, 1983). While this had the advantage of simplicity and uniformity, the methodology did not take account of many important processes that occur within the irrigated areas. For example, the procedures do not consider the potential for sub-surface water-logging, nor do they consider the potential for conjunctive use of surface and ground water irrigation within the command areas. The procedures also assumed a relatively low crop intensity of around 90–100%. Crop intensity is defined as the sum of areas cropped in all the three seasons, divided by the net sown area, whether single, double or triple cropped.

During the construction of the canal systems, there was initially surplus irrigation water available in the head reaches while the downstream canal system was under construction. Most irrigators took the opportunity to increase crop intensities to 150% or more. As the systems expanded, the irrigators in head areas were reluctant to reduce irrigation intensities back to design levels. As a result, there was inadequate water to supply tail areas, and irrigators in these tail areas gained no advantage from the works. Irrigators in many of the head areas noticed that groundwater levels were rising to within 2–3 m of the surface in the post-Monsoon, salinity and sodicity problems were increasing and crop production was decreasing (UPGWD 1996). Some areas near the heads of canals had to be abandoned due to the effects of water logging, while tree crops and salt-sensitive crops could not be grown in many areas with shallow water tables.

4.1 The Decision Support System Project

4.1.1 Background

By the 1990s, it was noticed that crop production in the state was stagnant or even decreasing, and improved management of groundwater was identified as a key factor in addressing this situation. In 1994 a committee of experts from a range of Central and State government agencies investigated this issue and released a report of their findings. The committee concluded that "The development of ground water resources will ensure dependable irrigation and that will bring about a sustained increasing agricultural production in the State". The committee recommended as follows:

> "The conjunctive use of surface and groundwater has to be encouraged in a scientific manner to control the water table which has to be kept between 6 and 10 m below land surface for best agricultural crop yield. ... The water-logged areas are mostly located in canal commands where limited groundwater exploitation is presently being done. A three-pronged action has to be taken a) strictly to restrict the canal water supply to the areas where the ground water table is less than 10 m below land surface, b) give all out assistance to develop the ground water resources in those shallow water table areas to supplement irrigation needs of the farmers and c) develop strong extension services in the water-logged areas to make farmers aware of the bad effects of excess canal water irrigation and the rising ground water table and its effects on declining crop yields, and good effects of timely and economic ground water use which is a step forward to increase their crop yields. The surface drainage has to be made fully efficient in the water-logged areas."

In 1995 the Central Ground Water Board released the results of a study of conjunctive use of surface and groundwater resources in one part of the Sarda Sahayak irrigation scheme that included the study area (CGWB 1995). The report noted the trend of rising groundwater levels in areas near heads of canals over the period 1974–93, while groundwater levels had fallen in tail areas of canal commands. The study used a simple water balance model to forecast future groundwater levels for a range of cropping intensities, used canal flow data for 1991–92 and estimated groundwater usage from the number of shallow tube wells and an assumed pumping rate. The study concluded that areas of critical water logging (0–2 m below ground level) would increase from 27% to 55% of the area over 20 years if there was no change in groundwater usage. However, if the proportion of groundwater used for irrigation was increased from the estimated current usage of 35% of total irrigation usage to 52% then the area of critical waterlogging would decrease from 27% to just 5% over a 30 year period.

In 2004 a concept paper was released by the Irrigation Department titled: "Water logging, Land Degradation and Drainages in Agriculture" (UPID 2004). This paper was prepared by a committee established by GoUP in 2003 to examine the problem and recommend actions. The committee found that, taking data for year 2000, the area with water table less than 3 m below ground level was about 9% of the plains area of Uttar Pradesh pre-monsoon and 40% post-monsoon.

The Concept Paper makes numerous recommendations for addressing the identified problems. The following is a summary of recommendations that are particularly relevant to the development of the DSS and Basin Plans for the study area:

- areas affected by waterlogging, either surface or sub-surface, should be mapped.
- surface drainage works should be constructed in surface and sub-surface waterlogged areas where the October/November ground water level is less than 5.0 m below land surface. In other areas, there should be no construction of surface drainage, rather check dams and water pools should be used to recharge groundwater.
- in critical and semi-critical waterlogged areas, free boring schemes should be provided to all farmers and the subsidy for installation of diesel pump sets should be increased from 25 per cent to 50 per cent to all farmers.
- there should be intensive pumping of groundwater in flood prone and surface waterlogged areas in Winter and Pre-monsoon seasons using farmer-owned shallow and/or deep tube wells to support increased crop production and to create space for recharge of floods and rainwater.
- supply of canal irrigation to areas that are waterlogged or potentially waterlogged should be restricted to promote use of shallow tube wells for irrigation. Complete closure of canal supplies for several years in the most severely waterlogged areas is suggested.
- prices of water from canals and shallow tube wells should be rationalized. Charges for water from canals should be made on a volumetric basis to promote conservation and help distribute the available water more equitably.
- ponds, lakes and marshy lands should be conserved as wetlands and developed for multiple integrated activities such as aquaculture, aromatic plants, vegetables or raising of pigs, ducks and other poultry.

4.1.2 Description of the Decision Support System

A spatial Decision Support System was developed to map and model the study area. The mapping used Arc-GIS software, while the Arc-Hydro extension was used to store the hydrological data base. A water-balance model operating at a daily time step was developed to simulate water movement within each of the spatial units. The results of the simulation were able to be mapped using the GIS software, while tabular outputs were also provided for analysis.

Mapping

The spatial Decision Support System was developed for the project using Arc-GIS as its core. This allowed the powerful capabilities of Arc-GIS to be applied to data analysis and presentation. The GIS included layers covering:

- physical features such as topography, soil texture and characteristics, canal and drainage infrastructure, roads, towns/villages
- population and socio-economic data based on 2001 census, down to the village level where available
- groundwater data including depth to groundwater for post-monsoon 2002, numbers of shallow tube wells per village (MID 2001)
- cropped areas and crop yields from Ministry of Agriculture data.

Mapping of a range of parameters allowed the relationship of various characteristics to be investigated.

Soils: Key physical parameters for the study area including sub-surface soil texture, soil drainage, slope, soil calcareousness, soil salinity/sodicity and erosion were assessed and mapped.

By comparing the mapped parameters, relationships between parameters were identified. The relationships between drainage, calcareousness, and salinity/sodicity are summarized in Table 6.7.

There is also an association between soil drainage, calcareousness and depth to ground water. Based on the post-monsoon 2002 data, the median ground water depths are 3.89 m for imperfectly drained soils, 3.94 m for moderately well-drained soils and 6.11 m for well-drained soils. Calcareous soils are associated with shallower ground water compared with non-calcareous soils. The post-monsoon 2002 data shows the median ground water depths are 4.67 m for calcareous soils and 6.09 m for non-calcareous soils.

Table 6.7 Relationships of soil parameters.

	Percent of study area covered				
	Calcareous			Non-calcareous	
	Imperfectly drained	Moderately-well drained	Well drained	Moderately-well drained	Well drained
Not saline/not sodic	1.4	-	25.3	3.4	33.8
Slightly saline/slightly sodic	1.5	5.8	-	-	-
Moderately saline/moderately sodic	19.4	8.3	-	-	-
Moderately saline/strongly sodic	2.5	-	-	-	-

Groundwater: The quality of groundwater in the study area is generally good for irrigation with low levels of salinity, although there are some areas of elevated fluoride concentrations. Transmissivity of the soils in the Gangetic Plain varies from 250 to 4,000 m^2/d and hydraulic conductivity from 10 to 800 m/d. Well yields range up to 100 l/s and more but yields of 40–100 l/s are common (GOI 1996). In some areas there is a build-up of "kankar" (calcium carbonate nodules) in a layer below the surface which becomes almost impermeable. This seriously reduces crop growth and can result in surface ponding of water, with precipitation of salt as the surface water evaporates in the hot season.

Groundwater is pumped from shallow tube wells for irrigation of crops. The density of shallow tube wells within each soil type was mapped. The map showed that shallow tube well densities are quite low (less than $10/km^2$) in the head areas of the canal command and are generally much higher ($20–40/km^2$ or more) in the tail areas of the canal command. An exception to this pattern was observed along the main canal in the central-north of the study area, where tube well densities are quite high even in the head reaches of the canal command. The reason for this exception was not determined.

Agriculture: The relationship between agricultural production and soil parameters was investigated. The data for crop yields based on crop cuttings were available for rice (paddy) for Monsoon Season 2004 and for wheat for Winter Season 2004–05. Using the GIS, the average yield of paddy and wheat was computed for each of 20 soil types mapped in the study area. Key soil parameters (drainage, subsurface texture, calcareousness, salinity-sodicity and erosion class) were input to a regression analysis, with crop yield as the dependent variable. The crop yields for each soil estimated by regression are compared with observed crop yields for each soil type. Paddy yield was quite variable over the various soil types, and a large part of this variability appears related to the soil parameters. Wheat yield is far less variable over the soil types, but this small variability also appears to be related to soil parameters.

Cropped areas were mapped for 2004–05 from LISS-III satellite imagery as part of the project. The areas cropped as a percentage of gross area were computed for each mapped soil type. A regression with soil parameters (codes representing drainage, subsurface texture, calcareousness, salinity-sodicity and erosion) as independent variables and percentage of area cropped as the dependent variable showed that there was very little correlation between soil type and percent of area cropped.

The relationship between depth to groundwater (post-Monsoon 2002) and agricultural productivity was also investigated. Using GIS tools, the percent of area cropped and the crop yields were computed for a range of groundwater depths. There was a clear relationship between depth to groundwater and crop yield, with paddy yield decreasing and wheat yield increasing as depth to groundwater increases. The area cropped also shows a relationship with depth to groundwater, with paddy area being very low for shallow groundwater (less than 2 m post-monsoon), peaking for groundwater depths from 3–5 m, then decreasing again as depth to groundwater increases. Wheat yield shows a similar pattern although less pronounced, with yield peaking for depth to groundwater of 4–6 m. Agricultural production was computed by multiplying the percentage of area cropped by the crop yield, to get crop production per gross hectare for the various groundwater depths. For paddy, productivity is low for shallow groundwater, peaks at groundwater depth of 4–5 m, then decreases again for deeper groundwater. Wheat shows a similar pattern although the peak occurs at groundwater depths of about 5–8 m, with a gradual decline for increasing groundwater depth, and a much more rapid decline as groundwater depth decreases below 5 m.

The mapping of soil parameters, surface slope and groundwater depths show that there are many associations. For example, areas of low slope tend to have greater calcareousness, salinity-sodicity and shallower groundwater. These areas also tend to have the greatest supply of canal water and highest canal seepages. From an analysis of the data it is not possible, however, to determine causative linkages between soil parameters, groundwater depths and crop productivity.

Additional data was collected to gain a better understanding of possible processes that may link soils, groundwater and crop production. Five piezometers installed at locations in the study area around 2004–05 as part of the World Bank project were identified for use in the study (Fig. 6.4).

These five piezometers represented different groundwater behaviors. The five locations (Locations 26, 27, 28, 29 and 37) were visited in January 2007 (the middle of the Winter crop season) by a team including specialists in groundwater, agriculture and water resources management. For Location 26 the piezometer plot shows the groundwater reaching within 1 m of the surface between early August and mid-October, then decreasing to mid-December, and staying roughly constant about 2 m below the surface through to early April, and then falling further to around 4 m below the surface by late June. The area around the piezometer shows significant areas of barren land with a scattering of trees of relatively low economic value, particularly trees tolerant to waterlogging and salinity.

For Location 28, the piezometer plot shows a similar pattern to Location 26 but between about 0.5 m and 2 m deeper. The groundwater rises from mid-July reaching a peak within 1.5 m of the surface around mid-September. Two distinct rises in groundwater during this period probably represent canal irrigation applications. The groundwater then drops reasonably

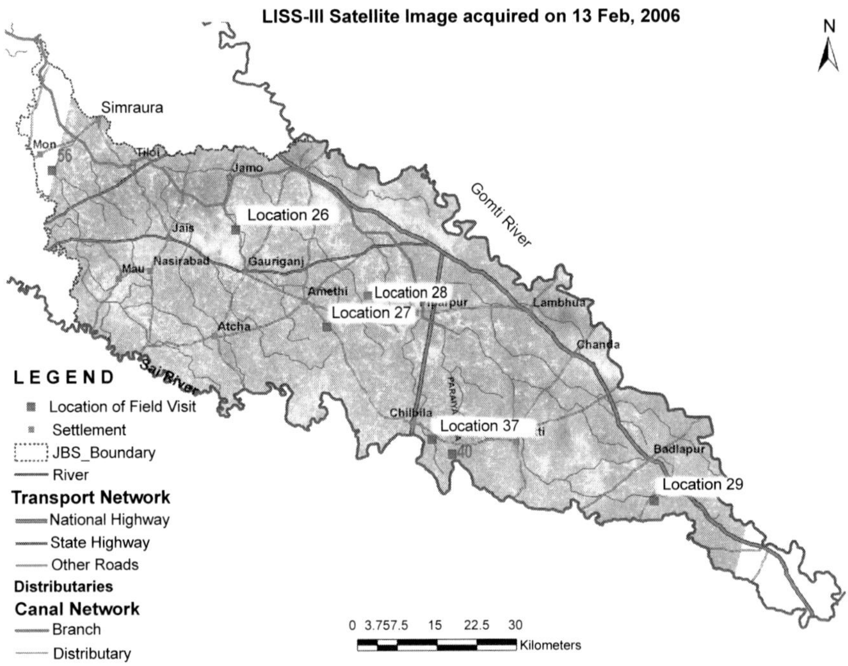

Figure 6.4. Distribution of peizometers within the study area.

steadily from mid-September to mid-December, stays roughly static from mid-December to mid-January, then steadily declines through to the end of June. The vertical lines in the plot from January through to mid-May probably indicate groundwater pumping from a shallow tube well which was observed close to the piezometer and show that the draw down from pumping recovers very rapidly. By mid-June the groundwater level appears to stabilize about 4.8 m below surface. The area around piezometer 28 shows many large economically-valuable trees and a diversity of crops.

While the comparison of only two locations (piezometers 26 and 28) cannot be considered to provide conclusive evidence, it is plausible that the better health of the soil and vegetation at Location 28 compared with Location 26 is the result of the greater usage of groundwater for the Winter crop at Location 28. By the start of the hot season about early April, the groundwater at Location 28 was at least 1.5 m lower than at Location 26. The groundwater level at Location 26 drops nearly 2 m from early to mid-June while at Location 28 it drops only about 0.8 m. There are several obvious rises in the groundwater at Location 28 between January and early-April that are likely to correspond to canal irrigation applications. Smaller rises (see Fig. 6.4) are also visible at Location 26, at about the same times. These are also likely to reflect canal irrigation applications, probably from a more

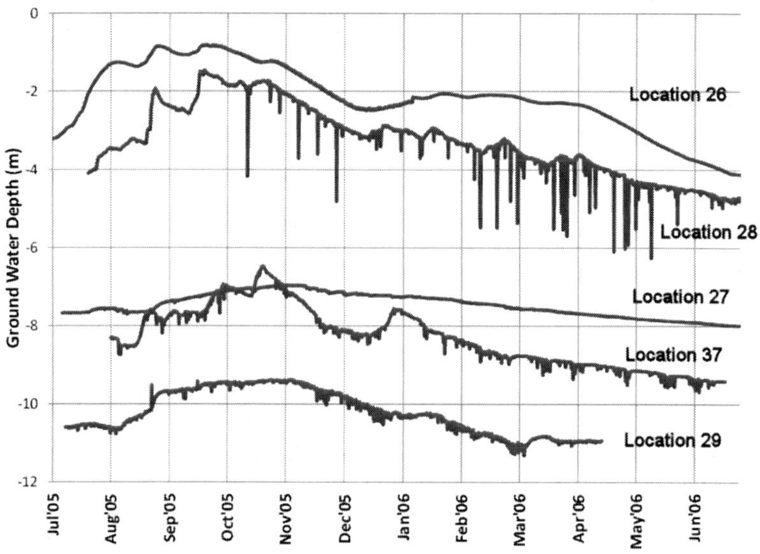

Figure 6.5. A time series of piezometer readings over a 12 month period at different locations within the study site.

distant source. The difference in the piezometer traces between Locations 26 and 28 from early April to late June may provide a clue as to possible mechanisms for the build-up of salinity-sodicity and "kankar" in the root zone of crops in Location 26 that does not occur in Location 28. At the start of the hot season (early April), the groundwater at Location 28 is about 3.8 m below surface, while at Location 26 it is only about 2.3 m below surface. The capillary rise in these soils is around 1.5–2.0 m, so groundwater at Location 26 can rise into the root zone from where it can evaporate through the soil pores or cracks in the soil. If this occurs, salt from the groundwater will be deposited in the crop root zone. At Location 28, the groundwater is about 1.5 m lower, so that transport of salts into the root zone is unlikely to occur. Over the years, the build-up of salts in the root zone at Location 26 would reduce crop health and degrade soil structure, including the reduction of soil permeability.

This comparison of Locations 26 and 28 does not provide conclusive evidence of a mechanism linking shallow groundwater to salt build-up in the crop root zone, but it does indicate a possible mechanism. This should be investigated in much more detail in the future because it is important that development of management strategies should be based on a good understanding of the mechanisms of soil degradation in the study area.

The piezometer traces at Locations 27, 29 and 37 (see Fig. 6.4) show groundwater being 6 m or more below the surface throughout the year. As a result, there would be little migration of salts from the groundwater into the root zones, even of deep-rooted vegetation such as trees. Crops and trees at all three locations appeared healthy, and farmers interviewed reported good crop yields although they mentioned the high cost of groundwater pumping. There was no evidence at any of these locations of soil salinity-sodicity, or areas of barren infertile land.

Modeling

The water balance model that was developed used a daily time step and kept an account of water ponded on the fields, percolation to the crop root-zone and storage within that zone, percolation to groundwater and the rise or fall of the water table. Canal seepage was assumed to percolate into the modeling units adjacent to the canal. The model included a parameter to indicate whether groundwater or surface water should be given preference for irrigation use, and allowed this preference to be changed dynamically within a simulation depending on the depth to groundwater. The model was validated against a 5-year period, although the lack of reliable data on surface water supply and drainage flow meant that the key parameter used for validation was the rise or fall in groundwater over the modeling period. The model allowed for cropping intensity and crop selections to be varied and for canal leakage rates to be varied to represent canals being lined or un-lined. The model kept an account of groundwater pumping and computed the cost of pumping based on the price of diesel fuel. The model also computed the value of crop production and used this to compare the economic performance of various management strategies.

4.1.3 Options

A major goal of the GoUP is to increase the value of agricultural production by increasing the intensity and diversity of crops. Agriculture specialists on the DSS team developed cropping patterns that would meet the subsistence needs of the population as well as providing cash income where possible. The cropping patterns rely on the availability of ground water from shallow tube wells throughout the crop calendar, and also assume that sub-surface water-logging has been addressed, and where necessary, sodic or saline soils have been reclaimed. The cropping patterns developed, therefore, represented a future sustainable agricultural system following any necessary soil reclamation works. Cropping patterns also assume that legume productivity is improved due to reductions in soil salinity/sodicity. Legumes are a much better source of protein than cereals and also have a lower irrigation requirement, so are beneficial for subsistence farmers who may have their access to groundwater irrigation restricted by their limited access to financial resources especially credit.

The planned cropping patterns include cash crops such as vegetables and menthe. The quantity of vegetables that can be grown is limited by market access. Where there is access to reasonable roads, there is potential to market the vegetables to urban centers, even as far away as Delhi. In much of the study area, roads are poor and most vegetables must be consumed locally, so there is a limit to the quantity that can be marketed. Where transport and marketing infrastructure is poor, menthe is a good crop, because the peppermint oil can be obtained by distillation within the villages, and the relatively small quantity of product can easily be stored and transported to market.

The lowering of the water table will also allow tree crops to be grown, such as mangoes and citrus. There will be a relatively long lead time for the establishment of tree crops, so these have not been included in the economic analysis.

The analysis assumes that diesel powered pumps will continue to be used for the shallow tube wells rather than electrically-powered pumps. The option of converting to electrically powered pumps was found to be problematic for several reasons. First, if all the 3.5 million pumps in UP (at the time of the study) were to be replaced by 3 kW electric pumps, then the peak demand when all pumps operated simultaneously would be about 10,000 MW. While it is unlikely that all pumps would be used simultaneously, it is possible that a large proportion of the pumps (say, 50%) would be operated simultaneously if there is a dry period during the peak irrigation season. Reliable provision of this power would require the building of several new power stations plus a huge distribution network. The power generation and distribution system required to meet peak pumping demands would be used infrequently, and is therefore unlikely to be economical. If electrical pumps are used, there will also be problems with metering usage, collecting fees and preventing unauthorized connections. Based on these considerations, it appears preferable to support continued usage of diesel powered pumps for ground water irrigation as there is already a well-established distribution system for diesel fuel. Increases in electrical power generation capacity in the state could be directed to urban areas to reduce the use of petrol and diesel fuelled generators. In these locations, the provision of distribution networks, prevention of unauthorized connections and collection of revenue is more easily achieved than in remote rural areas.

The modeling performed as part of the DSS project showed that annual crop intensity could be increased to 230% (88% in Monsoon season, 93% in Winter season and 49% in Pre-monsoon season), with increased crop diversity. This would provide improved nutrition for the residents, cash incomes to farmers and increased employment opportunities for rural labor.

The economic analysis of the various possible future scenarios (Table 6.8) shows that the Gross Margin per hectare of Culturable Command Area (CCA) can be increased substantially by adopting the proposed 230% cropping intensity, provided that additional irrigation supply is made available by desilting of the main canals and conjunctive use of surface and ground water. Note that the Gross Margin is calculated by subtracting the variable input costs for cropping from the value of the crop produced. The Gross Margin is not generally the same as farm income, because there are likely to be fixed input costs that are not included in the calculation. The table also shows the value of labor inputs that are required for cropping intensities of 149% and 230%.In calculating the Gross Margins, the labor input by the farmer's family or by other farm laborers is costed at market rates, based on estimates prepared by the GoUP. The added value of labor represents increased income for agricultural laborers in the study area, whether they are land holders or landless laborers. The table shows that total income (Gross Margin plus Labor input) is more than doubled in Scenario 3a compared with the Scenario 0 which is the existing situation.

Sensitivity testing showed that if diesel prices doubled, the Gross Margin for Scenarios 3a–3c would be reduced by an estimated INR 3600/ha of CCA.

The modeling also showed that the area of water-logging (post-monsoon ground water depth less than 3 m) would reduce from around 30% for Scenarios 0, 1a and 1b to under 4% for Scenarios 3a–3c, principally as a result of increased ground water usage for the higher crop intensity.

Table 6.8 Results of Scenario Comparison.

	Scenario					
	0	1a	1b	3a	3b	3c
Crop Intensity (%)	149	149	149	230	230	230
Conjunctive use	No	Yes	Yes	Yes	Yes	Yes
Canal desilting	No	Main canals	Main canals	Main canals	Whole system	Whole system
Canal lining	No	No	No	No	No	Yes
Gross Margin (INR/ha CCA)	8 550	9 800	9 950	21 100	22 400	22 450
Labor input (INR/ha CCA)	6 550	6 550	6 550	10 100	10 100	10 100

Note: In 2013, the exchange rate was INR 1000 = USD 14.9

4.1.4 The critical role of farmers

There is evidence from research (Pant 2004; Shah 2001) that farmers in eastern Uttar Pradesh, given access to adequate inputs, maximize their social and economic benefits from their land. The modeling adopted in this study assumes that farmers manage irrigation to meet their economic objectives by providing adequate water to their crops while also maintaining the productivity of their soils over the long term by keeping ground water levels more than 6 m below the surface through conjunctive use of surface and ground water.

A typical shallow tubewell installation might include a surface mounted diesel engine driving a pump located in a shallow excavation, with irrigation delivered through a flexible hose (see Fig. 6.2).

The particular characteristics of the shallow aquifer system allow groundwater to be managed locally at the village level. This is possible because the relatively low permeability of the aquifer means that the cone of depression around a tubewell extends only 100–200 m from the tubewell, and affects only the farmers in the immediate locality. Therefore, management of usage is most effective if performed locally through collaboration at the village level. In contrast, the canal system that delivers surface irrigation is huge and must be managed as an entire system. The canal system is a run-of-river system with negligible storage so it is not possible to operate the system flexibly in response to different rainfalls in different areas, or to differences in crop calendars. The key role of the canal system is to provide additional water to the agricultural areas to supplement rainfall and so allow an increase in cropping intensity. The canal system is also well-suited to providing much of the irrigation required for establishing the flooded paddy fields. Storage of irrigation water is most effectively achieved through use of the shallow aquifer which is recharged by rainfall and canal irrigation and drawn down by shallow tubewells. Storage of water underground reduces evaporation losses to some extent.

As recommended in the 1994 and 2004 reports that examined water-logging and soil degradation, the government should play an enabling role in the implementation of conjunctive use of surface and ground water irrigation. The 1994 report provided very clear recommendations for government action:

A three-pronged action has to be taken a) strictly to restrict the canal water supply to the areas where the ground water table is less than 10 m below land surface, b) give all out assistance to develop the ground water resources in those shallow water table areas to supplement irrigation needs of the farmers and c) develop strong extension services in the water-logged areas to make farmers aware of the bad effects of excess canal water irrigation and the rising ground water table and its effects on declining crop yields, and good effects of timely and economic ground water use which is a step forward to increase their crop yields.

4.1.5 On-going activities

In October 2013, a new credit agreement was signed between the GoI and the World Bank to help build the institutional capacity needed to increase agricultural productivity in Uttar Pradesh. The activities covered by the agreement will support a more holistic strategy for aquifer management, improving governance through water user associations, and focusing on rehabilitating and modernizing existing irrigation systems in critical areas. The GoI has identified these components and activities as directly supporting the government's efforts to improve management and efficiency of water resources, and to help improve agricultural growth in the state.

5 Conclusions

Large parts of the study area are experiencing decreasing crop productivity and crop diversity due to water-logging which under certain conditions leads to soil salinity and sodicity. In 1994 and again in 2004 multi-disciplinary committees were established by Government to investigate water-logging and associated soil degradation and to recommend actions to be taken to reduce the problems. These committee endorsed the conjunctive use of surface and ground water for irrigation as a key component of the water resources management.

The DSS Project commissioned in 2004 under the World Bank funded UP Water Sector Restructuring Project was given the task of investigating water-logging and agricultural production in a 540,000 ha study area, and assessing the potential for applying the actions recommended by the government committees.

By mapping, and development and application of a water balance model, the DSS Project was able to demonstrate that the actions recommended by the government committees could result in substantial increases in crop intensity and diversity in the study area, with substantial increases in income for farmers and landless laborers. The key action was to increase conjunctive use of surface and ground water to lower the ground water level in water-logged areas while also increasing the water available for irrigation of crops. This would allow farmers to have greater control of their irrigation scheduling, and allow higher value crops such as vegetables and cash crops (such as menthe) to be grown.

The major challenge in the implementation of the proposed changes in agriculture and irrigation management is the difficulty of educating farmers to use the more expensive ground water rather than the highly subsidized canal water in order to lower ground water levels and make more canal water available to the tail ends of canals.

In October 2013, a new agreement between the government of India and the World Bank was signed, providing for further institutional development of irrigation management together with rehabilitation and modernization of existing irrigation systems.

References

CGWB. 1995. Studies on Conjunctive Use of Surface Water and Ground Water Resources in the Sarda Sahayak Irrigation Project, U.P. Directorates of Northern Region, Lucknow and Conjunctive Use Projects, Faridabad, Central Ground Water Board, Ministry of Water Resources, GOI.

Danton, D. and A. Marr. 2005. Conjunctive use of surface and ground water in a water-abundant river basin. 8th International River Symposium, 5–9 September 2005, Brisbane, Australia.

GOI (Government of India). 1996. Groundwater Resources of India. Central Groundwater Board, New Delhi.

GoUP (Government of UP). 1996. Groundwater resources of UP. Groundwater Department, Lucknow.

MID. 2001. Minor Irrigation Department Census 2001.

MOA. 2004. Crop Summaries. Ministry of Agriculture, Uttar Pradesh.

Pant, N. 2004. Trends in Groundwater Irrigation in Eastern and Western UP. Economic and Political Weekly, July 31, 2004.

RSAC. 2002. Groundwater Depth Survey in JBS, Post-monsoon 2002.

Shah, T. 2001. Wells and welfare in the Ganga Basin: public policy and private initiative in eastern U.P., India. Research Report 54, International Water Management Institute (IWMI), Colombo, Sri Lanka.

UPGWD. 1996. Analysis Report on Census of Minor Irrigation Works 86–87: Ground Water Resource and Its Stage of Development. Uttar Pradesh Ground Water Department.

UPID. 1982. Planning total water management for Sai-Gomti Doab. Irrigation Department, Uttar Pradesh.

UPID. 1983. Hydrological Report of Sai River, Investigations & Planning (Water Resources) Division, Irrigation Department, Uttar Pradesh, Lucknow, October.

UPID. 1985. Sarda Sahayak Pariyojna Revised Project—1985 (Volume 1) Reports and Statements, November 1985. Irrigation Department, Uttar Pradesh.

UPID. 2004. Water-logging, land degradation and drainages in agriculture—concept paper. Uttar Pradesh Irrigation Department, November.

PART 2

Transboundary River Management and Politics

No problem is so vexed as how to manage rivers that pass through several jurisdictions, especially international transboundary rivers. It has been predicted that access to water will create conflict between countries, even if initial conflict eventually leads to heightened cooperation. In Africa, central Asia, west Asia and the Americas, some countries are already arguing fiercely over access to rivers and inland seas, and confrontations could arise as water shortages grow. Countries currently or potentially involved in international disputes over access to river water and aquifers include: Turkey, Syria and Iraq (the Tigris and Euphrates rivers); Israel, Jordan, Syria and Palestine (the Jordan River and the aquifers of the Golan Heights); India and Pakistan (the Punjab rivers); India and Bangladesh (the Ganges and Brahmaputra rivers); China and South-East Asian countries (the Mekong River); Tajikistan, Kyrgyzstan and Uzbekistan (the Amu Darya and Syr Darya rivers); Ethiopia, Sudan and East African riparian countries, including Kenya, Tanzania, Rwanda, Burundi, Uganda and Egypt (the Nile River); and Iran and Turkmenistan over the Atrek river and Caspian sea.

The 4 chapters here draw on examples from many countries. ***Hassenforder*** **and** ***Noury*** examine 8 case studies on transboundary water management projects drawn from their work in 4 continents, ***Kibaroglu*** **and** ***Ahmetova*** address the real life issues in the Tigris-Euphrates river basin, ***Sullivan*** deals with the largest river in Southern Africa that rises in Lesotho, flows across south Africa and enters the Atlantic ocean via Namibia. Water management policy and practice in the Nile river system receives scrutiny from ***Riddell*** **and** ***Thuo***.

7

Managing Transboundary River Basins

Emeline Hassenforder[1,*] and Benjamin Noury[2]

SYNOPSIS

Managing transboundary river basins is a complex endeavor that requires specific approaches and methods of management and governance. Part 1 of this chapter introduces transboundary river basins and their management: why they are now widely used as a unit of management, potential linkages with the management of other natural resources and with national and regional development planning efforts. It highlights the difference between operational management, strategic management and governance while introducing river basin organizations and their roles. Finally it paves the way to an open interpretation of the term "transboundary".

Part 2 introduces the notion of social-ecological systems and lists a number of factors, explaining their complexity and the inherent difficulties in managing them. We advocate in this part that there is a need to shift from an infrastructure-oriented approach, which focuses on supply-side management and was prevalent in the 1970's and 1980's, to more integrated and adaptive approaches considering water demand management, and environmental and social benefits. Based on a comparative analysis of 8 projects in seven transboundary basins, Part 3 shows concrete examples highlighting obstacles and key success factors in implementing these new approaches in practice.

[1] AgroParisTech, IRSTEA G-EAU, France/Australian National University, Canberra, Australia.
Email: emeline.hassenforder@irstea.fr

[2] Consultant, Transboundary water issues, Oxyo-water, France.
Email: benjamin@oxyo-water.com

* Corresponding author

Part 4 starts by advocating that an additional factor compounding the difficulty in implementing these new approaches is the apparent incompatibility between the notions of planning, integration and adaptation. Authors in the management science literature uphold that this argument is based on a very old-fashioned notion of planning and suggest using a more adaptive and integrated planning approach based on multi-scale, trans-disciplinary and participatory aspects. Finally, the conclusion questions the desirability of resilience and adaptation for planning purposes and concludes by highlighting the importance of thorough monitoring and evaluation as the unique means for responding to this question.

Keywords: cross linkages, holon, resource frontiers, intranational/interstate, social-ecological systems, polycentricism, complexity, multi-scale, interdependencies, participation, planning, integration, adaptation, robustness, sustainability, monitoring, Zambia, India, Australia, China, USA, Mali, Danube, Tigris, Euphrates, Jordan, Nile, Mekong, Guarani aquifer, Okavango

1 On the Notion of Transboundary Cooperation

1.1 The river basin as a unit of management

Today's greatest challenges occur in the world's major river basins: from development of agriculture and industry to urban development, as well as increasing population densities and political, cultural and social changes. It is now increasingly recognized that a coherent unit, the "holon",[1] for managing the river along with its surrounding resources, is the river basin (Luca and Glavany 2011; Pegram et al. 2013), although this unit can be difficult to define as will be discussed (see also Daniell et al., this volume).

A river basin is the area of land from which all surface run-off flows through a sequence of streams, rivers and, possibly, lakes into the sea or a lake at a single river mouth, estuary or delta (European Environment Agency, n.d.). Each river basin can be subdivided into a certain number of elementary basins, sometimes called "sub-river basins", corresponding to the run-off surface of the tributaries flowing into the main stream.

This division holds several advantages regarding water management. On the one hand, the basin scale is relevant for basin authorities for developing an overall vision and strategy for the territory by considering upstream-downstream hydrological and social interactions in a holistic manner. Such scale allows for a systemic approach (Hassenforder and

[1] A **holon** is a component or unit that is simultaneously a whole and a part (Lankford 2008).

Noury 2012a). On the other hand, since some river basins are too large and diverse for an efficient implementation of these strategies across the entire territory at once, the sub-basin scale is relevant for water managers and professionals (technicians, politicians, etc.) to carry out concrete and pragmatic activities on the landscape.

But this division also has some limitations:

- Basin boundaries and water management boundaries are not necessarily congruent: for example, urban and agricultural water supply systems do not necessarily follow basin boundaries (Pegram et al. 2013);
- In addition, even if the 1997 convention broadens the definition of a river basin by integrating underground waters (International Law Commission 1997), it is a unit that takes into account mostly water and not other resources for which the most coherent management unit might not be the river basin.

One can, therefore, wonder what dimensions and limitations of management should be considered. In other words, when managing the environment, should we focus only on water or should we integrate other resources, and if so, which ones and how? For instance, is it possible to integrate some elements of land management and if so which ones? Following the same line of reasoning, should we focus only on certain dimensions (such as technical or legal aspects for instance) or should we integrate others, and if so, which ones and how?

These issues are at the core of current debates. However, certain elements can already enlighten the discussion. Water is increasingly recognized as the crucial link between food, energy and the environment. It is increasingly considered as cross-cutting in nature (GWP and AMCOW 2010a). Therefore people tend to consider that water is a good starting point for managing various resources and sectors and establishing cross-linkages. This is reflected for instance in the principles for integrated water resources management (IWRM) stating, among other aspects, that: "A basin-level perspective allows water managers to address the linkages between water resources management and the management of land and other related resources" (UNESCO 2009).

However, these cross-linkages are not easy to implement in practice. One can wonder for instance which are the most relevant organizations and institutions that would be cross-sectoral enough to be able to deal with water management and other resources. About this IWRM coordination, the Global Water Partnership (GWP) states that "it is not effective to rely on one line ministry (i.e., water) to coordinate others. For some countries, a central 'apex' body that is independent of sectoral pressures and with the convening powers to bring together sectors (such as Ministry of Finance/Economic

Development Planning, Prime Minister's or Vice President's office) may be appropriate in facilitating coordination". In Zambia for instance, the Sixth National Development Planning (SNDP) process was coordinated by the Ministry of Finance and National Planning, working alongside other line ministries. Sector strategies were established with cross-linking input from other sectors (GWP and AMCOW 2010a).

This raises the question about the attempted linkage between water management, or natural resources management (NRM) in general, and national and regional development planning. This is particularly relevant in transboundary river basins where natural resources management and planning can be applied at the transboundary level even though countries and lower administrative units have their own objectives and plans.

Even though planning actions in a transboundary basin may take various forms and names, it often includes two main components (INBO and GWP 2012):

- The **strategic action plan**: (in the long term, 20–30 years) which is generally consensually developed by the transboundary basin organization (when it exists);
- The **action plan**—or detailed program of measures—(in the short term, 3–5 years) which must be formally approved by the basin States to give it strength in enforcement.

The development of these plans should as much as possible include not only the central government of each state in the river basin but also lower levels of government (line ministries, municipal administrations, etc.); other transboundary bodies (donor agencies, Regional Economic Communities (RECs), other River and Lake Basin Organizations, etc.); local bodies and agents (farmers, water users, businesses, etc.) and civil society at various levels (NGOS, community based organisations, local associations, etc.). It should be inclusive both vertically (cross-level) and horizontally (cross-sectoral). This participatory approach should increase compliance between the basin development plan and other national and regional plans.

Figure 7.1 provides an illustration showing the interlinkages across sectors and levels in natural resources planning.

1.2 Introduction to international transboundary river basins and their management

The notion of river basin itself implies that attention needs to be paid to natural boundaries rather than administrative ones: taking into account the famous expression in transboundary water management, "water does

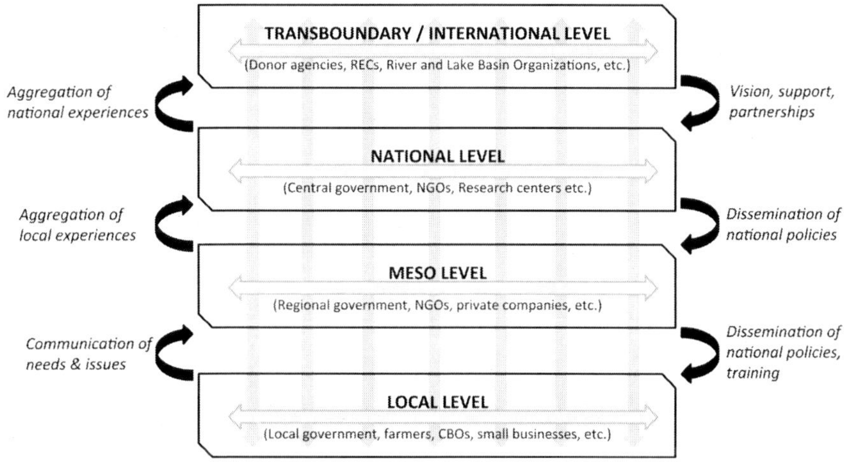

Figure 7.1. Interlinkages across sectors and levels in natural resources planning.

not stop at the border". Being defined by their natural boundaries, many river basins encompass several countries. According to the 1997 convention, an international watercourse is a watercourse, part of which is situated in different states (International Law Commission 1997). One can also talk of a transboundary watercourse.

The river basin includes not only the main watercourse but also the entirety of its hydrographic network. The Danube River for instance counts 10 states that are riparian of its main watercourse but 19 states have some area within the whole river basin ("ICPDR website", n.d.).

276 transboundary water basins have been inventoried in the world (UN Water, n.d.). They cover around 45% of the globe's surface, correspond to around 60% of superficial freshwater and gather 40% of the world's population (Luca and Glavany 2011) (cf. Fig. 7.2).

There were 445 transboundary aquifers at last count (IGRAC 2012). Census and mapping is still ongoing (cf. Fig. 7.3).

Since international transboundary river basins are shared among several countries, it is necessary, in order to manage them equitably and reasonably, to establish joint management and governance structures. Even though this chapter focuses on international transboundary river basins and aquifers, it must be noted that some of the most contentious water management issues concern the management of river basins within nations—India, China, USA, and Australia—where states or independent provinces must share the resources within river basins. For such intranational/interstate transboundary river basins, joint management and governance structures are equally necessary.

Figure 7.2. International Transboundary river basins in the world (Wolf et al. 1999).

Governance refers to "...the whole of public as well as private interactions taken to solve societal problems and create societal opportunities." It includes the formulation and application of principles guiding those interactions and care for institutions that enable them (Kooiman and Bavinck 2005). Environmental governance is a specific form or subcategory of this broader governance concern (Plummer and Armitage 2010). Water governance is specifically defined by the United Nations Development Programme (UNDP) as "the political, social, economic and administrative systems that are in place, and which directly or indirectly affect the use, development and management of water resources and the delivery of water service delivery at different levels of society" (UNDP et al. 2013). Water governance deals with the way decisions about water are made: how, by whom and under what conditions. It concerns both this decision-making aspect and the formal and informal institutions by which authority is exercised (Akhmouch 2012).Water management, by contrast, includes everyday management of the resource: analysis, monitoring, implementation of measures, administrative management, etc. "Governance sets the rules under which management operates" (Pahl-Wostl 2011).

In many transboundary basins, river basin organizations exist that deal with either or both water management and governance. The International Network for Basin Organizations gives the definition of a basin organization as "any formal or informal entity that manages water resources at the basin scale" (INBO and GWP 2009). This definition includes a broad range of formal and informal organizations covering various functions for both basin management and basin governance.

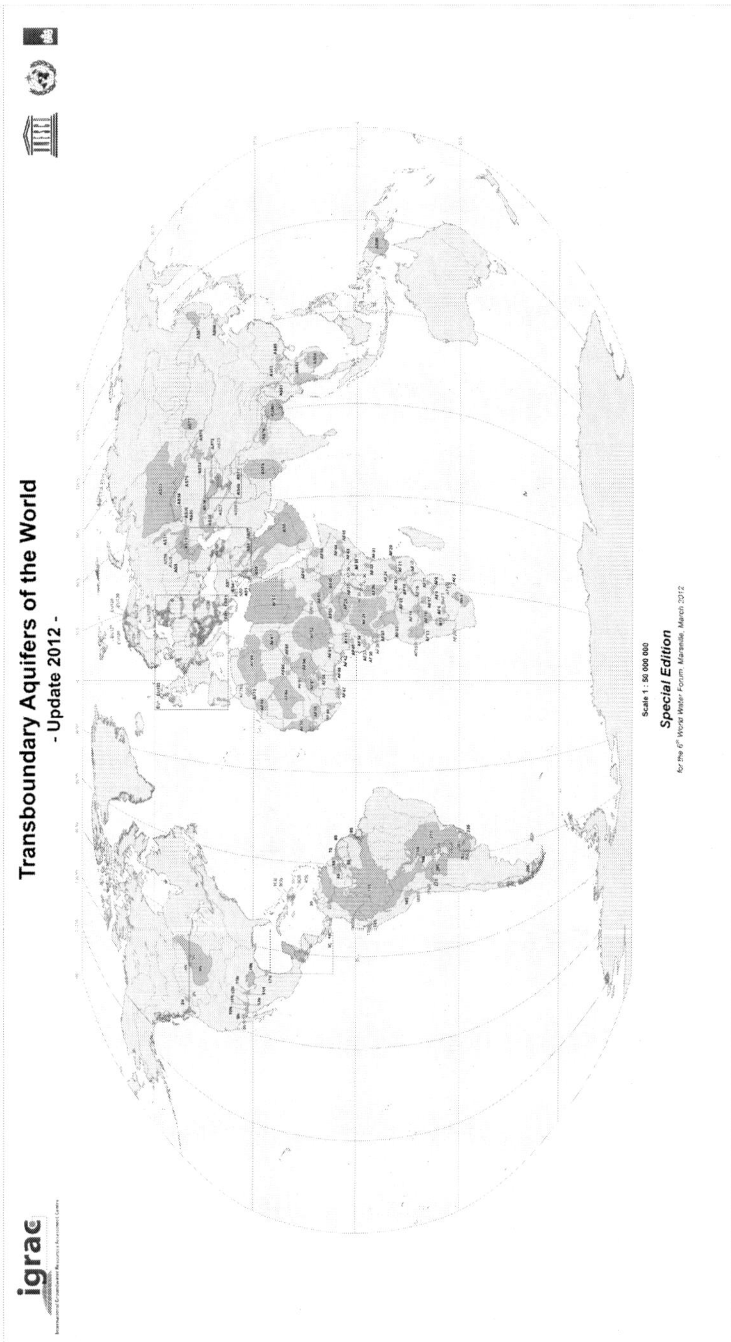

Figure 7.3. Transboundary Aquifers of the World 2012 (IGRAC 2012).

1.3 Roles and actions within transboundary basins

The roles of a river basin organization can be of different natures, including the management of:

- *Information*: Data collection, compilation and evaluation, information system updating, etc.
- *Consultation*: Identification of the various stakeholders, communication with individuals and groups, organization of participatory meetings, forums, etc.
- *Decision*: Definition of strategic orientations of the river basin, including river basin planning and project development.
- *Implementation of projects and programs*: Field work, project operations and contracts, project evaluation, etc.
- *Resource protection*: Water quality monitoring, alert systems, pollution source inventories, etc.
- *Arbitration*: Conflict management and resolution.

Here it can be considered that decision and arbitration are functions of governance, whereas other functions are more managerial. This list is, of course, not exhaustive. According to the river basins concerned, existence or not of these functions depend on numerous factors such as the needs and skills of the staff and beneficiaries, institutional capacities of the riparian countries and the tools and methods available.

Actions implemented through projects and programs can be of different natures ranging from institutional or capacity building actions to more technical and infrastructure actions (hydropower generation, irrigation, navigation, etc.) including legal and economic ones.

It is important at this stage to distinguish between:

- The **management of the River Basin Organization itself**: statutes, functioning of the secretariat, accounting, financing system for operation, human resources management, staff training, etc.
- The **operational management of the river basin**: water quality and quantity monitoring, information system, alert system, etc. It integrates all activities that must be undertaken on a regular or daily basis. Operations are ongoing and deliver repetitive products, services or results.
- What is often called the **strategic management of the river basin**: this strategic part relates to the basin planning which defines the direction for actions in the basin, the vision, projects and programs which must follow and help in order to fulfill the basin plan.

It is this latter part that interests us most and is investigated more thoroughly in the rest of this chapter.

1.4 The multiplicity of boundaries

We can consider that a transboundary river basin is one that is shared among several countries. Although as previously explained, there can also be transboundary issues between jurisdictions within a nation. It is the case for instance with the conflict over the upstream-downstream allocation of the Cauvery water between the States of Karnataka and Tamil Nadu in southern India. In that sense, we are talking of "interstate" boundaries, or administrative boundaries at a lower scale. The degree of independence of these administrative units in the management of natural resources can vary widely from one country to another. In countries like India and Australia, states have, under the national constitution, the right to use and manage natural resources within their boundaries. Provinces in China also have this independent power. The degree of financial independence and power is an important factor in achieving coordination.

But a boundary is not necessarily administrative or geographical; it can also be sectoral, cultural, historical, or social. We can take here as an example the great migration routes of Peul people across Africa: certain pastoralist ethnic groups from Mali are said to have crossed the continent in order to buy cattle in Ethiopia without being hindered by administrative or topographic boundaries. They had "laissez-passer" ("right to passage") allowing them to cross lands, which illustrates the notion of "mixed" boundary.

In her book "Friction, an ethnography of global connection" (Lowenhaupt Tsing 2004), Anna Lowenhaupt Tsing, an American anthropologist, describes "resource frontiers" as "frontiers between nature and local ecologies/livelihoods [...] between proliferation/out-of-control capitalist expansion vs. local systems of human access and livelihood/ ecological dynamics of replacement and replenishment" (p. 28). For her, the frontier is a shifting terrain between legality and illegality, public and private ownership or disciplined and wild territories.

This broadens our vision of a "transboundary" issue, showing that, even if this chapter focuses on boundaries in geographical and administrative terms, most of the issues, challenges and propositions mentioned thereafter can be conceptualized and transposed for systems with other types of boundaries.

2 The Complexity of River Basin Management

2.1 Complex properties of social-ecological systems

River basins can be considered as social-ecological systems (SES). The term "social-ecological" started to be used by Berkes and Folke (1998) to emphasize the intricate linkages between humans and nature and to stress the artificial and arbitrary delineation between social and ecological systems (Folke et al. 2005). Indeed, many human activities take place in and are dependent on river basins.

SES are by definition complex: they involve a multitude of actors, sectors and scales which are interdependent and their evolution over the coming decades is uncertain, which makes it hard to have a global understanding of systems and changes. This complexity inherent in SES increases the complexity in their management and governance.

Global and climate change

Rapid technological, social and economic changes drive increased complexity by applying growing pressure on water and other natural resources. Rapid population growth, increased economic activity and improved standards of living typically lead to expanding pollution and a rise in water demand. In order to meet it, water resources are more intensively developed; but in water-scarce countries, water supply can often be insufficient to cover all needs, leading to competing uses and consequently to the emergence of conflicts for water appropriation. Climate change increases this challenge by adding further uncertainty and variability in the occurrence of water leading to time and space shifts in rainfall patterns (increased rainfall variability) (UNDP 2006) and to more frequent and intense extreme events (e.g., floods and droughts) (IPCC 2007; Hassenforder et al. 2012).

Imperfect knowledge

The aforementioned challenges are all the harder to manage as scientific and technical data about the resource is often lacking, unreliable, incomplete or inconsistent (UN WWAP 2006). For example, there is a lack of comprehensive scientific knowledge of the water cycle, environmental impacts and behavioral variables in certain areas (Malano 2010). On the Senegal River for instance, an extensive amount of water quality data remains missing. In addition, the Organisation pour la mise en valeur du fleuve Sénégal (OMVS) has almost no information on the Guinean part of the river located upstream of Mali, Mauritania and Senegal (Hassenforder and Noury 2012b). There are various reasons for why such information is not collected or made available including lack of political will, cost, available expertise and technology, confidentiality concerns, and conflict avoidance.

Risks, hazard, vulnerability and uncertainty

Authors working on risks and hazards (Burton, Blaikie, Cutter, Brooks, Dzyton-Johnson, Wisner, Haque among others, cited in Plummer and Armitage 2010) focus on assessing the risk of a social-ecological system from exposure to a hazard. Risks are probabilities of losses or damageable consequences. The risks for a community to undergo losses or damages result from the interaction between possible natural or anthropogenic hazards and the vulnerability of the community towards these dangers. Most simply risks to particular populations can be schematized by:

Risk = Hazards x Vulnerability
Formula 1 (UNISDR 2004)

Since hazard is exacerbated by climate change, so is risk. There is a need to improve understanding of these risks and changes and to develop an adaptive and anticipative behavior.

However, certain threats or natural disasters can hardly be anticipated. These perturbations are called uncertainties. Uncertainty can be defined as an unidentified and unexpected threat which may in some cases result from a catastrophic accident. It often manifests as a 'surprise' and requires a different management approach to traditional risk management (UNISDR 2002). Therefore, managing uncertainty often requires identifying "weak" signals to anticipate potential threats. Uncertainty is also strengthened by climate change since the latter increases unpredictability and nonlinearity: water managers cannot precisely know the consequences of their action (Pahl-Wostl et al. 2008). Certain mitigation actions can, therefore, have opposite and amplifying consequences.

Multiplicity of actors and sectors

In order to understand the increasing complexity of the environment in which water evolves, it is useful to distinguish two major interdependencies: among the increasing number of stakeholders involved in water management and between the various disciplines around water. Water and other resources are used by a wide range of activities belonging to various sectors themselves managed by a wide range of stakeholders. All of these stakeholders are dependent on the resources and concerned by its management.

This multiplicity of stakeholders and sectors makes water management more complex. Each has a particular interest towards the resources. These interests can be compatible but also contentious, for instance where there is competition among domestic, agricultural and industrial uses. In such cases, there is a need for trade-offs between alternative economic, social and environmental objectives and between available resources/supply and demand (see also chapters by Li & Squires, Loch et al. and Sullivan in this volume).

Additional interdependencies exist in transboundary basins whereby water creates an environmental, social, economic and political dependency and interconnection among countries. These interdependencies add to the complexity for resources management and interstate relationships.

Multiple scales

A lot of attention is given to multi-level governance and cross-scale interactions in relation to social-ecological systems and adaptive co-management (see notably Costanza, Ostrom, Young, Folke, Pritchard, Berkes, Svedin, Riordan, Jordan, Dolsak cited in Folke et al. 2005). Different scales and dimensions, along with their interactions, are being mobilized in the governance of natural resources:

- Spatial and temporal scales,
- Social and ecological systems, the interactions between which were presented earlier in this chapter,
- The dimensions of the social system: institutional, juridical, regulatory, aesthetic, cultural, economic, etc. (Cash et al. 2006),
- The various dimensions of the ecological system: biotic-abiotic, biodiversity, soil water content, etc. (Görg 2007).

Each of these dimensions may be considered on several levels (Cash et al. 2006).

This multi-scalar vision is entirely linked to complexity. For a long time mono-centric centralized governance, based on centralized coordination, command-and-control and hierarchical regulation by a national resources agency, was considered to be the best approach to ensure fair allocation and supply and avoid the destruction of the resources (Cash et al. 2006). In the last 30 years, the acknowledgement of the difficulties of hierarchical coordination mechanisms on which is based mono-centric governance has led to the development of two other theoretical models of governance: market coordination and coordination through negotiations, mobilization of consensus and building of mutual understanding (Jessop 2002). This shift led to the development of the 'polycentric governance model' (Marshall 2008) which can be defined as systems involving higher levels of government as well as local systems (Ostrom et al. 1961; Ostrom 1997 cited in Ostrom 2005).

This approach has been judged by some authors as inefficient because they look messy and are non-hierarchical in structure, but they can be drawn on by resource users at multiple levels to aid in the crafting of new institutions that cope with changing situations (Burger et al. 2001; Ostrom 2002). According to Folke et al., the real challenge is dealing with systems that are not only cross-scale but also dynamic, whereby the nature of cross-scale influences in the linked SES changes over time (Lowell Pritchard and Sanderson 2002 cited in Folke et al. 2005).

All these aspects make it difficult to comprehend the system as a whole and highlight the need for adapted models and tools. The multiplicity of these elements, along with their interactions, are factors which increase complexity in basin management and governance. Each element is in synergy with other elements of the system and can therefore impact the system as a whole. In this context, developing more integrated approaches is fundamental (Hassenforder and Noury 2012a).

2.2 The need to shift to a new management approach

In the 1970's and 1980's, it became increasingly recognized by the water development community that purely technical and engineering solutions were no longer adequate to address the multifaceted and interconnected problems of basin management. There was a need to address a wider range of issues and challenges: the focus moved from infrastructure management to a wider range of management foci at the basin scale, including the environmental and social importance of water, the participation of a wide range of groups in decision-making and attempts to manage the use of water as well as augment supply (Pegram et al. 2013).

This shift laid the basis for the emergence of the concept of Integrated Water Resources Management (IWRM) in the 1990's. A number of countries undertook thorough reforms of their water policy and legislation, and incorporated new basin-scale management and institutional arrangements into their legal frameworks, often based on IWRM principles. This decade saw an increasing shift from management for optimal use and control of resources to a management of both humans and ecosystems through approaches such as IWRM, adaptive management, and ecosystem-based management.

Adaptive management is an approach to managing natural resources that encourages "continually improving management policies and practices by learning from the outcomes of implemented management strategies" (Pahl-Wostl 2006). The goal of adaptive water management is to increase the adaptive capacity of the water system in a river basin. It considers implementing policies as experiments (Bormann et al. 1994) "Adaptive management is learning to manage by managing to learn" (Allan and

Curtis 2005; Kashyap 2004 cited in Lankford 2008; Pahl-Wostl et al. 2005). Adaptive management is often put forward as a more realistic and promising approach to deal with SES complexity (Gunderson 1999) than management for optimal use and control of resources (Folke et al. 2005; Holling and Meffe 1996) (see Table 7.1).

Concrete examples of successes and challenges in the implementation of these integrated and adaptive approaches are illustrated through eight case studies in the following part of this chapter.

Table 7.1 Strategic/adaptive model of management vs. Traditional/linear model of management.

Strategic or adaptive management	Traditional/linear model of management
People management	Results-based management
Bottom-up	Top-down
Modeling/Adaptive planning	Planning (predefined goals, objectives and outcomes)
Learning by mistakes	Control mistakes
Risks and uncertainties follow-up	Scope, time and cost follow-up
MANAGEMENT	COMMAND-AND-CONTROL

3 Learning from Past Management and Mis-management Practices

3.1 Difficulties in implementing this new approach: conclusions from the comparative study of 8 transboundary cooperation water projects

The complexity of SES highlighted in the previous part makes it difficult to implement these "new" integrated and adaptive methods of management in practice. This part will draw upon independent research making a comparative analysis of eight cooperation projects implemented in seven transboundary river basins (Danube, Tigris & Euphrates, Jordan River, Mekong, Guarani aquifer, Okavango and Nile). The objective of this research was to analyze the management of these projects, identify their success and challenges from a project management perspective, in order to draw principles and suggest practical tools for the management of complex water projects.

Data collection took place between November 2008 and June 2010. The study draws upon a mix of dominantly qualitative data from 273 interviews, including focus group discussions. It was completed by field studies of relevant sites as well as documentary research. A decision and management model called Development Modeling® was used for data analysis. Case studies were compiled on each project (based on the principle of theory building from cases from Eisenhardt and Graebner 2007; Eisenhardt 1989) and a comparative analysis made. This part offers a synthesis of relevant items in regards to integration and adaptation. For full details on collected data and results, see Hassenforder et al. (2012).

The level of effectiveness of these projects with regard to the goal setting and to their level of integration and adaptation is variable.

Based on the literature, some obstacles to the implementation of integrated and adaptive approaches can already be highlighted (Hassenforder and Noury 2012a; Pegram et al. 2013):

- IWRM and adaptive management are conceptual frameworks that present practical difficulties in implementation. At the micro scale, it is possible to develop an integrated approach. But as soon as one deals with larger scales such as a region, a country or a river basin (meso and macro), the number of parameters that need to be taken into account is large, making implementation much more complex (Biswas 2012). "IWRM is a principle and not an objective: it explains what needs to be done but not how to do it" (Froebrich 2012).
- IWRM implementation is often constrained by a lack of political will: proposed solutions are often technical while the main obstacle to IWRM implementation is political. There is often an inadequate leadership to drive implementation and allocate resources (particularly to collaborating and related agencies), preventing other stakeholders from adopting the necessary changes.
- Implementation may be limited by institutional mandate and capacity. Key institutions need to have the power and capacity to give effect to the strategic actions of the plan. This can require policy and legislation to be in place, as well as institutional strengthening and capacity-building. In addition, it can be difficult to implement transversal and cross-sectoral management of resources when administrative organization is sectoral.
- IWRM and adaptive management imply participatory processes and stakeholder involvement. Failure to involve legitimate stakeholders and obtain their cooperation can be hindering factors in implementing integrated and adaptive plans.
- Poor information and communication and a lack of data transparency can prevent information sharing to support action. Sound monitoring systems, communications strategies and stakeholder engagement mechanisms are necessary tools and methods for sound IWRM.

Adaptation and integration are not additional sets of actions to be added in NRM plans but a totally different approach to planning that must be ingrained into management processes and the plan itself. Unfortunately, so far much of the effort in adaptation, especially but not only in regards to climate change, has set a separate agenda for adaptation activities and programs, and in some cases separate budget lines (GWP and AMCOW 2010b). Too often, traditional approaches to management and planning prevail which try to predefine goals, objectives and outcomes and correct any deviation from the planned trajectory, considering such deviation as a failure rather than an opportunity to learn.

There is also a lack of learning from past experiences which is closely linked to limited efforts and investments in monitoring and evaluation (M&E). Development and implementation of M&E procedures has become customary in development projects and is yet too often seen as a requirement/control by the donor rather than as a key part of the project cycle (Couix 1997). Differences between sound evaluation and control are summarized in Table 7.2.

These are all hindering factors to a successful shift to more integrated and adaptive approaches.

Table 7.2. Differences between evaluation and control (adapted from Couix 1997).

	Control	**Evaluation**
Objective	Check conformity of a phenomenon or item to a norm or a reference	Information elaboration process in order to attribute" values" to an action (political, economic, philosophic, etc.)
	Make sure that predefined objectives are achieved for a specific action	Build an estimation of the phenomenon
	Assess the gaps between targeted and achieved results	Questioning the meaning and interpretation of actions and unexpected consequences
Type of information to be elaborated	Quantitative	Qualitative
Approach	Analytic	Hermeneutics
	Explain results ("why")	Understand and interpret (« for what » and « how ») (evolution, history, effects, context)
Focus	Achievement of predefined objectives	Relevance of predefined objectives
	Explicit (official discourses)	Implicit (vision, intention, perceptions, interpretations
Main assumptions	Existence of a preset referent, unique and constant	Elaboration/construction of a conceptual model, not provided a priori but evolving throughout the evaluation process
	Measure and comparison to standards	Make sense
	Knowledge of the "good" implementation of action	Multiple referent (each action can be considered from various points of view)
Rationality	Substantial	Procedural
Norm	Contractual (established out of the piloting process)	Emerging (established at the core of the piloting process)
Paradigm	Paradigm of measurement	Paradigm of diagnostics
Conclusions	Objective conclusions	Subjective conclusions

Folke and Olsson identify two main factors that can foster successful transformations towards adaptive management (Folke et al. 2005): precedence of the emergence of informal networks and orchestration by key individuals that help facilitate information flows, identify knowledge gaps and create nodes of expertise. Based on this acknowledgement, they list constituencies they believe necessary for institutions and organizations to be viable, adaptive, flexible, to change rapidly and cope with uncertainty:

- Polycentricism,
- Large multi-stakeholder organizations or encompassing organizations (McCay 2002). Berkes goes even further saying that real co-management, including shared management and shared power, is required (Berkes and Folke 2002),
- Multi-scale or cross-level (i.e., involve stakeholders operating at different levels),
- Develop strategies that prepare for uncertainty and surprise,
- Ongoing dynamic knowledge acquisition and learning over decades,
- A civic society with a certain level of social capital,
- Accumulation of experience and social memory,
- Framed creativity, creative teams and actor groups,
- Key individuals with strong leadership able to direct change and transform governance (transformational leadership),
- Political independence,

Organizations gathering these qualities are what can be called "bridging organizations" (Folke et al. 2005).

The following parts will provide concrete examples of success and challenges faced in the 8 projects while attempting to implement these approaches. These experiences are based on water project management but they can similarly be used as more general examples for basin and natural resources management.

3.2 Vision

One of the strategic roles of project decision-makers is to establish development orientations and priorities for the basin. Making sure that this "desired vision for the future" is clear, that projects and programs are aligned to it, and that each project will deliver results that are in line with the strategic expectations, is key to integrated and adaptive management.

Goal

Ideally, the goal must be consensual and acceptable for all stakeholder groups. Project team members and local stakeholders should play leading roles in defining the project goal in order to ensure ownership and commitment (European Commission, n.d.). Defining the goal provides an opportunity for the stakeholders involved in the project to meet altogether, to share their experiences and opinions, and to be aware of their respective interests and potential contradictions.

Some stakeholders can have interests that do not necessarily concur with those of other stakeholders (Cleland and Ireland 2002). These potentially conflicting views need to be formulated and considered to establish a clear goal for the project (Turner and Simister 2000). The goal should not hide the contradictions among the stakeholders, but rather highlight them. Setting the goal is an opportunity to reveal and eventually resolve conflicting interests early in the project development phase, thus shielding project managers from later conflicts during the implementation phase which potentially may have broader negative impacts. *The Red Sea—Dead Sea Water Conveyance Study Program* between Jordan, Israel and the Palestinian Territories illustrates this point. The overall goal of the project was "to investigate the feasibility of reversing the environmental degradation of the Dead Sea Region by transferring water from the Red Sea" ("World Bank website", n.d.). However, the following events revealed that the three parties had undisclosed interests: Jordan needed additional water supply for its capital Amman, Israel wanted to be part of a "peace project" and the Palestinian territories hungered for being recognized as riparian to the Dead Sea and therefore as a country. These disagreements generated several twists and turns in the project development (Hassenforder et al. 2012).

Indicators

The medium and long-term success of a project must be defined and debated in the early stages of the project. The very definition of the notion of success must be discussed by the stakeholders: "when will we be able to consider that we will have reached our goal?". Focusing stakeholders' attention on the expected long-term performance (and the required indicators to measure it) is a central element for discussion and negotiation during the project design stage. The definition of indicators is a way of focusing everybody's energy on practical questions. The choice of criteria is critical and depends on the purpose of the evaluation. These indicators and criteria can be modified in the course of the evaluation "*chemin faisant*". Evaluation also questions the relevance of criteria (Couix 1997).

Since the goal defines a long-term end-state to achieve, it often makes it complicated to identify and select the appropriate indicators and targets (World Bank 2005). Since it is a difficult exercise, indicators at the strategic level of the project (goal and objectives) do not always exist. Yet it is a necessary effort in order to be able to evaluate what is considered by the stakeholders to be the success of a project. Indicators which are too vague make it easier for "everybody" to post-evaluate the project as a success. However, it doesn't help the project decision makers monitor the long-term and sustainable changes resulting from the project implementation. Goal indicators should also be measured several years after the end of the project in order to assess the sustainability of the project efforts. Unfortunately, the question of who should fund this post-evaluation remains an unanswered question in most basins.

Dimensions

The analysis of the eight transboundary water projects revealed that technical, political and social dimensions are often given a lot of importance while financial and legal dimensions are often underestimated at the design stage. Financial and legal aspects are often considered as "supportive functions", not really part of the core dimensions of the project. However, such dimensions are necessary to maintain the project sustainably in the long run. Failure to integrate this multidimensional perspective into a project early enough generally leads either to later modifications of its design, or to poor sustainable performance due to underestimated or forgotten dimensions. The Guarani Aquifer System (GAS) Project initially planned to develop a single deliverable in the legal dimension: a multicountry agreement. During implementation, the project team realized that many other legal agreements were needed to support the project development. Laws on wells protection or memoranda of understanding in pilot areas were therefore formalized later on. None of them were forecasted initially in the project design.

Final deliverables

Final deliverables, also called outputs, should be long-term oriented; either the deliverables themselves, or their benefits, should last after the end of the project. Final deliverables are future resources and expected results that will be used in the long run. For instance, a "monitoring system" is a deliverable that is long-term oriented, since regular monitoring of the quality or quantity of water in the basin will continue after the end of the project. Training, technical analysis and awareness-raising documents are other examples. Even if they are produced during the project, they generate or gather knowledge that will last after the end of the project implementation phase.

Determining whether deliverables or their benefits are going to last is a matter of project sustainability. Our analysis of the transboundary water projects revealed that the design of complex water projects is often very weak in that domain. Many deliverables considered as "final" are not actually "sustainable"; once the project ends, they are not maintained.

All complex water projects should include a phase-out plan, also sometimes called exit strategy, that includes three major aspects:

- The selection of the project final deliverables that will need to be sustained in the long term,
- The list of the stakeholders who will be responsible for them and
- The agreement of the stakeholders who will pay for them.

The *Danube Regional Project (DRP)* for instance, implemented a phase-out plan. The phase-out plan identified a list of core and additional activities. For each core activity to be transferred, recommendations were provided on how they could be carried forward. For each category, a table was drafted as in Table 7.3.

Table 7.3 Framework of The Danube Phase-Out Plan (adapted from Phase—out of the DRP Project 2006).

Activity	Anticipated costs €	Potential sources
Deliverables to be maintained. Example: Information system, Danube Day, Danube Watch Newsletter, etc.	Costs for maintaining the deliverable	Stakeholders who can take over the responsibility for managing and funding the deliverable

The conclusion of those projects is that the key to the financial sustainability of the project activities is to start the process early. Phase out plans should be considered right at the drafting stage of project documents while implementation of these plans should start at the mid-way point of the project. This would allow sufficient time for the phase down process to take effect and the countries to budget for their increased responsibilities (Final Evaluation of the DRP Project 2007). This process is often initiated at the very end of the project which does not leave enough time to create a strong country buy-in and an easier transition to post-project self-sufficiency.

3.3 Implementation

Integration and adaptation are equally important when implementing activities.

Uncertainties during project implementation

Since the contexts of SES are continually evolving, management of resources takes place in situations that are not totally unknown but show unexpected characteristics (Couix 1997). The project implementation phase is the transitory period when all the final deliverables are being produced, and eventually delivered. During the project life cycle, many different actions are implemented, and many different "events" happen and influence the final deliverables production process. Some of them have small impacts that do not influence the long-term expectations for the project; some others impact irreversibly the expected performance of the project. The project implementation phase is very sensitive to the complexities generated by the multiple stakeholders involved in the project, by changing interactions, exogenous events, etc.: the required authorization was not granted, the project funding was cut by half, two private companies shared the job instead of one, etc.

In 2004, the United Nations Office for Project Services (UNOPS) decided to use a new Enterprise Resource Planning (ERP) software called Atlas. This modification had a major impact on many projects, among which were the *Danube Regional Project* and the *Nile Transboundary Environmental Action Project*. The use of new financial procedures modified many local activities and by consequence delayed or cancelled some of the final deliverables.

These changes might require readjustments to be made (Couix 1997) on the:

- Objectives of the strategy or the program: are objectives relevant in regards to the situation at stake?
- Means and resources: are available means adequate?
- Strategy and actions: are actions relevant to reach these objectives?

That is why it is necessary to continuously diagnose both the context and the project functioning in order to adopt adaptive behaviors and decisions in the face of unexpected constraints. In that way, unpredictable constraints can lead to new opportunities.

Experimentation

Projects are means of experimentation, they provide good opportunities to test actions. This aspect is key to adaptive management. It is emphasized by Kato and Ahern (2008) who consider that one of the key concepts and principles of adaptive management is conceiving management actions as experiments. Adaptive management conceives management actions as experiments with testable a priori hypotheses (Lee 1993; Rutledge and Lepczyk 2002). On the contrary, traditional or reactive management hesitates to apply new policy decisions until proof of efficacy is obtained through short and long-term empirical studies. Adaptive management is a proactive method, where policy decisions are used as 'experimental probes', enabling the manager or planner to learn by doing (Holling 1978).

3.4 Organization

An in-depth analysis of the main issues faced by the eight transboundary projects showed that most issues in implementation came from stakeholders; highlighting the importance of considering human aspects in NRM.

Networks

Hierarchical models correspond perfectly to routine and organized structures. But when it comes to complex projects and SES, organizations are multiple and not ruled by clear power liaisons: the hierarchical model becomes inappropriate. Organization is understood as "political, economic, social and educational bodies. They are groups of individuals bound by some common purpose to achieve objectives". "Organizations [are] agents of institutional change" (North 1990). An organization is, thus, a body of agents, a group of individuals, even where these individuals may be within organizations/agencies/departments/corporations/etc. Considering organizations in terms of networks rather than hierarchies helps to adopt a more adaptive and integrated approach.

Power

By definition, relations among stakeholders (whether individuals, organizations or countries) are asymmetric. Adaptive management and integrated water resources allocation and management take place in circumstances of asymmetric power (Warner et al. 2013). Often this asymmetry and the presence of power relations are underestimated in water management. They need to be recognized and taken into account.

It is not the purpose of this chapter to enter into too much detail about power relations, but it is worth mentioning that the binary conception of dominator versus dominated has long prevailed. Long (1999), however, suggests that power is not either-or, but an ongoing, fluid interaction between compliance, submission, resistance and defiance in our everyday lives (cited in Wong 2012). In that sense, power can be difficult to fully comprehend. Its interactive nature must nevertheless be given consideration when managing natural resources.

Champions

For effective project implementation, champions are needed at various levels. In transboundary projects, champions at the top government level are very important to achieve cooperation and coordination among government agencies. But local champions are equally important for such projects. Identifying and working with people coming from the communities who are trustworthy, willing to work on the project activities and to know more about their area is crucial to public participation. The fact that they come from the community is an invaluable asset for their knowledge of local and traditional practices, traditions, customs and behaviors. On the transboundary *Good Water Neighbors Project* on the Jordan River, a field activity had been planned in the community for months. A week before the event, the dean of the village died and his funeral was arranged for the day of the activity. The field researcher knew about it and was able to postpone the activity since he was from the village. If he had not been from the village, the activity would have been maintained, with no one participating. The "champions"' role can also be considerably important for communication particularly in countries where many dialects coexist. Thirdly, "champions" can facilitate logistical and administrative constraints. Particularly for transboundary water projects, logistics can constitute a real burden. Transportation costs, visas and communication expenses add to project management burden and constitute obstacles to good communication. On the Jordan River, the situation is very complex: Palestinians and Israeli can cross to Jordan through certain borders only. In the *Good Water Neighbors Project* case, each office in Amman, Tel Aviv and Bethlehem was in charge of the administrative papers of its residents. Through effective work by the communities' "champions", the project team was able to organize several regional meetings.

Many authors emphasize the role of leadership, or champions, in building resilience (Folke et al. 2005; Gupta et al. 2010; Herrfahrdt-Pähle and Pahl Wostl 2012).

4 Food for Thought

4.1 Planning, integration and adaptation: are they really incompatible?

We have seen in the preceding parts that a new management approach is needed based on the principles of integration and adaptation. We have also seen that this approach is not easy to implement in practice and faces major difficulties and challenges. This is also due to the fact that many people consider that the idea of planning activities, which is inherent in project design, is in nature opposed to the concept of adaptation. They believe that since planning aims at deciding the goal, objectives and actions at, or before, activities commence, it prevents any form of adaptive behavior.

We advocate here that this argument is based on a very old-fashioned and outdated notion of planning. In 1994, Mintzberg defined planning as "a formalized procedure to produce an articulated result, in the form of an integrated system of decisions" (Mintzberg 1994). According to Avenier, this "ballistic" conception, according to which a plan or a strategy is a trajectory to reach a certain target, is not appropriate in a context perceived as complex (Avenier 1997). Instead, a more "adaptive" concept of planning can be considered, which is often called "strategic planning". According to Simon, the task of strategic planning is to assure a stream of new ideas that will allow the organization to continue to adapt to its uncertain outside world (Simon 1993). Strategy is the art of using information arising during action, integrate it and quickly formulate action schemes (Avenier 1997; Morin 1990).

This notion of planning is compatible with the notions of adaptation and integration. It implies being open to potential endogenous and exogenous changes in the environment and to adapt to them. Still one needs to define the vision, indicators and outputs to be produced but the planner should be aware that these might need to be reviewed and amended when conditions or knowledge changes. For instance, an objective of the *Guarani Aquifer System* (GAS) Project implemented by Brazil, Argentina, Uruguay and Paraguay was originally to evaluate the geothermal potential of the aquifer. The studies concluded, however, that contrary to earlier expectations, the temperature of the GAS was too low to produce energy and resources and planned effort for this activity were reallocated to other project activities (Hassenforder et al. 2012).

In order to adopt an adaptive management approach, environmental managers need flexible decision support and planning tools which suit a strategic approach. However, planning tools suggested by donors are often difficult to use and seen as constraints rather than decision support tools. We advocate that much simpler tools are needed to provide decisions in a fast and answer-oriented manner. Three major notions need to be taken into account for a more adaptive and integrated planning: multi-scale, trans-disciplinary and participatory processes.

4.2 Multi-scale

The multiplicity of scales and levels in transboundary river basins is both a factor of complexity and a window of opportunity. We have discussed the definition of multi-scale and its implications for complexity in the first part of this chapter. Now, what does it imply in terms of planning, integrated and adaptive management?

Multi-scale is a complex notion to understand because one cannot categorize a stakeholder as being from the "local" or "national" scale simply by his or her role in a process. A stakeholder can work in a local administration and at the same time be a farmer, be involved in a local association and have a brother who works in the ministry.

Since resource systems are multi-scale, they might require different management approaches at different scales simultaneously (Berkes 2002). Usually multi-level planning processes involve simultaneous planning processes with various social groups and at various scales (from the particular to the whole). Plans produced at the various levels need to be nested: meaning that local plans have to fit into global plans and that there needs to be interaction among stakeholders.

For example, it is increasingly common to see development projects advocate that work be carried out at a meso-scale. The meso-scale can be defined as "an intermediary level between local (or community) level and national level. At this level national policies are disseminated or interpreted for lower level implementation ("scaling down") and/or local experiences aggregated and disseminated ("scaling up" and "scaling out")". Meso-scale institutions can also have their own perspectives and can have specific functions such as conflict management, support for operational management (such as planning), and as well, are effective in scoping and framing issues (AfroMaison, n.d.).

4.3 Trans-disciplinary

Trans-disciplinarity emphasizes the importance of involving people from various disciplines in the planning process. Jean Paul Billaud, in a paper on trans-disciplinarity, traces back the origins of trans-disciplinary endeavors in French research programs dealing with nature/society interactions (Billaud 2003). Viewed from this perspective, the history of trans-disciplinarity, in western countries at least, can be compared to the history of evolution of natural resources management evoked in the first part of this chapter. From around the 1970's, the majority of environmental research was largely framed by a conception of natural science that aimed to embrace complexity through computing and mathematics, synthesis of inputs from various disciplines and ultimately produce knowledge for action. In the 1990's, assessment of 20 years of trans-disciplinary programs showed that this "bland" scenario of trans-disciplinarity was flawed. Trans-disciplinary practice faces numerous challenges, among which are difficulties in coordinating disparate groups, communication and language barriers and lack of tools and instruments specifically dedicated to trans-disciplinarity. Each discipline has its own rationality and criticizes those in other disciplines of irrationality because they do not understand or accept that rationale. Pioneers of trans-disciplinarity are looking for a "meta-science" at the interface between nature and society, which would align the various disciplines on one rational logic that would integrate the various knowledge pools.

Nowadays, in the context of increasing uncertainty and complexity, especially social and scientific (see the earlier comment on climate change), improved trans-disciplinary approaches are all the more required, triggered by a new relationship between science and society: with a shift from rigid planning to greater interactions among disciplines and with the broader society. There is a modification in the conditions for the production of knowledge that is no longer based only on "pure science" but need to include participatory approaches, and community and 'traditional' knowledge. Under these circumstances, knowledge is unstable, uncertain and negotiated. To implement a trans-disciplinary approach, researchers do not only need to negotiate with each other but also with managers and other citizens. In this uncertain context, Billaud suggests to appeal to new instruments to build bridges among disciplines. He suggests that language be used not only as a communication tool but also in order to build common representations of the world. He also suggests reconsidering and reaching agreement on "intermediary items", i.e., items for which each discipline gives a different definition, which can be meeting and confrontation points among disciplines. According to him, no generalization can be drawn from trans-disciplinary projects as long as the various disciplines disagree on key words and concepts such as "sustainable development", "rational and equitable use" or "integrated management" (Billaud 2003). In that respect, the long negotiations around articles 5 and 7 of the 1997 Convention on the Law of the Non-navigational Uses of International Watercourses about the definition of the terms "equitable and reasonable utilization" and "obligation not to cause significant harm" are yet another example of potential negative impacts of trans-disciplinary tensions and disagreements (Dinar 2009). We, therefore, advocate not only integrating people from various disciplines when trying to implement these new approaches but also agreeing on a common language and key concepts.

4.4 Participatory

Participation is closely linked to the two preceding notions. It is now commonly acknowledged that development projects tend to be more adaptive and their final deliverables more sustainable when they integrate representative stakeholders and beneficiaries. This assumption is even truer for planning processes. As highlighted by Hamel and Prahalad: cognizant of the limits of individual cognitive capacities, "it is difficult to produce creative strategies when its formulation is the responsibility of elites only" (Hamel and Prahalad 1990).

"Public participation may be defined at a general level as the practice of consulting and involving members of the public in the agenda-setting, decision-making, and policy-forming activities of organizations or institutions responsible for policy development" (Rowe and Frewer 2004). Participatory approaches often require a high level of initial investment both in terms of time and budget, even if it is not always the case, as shown in Daniell (2008). However, the benefits for the management of water and its linked social-ecological systems are, among others: improvement of ownership and empowerment of the stakeholders involved; identification of local problems, causes and potential solutions for these problems; and anticipation of potential conflicts and disagreements (Lotfy 2008).

In order to implement such processes, identifying key stakeholders is an essential requirement (INBO and GWP 2009). "A stakeholder is an individual or group influenced by—and with an ability to significantly impact (either directly or indirectly)—the topical area of interest" (Pahl-Wostl et al. 2005). In the field of water and natural resources, this notion of stakeholders encompasses a wide range of people. Potential groups concerned can come from various scales, disciplines or fields of work (Hassenforder and Noury 2012a).

These three multi-scale, trans-disciplinary and participatory aspects can seem conflicting at first: would one want to involve both people from the various scales and from various disciplines in a transboundary endeavor, he might end up with a planning process including thousands of people. This highlights the need for a thorough initial stakeholder analysis which remains flexible throughout the process to the potential of including stakeholders from other scales or disciplines and to adapt to uncertainties, conflicts or surprises (Von Korff et al. 2010). Integrating these three approaches should be possible. Success can be maximized by building on lessons learnt from past projects such as using network rather than hierarchical organizations, identifying champions, taking into account power relations and maintaining a strategic direction while "learning by doing" and adapting *chemin faisant*.

5 Conclusion

5.1 Resilience and adaptive capacity are neither necessarily good nor desirable in all situations

We have advocated throughout this chapter the need to move to a more adaptive and integrative management of water resources, especially when it comes to transboundary river basins. After stressing the difficulties associated with such transition, we have emphasized that, in order to face increasing complexity, this management should be, as much as possible, multi-scale, trans-disciplinary and participatory. We will conclude this chapter with a more polemic issue wondering whether adaptive and integrative planning and management are "good" per se. Environmental scholars, including ourselves, tend to focus on adaptive and integrated institutions and organizations as one, if not the, main means of working towards sustainability and resilience in our social-environmental system, sometimes forgetting that adaptive and integrative capacity are neither necessarily good nor desirable in all situations. As for example, some of the most despotic political regimes are resilient and adaptable.

Hence one, if taking an extreme position, could argue that certain places might be better off without people. Of course, such a statement would mean ignoring the unswerving connection between people, place and the governance of those places. Resources cannot be considered separately: there is interdependency between land, water and other resources. In the same way, people cannot be separated from place: as outlined in the introduction of this book, "goods and services derived from the waterways and the surrounding land are vital to human welfare". Even places, after decades of human intervention, might not be able to return to a sustainable ecological state if the people were removed. The need for resilience and the desirability of sustainability also arise from the characteristics of natural systems: natural communities cannot withstand rapid and continuous change, and climate change uncertainty requires adaptability.

After all, we should not forget that our main aim is not the sustainability and resilience of our institutions but of our social-environmental system. Therefore, the most important shift needed might not be from command-and-control to integrated and adaptive management but from having governance "making the most *of* the environment" to "making the most *for* the environment".

5.2 The importance of monitoring

In order to end this chapter on a more optimistic tone, we would like to emphasize the fact that, however dependent on the environment, in order to be able to know whether these new approaches and concepts are "good" or "bad", we need to be able to monitor their impacts. There is a need to continue monitoring and evaluation, and provide effective feedback to stakeholders, especially decision-makers (Kato and Ahern 2008), for several years after the end of development projects. Such monitoring should be integrated, adaptive, trans-disciplinary, multi-scale and participatory.

Acknowledgements

We would like to thank all the people: farmers, fishermen, civil servants, decision-makers, human rights and environmental believers who took time to meet us during the "*Entre Deux Eaux*" research project and to answer our questions. We are grateful to Katherine Daniell, Victor Squires and Hugh Milner for giving us the opportunity to contribute to this important book.

References

AfroMaison. (n.d.). AfroMaison Grant agreement Annex 1 "Description of Work" part B1 (p. 5).

Akhmouch, A. 2012. Water Governance in Latin America and the Caribbean: A Multi-Level Approach. Retrieved from http://dx.doi.org/10.1787/5k9crzqk3ttj-en.

Allan, C. and A. Curtis. 2005. Nipped in the bud: why regional scale adaptive management is not blooming. Environmental Management 36(3): 414–425. Retrieved from http://www.ncbi.nlm.nih.gov/pubmed/16132442.

Avenier, M.-J. 1997. Chapitre 1 Une conception de l'action stratégique en milieu complexe. *In*: La stratégie chemin-faisant (Economica., p. 393).

Berkes, Fikret. 2002. Cross-scale institutional linkages for commons management: perspectives from the bottom up. pp. 293–321. *In*: E. Ostrom, T. Dietz, N. Dolsak, P. Stern, S. Stonich and E.U. Weber (eds.). The Drama of the Commons. National Academy Press, Washington DC.

Berkes, F. and C. Folke. 1998. Linking social and ecological systems: management practices and social mechanisms for building resilience. pp. 151–153. *In*: Berkes, Fikret and C. Folke (eds.). Ecological Economics, Vol. 28. Cambridge University Press. Retrieved from http://books.google.co.za/books?id=XixuNvX2zLwC&dq=Linking+Social+and+Ecological+Systems:+Management+Practices+and+Social+Mechanisms+for+Building+Resilience&printsec=frontcover&source=bn&hl=en&ei=5gdETO3-A9mVOKuJiOMP&sa=X&oi=book_result&ct=result&resnum=4&ved=0CCwQ6AEwAw#v=onepage&q&f=false.

Berkes, Fikret and C. Folke. 2002. Back to the future: ecosystem dynamics and local knowledge. pp. 121–146. *In*: L. Gunderson and C. Holling (eds.). Panarchy: Understanding Transformations in Human and Natural Systems. Island Press, Washington, DC.

Billaud, J.P. 2003. De l'objet de l'interdisciplinarité à l'interdisciplinarité autour des objets. Nature Sciences Sociétés.

Biswas, A.K. 2012. Enhancing good governance in the water sector, setting the scene. *In*: 6th World Water Forum. Marseille, France.

Bormann, B., P. Cunningham, M.H. Brooks, V.W. Manning and M.W. Callopy. 1994. Adaptive ecosystem management in the Pacific Northwest. General Technical Report, PNW-GTR(341), 32. Retrieved from http://www.citeulike.org/group/14168/article/8039911.

Burger, J., E. Ostrom, R. Norgaard, D. Policansky and B. Goldstein. 2001. Protecting the Commons. A Framework for Resource Management in the Americas (Island Press, Ed.). Washington DC, USA.

Cash, D.W., W.N. Adger, F. Berkes, P. Garden, L. Lebel and P. Olsson. 2006. Scale and cross-scale dynamics: governance and information in a multilevel world. Interactions 11(2): 8. doi:10.1111/j.1742-4658.2008.06281.x.

Cleland, D.I. and L.R. Ireland (eds.) 2002. Project Management: Strategic Design and Implementation, 4th Edition. McGraw Hill, New York, p. 656.

Couix, N. 1997. Chapitre 6 Evaluation « chemin faisant » et mise en acte d'une stratégie tâtonnante. *In*: Economica (ed.), La stratégie chemin-faisant (p. 393).

Daniell, K. 2008. Co-engineering participatory modelling processes for water planning and management. Australian National university, AgroParisTech.

Dinar, S. 2009. Treaty Principles and Patterns: Negotiations Over International Rivers (p. 452). ProQuest. Retrieved from http://books.google.com/books?id=zSvObjuN8wYC&pgis=1.

Eisenhardt, K.M. 1989. Building theories from case study research. Academy of Management Review 14: 532–550. doi:10.5465/AMR.1989.4308385.

Eisenhardt, K.M. and M.E. Graebner. 2007. Theory building from cases: opportunities and challenges. Academy of Management Journal 50(1): 25–32. doi:10.2307/20159839.

European Commission. (n.d.). Project Cycle Management Guidelines. Aid Delivery Methods, 158 p.

European Environment Agency. (n.d.). European Environment Agency Water WISE Glossary definitions. Retrieved September 02, 2013, from http://glossary.eea.europa.eu/EEAGlossary/R/river_basin.

Folke, C., T. Hahn, P. Olsson and J. Norberg. 2005. Adaptive governance of social-ecological systems. Annual Review of Environment and Resources 30(1): 441–473. doi:10.1146/annurev.energy.30.050504.144511.

Froebrich, J. 2012. IWRM for all: make water resource planning a reality by adopting IWRM Plan. *In*: 6th World Water Forum. Marseille, France.

Görg, C. 2007. Landscape governance: the "politics of scale" and the "natural" conditions of places. Geoforum 38(5): 954–966. Retrieved from http://www.sciencedirect.com/science/article/B6V68-4NC5V82-1/2/d79a210d0a680d937ddc5ea1acc60ca0.

Gunderson, L. 1999. Resilience, flexibility and adaptive management: antidotes for spurious certitude? Conservation Ecology 3(1): 1–10.

Gupta, J., C. Termeer, J. Klostermann, S. Meijerink, M. Van Den Brink, P. Jong and E. Bergsma. 2010. The adaptive capacity wheel: a method to assess the inherent characteristics of institutions to enable the adaptive capacity of society. Environmental Science Policy 13(6): 459–471. doi:10.1016/j.envsci.2010.05.006.

GWP and AMCOW. 2010a. Policy Brief 2 Building on the Foundations of Integrated Water Resources Management. Retrieved from http://www.gwp.org/Global/WCDP Files/WACDEP Publications/WaterSecurity_Brief2_WEB.pdf.

GWP and AMCOW. 2010b. Policy Brief 3 Ensuring Adaptation At All Levels. Retrieved from http://www.gwp.org/Documents/WACDEP/WaterSecurity_Brief3_WEB[1].pdf.

Hamel, G. and C.K. Prahalad. 1990. Les stratèges du soleil levant. Harvard-L'Expansion.

Hassenforder, E. and B. Noury. 2012a. Gestion des bassins versants transfrontaliers. Techniques de l'Ingénieur, 42506210 (Eau: propriétés, qualité, valeurs d'usage). Retrieved from http://www.techniques-ingenieur.fr/base-documentaire/environnement-securite-th5/eau-proprietes-qualite-valeurs-d-usage-42506210/.

Hassenforder, E. and B. Noury. 2012b. La gestion des projets de cooperation dans les bassins transfrontaliers: vers un nouveau paradigme. Géoéconomie 60: 152.

Hassenforder, E., P. Daniel and B. Noury. 2012. New Perspectives for the Management of Water Projects: Models and Applications from 8 Transboundary Projects. La Guilde des créateurs de Mondes Ed., Lille, France, 200 p.

Herrfahrdt-Pähle, E. and C. Pahl-Wostl. 2012. Continuity and change in social-ecological systems: the role of institutional resilience. Ecology and Society 17(2): 8. Retrieved from http://dx.doi.org/10.5751/ES-04565-170208.

Holling, C.S. 1978. Adaptive Environmental Assessment and Management (C. S. Holling, ed.). Proc Gulf of Mexico coastal ecosystem workshop Sept 1979 Port Aransas Texas (Vol. 3, p. 377). International Institute for Applied Systems Analysis. Retrieved from http://www.scopus.com/inward/record.url?eid=2-s2.0-0019114062&partnerID=40&md5=0fd8b3a100e424622aa5240cc241aa55.

Holling, C.S. and G.K. Meffe. 1996. Command and control and the pathology of natural resource management. Conservation Biology 10(2): 328–337. doi:10.1046/j.1523-1739.1996.10020328.x.

ICPDR website. (n.d.). Retrieved March 23, 2012, from http://www.icpdr.org/.

IGRAC. 2012. Transboundary aquifers of the world. Retrieved from http://www.un-igrac.org/publications/456#.

INBO and GWP. 2009. A Handbook for Integrated Water Resources Management in Basins.

INBO and GWP. 2012. The handbook for Integrated Water Resources Management in transboundary basins of rivers, lakes and aquifers. Retrieved from http://www.gwp.org/Global/About GWP/Publications/INBO-GWP Transboundary Handbook/MGIREB-UK-2012_Web.pdf.

International Law Commission. Convention on the Law of the Non-Navigational Uses of International Watercourses New York (1997). Retrieved from http://treaties.un.org/doc/Treaties/1998/09/19980925 06-30 PM/Ch_XXVII_12p.pdf.

IPCC. 2007. FAQ 10.1: Are Extreme Events, Like Heat Waves, Droughts or Floods, Expected to Change as the Earth's Climate Changes? http://www.ipcc.ch/publications_and_data/ar4/wg1/en/faq-10-1.html.

Jessop, B. 2002. Governance and Metagovernance: On Reflexivity, Requisite Variety, and Requisite Irony. *In*: H. Heinelt, P. Getimis, G. Kafkalas, R. Smith and E. Swyngedouw (eds.). Participatory Governance in Multi-level Context. Concepts and Experience (pp. 17–32). Opladen: Leske & Budrich.

Kashyap, A. 2004. Water governance: learning by developing adaptive capacity to incorporate climate variability and change. Water Science and Technology 49(7): 141–146.

Kato, S. and J. Ahern. 2008. "Learning by doing": adaptive planning as a strategy to address uncertainty in planning. Journal of Environmental Planning and Management 51(4): 543–559. doi:10.1080/09640560802117028.

Kooiman, J. and M. Bavinck. 2005. The governance perspective. pp. 11–24. *In*: J. Kooiman (eds.). Fish for Life, Interactive Governance for Fisheries. Amsterdam University Press,Amsterdam, Netherlands.

Lankford, B.A. 2008. Integrated, adaptive and domanial water resources management. pp. 39–59. *In*: Pahl-Wostl, Claudia, P. Kabat and J. Möltgen (eds.). Adaptive and Integrated Water Management: Coping with Complexity and Uncertainty. Springer, Berlin Heidelberg.

Lee, K.N. 1993. Compass and Gyroscope: Integrating Science and Politics for the Environment. Conservation Ecology, Vol. 3. Island Press, Washington D.C., USA, p. 255. Retrieved from http://books.google.com/books?hl=en&lr=&id=nJJ8irPTTiwC&oi=fnd&pg=PT4&dq=Compass+and+gyroscope+:+integrating+science+and+politics+for+the+environment&ots=S2v_JVGip2&sig=--PO9bG37Osvmt1HE2qaYQzzHq8.

Long, N. 1999. The multiple optic of interface analysis. UNESCO Background Paper on Interface Analysis.

Lotfy, H.R. 2008. Involving stakeholders in transboundary water management in Southern Africa. *In*: Vth International Symposium on Transboundary Waters Management. Thessaloniki, Greece.

Lowell Pritchard, J. and S.E. Sanderson. 2002. The dynamics of political discourse in seeking sustainability. pp. 147–169. *In*: L.H. Gunderson and C.S. Holling (eds.). Panarchy Understanding Transformations in Human and Natural Systems. Island Press, Washington, D.C., USA. Retrieved from http://isbndb.com/d/book/panarchy_understanding_transformations_in_systems_of_humans_.html.

Lowenhaupt Tsing, A. 2004. Friction, An Ethnography of Global Connection, Princeton University Press, Princeton, USA, p. 344.

Luca, L. and J. Glavany. 2011. La géopolitique de l'eau. Paris, France. Retrieved from http://www.assemblee-nationale.fr/13/rap-info/i4070.asp.

Malano, H.M. 2010. Modelling and decision-making in water resource management. pp. 111–126. *In*: Khan, Shahbaz Savenije, Huber Demuth, Siegfried Hubert and Pierre (eds.).

Hydrocomplexity: New Tools for Solving Wicked Water Problems, proceedings of the Kovacs Colloquium. IAHS Publication, Paris, France.

Marshall, G.R. 2008. Nesting, subsidiarity, and community-based environmental governance beyond the local level. International Journal of the Commons 2(1): 75–97.

McCay, B.J. 2002. Emergence of institutions for the commons: contexts, situations, and events. pp. 361–402. *In*: Ostrom, Elinor, T. Dietz, N. Dolšak, P.C. Stern, S. Stonich and E.U. Weber (eds.). The Drama of the Commons. National Academy Press, Washington D.C., USA. Retrieved from http://books.nap.edu/openbook.php?record_id=10287&page=361.

Mintzberg, H. 1994. The Rise and Fall of Strategic Planning. (Free Press and Prentice Hall International, Ed.) (p. 458). New York, USA: Simon and Schuster.

Morin, E. 1990. Introduction à la pensée complexe. Edition du Seuil. ESF, Paris, France. Retrieved from http://meatschool.free.fr/R%E9f%E9rences/Introduction %E0 la pens%E9e complexe.pdf.

North, D.C. 1990. Institutions, Institutional Change and Economic Performance. p. 152. Retrieved from http://books.google.com/books?hl=fr&lr=&id=oFnWbTqgNPYC&pgis=1.

Ostrom, Elinor, T. Dietz, N. Dolsak, P.C. Stern, S. Stonich and E.U. Weber. 2002. The Drama of the Commons (Elinor Ostrom, T. Dietz, N. Dolsak, P.C. Stern, S. Stonich and E.U. Weber, eds.). Social Sciences (p. 521). National Academy Press,. Retrieved from http://site.ebrary.com/lib/nuls/docDetail.action?docID=10032451.

Ostrom, Elinor. 2005. Understanding Institutional Diversity (I. University, Ed.). Public Choice (Vol. 132, p. 376). Princeton University Press. doi:10.1007/s11127-007-9157-x.

Ostrom, V. 1997. The Meaning of Democracy and the Vulnerability of Democracies: A Response to Tocqueville's Challenge. C.U. The University of Michigan Press, Ed. Ann. Arbor, USA.

Ostrom, V., C.M. Tiebout and R. Warren. 1961. The organization of government in metropolitan areas: a theoretical inquiry. American Political Science Review 55: 831–842.

Pahl-Wostl, C. 2006. Transitions towards adaptive management of water facing climate and global change. pp. 49–62. *In*: E. Craswell, M. Bonnell, D. Bossio, S. Demuth and N. van de Giesen (eds.). Integrated Assessment of Water Resources and Global Change: A North-South Analysis. Reprinted from Water Resources Management 21/1 (2007) 2007, VI, 373 p.

Pahl-Wostl, Claudia. 2011. Water Governance Meeting the Challenges of Global Change. *In*: Proceedings of the Water Governance Conference. Obergürgl, Austria.

Pahl-Wostl, Claudia, T. Downing, P. Kabat, P. Magnuszewski, J. Meigh, M. Schlueter and S. Werners. 2005. Transition to adaptive water management: the NeWater project Working Paper 1. Water 23. Retrieved from www.newater.info.

Pahl-Wostl, Claudia, P. Kabat and J. Möltgen. 2008. Adaptive and Integrated Water Management: Coping with Complexity and Uncertainty, Vol. 89, p. 306. Springer, Dordrecht, The Netherlands. doi:10.1029/2008EO330009.

Pegram, G., Y. Li, T. Le Quesne, R. Speed, J. Li and F. Shen. 2013. River Basin Planning. Paris. Retrieved from http://www.iadb.org/intal/intalcdi/PE/2013/12283.pdf.

Plummer, R. and D. Armitage. 2010. Integrating perspectives on adaptive capacity and environmental governance. pp. 1–19. *In*: R. Plummer and D. Armitage (eds.). Adaptive Capacity and Environmental Governance. Springer, Berlin Heidelberg.

Rowe, G. and L.J. Frewer. 2004. Evaluating public-participation exercises: are search agenda. Science Technology and Human Values 29(4): 512–556. doi:10.1177/0162243903259197.

Rutledge, D.T. and C.A. Lepczyk. 2002. Landscape change: patterns, effects, and implications for adaptive management of wildlife resources. pp. 312–333. *In:* Liu, Jianguo and Taylor, William W. (eds.). *In*: Integrating Landscape Ecology into Natural Resources Management. Cambridge University Press.

Simon, H.A. 1993. Strategy and organizational evolution. Strategic Management Journal 14(S2): 131–142. doi:10.1002/smj.4250141011.

Turner, J.R. and S.J. Simister. 2000. Gower Handbook of Project Management 3rd Edition. p. 847.

UN International Secretariat for Disaster Reduction. 2004. Terminology: Basic terms of disaster risk reduction. http://www.unisdr.org/2004/wcdr-dialogue/terminology.htm.

UN Water. (n.d.). UN Water statistics, transboundary water. Retrieved September 02, 2013, from http://www.unwater.org/statistics_trans.html.

UN WWAP. 2006. World Water Development Report 2nd edition Water, a Shared Responsibility.

UNDP. 2006. United Nations Human Development Report. Beyond scarcity: power, poverty and the global water crisis. New York, USA. Retrieved from http://hdr.undp.org/hdr2006.

UNDP, GEF and ICPDR. 2006. Phase—out of the DRP Project (p. 14). Vienna, Austria. Retrieved from http://iwlearn.net/iw-projects/1460/reports/capacity-building-and-transboundary-cooperation-for-the-danube-basin/phase-2013-out-of-the-danube-regional-project.

UNDP, GEF, & ICPDR. 2007. Final Evaluation of the DRP Project (56 p). Vienna, Austria. Retrieved from http://aws.undp-drp.org/pdf/3.3_Communication and Public Participation - phase 2/DRP Draft Final Report_June 12.pdf.

UNDP, Water Governance Facility and SIWI. 2013. What is water governance. Retrieved September 02, 2013, from http://www.watergovernance.org/whatiswatergovernance.

UNESCO. 2009. IWRM guidelines at river basin Level—Part 1: Principles (p. 24). Retrieved from http://www.unesco.org/water/news/pdf/Part_1_Principles.pdf.

UNISDR. 2002. Background document on natural disasters and sustainable development. Genève, Suisse.

Von Korff, Y., P. d'Aquino, K.A. Daniell and R. Bijlsma. 2010. Designing participation processes for water management and beyond. Ecology and Society 15(3): 1. http://www.ecologyandsociety.org/vol15/iss3/art1/.

Warner, J., M. Zeitoun and N. Mirumachi. 2013. How "soft" power shapes transboundary water interaction. Retrieved from http://www.globalwaterforum.org/2013/06/03/6928/.

Wolf, A., J. Natharius, J. Danielson, B. Ward and J. Pender. 1999. International river basins of the world. International Journal of Water Resources Development 15(4): 387–427.

Wong, S. 2010. Elite capture or capture elites? Lessons from the "counter-elite" and "co-opt-elite" approaches in Bangladesh and Ghana (No. 82). Helsinki, Finland. Retrieved from http://hdl.handle.net/10419/54126.

World Bank. 2005. The Logframe Handbook—A Logical Framework Approach to Project Cycle Management. Washington DC, p. 113.

World Bank website. (n.d.). Retrieved April 12, 2011, from http://web.worldbank.org/rds.

8

The Privatization Initiatives in the Turkish Water Sector: Reflections on the Transboundary Water Politics in the Euphrates-Tigris River Basin

Aysegul Kibaroglu[1,*] and Radkha Ahmetova[2]

SYNOPSIS

Turkey's proposed privatization plan, notwithstanding the fact that it is not yet implemented, is the first case of a planning of a large-scale privatization of (parts of) the transboundary rivers, namely the Euphrates and Tigris Rivers. This chapter aims to articulate a number of risks associated with any privatization in the Euphrates-Tigris transboundary river basin. To this end, the first section expresses major policy configurations in the water sector by following a historical line of reasoning. A wide range of literature outlining arguments for and against privatization in the water sector is provided as a ground for debate. The second section displays the Euphrates-Tigris River basin's characteristics, hydropolitical relations and possible outcomes resulting out of the privatization plan. Additionally, the chapter provides considered recommendations that can be used in near future as the Turkish government continues to support the private sector, privatization projects in particular.

[1] Department of International Relations, Okan University, Istanbul, Turkey.
Email: aysegul.kibaroglu@okan.edu.tr
[2] The Boston Consulting Group, Istanbul, Turkey.
Email: aysegul.kibaroglu@okan.edu.tr
* Corresponding author

Keywords: Privatization, common good, transboundary river basin, Euphrates and Tigris, Iraq, Syria, Turkey, Public Private Partnerships, Build-Operate-Transfer, hydropolitics, water allocation and sharing, GAP project, conflict and cooperation

"Watersheds come in families; nested levels of intimacy. On the grandest scale the hydrologic web is like all humanity—Serbs, Russians, Koyukon Indians, Amish, the billion lives in the People's Republic of China—it's broadly troubled, but it's hard to know how to help. As you work upstream toward home, you're more closely related. The big river is like your nation, a little out of hand. The lake is your cousin. The creek is your sister. The pond is her child. And, for better or worse, in sickness and in health, you're married to your sink."

—Michael Parfit, National Geographic[1]

Introduction

For a long time, water was considered as a specific common good[2] which needs to be managed collectively and from the utilization of which no one was to be excluded (Hochuli 2004). Thus, it was generally assumed that a state-run system is the usual way of provision and distribution of common goods. This trend, however, has started to change with the introduction of the so called Washington Consensus which is based on the understanding that privatization measures articulated in the 'modernisation' agendas of the governments in the 1980s pave the way for more efficient state organization and faster economic growth. Thereby, following the efforts of global actors such as the World Bank and the International Monetary Fund, the water sector has experienced a massive wave of privatization covering different regions of the world, especially during the years after 1989. Consequently, the private sector participation in the water industry has become an important tool to finance water related projects in Turkey.

From the 1980s on, Turkey has experienced extensive privatization measures in a number of sectors including water and energy distribution, hydropower infrastructure, telecommunication and health care. The rise of

[1] As it is cited in Barlow, M. 2001. Blue Gold: The Global Water Crisis and the Commodification of the World's Water Supply, available at http://www.ratical.org/co-globalize/BlueGold.pdf.

[2] It is important to state that common goods such as water can refer to a number of concepts including the above definition which describes the common good as a good of community that is shared and provides benefits for all members.

neoliberal policies together with a political and economic crisis paved the way for the transformation of the Turkish economy under the leadership of Prime Minister and former IMF consultant TurgutOzal (Pamuk 2008; Kibaroglu et al. 2009). The government was looking for alternative ways to channel professional expertise through numerous forms of Public Private Partnerships (PPPs) which were/are expected to contribute to achieving national goals in affordable ways. Build-Operate-Transfer (BOT)[3] was one of the first public private partnership models implemented in various large scale public facilities in Turkey. The BOT model is seen as an effective way to generate a partnership between public and private sectors and to join their financial resources, know-how and expertise to meet the challenges facing service provision systems normally supplied by governments. However, the privatization process including the various public private partnership models in the water sector has resulted in mixed and mostly unsatisfactory results. Despite these disappointing results, the Justice and Development Party (AKP, in Turkish acronym) continues to favor the partnerships established between the public and private sectors as a way to contain efficiency problems and financial dead-locks of large scale development facilities. In this line of reasoning, the second AKP government announced a new solution plan[4] in 2007, which is the privatization of more than 12 water resources across the country including the Euphrates and the Tigris rivers, to tackle severe drought problems in the regions that lack effective irrigation systems. In other words, the rivers and lakes subject to the proposed privatization plan will be transferred to the private sector for a period of 49 years under the Build-Operate-Transfer (BOT) model. According to the plan, the privatization of the Euphrates and the Tigris rivers is the first case of a large scale privatization of (parts of) trans-boundary watercourse systems in the world, thereby, sets a new precedent in this field.

Thus, the aim of this chapter is to answer the following question: What are the possible consequences of the recent privatization plan of water resources in Turkey given the transboundary nature of the Euphrates-

[3] Among the many different definitions of BOT, the following definition is chosen to fulfill the purpose of this chapter: 'A private party retains a concession for a fixed period from a public party, called principal (client), for the development and operation of a public facility. The development consists of the financing, design and construction of the facility, managing and maintaining the facility adequately, and making it sufficiently profitable. The concessionaire secures return of investment by operating the facility and, during the concession period, the concessionaire acts as owner. At the end of the concession period, the concessionaire transfers the ownership of the facility free of liens to the principal at no cost.' in Menheere, S.C.M and S.N. Pollalis. 1996. Case Studies on Build Operate Transfer, (1. Build Operate Transfer 2. Case Studies 3. International Project Management), Project Management and Real Estate Development, Delft University of Technology-The Netherlands, p. 8.

[4] The plan was announced by the-then Turkish Minister of Energy and Natural Resources, HilmiGuler, which is known to last for a period of 49 years.

Tigris river basin? The literature predicts growing pressures over water resources across different parts of the world which result from a number of factors including the drastic increase in population, urbanization, industrialization combined with the rapid speed of environmental degradation and deterioration. Thus, the management of transboundary waters is becoming an ever important issue as water shortages are very likely to result in international disputes. In this sense, the privatization plan that envisages to delegate 'some power' to private water companies on shared water resources complicates the management of transboundary waters by introducing new actors such as corporations and legal entities.

The first part of the chapter seeks to introduce a historical line of reasoning in order to disclose the major developments in the water industry. This will be assessed in the context of changing macro-economic configurations throughout the 20th century to provide a broader international political economy framework. Then, it aims to highlight a wide range of literature regarding the process of privatization in the water sector. Thus, major arguments for and against the merits of privatization will be outlined to capture the dominant trends in the water sector. The inclusion of private actors in the water sector, however, has paved the way for 'clash of interests' (in most of the cases) between public and private sectors which in turn translated into greater numbers of legal disputes. By taking into account disputes resulting from competing demands between private and public actors, the chapter will show the current trends in international investment and trade agreements related to water resources which are generally inclined to protect the rights of foreign investors and guard against their possible implications for governments.

As the recent privatization plan relies on the strength of the BOT model's premises, the chapter will bring out general characteristics and the rationale behind using public-private-partnerships (PPP) in the water sector given a popular belief that the PPP models such as BOT are a well balanced strategy to tackle the challenges of the service provision systems previously exercised by the public sector. The first section of this part also incorporates the details of the announced privatization plan of water resources in 2007 through which the contemporary debates and developments have become constituted.

Finally, the last part will delve into the hydro-politics of the Euphrates-Tigris river basin to outline the changing dynamics among the riparian countries—Turkey, Syria and Iraq. In this regard, the ET's twentieth century history and geography present a clear perspective with respect to the political-geographic power relationships in the region. Lastly, it explores the possible outcomes of the privatization process on the transboundary watercourses, the Euphrates-Tigris river basin and provides considered recommendations based on the flows in the proposed privatization plan.

1 Understanding the Process of Privatization in the Water Sector

1.1 Historical accounts

Following the post-First World War era, the water industry has long been considered as a vital part of growing national concern. In this process, the nation-states became the primary actors of water services provision in line with Keynesian state-led social and economic policies. Thus, investments in grand infrastructure works such as dams and canals were largely attributed to state-driven efforts that aimed to generate economic growth and sustain development projects, especially in the developing parts of the world (Tekeli and Ilkin 1977; Tezel 1986; Pamuk 2008). It would appear that a similar trend was being followed in Turkey through the operation of State Economic Enterprises (SEE).[5] From the 1930s, the state played an ever increasing role in the main industrial sectors, particularly in financing infrastructure projects (Boratav 1981). Starting with the 1950s, Turkey has expanded its scope of water allocation and planning policies to include social aspects of development such as the resolution of regional economic and social imbalances aiming to raise the living standards of the population, especially the least developed parts of Turkey. In this regard, the Euphrates and the Tigris rivers were chosen to be the backbone of socio-economic policies with respect to water development in Turkey. In this sense, the Southeastern Anatolia Project, GAP in Turkish acronym, is a grandiose development project which envisages the construction of 22 dams and 19 hydroelectric power plants on the Euphrates and the Tigris rivers. For many observers, the GAP Project is one of the world's largest and most ambitious regional development projects comprising not only a giant water resources development plan, but also socio-economic projects in areas such as agriculture, energy, transportation, telecommunications, health care, education as well as urban and rural infrastructure (Carkoglu and Eder 2005).

In the 1980s, the global recession of the late 1970s paved the way for the subsequent transition to neo-liberal economic policies, which resulted in the demise of state-led economic growth in the water sector. In this respect, the emerging structure of international political economy in the 1980s called for 'structural adjustment' policies to bring long-lasting solutions to the

[5] For more information regarding economic policies of Turkey in this era, see Boratav, K. 1981. 'Kemalist economic policies and etatism'. pp. 172–189. *In*: A. Kazancıgil and E. Özbudun (eds.). *Atatürk: Founder of a Modern State* (Turkish accounts) Tekeli, I. and S. Ilkin. 1977. '1929 Dünya Buhranı'ndaTürkiye'niniktisadipolitikaarayışları, Ankara: Orta Doğu Teknik Üniversitesi; Tezel, Y. 1986. Cumhuriyet Dönemininiktisaditarihi (1923–1950), 2nd edn. Ankara: Yurt Yayınları, pp. 98–106.

economic crisis of the previous decade. The similar trend was observed in Turkey through the neo-liberal economic policies that have been initiated by high-level politicians, in particular TurgutOzal.[6] The central objective of these neo-liberal economic policies is to minimize the role of the public sector in economy while maximizing the private sector involvement based on various assumptions such as greater efficiency gains and greater productivity. In addition to the effects of macro-economic policy configurations, other factors such as low prices, subsidized water investments and aging water infrastructure together with an ever increasing water demand have changed the balance in public/private interplay in the water sector. The Washington-based institutions, particularly the World Bank (WB) and the International Monetary Fund (IMF), have taken significant parts in the formulation of the so called Washington Consensus with varying 'conditions' attached to loans.[7] In addition to the World Bank and the IMF, other actors such as the Organization for Economic Cooperation and Development (OECD) and the European Union (EU) have entered to the domain of public/private transition policies (Kibaroglu et al. 2009).

However, in recent years, a growing amount of literature, including a number of World Bank publications,[8] have started to articulate the short-comings of the privatization processes of the 1980s and 1990s:

'As developing and transition economies began restructuring and privatizing their infrastructure. ... Under pressure from international agencies, investment banks, and financial advisers, many of these countries have hastily adopted regulatory templates from industrial countries, especially the United Kingdom and the United States. But these models have rarely been adapted to the political and institutional features common to poorer countries, including lack of checks and balances, low credibility,

---

[6] TurgutOzal's influence in Turkey's political economy in the 1980s, particularly his efforts in transforming the Turkish economy through neo-liberal economic policies have paved the way for a wide literature. It is worth nothing that most of the literature is written in Turkish, for more information see ZiyaOnis. 2004. TurgutOzal and His Economic Legacy: Turkish Neo-Liberalism in Critical Perspective, Middle Eastern Studies, vol. 40, no 4; Ihsan Sezai and Ihsan Dagi (eds.). Kim Bu Özal? Siyaset, İktisat, Zihniyet. Istanbul: BoyutYayıncılık. 2001. FerideAcar, TurgutOzal: Pious Agent of Liberal Transformation, in MetinHeper and SabriSayarı (eds.). Political Leaders and Democracy in Turkey (Lanham, MD: Lexington Books (2002), pp. 163–180.

[7] To illustrate, the World Bank prompted the irrigation management transfer (IMT) in Turkey through a loan that led the country to facilitate a rapid transformation in the irrigation practices.

[8] World Bank. 2003a. Infrastructure Action Plan, 8 July; World Bank. 2003b. Implementing the World Bank Group Infrastructure Action Plan, DC2003-0015,13 September; World Bank. 2004a. Reforming Infrastructure Privatization, Regulation, and Competition, New York, NY: World Bank/OUP; World Bank. 2004b. 'Credible regulation vital for infrastructure reform to reduce poverty', pressrelease, 14 June. This issue has also been elaborated in Wall Street Journal. 2003. The World Bank as privatization agnostic, 21 July.

widespread corruption and regulatory capture, limited technical expertise, and weak auditing, accounting, and tax systems. ... As a result such efforts have had limited success- or been outright failures.'[9]

It is worth stating that a majority of privatization cases seem to justify the objections put forward by the privatization opponents which show that the perceived efficiency of the market has been less than successful (Vining et al. 2006). In this respect, profit as being the major drive for the operations of water companies has been prioritized over numerous issues such as maintenance and construction of necessary infrastructure for water development and conservation projects, public interest human rights matters, and degradation and deterioration of natural resources. In Canada, for instance, the faith in markets paved the way for the privatization of the City of Hamilton's water and wastewater services based on the belief known as 'non-market' failure which is strongly articulated by Charles Wolf.[10] The City of Hamilton's experience, however, demonstrates the private sector's inability to deliver services in the manner articulated by advocates of privatization (Ohemeng and Grant 2008).

1.2 Arguments for and against privatization

The trend which is associated with growing numbers of privatization failures in the water sector, starting in the 1990s, paved the way for a wide range of literature about both for and against the merits of privatization. In this regard, a number of well-known scholars have analyzed the privatization practices from different points of views (Birdshall and Nellis 2002; Dore et al. 2004; Coghill and Woodward 2005). Within this framework, authors such as Coghill and Woodward strongly state that privatization provides benefits if the assumption that the private sector has distributional superiority proves right (2005). In line with this argument, Nellis claims that 'the positive distributional impact of increased access to privatized utility services far outweighs any negative impact of increased tariffs' (Birdshall and Nellis 2002, p. 18). Barlow's observations, however, show that increased access of distribution has benefited only a small fraction of people especially when these numbers compared to growing negative impacts such as drastic price hikes that affect the whole population (Barlow 2001, p. 58–92).

[9] World Bank. 2004a. Reforming Infrastructure Privatization, Regulation, and Competition, New York, NY: World Bank/OUP.

[10] One of the preeminent scholars who have analysed the privatization of public services, Charles Wolf, has used the theory of non-market failure to underline his views stating that the public sector lacks performance indicators which enable governments to determine consumer behaviour and profit-and-loss. For for more information, see Frank K. Ohemang and John K. Grant. 2008. When markets fail to deliver: An examination of the privatization and de-privatization of water and wasterwater services delivery in Hamilton, Canada, Canadian Public Administration, pp. 475–499.

In this respect, the literature outlining positive aspects of privatization is mostly centered on economic, political and social grounds (Ohemeng and Grant 2008). The most prominent economic assumption is that privatization results in greater efficiency and productivity gains (Osborne and Gaebler 1993). Following this line of reasoning, over the last two decades or so, the water sector has become one of the major testing grounds for the implementation of Public Private Partnerships (PPP) as governments have been under severe pressure to provide alternative approaches to the provision of essential infrastructure. In this regard, PPPs are seen as an effective solution to establish new ways to the delivery of economic infrastructure as alternatives to the traditional state-led development projects. Within the boundaries of this argument, the public sector seeks to benefit from commercial dynamism, innovation and efficiencies, harnessed through the introduction of private sector investors who are eager to provide their financial resources, skills and expertise.[11] In terms of political impacts of privatization, advocates claim that the process of privatization saves lots of money for taxpayers since governments delegate the delivery of public services to more 'efficient' actors, the private sector (Ohemeng and Grant 2008). Additionally, socially, privatization enables governments to transform some of their obligations into the realm of democratic governance which brings together many different social actors including the private for-profit sector, the private not-for-profit sector and the community at large (Sclar 2001; Hrab 2004; Ohemeng and Grant 2008).

Despite its highly articulated advantages as outlined above, many observers bring out their strong reservations regarding the matters of privatization (Starr 1987; Schmidt 1996; Hart et al. 1997; Barlow 2001; Sclar 2001). One of the most cited oppositions revolves around the threat of international trade and investment agreements. According to advocates of privatization, the private sector, through 'legalistic contractual agreements', coordinates various aspects of service delivery in more efficient terms while the same process becomes highly complicated if the public sector is both the owner and the producer/deliverer of a service (Ohemeng and Grant 2008, p. 478). This argument, however, lacks its practical grounds since most of contractual arrangements are incomplete (Schmidt 1996; Hart et al. 1997; Barlow 2001). This implies that the public sector, in a majority of privatization cases, is not aware of these agreements' contents and their

[11] For more information see, Andrew Smith, Forward by the Chief Secretary, "Public Private Partnerships, The Government's Approach," Published with the permission of HM on behalf of the Controller of Her Majesty's Stationery Office (2000), 5.

possible shortcomings[12] which pave the way for a process in which these international trade and investment agreements are gaining in power and scope (Barlow 2001).

1.3 International investment and trade agreements

It should be noted that international investment law is mostly inclined to protect the profit-driven interests of foreign investors as various studies conducted in this field suggest. Furthermore, these agreements have not been justified on value-driven measures as the private sector is hesitant to incorporate issues such as equity, public interest, justice, human rights as well as environmental standards. Consequently, an increasingly large number of private water corporations tend to rely on the underlying benefits of the investor-state panels.

It is worth stating that only a few governments have laws and regulations related to their water provision systems and majority of governments have not even begun to address the issues regarding water such as privatization, commercialization (also known as commodification) and trade in water (Barlow 2001). On the other hand, an increasing number of governments in different regions of the world are in the process of signing international investment and trade agreements, thereby, leaving their water resources and local people unsheltered by legislation. For instance, if local farmers' rights regarding access to water are not outlined within the sphere of national law, foreign investors are likely to use their international investment rights to displace the claims of local farmers in international investment panels.

It should be pointed out that these international investment and trade agreements supersede national law which provides the necessary ground for profit seeking investors, especially since the rights of (foreign) investors are heavily protected during the process of so called state-investor arbitration process. The underlying premises of this understanding are based on 'legitimate expectations' of investors since it is the private sector

[12] 'The Attorney General of Pakistan, Makhdoom Ali Khan, speaking at a colloquium hosted by the International Centre for Settlement of Investment Disputes (ICSID), cautioned States to scrutinize closely any international investment treaties which they conclude with other governments (Peterson 2006c). Speaking of his own country's experience, he noted that Pakistan long treated such treaties as 'photo-op' agreements, which could be signed hastily, with little consideration of their concrete legal consequences: "Because someone is going visiting someplace and wants to sign an 'unimportant' document; or someone is coming over for a visit and an 'unimportant' document has to be signed. And a ... "[bilateral investment treaty] until very recently regarded as one such an (unimportant) document. ... These are signed without any knowledge of their implications. And when you are hit by the first investor-state arbitration you realize what these words mean" in Makhdoom Ali Khan (2006). In Pakistan's case the first arbitration to arise under one of its investment treaties was filed by a Swiss multinational, Societe Generale de Surveillance (SGS) in 2001.

that undertakes the financing of large-scale projects and aims to provide distributional superiority and efficiency gains. Therefore, international dispute settlement panels are biased towards the rights of investors. What's more, in legal terms, international law prevails with respect to a conflict between national law and international law, which is also in line with the Vienna Convention on the Law of Treaties. The Convention states that the content of national law is not a legitimate excuse for breaching international commitments (Mann 2006, p. 30).

These investment arrangements incorporate matters related to water, particularly trade in water, and some of them grant water rights to the private sector in explicit terms such as the North American Free Trade Agreement (NAFTA)[13] signed by Canada, the United States and Mexico (Barlow 2001). NAFTA's approach to water privatization revolves around the notion that water is a tradable commodity, a service as well as an investment which outlined as:

'Chapter 3 of NAFTA establishes obligations regarding the trade in goods. Using the General Agreement on Tariffs and Trade (GATT) definition of a "good" which clearly lists "waters, including natural or artificial waters and aerated waters", NAFTA adds in an explanatory note that "ordinary natural water of all kinds (other than sea water)" is included. Chapter 12 sets out a comprehensive regime to govern trade and investment in the service sector, including water services. Chapter 11 establishes an extensive array of investor rights, including investors in water goods and water services. Thus, under NAFTA, water is a commercial good, a service and an investment.'[14]

Furthermore, the proliferation of bilateral investment and trade agreements, in recent years, has enabled the private sector to undertake investments in various sectors including the water sector. Once again, it should be outlined that very few governments protect their water systems through law and regulations which leaves the rest unprotected from foreign investors and threats associated with privatization. One of such threats is the clause defining investor-state arbitration process, very similar to Chapter 11 of NAFTA. As stated above, this clause allows investors of the signatory countries to take governments to court for the so called expropriation compensation (Barlow 2001).

[13] The North American Free Trade Agreement (NAFTA) was implemented on January, 1, 1994. It is designed to remove tariff barriers between the US, Canada and Mexico and considered as the world's largest free trade area, for more information http://export.gov/FTA/nafta/index.asp, http://www.law.duke.edu/lib/researchguides/nafta.html.

[14] For more information regarding water and NAFTA, see Barlow, M. 2001. Blue Gold: The Global Water Crisis and the Commodification of the World's Water Supply, available at http://www.ratical.org/co-globalize/BlueGold.pdf.

Moreover, possible changes to the terms of contract (or/and national laws) go through a very difficult and expensive path after an agreement being signed between the parties of public private partnership (Mann 2006). For example, possible changes regarding environmental standards to national laws that might affect the profit calculations of foreign investors enable the private sector to claim compensation, mostly in the form of financial compensation, under the rights provided by international investment agreements-expropriation protections of investors. There is also an increasing trend in which threats of using international investment panels are becoming a strong vehicle, in the hands of investors, against proposed changes to national laws and the terms of contract. Thus, there is no doubt that investors will exercise this threat when it is necessary indicating a process where governments are subject to an ever increasing pressure.

1.4 Oligopolistic market structure

Another problem with privatization is the issue related to competition since the private sector involvement in the water sector has resulted in local monopolies rather than an environment of competition as emphasized by advocates of privatization. It should be noted that despite its perceived/expected efficiency gains resulting out of competitive market conditions, privatization of public delivery systems has paved the way for an oligopolistic market structure with minimal competition among very few firms. In this regard, the largest transnational corporations which dominate the world's privatized water markets have mostly benefited from the process of privatization given their vast experience, financial resources as well as lobbying activities (Hall et al. 2003). More specifically, the French water companies such as Ondeo (Suez) and Vivendi (now Veolia) are controlling more than two-thirds of the global privatized water market, with Thames Water (part of the German multi-utility RWE) and SAUR holding substantial share but coming behind the former two French establishments (Swyngedouw 2005). One of the drastic consequences most likely to occur as a result of the oligopolistic market structure is water rate increase which was evident in the UK's privatization case. This case, as being the one of the first privatization examples, sheds the lights on future privatization projects and provides a clear line of reasoning with respect to the dangers associated with oligopolistic market conditions, if not regional monopolies. In the UK, 'the water price immediately increased significantly, and many non-paying households were cut off, while companies and their shareholders gained considerable profits' (Bakker 2001). By following negative outcomes resulting out of the privatization process in the water sector, it becomes clear to state that the private sector tends to heavily rely on its profit-driven interests. The same process of profit maximization, however, is very likely to endanger most of the inherent objectives of public services systems. For instance, Turkey's GAP Project, as it was outlined in the previous section, aims to eliminate regional development disparities for the people living in the Southeastern Anatolia region of Turkey. However, profit-driven interests of investors, in the region, can easily halt the socio-economic development objectives of the GAP Project leaving the local population vulnerable to the private sector and its initiatives.

2 Moves Towards the Privatization of the Euphrates-Tigris River Basin

The privatization of water is a trend that takes its origins from the so called success of water privatization in developed countries such as England and Wales which encouraged many developing countries including Turkey to undertake privatization measures in the water sector. In this regard, given the common good nature of natural resources in the public eye, privatization in the water sector has mostly taken the form of Public Private Partnerships (PPP).[15] Turkey is considered as one of the leading actors in the field of cooperative arrangements established between the public and private sectors with a number of privatization models backed by the international finance institutions together with the political support of successive governments (Cinar 2009). In this regard, the Build-Operate-Transfer (BOT) model is one of the major delivery ways implemented in a number of large scale projects in Turkey.[16]

2.1 Turkey's recent privatization plan

Although the trend towards water privatization in Turkey dates back to the 1980s, as outlined in the previous section, an accelerated privatization of the

[15] This terminology aims to confront the critiques of privatization such as the 'commodification of natural resources' debates by bringing the public domain into play while also appearing more acceptable to society, for more information see, Franceys, R. 2008. GATS, privatization and institutional development for urban water provision: future postponed? Progress in Development Studies 8(1): 5–58.

[16] The water privatization measures in Turkey have been implemented mainly at the level of municipalities such as the Antalya case and Yuvacik Dam project, combined with mixed and mostly unsatisfactory results. Izmit, as being an important industrial city, has been subject to a growing water demand which forced the government to launch 'Izmit Urban and Industrial Water Supply Project' in the mid-1980s. The project is known as the largest privately financed water-supply scheme in the world costs approximately US$ 900 million (Bennett et al. 1999). The project, however, raised issues related to the private sector financing. The Antalya case displays dangers with respect to international arbitration process since the private entity decided to seek international arbitration and demanded compensation of US$30 million.

water sector has expanded its scope to include the privatization of water resources including the transboundary watercourses, the Euphrates and the Tigris Rivers. The plan was announced by Turkey's Minister of Energy and Natural Resources, HilmiGuler, in 2007, and the objective central to the proposed plan is to privatize rivers and lakes as an action plan to fight severe drought that occurred in various areas across the country during the summer of 2007.[17] The implementation stage of the proposed privatization plan relies on the premises of the BOT model, which is expected to last for 49 years. In this respect, BOT is considered to be a successful privatization approach which enables the private actors to engage in the process of development and organization of a facility normally implemented by the public sector. As a result of the privatization plan, Turkish officials expect greater efficiency gains and productivity, particularly in the regions that are subject to the conditions of severe drought such as eastern and southeastern parts of Turkey. It should also be noted that, the government expects a profit of US$3.1 billion, resulting out of the proposed privatization of more than 12 water resources across the country.[18]

However, in spite of such expected benefits, mounting concerns (following the announcement) with respect to the proposed privatization plan demonstrate strong hesitation as well as growing uneasiness regarding the matters of privatization.[19] In this respect, as the critics had reached a critical mass, Turkish officials addressed the issue with a statement indicating that not the river itself but the operation and the sale rights are intended to be the subjects of privatization.[20] Within this wave of concerns that outline serious reservations regarding the proposed privatization plan, questions with respect to the privatization of shared watercourses, (parts) of the Euphrates and the Tigris Rivers and its potential international impacts remain intact. In this regard, given a clear trend of international water disputes which contains new claims, claimants and dispute settlement institutions where the parties are no longer one state against another state (Salman 2006), the following part aims to articulate a number of

[17] TDN 01.08.2007.

[18] According to the numbers given by the General Directorate of State Hydraulic Works (DSI), a 29-year sale price of the Euphrates is $950 million while than of the Tigris is $650 million. Turkish Daily News. 2007. 1 August.

[19] Many have expressed serious reservation regarding the sale of water resources, civil society organizations, in particular, such as the Union of Chambers of Turkish Engineers and Architects (TMMOB, in Turkish acronym) and one of the pre-eminent environmental foundations, TEMA.

[20] The then newly appointed Minister of Environment and Forestry, Veysel Eroglu, corrected the former Minister's (HilmiGuler) statement by denying any attempts with respect to the sale of natural resources. TZ 14.09.2007, New Minister Objects to Privatization of Water Resources (http://todayszaman.com/tz-web/detaylar.do?load=detay&link=122100).

possible outcomes that are bound to happen after the privatization of the Euphrates and the Tigris rivers. The outcomes that are likely to happen in the basin area since the privatization process opens up doors for long term public-private arrangements, i.e., build-operate-transfer model, where multinational corporations may become new actors on matters related to the Euphrates and the Tigris rivers. It starts with the Euphrates-Tigris River basin's characteristics and then illustrates milestone events of the basin area within a historical line of reasoning. Finally, this part will introduce a number of possible dangers resulting out of the privatization process as well as considered recommendations since the transboundary nature of the Euphrates and the Tigris rivers have not been explored properly, if not at all, within the proposed privatization plan.

2.2 The Euphrates-Tigris river basin

The Euphrates-Tigris River basin is one of the most important transboundary watercourses in the world which is shared by Iraq, Syria and Turkey as its major riparian states.

It is important to underline that both rivers originate in Turkey, making Turkey an upstream riparian country with respect to the basin related issues. The Euphrates and the Tigris rivers start to flow scarcely 30 m from each other in the mountains of eastern Turkey, and travel southeast through Syria and Iraq to the Persian Gulf (Kibaroglu 2008). The rivers are subject to seasonal and multi-annual hydrological changes of water levels compounded with severe drought and flooding problems in the basin area. In addition to the hydrological factors of the basin, the riparian states' water development projects on the Euphrates and the Tigris rivers which started in the 1960s have complicated relations among Turkey, Syria and Iraq. In this regard, the major factor behind the projects was increasing demand for electricity and irrigation facilities given the drastic surge in population and the efforts of economic development in the second half of the 20th century. The damming activities on the Euphrates-Tigris River basin, therefore, accelerated thereafter resulting in 'a series of dams, first in Iraq, then in Syria and Turkey to provide irrigation water and to generate hydropower' (Kibaroglu and Scheumann 2010, p. 5).

For instance, the largest of these water development projects is the Southeastern Anatolia Project (GAP), as outlined in the previous section, was a major outcome of Turkey's State Planning Organization in the 1950s. Turkey's highly ambitious GAP Project, however, raised eyebrows among the downstream riparian states of the Euphrates-Tigris river basin, Syria and Iraq in particular.

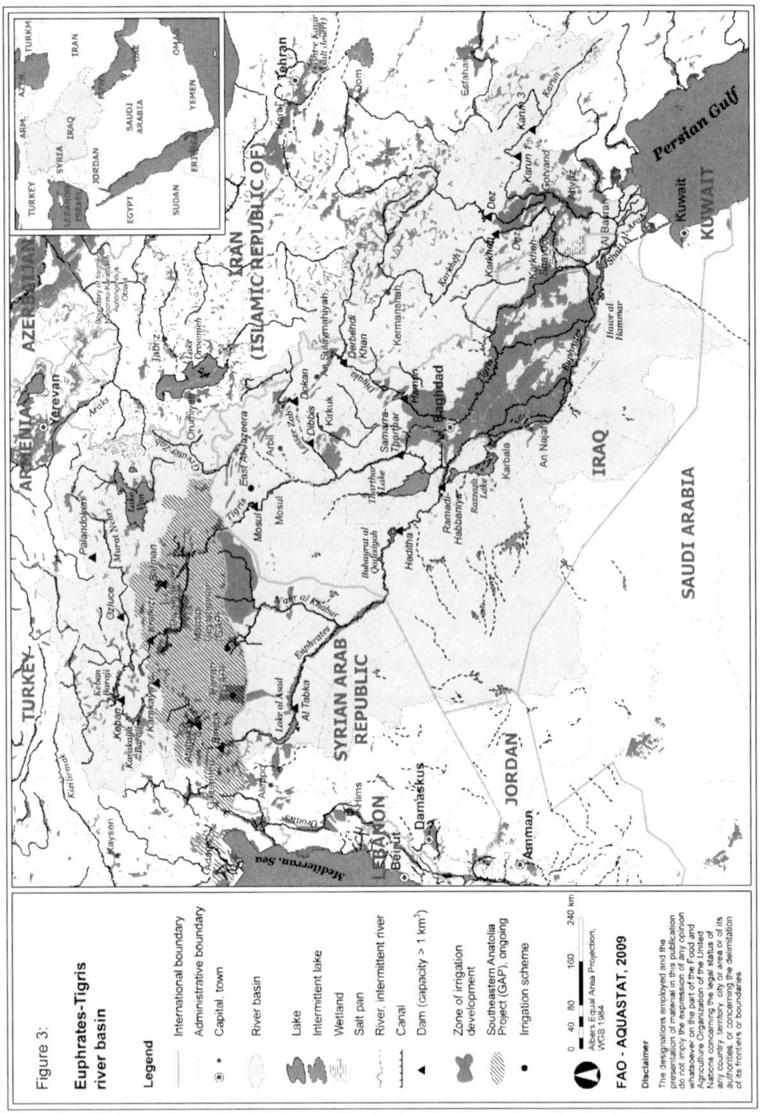

Figure 8.1. The Tigris Euphrates River Basin and the three riparian states (FAO AQUA STAT 2009 used with permission).

2.3 The crises in the Euphrates-Tigris river basin

Following the riparians' uncoordinated and unilateral water development projects, the Euphrates-Tigris River basin witnessed a number of crises during the 1970s, 1980s and 1990s. The first incident among the riparian countries was over the issue of impounding of both Turkey's Keban reservoir and Syria's Tabqa Dam in the 1970s which paved the way for the escalation of tension in the region. The Iraqi government blamed Syria for reducing the Euphrates' water level, while Syria was busy with placing the burden on the upstream riparian state, Turkey. The crisis on the river basin which eventually turned into mutual threats was halted over the mediation efforts of Saudi Arabia. In the 1980s and 1990s, the crises on the basin area moved into the realm of high politics as the riparian states preferred to rely on their coercive foreign policy tools. The 'hawkish' foreign policy, thereby, left almost no room for the riparian states to accommodate each other's interests and concerns with respect to transboundary water issues. In the early 1990s, for example, both Syria and Iraq made official complaints about the impounding of Turkey's Ataturk Dam which sits on the Euphrates River (Kibaroglu and Scheumann 2010). Despite notifications being sent to the governments of Syria and Iraq, Turkey was accused of reducing the river's flow and consequently the downstream riparian states demanded an agreement to share the waters of the Euphrates (Kibaroglu 2008).

The rapidly growing populations and ambitious development plans compounded with the severe drought conditions in Syria and Iraq added an extra layer of frustration to the water development plans in the upstream country, particularly the GAP Project which was evident once more when Turkey started to construct the Birecik Dam on the Euphrates River in 1996 (Kibaroglu and Scheumann 2010). In addition to the official complaints underlining the downstream countries' objections to the construction of the dam which revolve around the issues of water quantity and quality, both Syria and Iraq asked the Arab League countries to halt financial aid to Turkish projects and to boycott European companies which undertook financing of the dam, thereby, making the issue to become an international affair (Scheumann 2003; Kibaroglu and Scheumann 2010).

2.4 Cooperation platforms

At the same time, the riparian states of the Euphrates-Tigris River basin managed to establish a number of negotiation platforms to enable cooperation through technical negotiations. The major aim of these technical negotiations starting in the 1960s was to enable information sharing among the riparian states. In 1965, for instance, the three riparian countries gathered in Baghdad for the first time to exchange technical data on the

Haditha (Iraq), Tabqa (Syria) and Keban (Turkey) dams (Kibaroglu and Scheumann 2010). In 1980, upon the proposal of the Iraqi government, a permanent Joint Technical Committee (JTC) was established at the end of the first meeting of the Joint Technical Commission between Turkey and Iraq to discuss and finalize the water issue among the riparian states. After three years, Syria decided to join the cooperative initiative and during the years between 1980 and 1993, the JTC was able to gather technocrats from Turkey, Syria and Iraq for a total of 16 times (Kibaroglu 1996). Despite its initial intent to establish an environment of cooperation, the JTC could not perform its designated role as the riparian states continued their unilateral water development plans. It is worth noting that there was no agreement at the end of the meetings on sharing waters as it was heavily demanded by the governments of Syria and Iraq. However, two bilateral accords-that are acknowledged as being interim agreements by all riparian states-were signed, in 1987 and 1990, resulting out of a number of high-level meetings of top officials in the region (Kibaroglu and Scheumann 2010). The Turkish-Syrian Protocol of 1987 is important since it underlines Turkey's commitment in the Euphrates-Tigris River basin as to release 'a yearly average of more than 500 m^3/s at the Turkish-Syrian border and in cases where monthly flow falls below the level of 500 m^3/s, the Turkish side agrees to make up the difference during the following month' (Article 6 of the Protocol). Furthermore, the Syrian-Iraqi water accord of 1990 states that the Syrian share of the Euphrates flows is 42 percent and the remainder quantity 58 percent as fixed annual total percentage (The Joint Minutes [1] of the Syrian-Iraqi water accord of 1990).

Beginning in the late 1990s, however, the opportunity for a lasting solution over the Euphrates and the Tigris has been an emerging trend as cooperation, mainly in the security domain, has intensified between two of its major riparian states, Turkey and Syria (Kibaroglu 2008). It is important to note that, in 1998, Turkey and Syria appeared to be on the verge of war as Turkey threatened to use military force against Syria. For many observers, Turkey's coercive attitude was attributed to the country's foreign policy goals that were centered on its geopolitically important geography in the post-Cold war years which transformed into more cooperative and collaborative tone in the 21st century. The reasons of this drastic transformation are beyond the scope of this chapter, but the constructive foreign policy of Turkey in the late 1990s and early 2000s has clearly contributed to the emergence of cooperative frameworks in the Euphrates and Tigris River basin. After escalating at the peak of the political confrontation, Turkey and Syria signed the Adana Accord on 20 October 1998, a framework security agreement, which marked the beginning of a new era that is based on more cooperative initiatives of interest to both sides (Kibaroglu 2008). One of the first cooperative initiatives was a joint communiqué signed between

Southeastern Anatolia Project Regional Development Administration (GAP RDA) from Turkey and the General Organization for Land Development (GOLD), a part of the Syrian Ministry of Irrigation, in 2001. The underlying premises of this cooperation are defined as providing sustainable utilization of land and water resources in the region (Kibaroglu and Scheumann 2010). The cooperative relations that have been built on a number of cooperative arrangements between Turkey and Syria, among which the signing of Free Trade Agreement in 2004 has been considered as reaching a milestone, have continued to flourish in the 2000s.

It is worth stating that the relations among the riparian states of the Euphrates-Tigris River basin have been mostly governed on the levels of diplomatic correspondence, technical negotiations and meetings of top officials. In other words, interactions between the riparians were conducted on state-to-state basis, mostly in the form of bilateral high-level exchange between state officials, where the driving rationale has been the provision of strategic interests of Turkey, Syria and Iraq. However, the disputes that may arise out of or in connection with the proposed privatization plan of the transboundary watercourses, the Euphrates and the Tigris rivers, will most likely to reveal a new set of claimants (multinational corporations) and claims such as claims for compensation (financial/expropriation compensation). Thereby, transboundary water resolution of the Euphrates-Tigris River basin goes beyond traditional practice of one state against another state indicating complexity and expansion of actors in relation to possible water related disputes in the basin area.

Discussion

In this regard, by taking into account the historic pattern of political confrontations and escalating transboundary water disputes among Turkey, Syria and Iraq in the second half of the twentieth century, the Turkish officials should take a cautious approach with respect to the issues related to the privatization process of (parts of) the Euphrates and the Tigris rivers. Thus, it is vital for the officials of Turkey to understand the complexity of the proposed privatization plan as the playing field in the basin area will be expanding with new claimants such as multinational corporations. It is worth stating that if the privatization plan which delegates the release of the rivers' water to the private sector continues to ignore the novelty of issues related to the privatization of (parts of) the Euphrates and the Tigris rivers, 'blue-gold' is very likely to be one of the major points of contention among the riparian countries.

At this point, the downstream riparian countries namely Syria and Iraq may demand a specific form of agreement from the upstream riparian country, Turkey, to guarantee a certain amount of water to be released from the Euphrates and the Tigris rivers. It is worth underlining the fact that there is no permanent agreement between the riparian states of the Euphrates-Tigris River basin which guarantees fixed water quotas from the upstream riparian country to the downstream riparian states. However, Turkey, as being the upstream country, has certain obligations to the downstream riparians which were underlined by bilateral accords concerning the Euphrates River. As indicated earlier, Turkey ensures release of a yearly average of more than 500 m^3/s at the Turkish-Syrian border and Syria commits itself to share the water of the Euphrates River (500 m^3/s) with Iraq, 42 percent and 58 percent, respectively. Thus, Turkey's obligations resulting out of temporary bilateral accords with its riparian states should be well understood prior to any possible privatization arrangements between public and private sectors. Specifically, new actors of the basin area such as multinational corporations should grasp the fundamentals of these agreements since Turkey continues to remain accountable to its downstream neighbors. In this sense, Turkey should ensure that the specific requirements and unique characteristics of the Euphrates and the Tigris rivers along with the river basin's complex history are communicated thoroughly to the basin area's new actors, private corporations. A thoughtful and detailed explanation will avoid the possibility of future disputes between the public and private actors arising out of Turkey's obligations in the region. Consequently, Turkey needs keeping an eye on private operators to guarantee certain water volumes. However, details regarding such obligations under any particular surveillance mechanism have not been released yet.

Additionally, a number of interviews[21] conducted with officials from Turkey's Ministry of Foreign Affairs and various organizations such as State Hydraulic Works (DSI) demonstrate Turkey's lack of interest to grasp possible outcomes as well as shortcomings of the proposed privatization plan including the role of the private sector in meeting the terms of bilateral accords related to the Euphrates River. These interviews illustrate the absence of any advisory mechanism established prior or during the announcement of the large-scale privatization plan. A careful reading of the interviews that were conducted in 2009 display the fact that there were no arrangements in place or preliminary studies undertaken by Turkey's officials with respect to the privatization of (parts of) the Euphrates and the Tigris rivers. Despite the plan's ambitious rhetoric which aims to curtail

---

[21] These interviews are based on the questions of Prof. Aysegul Kibaroglu (conducted over email), that aim to portray the lack of information regarding the proposed privatization plan.

drought problems in Turkey, it seems, however, that Ankara's privatization project of more than 12 rivers has been undertaken without a strategic plan and with little integration of water issues into Turkey's overall foreign and security policies.

As outlined in the previous section regarding the threat of international investment and trade agreements, 'if contracts and extant laws and regulations do not expressively recognize and give priority to the rights and needs of local citizens, or are not sufficient to ensure long term water quality management, the existing international trade and investment rules will reinforce any weaknesses and imbalances by ensuring the investor's rights can be enforced under international law, and outside of national legal systems' (Mann 2006). Thus, this chapter strongly suggests that Turkey must study the long-term implications of the proposed privatization plan of 2007 to prepare for any consequent dangers in future.

In this line of thinking, one of the most dangerous scenarios that Turkey could face as a result of the privatization plan of the Euphrates and the Tigris rivers is a drastic deterioration of the current positive relations among the riparian states. In other words, the process of privatization of the Euphrates and the Tigris rivers is very likely to endanger the cooperative approach of the riparian countries with respect to the matters of the Euphrates-Tigris River basin.[22] For instance, in recent years, one of the most significant developments related to the basin area is the Euphrates-Tigris Initiative for Cooperation (ETIC) established in May 2005. ETIC aims to translate its cooperative vision to technical, social and economic development in the Euphrates-Tigris River basin which is primarily driven by the efforts of scholars and professionals from the three major riparian countries—Turkey, Syria and Iraq (Kibaroglu 2008). In addition to such initiatives coordinated through a network of experts, encouraging developments in the basin have taken place on the highest political levels to reach a consensus with respect to the region's political economy and social-cultural affairs. Turkey's cooperative and constructive foreign policy, in the 2000s, has paved the way for broader cooperation with southern neighbors which is also reflected in cooperative initiatives related to transboundary water development and management of the Euphrates, Tigris and the Orontes Rivers such as the High-Level Strategic Cooperation established between Turkey and Iraq committing the parties to cooperate on matters of politics, economy, water, culture and security as well as the Turkish-Syrian High-Level Strategic Cooperation Council Agreement signed on 13 October 2009 which aims to

---

[22] For a detailed discussion see Kibaroglu, A. and W. Scheumann. 2010. 'Euphrates-Tigris River System: Political Rapproachment and Transboundary Water Cooperation which explores the details regarding the basin area as well as hydropolitics of the basin area.

underline the commitment of both sides to promote solutions for regional and international problems (Kibaroglu and Scheumann 2010).

The above initiatives and cooperative frameworks established mostly through Turkey's active foreign policy, starting in the late 1990s, using both diplomacy and trade, indicates a clear trend of positive relations that possibly go beyond good wishes and more towards institutional settings in the region. On the other hand, as water becomes an economic commodity, questions of supply, access and management, both in quantitative and qualitative terms, emerge as key issues both for private and public actors. In this line of reasoning, private actors' revenue calculations combined with severe drought problems of the region are very likely to intensify the number of concerns related to relations of the riparian states, most importantly as a result of possible disruptions on the Euphrates and the Tigris rivers' flows. In other words, a deterioration of relations among the riparian states of the Euphrates-Tigris River basin becomes a highly possible incident as the historical record of the riparian relations was, and continues to be, the guiding vehicle to understand a number of possible outcomes.

As indicated earlier, historical developments of the riparian relations and concerns related to cooperation and conflict on the Euphrates-Tigris River basin present a rather interesting and mostly mixed picture. The impounding actions at the Ataturk Dam and many similar interruptions to the flow of the Euphrates resulting from the Southeastern Anatolia Project (GAP), for instance, paved the way for worsening of the riparian relations as both Syria and Iraq perceived these disruptions as being against their own development interests. The risk associated with the announced privatization plan, as outlined above, could endanger the current positive developments in the basin area. Furthermore, possible disruptions of water flows in the basin due to activities of private actors may pave the way for an international investment panel where foreign investors can claim financial compensation under expropriation protections of investment agreements. In other words, the private actors can refuse to ensure Turkey's obligations to its downstream riparian states given investors' profit-based calculations in the region and take the issue to international arbitration panels. In this respect, the absence of any sharing agreements between the riparian states in relation to the Euphrates and the Tigris rivers as well as lack of information and consultation provided to foreign investors prior to any contract signed between public and private actors will complicate the picture while shifting the balance in favor of foreign investors. In fact, this would put a heavy financial burden on Turkey's shoulders and affect Turkey's relations with its downstream riparian neighbors given the obligations and rights of the states in the river basin.

Additionally, Turkey's state actors and those undertaking the privatization process of natural resources should take into account the public interest as it is stated in the Article 43 of the Constitutional law which reads: 'The public good is the primary consideration in respect to benefiting from the shores of seas, lakes and rivers, and the coast line surrounding seas and lakes.'[23] However, the centrality of foreign investors' so called legitimate expectations (profit-driven calculations) in front of international investment agreements becomes one of the major obstacles to tackle for the public sector. Having said this, this chapter claims that the wording of any such arrangements should openly reflect the public interest while defining the role and the limitations of the private sector with clear terms and conditions.

The involvement of foreign investors in water disputes, especially in the case of transboundary watercourses, adds another layer of difficulty and complexity to such disputes. Less noticed is the fact that international investment agreements on matters of water privatization are being kept secret for various reasons such as the political struggles around water privatization projects and the popular resistance against the privatization of natural resources given the common good nature of water. It should be clarified that the investor-state panels operate in almost total secrecy and 'at least four water related cases are known to have been started in the past five years under investor-state arbitration procedures, and it is suspected others have already begun or will begin soon, but remain shrouded in secrecy' (Mann 2006). As such those cases do fall within the scope of this chapter, it is worth elaborating on the Bayview Irrigation District V. United Mexican States case which is one of the few known cases related to the complex and contentious issue of privatization of the transboundary watercourses and the role of the private sector during the process.[24] In this case, the coalition of claimants from Texas (Bayview Irrigation District and others) submitted a request for arbitration to International Centre for Settlement of Investment Disputes (ICSID) which is filed under the provisions of Article

[23] Turkish Daily News. 2007. Rivers to be privatized as a solution to water crisis, 1 August 2007.

[24] Fernando Cabrera. 2007. NAFTA Tribunal lacks jurisdiction to hear Texans water dispute with Mexico, Investment Treaty News, 12 July, International Institute for Sustainable Development (IISD), http://www.iisd.org; Fernando Cabrera and Eric Luke Peterson (2006), US makes last-minute filing in arbitration between Texan farmers and Mexico Investment Treaty News, 19 December, IISD; Eric Luke Peterson (2004), US water rights-holders sue Mexico under NAFTA, Investment Law and Policy Weekly News Bulletin, 1 June, IISD in Miguel Solanes and Andrei Jouravlev (2007), Revisiting privatization, foreign investment, international arbitration, and water, United Nations Publication, p. 16. For more information about the case see also: International Arbitration-ICSID Jurisdiction Under NAFTA-*Bayview Irrigation District V. United Mexican States*, ICSID Case No. ARB(AF)/05/1, AWARD 19 June 2007.

1120 (1) (b) of the North American Free Trade Agreement (NAFTA) and is based on Mexico's debt. This is an extremely interesting case which goes on to highlight that 'each claimant is an investor and owner of an integrated investment which includes rights to water located in Mexico' which considers such rights as property rights (Salman 2006). The claimants, in this case, chose to invoke NAFTA and 'allege that Mexico has violated Articles 1102, 1105, and 1110 that provide that NAFTA Parties shall treat investors of another party no less favorable than they treat their own investors with respect to the establishment, acquisition, expansion, management, conduct, operation and sale, or other disposition of investments' (Peterson 2004; Cabrera and Peterson 2006; Salman 2006; Cabrera 2007). In this respect, taking into account the Bayview case, the commodification process of common goods such as water presents extremely novel claims based on the understanding that water is a tradable commodity and related investments on water resources subject to a number of rights including property rights. Based on this, it is important to underline that Turkey should consider various aspects of the privatization process of the Euphrates and the Tigris Rivers since the basin area is shared by the riparian states, Turkey, Syria and Iraq. Therefore, it should be pointed out in this regard that, the proceedings of the Bayview Irrigation District V. United Mexican States may provide guidance for similar situations in future, particularly with respect to the privatization of (parts of) the Euphrates and the Tigris Rivers.

Conclusion

The privatization of the Euphrates and the Tigris rivers, as a result of Turkey's proposed privatization plan that covers more than 12 water resources, is the first case of a large-scale privatization of (parts of) the trans-boundary rivers. Therefore, this chapter aimed to articulate a number of dangers associated with any privatization in the Euphrates-Tigris River basin. To this end, the first section of this chapter articulated major policy configurations in the water sector by following a historical line of reasoning. A wide range of literature outlining arguments for and against privatization in the water sector provided a ground for debates in the last section. The second section displayed Turkey's experience with respect to water privatization as well as the details of the proposed privatization plan which paved the way for the last part that brings out the Euphrates-Tigris River basin's characteristics, historical record and possible outcomes resulting out of the privatization plan. Additionally, this chapter provided considered recommendations that can be used in near future as the AKP government continues to support the private sector, privatization projects in particular.

In this regard, as corporations emerge as new actors within the transboundary water politics, it is important that governments, particularly Turkey as being the upstream country, do understand the process of privatization and consequent dangers of the process. As indicated in the various parts of this chapter, foreign investors are very likely to treat water as a basic requirement which enables the private sector to manage water as a commercial good. This line of reasoning, in turn, could endanger Turkey's volatile relations with its southern neighbors, as foreign investors may pursue their interests in matters vital to the transboundary water politics such as possible disruptions to the flows of the Euphrates and the Tigris rivers. Furthermore, in the absence of a thorough and thoughtful explanation with respect to the river basin's special characteristics and Turkey's specific obligations to its downstream riparian countries that are communicated to the private sector prior to any PPP arrangements (foreign), investors are in the position to demand financial compensation which mostly results in high costs to the public sector. One of the lessons that this chapter provides is that the proceedings of the proposed privatization plan should be followed very carefully including the issues related to the rights of investors which requires Turkey to consider the plan's long term implications and possible shortcomings. In the context in which foreign investment in the water sector is becoming an essential tool in the hands of policy makers, this chapter strongly claims that it is very important that the public sector does understand various outcomes of international investment agreements including their international implications in the trans-boundary river basins such as the Euphrates-Tigris river basin and the so-called investor-state arbitration process.

Thus, Turkey needs to articulate an over-arching privatization plan which outlines Turkey's strategic position in the region as well as its obligations to the downstream riparians while placing the plan within Turkey's overall foreign and security policies. To this end, Ankara needs to improve coordination among government agencies in all areas related to water and between private investors undertaking the privatization of rivers and the foreign policy and security ministries.

References and Further Readings

Bakker, K. 2001. Paying for water: water charging and equity in England and Wales. Transactions of the Institute of British Geographers New Series 26: 143–164.

Barlow, M. 2001. Blue Gold: The Global Water Crisis and the Commodification of the World's Water Supply. Available at http://www.ratical.org/co-globalize/BlueGold.pdf.

Bennett, E., P. Grohmannn and B. Gentry. 1999. Public-private Partnerships for the Urban Environment: Options and Issues. PPPUE Working Chapter Series, Volume 1. UNDP and Yale University, New York.

Birdshall, N. and J. Nellis. 2002. Winners and losers: assessing the distributional impact of privatization. Center for Global Development Working Chapter No. 6. Center for Global Development, Washington, DC.

Boratav, K. 1981. Kemalist economic policies and etatism. pp. 172–189. *In*: A. Kazancıgil and E. Özbudun (eds.). Atatürk: Founder of a Modern State, London, C. Hurst & Co., 1981.

Cabrera, F. 2007. NAFTA Tribunal lacks jurisdiction to hear Texans' water dispute with Mexico. Investment Treaty News, 12 July, International Institute for Sustainable Development (IISD), http://www.iisd.org.

Cabrera, F. and E.L. Peterson. 2006. US makes last-minute filing in arbitration between Texan farmers and Mexico. Investment Treaty News, 19 December, IISD.

Carkoglu, A. and M. Eder. 2005. Developmentalism a la Turca: The Southeast Anatolia Development Project (GAP). pp.167–184. *In*: F. Adaman and M. Arsel London, Ashgate (eds.). Environmentalism in Turkey: Between Democracy and Development. Ashgate Publishing.

Cinar, T. 2009. Privatisation of urban water and sewerage services in Turkey: some trends. Development in Practice 19(3): 362–363.

Coghill, K.A. and D.F. Woodward. 2005. Political issues of public-private partnerships. pp. 81–94. *In*: Hodge, Graeme and Carsten Greve (eds.). The Challenge of Public-Private Partnerships: Learning From International Experience. Edward Elgar Publishing Limited, Cheltenham, UK.

Dore, M.H.I., J. Kushner and K. Zumer. 2004. Privatization of water in the UK and France-What can we learn? Utilities Policy 12(1): 41–50.

Franceys, R. 2008. GATS, privatization and institutional development for urban water provision: future postponed? Progress in Development Studies 8(1): 5–58.

Hall, D., E. Lobina and R. de la Motte. 2003. Public solutions for private problems? Responding to the shortfall in water infrastructure investment. PSIRU, London, available at www.psiru.org/reports/2003-9-W-strats.doc.

Hall, D., E. Lobina and R. De la Motte. 2005. Public resistance to privatisation in water and energy. Development in Practice 15(3&4): 286–301.

Hart, O., A. Schleifer and R.W. Vishny. 1997. The proper scope of government: theory and an application to prisons. Quarterly Journal of Economics 112(4): 1127–1161.

Hochuli, M. 2004. Water is a Common Good, International Conference: The Public Eye on Davos (21 January 2004), available at http://www.evb.ch/it/p25002715.html. Accessed 27 August 2011.

Hrab, R. 2004. Private delivery of public services: public private partnerships and contracting out. Research Chapter 21, available at http://www.law-lib.utoronto.ca/investing/reports/rp21.pdf.

International Arbitration-ICSID Jurisdiction Under NAFTA-Bayview Irrigation District V. United Mexican States, ICSID Case No. ARB(AF)/05/1, AWARD 19 June 2007.

Kibaroglu, A. 1996. Prospects for cooperation in the Euphrates-Tigris Basin. pp. 31–39. *In*: P. Howsam and R. Carter (eds.). Water Policy: Allocation and Management Practice. E&FNSpon, London.

Kibaroglu, A. 2008. The role of epistemic communities in offering new cooperation frameworks in the Euphrates-Tigris river system. Journal of International Affairs, Spring/Summer 61(2):183–198.

Kibaroglu, A. and O. Ünver. 2000. An institutional framework for facilitating cooperation in the Euphrates-Tigris river basin. International Negotiation: A Journal of Theory and Practice, Dordecht, The Netherlands 5(2): 311–330.

Kibaroglu, A. and W. Scheumann. 2010. Euphrates-Tigris river system: political rapprochement and transboundary water cooperation. *In*: A. Kibaroglu, W. Scheumann and K. Kramer. Springer Verlag, 2011.

Kibaroglu, A., A. Baskan and S. Alp. 2009. Neo-Liberal transitions in hydropower and irrigation water management in Turkey: main actors and opposition. pp. 287–304. *In*: D. Huitema and S. Meijerink (eds.). Water Policy Entrepreneurs. A Research Companion to Water Transitions Around the Globe. Edward Elgar, Cheltenham, UK.

Mann, H. 2006. Implications of international trade and investment agreements for water and water services: some responses from other sources of international law, Social Vision of Water in the Andes. http://www.aguavisionsocial.org.

Menheere, S.C.M. and S.N. Pollalis. 1996. Case Studies on Build Operate Transfer. 1. Build Operate Transfer 2. Case Studies 3. International Project Management, Project Management and Real Estate Development Delft University of Technology, p. 8.

Ohemeng, F.K. and J.K. Grant. 2008. When markets fail to deliver: an examination of the privatization and de-privatization of water and wastewater services delivery in Hamilton, Canada. Canadian Public Administration 51(3): 475–499.

Osborne, D. and T. Gaebler. 1993. Reinventing Government: How the Entrepreneurial Spirit is Transforming the Public Sector. Plume, New York.

Pamuk, S. 2008. Economic change in twentieth century. Is the glass more than half full? *In*: R. Kasaba (ed.). Cambridge History of Modern Turkey. New York, Cambridge University Press. 266–300.

Peterson, E.L. 2004. US water rights-holders sue Mexico under NAFTA. Investment Law and Policy Weekly News Bulletin, 1 June, IISD.

Salman, M.A.S. 2006. International water disputes: a new breed of claims, claimants, and settlement of institutions. Water International 31(1): 2–11.

Scheumann, W. 2003. The Euphrates issue in Turkish-Syrian relations. pp. 745–760. *In*: H.G. Brauch, P.H. Liotta, S. Marquina, P.F. Rogers, M. El SayidSelim (eds.). Security and Environment in the Mediterranean, Conceptualizing Security and Environmental Conflicts. Springer, Berlin.

Schmidt, K.M. 1996. The cost and benefits of privatization: an incomplete contract approach. Journal of Law, Economics and Organization 12(1): 1–24.

Sclar, E.D. 2001. You Don't Always Get What You Pay For: The Economics of Privatization. Cornell University Press, Ithaca, New York.

Sclar, E.D., K.H. Schaeffer and R. Brandwein. 1989. The Emperor's New Clothes: Transit Privatization and Public Policy. Economic Policy Institute, Washington, D.C.

Smith, A. 2000. Public Private Partnerships, The Government's Approach (forward by the Chief Secretary) published with the permission of HM on behalf of the Controller of Her Majesty's Stationery Office.

Starr, P. 1987. The limits of privatization. Proceedings of the Academy of Political Science 36(3): 124–137.

Swyngedouw, E. 2005. Dispossessing H_2O: the contested terrain of water privatization. Capitalism Nature Socialism 16(1): 5.

Tekeli, I. and S. Ilkin. 1977. 1929 Dünya Buhranı'nda Türkiye'nin İktisadi Politika Arayışları, Ankara: Orta Doğu Teknik Üniversitesi.

Tezel, Y. 1986. Cumhuriyet Döneminin İktisadi Tarihi (1923–1950), 2nd edn. Ankara: Yurt Yayınları, pp. 98–106.

Turkish Daily News. 2007. Rivers to be privatized as a solution to water crisis, 1 August.

TZ 14.09.2007, New Minister Objects to Privatization of Water Resources available at http://todayszaman.com/tz-web/detaylar.do?load=detay&link=122100.

Vining, A.R. and D.L. Weimar. 1991. Government supply and government production failure: a framework based on contestability. Journal of Public Policy 10(1): 1–22.

Vining, A.R., A.E. Boardman and F. Poschmann. 2006. Public-private partnerships in the US and Canada: There are no free lunches. Journal of Comparative Policy Analysis 7(3): 1–22.

Wall Street Journal. 2003. The World Bank as privatization agnostic, 21 July.

World Bank. 2003a. Infrastructure Action Plan, 8 July, available at www.worldbank.org/infrastructure/files/InfrastructureActionPlan.pdf.

World Bank. 2003b. Implementing the World Bank Group Infrastructure Action Plan, DC2003-0015, 13 September, available at http://siteresources.worldbank.org/DEVCOMMINT/Resources/Fall-2003/DC2003-0015(E)-Infrastructure.pdf.
World Bank. 2004a. Reforming Infrastructure Privatization, Regulation, and Competition, World Bank/OUP, New York, NY.
World Bank. 2004b. Credible regulation vital for infrastructure reform to reduce poverty. Press release, 14 June, available at http://www.cefe.net/forum/CredibleRegulation.pdf.
Yeldan, E. 2006. Neoliberal global remedies: speculative-led growth to IMF-led crisis in Turkey. Review of Radical Political Economics 38(2): 193–213.

9

Challenges for Management of the Orange/Senqu River Basin

Caroline A. Sullivan

SYNOPSIS

The Orange river is one of the most important freshwater systems draining the Southern part of Sub-Saharan Africa. Transboundary as a result of both geography and politics, the river basin is shared between Lesotho, where it rises, South Africa, its greatest user, and two other riparian states, Botswana and Namibia. Flowing from the Maloti Mountains in Lesotho, where the river is known as the Senqu, high levels of convectional rainfall feed the thousands of waterways and wetlands in the high altitude uplands, where the river has carved a basin covering almost one million square kilometres, one of the largest in the world.

Inevitably, in an area this size, there is huge diversity of geographic, hydrological, ecological and socio-cultural conditions, placing great challenges on institutional arrangements for the management of the river and its tributaries. Varying political and economic drivers have given rise to great diversity in the way water is used across the basin, but without doubt, its waters have driven economic development for the region as a whole. Rainfall variability is a constant challenge across the basin, but at the international scale, ground-breaking transboundary cooperation has provided both water security for the industries of South Africa, and economic security for the tiny mountain Kingdom of Lesotho.

School of Environment, Science and Engineering, Southern Cross University, NSW, Australia.
Email: caroline.sullivan@scu.edu.au

Extraction of natural resources from the Orange basin has generated much of the financial wealth of the region, while inevitably depleting its natural capital. Large scale water abstraction for both agriculture and mining has created tensions between sectors and communities. Political influence and legacy issues have resulted in social tension, and climate change is highlighting the regions' vulnerabilities, and the need for better water sharing.

Massive variability in socio-economic condition is evident across the basin, with the majority of inhabitants living at low levels of income, in spite of the wealth that the river generates. This lack of equitable distribution of income is also reflected by inequitable distribution of water resources, and the range of water related vulnerability in different parts of the basin.

Keywords: Transboundary water management, Orange-Senqu basin, IWRM, Water Vulnerability Index, water poverty, water values, ORASECOM

1 Introduction

The Orange River Basin is a vast and complex river system draining the whole of the southern part of the African continent. In many areas, it is the key generator of accessible water for millions of people in Southern Africa, and for their industries and livelihoods. Rising in the Lesotho Highlands some 3,300 meters above sea level, the Orange, illustrated in Fig. 9.1, is known at that point as the Senqu. Fed by its many tributaries, the Senqu flows across extensive upland wetland systems and through the steep gorges and rocky crags of the upper Senqu basin. Reaching the flatter 'Lowveldt', it flows westwards into South Africa, known as the *Gariep* or Orange, from that point on (Diederichs et al. 2005; Earle et al. 2005).

The whole Orange River system drains a catchment area of 896,368 km^2, with a total length of 2,300 km. In a basin the size of the Orange, there are a number of important tributaries, each being a sizeable river in its own right (DWAF 2004). Of particular note are the Caledon, and the Vaal Rivers, in South Africa, the Fish, and the more ephemeral Nossop and Molopo Rivers in Botswana and Namibia. Although the flow of the ephemeral Molopo is now blocked by desert dunes from the Kalahari, it forms a sizeable border between Botswana and South Africa. The Caledon forms the Northern and Western border between South Africa and Lesotho, flowing first through wetlands and then on through productive pastoral and arable land. The Orange itself provides the border between Nambia and South Africa (NMET 2002).

There is a very diverse human population of 19 million people distributed across the Orange basin. While South Africa is by far the most

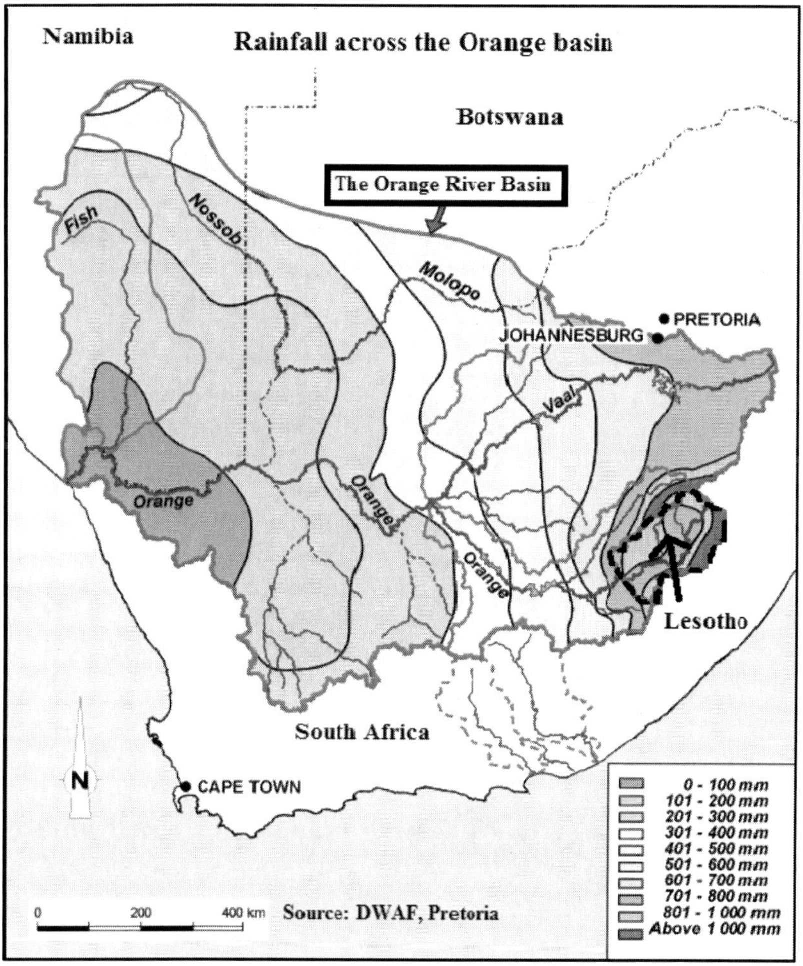

Figure 9.1. The Orange basin and its rainfall variability.

Color image of this figure appears in the color plate section at the end of the book.

populous, but within each country there is great diversity of population distribution, with multiple tribes and cultural groups being strongly represented. This diversity is further enriched by immigrants from many other African countries, as well as from elsewhere (Biggs et al. 2004). These migrants often supply cheap labour, especially in mining areas. Much of this mining labour force may be from countries within the basin, and this supports a significant level of remittances to families left at home. However, structural adjustment within the regional and world economy, and the continued impact of HIV-AIDS, means that this mobile population puts pressure on municipal social and health services (Ashton and Ramasar 2002).

In situations of such social diversity, it is often very difficult to achieve a genuine representation of stakeholders (de Coning and Sherwill 2004). What often happens is that both government and non-government organisations end up being identified as the stakeholders in major decisions, and this often leads to disenfranchisement of weaker social groups. Such diversity of population can create challenges in any river basin, and in one as large and as complex as the Orange, this is certainly the case (de Jonge Schuermans et al. 2004; Mysiak et al. 2010).

In terms of economic and political influence, South Africa is the powerhouse. Its economic might and related political power have been important drivers of regional development. There is no doubt that its impact not only includes areas of the wider Orange basin, but also in other large river basins in the region, such as the Limpopo and the Zambezi (Ashton and Turton 2004; Conley and van Niekerk 2000). With its apartheid history and continuing disadvantages for the majority of people, the largest country in the Orange basin is far from socially or politically stable (Turton 2003a). Any attempt to negotiate over water even within South Africa is fraught with tension, and in the semi-arid environment most communities have to contend with, water is never far from people's thoughts. When the diversity of views from the other riparian states is brought in, the problem becomes even more of a challenge (Falkenmark 1989; Solomon and Turton 2000).

2 Where does the Water in the Orange River Come from?

Within any large basin, spatial variability means that some places generate more water as run-off, than others. By comparing the size of these areas, with their specific run off rates, we can understand more about which parts of the basin are generating water benefits (run-off and river in-flow), and which parts are not (McKenzie and Craig 2001; Sene et al. 1998). A comparison of the area of the different parts of the basin, and the rate of runoff experienced there, is summarised in Table 9.1.

From Table 9.1 it is interesting to note that while the Senqu and its tributaries only include 2% of the area of the basin, that small area catches as much as 34% of the total run-off. Similarly, the Caledon covers just 2% of the catchment area, but records 11% of the total runoff. This is said to be the reason for the region being described as the 'water tower of Southern Africa' (Diederichs et al. 2005).

From an economic perspective, the most important tributary of the Orange is the Vaal, providing water for the industrial heartland of Gauteng, and the agricultural demand from the Free State and Mpumalanga. Another

Table 9.1 Comparing area and run-off in different parts of the Orange basin.

River section/tributary	% of Orange Basin Area	% of Total Run-off	Run-off per Area/ Tributary (Mm^3/a)
Vaal River basin	19%	36%	4,300
Senqu River to Lesotho/SA border	2%	34%	4,010
Caledon River Welbedacht Dam	2%	11%	1,240
Orange upstream to confluence with the Vaal	6%	11%	1,300
Lower Orange excluding Fish River basin	64%	4%	420
Fish River basin	7%	4%	480
Total Basin	100%	100%	**11,750**

major tributary is The Fish River, which drains a large part of Southern Namibia (Heyns 1995; NMET 2002). At its confluence with the Orange 112 km from the Atlantic, the Fish rarely currently delivers surface flow to the Orange, although it may indeed do so under different climatic conditions. There are also several other large tributaries which only flow during short periods of the year.

The Vaal is a very heavily used river. Almost 50% of the conventional electricity generated in the whole of Africa comes from the water of the Vaal, which meets 80% of South Africa's energy needs. In addition to this, its water supports production in some of the world's largest platinum and gold mines and huge coal reserves. As a result, very little water flows from the Vaal tributary into the Orange, except during flood events (Diederichs et al. 2005).

3 Water Use in a Transboundary Context

As is the case in many transboundary river systems, the way water is used varies considerably across the basin as a whole. Nevertheless, overall, agriculture is by far the biggest water user, accounting for as much as 80% of total water use.

Table 9.2 indicates how South Africa uses by far the most water from the basin, even after system losses through evaporation and environmental requirements are taken into account. It also indicates that just 23.5% of available water in the catchment is available for uses other than those already accounted for.

Withdrawing a total of 63% of freshwater available in the basin, South Africa is by far the largest water user in the Orange basin. Namibia, with its mining interests, comes in second, but its use amounts to only 1.3% of available water in the basin. Lesotho, with its plentiful supplies, only uses

Table 9.2 Who uses water from the Orange?

Orange River Basin	Quantity of Water (Mn m³ p/a)	% Resource Used/Available
Mean annual run-off	11,500	
System losses (evaporation)	–3,000	
Resource potential	*8,500*	*100%*
Less present utilisation		
Botswana	0	0%
Lesotho	–20	0.20%
Namibia	–110	1.30%
South Africa	–5,370	63%
Environment	–1,000	12%
Total utilisation	*–6,500*	*76.50%*
Remaining potential	**2,000**	**23.50%**

0.2% of the water of the Orange, and with just subsistence abstraction by a few small communities located in the Orange basin, Botswana uses a negligible amount nothing at all, at present (Diederichs et al. 2005).

In South Africa, along the main stem of the Orange River, irrigated agriculture dominates. In the Vaal tributary, mining and industry provide the biggest pressures, compounded by the inevitable[1] pollution from growing urban and peri-urban populations (Biggs et al. 2004). Groundwater is important, both in South Africa and in the other riparians (Bohensky et al. 2004). Important for domestic use in rural areas across the basin, groundwater resources in the northern part of South Africa, are fully allocated to existing use. Across the entire basin, water stress is widespread. According to the records of South Africa's Department of Water Affairs and Forestry, water stress is regularly reported in over half the water management areas, including those in the Orange basin (DWAF 2004).

In contrast to the whole of the Southern African Development Community (SADC)[2] region, Lesotho is a land benefitting from significant water resources, but ironically, in the capital Maseru, water stress is becoming a more frequent occurrence, due to population growth (Chakela 1997; Vörösmarty et al. 2010). While the water resources originate in the highlands, agricultural opportunities and industrial growth are occurring in

[1] According to the laws of thermodynamics, all life generates some form of pollution (waste).

[2] The Southern African Development Community comprises 15 member states: Angola, Botswana, Democratic Republic of Congo, Lesotho, Malawi, Mauritius, Mozambique, Namibia, Seychelles, South Africa, Swaziland, Tanzania, Zambia, Zimbabwe and Madagasgar whose membership currently suspended after the coup d'état led by the former mayor of Antananarivo Andry Rajoelina.

the lowlands. This requires various forms of water transfer to be considered, including the construction of the Metolong Dam. A recent development has been the opening up of the Letseng diamond mine, some 3,200 metres above sea level, which has started producing some of the largest diamonds in the world. This development, and potential pollution associated with it, will put additional stress on water resources at the local scale (Salman 2004).

Delivering water to the lowlands of Lesotho must be seen as an essential outcome of the Lesotho Highlands Water Scheme (Alexander 2005), as this is where the national beneficiaries of the project reside, as opposed to downstream in South Africa where beneficiaries are largely multinational mining or industrial corporations, or global agri-businesses. Very little water is used consumptively in Lesotho, and there is much scope for further development of water use for income generation, including electricity (Hoover 2001).

Botswana is a highly stressed water area, with demand doubling by about 2020, and increasingly likely to outstrip supply. Being hydraulically connected to the Orange basin, Botswana does have the potential to draw on this source if economic and social pressures could mobilise the required resources. To date there is little population development within the Botswana part of the basin, and no significant industrial development, although increasing interest in the Kudu gas field power generation is likely to change this in the future. Similarly, Namibia is highly dependent on groundwater, is significantly water stressed in some areas and has potential to demand additional water from the Orange basin (de Jonge Schuermans et al. 2004; Diederichs et al. 2005).

4 Water Poverty in the Orange Basin

A useful way to investigate the comparative state of water resource development and its impacts in a country is through the application of the *Water Poverty Index* (Sullivan 2002; Sullivan et al. 2003). This is a composite index approach taking into account five core dimensions of what makes up integrated water resources management. An example for the four riparian countries of the Orange is shown in Fig. 9.2.

Amongst a range of indicators, the calculation of the Water Poverty Index takes account of the population requirements for water. This is particularly clear when considering the values for *resources*, which also takes account of rainfall variability. From analysis provided by the Water Poverty Index, we can see that Lesotho (with a score of 43.2 index points)[3]

[3] These scores are based on the comparative scale of 143 countries across the world, see Sullivan et al. 2002.

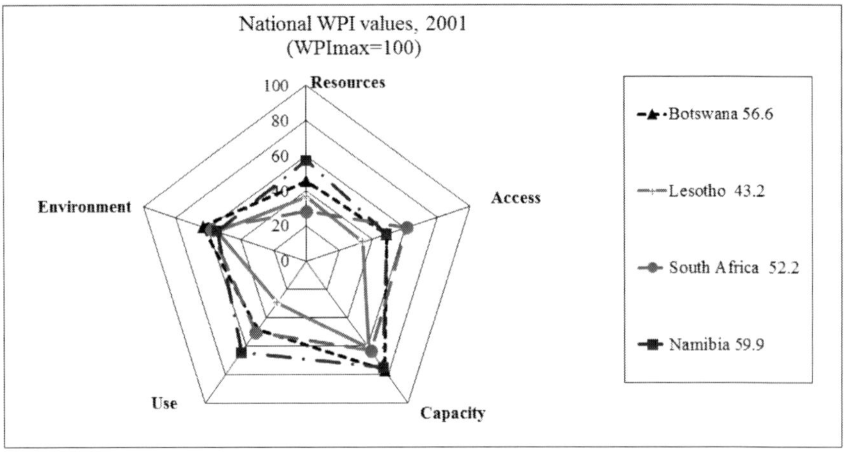

Figure 9.2. Comparing *Water Poverty* in the riparian states of the Orange basin.

Color image of this figure appears in the color plate section at the end of the book.

is the lowest scoring country in the group (and is thus the one with greatest water poverty). This is mostly due to very low access to water, low capacity to manage it, and inefficient use.

Due to their higher scores on capacity to manage, and efficiency of use, coupled with their low populations, both Namibia and Botswana score relatively well on this WPI measure. South Africa, in spite of its affluence and considerable infrastructure, scores significantly lower, due to its higher population, and its lower levels of capacity and use. It is notable, however, that access to water is actually better in South Africa than in the other riparians (mainly due the large volume of irrigation provision).

5 Agriculture Across the Orange Basin

Irrigated agriculture is of vital importance to the South African region as a whole (Salman 2004). Within the Orange basin, crops produced vary, with highest value crops (citrus, table grapes, pecans) being produced in the lower Orange basin, and cereals and fodder crops further upstream where water (thanks to irrigation and water transfers) is less of a limiting factor. Much small scale subsistence agriculture may be irrigated by hand, but the biggest volume of water use is taken by commercial-scale irrigated agriculture, as illustrated in Fig. 9.3 (DWAF 2004; Namibia MET 2002).

In other parts of the basin, irrigated agriculture is of less importance. In Lesotho, irrigation potential is limited by soils, slope geology, etc., and just 7% of the total land area is cultivated, with just over 1% being irrigated land. Main crops are maize, potatoes and other vegetables, and sheep and

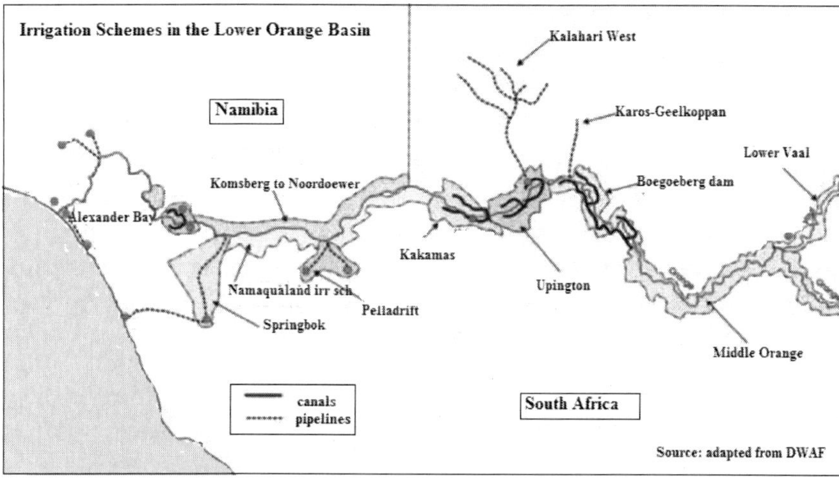

Figure 9.3. Irrigated agriculture in the South African part of the Orange basin.
Source: DWAF

Color image of this figure appears in the color plate section at the end of the book.

cattle are important, especially in highland communities where cropping is more difficult. In Namibia, due to geological and geographical factors, very little water is available from the Orange basin, and what little water used is used for high value crops Grapes. Nevertheless, its recognition as a riparian state gives it legal access to water resources from the Orange River (NMET 2002), which it is likely to develop in the future, as discussed in Section 10.

Like Namibia, Botswana is a water stressed region. It has highly variable ephemeral water resources and relatively unreliable groundwater sources. Little water is available for use in irrigation, and agricultural water use tends to be used for livestock. Cattle and goats are important for subsistence households, but this gives rise to serious problems of soil erosion. Being part of the Orange basin, it is likely that Botswana will also in future seek to receive some share of water from the basin, and being a member of ORASECOM, this is likely to be tabled in transboundary discussions in the future.

6 Industry and Mining in the Orange Basin

South Africa, Namibia and Botswana all generate the vast proportion of their wealth from the exploitation of geological resources such as zinc and diamonds. Within the Orange basin, only Lesotho is not well known for its mining operations, although this is now changing. In addition to mining operations, industry is also important in some parts of the basin. As a result of the economic and trade sanctions of the apartheid era, South Africa has a highly diversified economy, with the full range of industries present including engineering, food processing, textiles, car manufacturing, steel and aluminium plants, etc. Indeed, in many ways, Johannesburg is the economic engine of the whole of Africa, and it is almost completely dependent on water from the Orange River. Industry in Botswana and Namibia is limited and mostly related to mining (Biggs et al. 2004).

Electricity generation is an important activity in the basin, earning significant income for Lesotho and South African companies. However, in recent years, low rainfall conditions have given rise to a number of 'brown-outs' as a result of energy supplies not keeping up with demand.

In Lesotho, textiles and other forms of light industry development have become important, supported by large commercial investments from Taiwan and China. Many of these investments had been driven by the existence of the WTO Multi-fibre agreement, which allowed textile products from low income countries to benefit from negligible tariffs in rich markets. When the Multi-Fibre Agreement was brought to an end in 2009, a number of these operations closed down, creating unemployment in peri-urban areas of Maseru.

These industrial and mining activities create a strong demand for water resources, and many also have a significant polluting impact. During the WTO Multi-fibre agreement period, rapid growth in the textile industry in Lesotho, under conditions of low or poorly enforced environmental legislation, led to some significant pollution events which had knock-on effects downstream. The transboundary pollution incidents brought about external pressure for change in environmental legislation in Lesotho.

Ecotourism has evolved in several parts of the basin, ranging from pony trekking in the wilds of the high Maloti mountains of Lesotho, to the bird watching and wild flower viewing down in the lowlands towards the mouth. Within many parts of the basin, traditional livelihoods continue to dominate, with livestock being the cultural mainstay (Chakela 1997).

7 Water Storage Across the Orange Basin

To support both agricultural and industrial development, water storage is essential at all scales of human endeavour. To address the storage needs of water for the economy of South Africa, large dams have been constructed across the upper Orange basin, as shown in Table 9.3. The two dams in Lesotho listed in the table are part of the Lesotho Highlands Water Scheme and under the jurisdiction of the Lesotho Highlands Development Corporation.

Table 9.3 Major water storage infrastructure, Orange Basin, 2008.

Country	Dam name	Storage capacity, Mil m3	% ot total basin storage	Key functions
South Africa	*Gariep*	5,675.00	28%	Hydropower, transfer to Fish River, flow regulation, irrigation
	Vanderkloof	3,237.00	16%	Hydropower, flow regulation, irrigation
	Sterkfontein	2,617.00	13%	Hydropower, receives transfers from Tugela
	Vaal	2,122.00	10%	Supplies Gauteng industrial complex
Lesotho	*Katse*	1,950.00	10%	Main collector for transfer to Vaal for irrigation
	Mohale	938.00	5%	Transfer to Katse and then to Vaal
Total for these 6 dams		16,539 M m3	82%	

In the other riparian states of the Orange basin, there has been much less development of large scale water storage, as this type of infrastructure is usually driven by large scale industry, agriculture or energy generation, none of which exists to the same extent in any of the other countries. Furthermore, the meteorological conditions do not lend themselves to efficient above-ground storage.

8 The Importance of Wetlands and Biodiversity in the Orange Basin

The whole region of Southern Africa is an arid or semi-arid area characterised by scrubland vegetation, characteristic *Acacia* landscapes and mombio woodlands. In the Orange basin itself, rainfall can be below 100 mm per year in the lower Orange, or over 1000 mm per year up in the Lesotho Highlands. Temperatures across the basin are equally variable, with high diurnal variation as well as variation due to altitude. In spite of the well-known threats to key iconic megafauna such as the Rhino, and tragic stories of whole herds of elephants being sacrificed for their ivory tusks, wildlife and biodiversity in Southern Africa remains rich and globally important. Wetlands across the Orange basin play a vital role in water purification, water storage and water delivery, and their better management is vital to water security (Hirji et al. 2002; McKenzie and Craig 2001; Leiman et al. 2000).

9 Environmental Threats in the Orange Basin

There are numerous environmental threats in the Orange basin, but these are different, in both their nature and their impact, in the upper, middle and lower parts of the basin (Sullivan and O'Keeffe 2011). Pollution from mining is a continuing concern for regulatory authorities, although across the basin the strength of these institutions is quite variable. The non-point source pollution associated with large scale agriculture has had an impact in some areas, particularly along the mainstem of the Orange River itself, while conventional point sources such as Waste Water Treatment Plants and chicken farms have created pollution hotspots in some parts of the basin, particularly in the Vaal. This tributary of the Orange also has to contend with large scale industrial and other municipal effluent.

Significant levels of soil erosion occur in the upper part of the basin, both on the steep slopes of the Maloti Mountains in Lesotho, and on the uplands of South Africa where plantation forestry has been important (Rooseboom 1983; Sene et al. 1998). As a consequence, sediment loads are high in some places, threatening wetlands and waterways, as well as creating problems for water management infrastructure.

In both Lesotho and South Africa, there are thousands of wetlands of several different types, from upland seeps and permanently flooded marshes, down to the dry salt pans and estuarine wetlands at the river mouth, which is one of the 20 internationally recognised Ramsar sites in South Africa (Nel et al. 2004). In 2003, concerns were raised about the condition of the Orange delta, as a result of poor water quality, and more importantly, such low flows that the nature of the delta was under threat (Anderson et al. 2003). This has brought more pressure to reduce pollution impacts and regulate flow reduction, and while some licencing measures have been successful, the sheer pressure of population impacts in some areas continues to be an issue, particularly in drought conditions.

In the upper part of the basin, wetlands perform an important storage function, and are responsible for delivering good quality water downstream. Other benefits from wetlands are numerous, ranging from food and fibre provision for subsistence households across the basin, livestock watering, as a source of medicinal plants, and of course as biodiversity habitats and recreational sites. Better conservation of soil and water across the basin will help to protect wetlands from sedimentation, but many are largely held in private hands, and are often converted to other uses. One way in which wetlands may potentially be protected is through increasing use of schemes for *Payments for Ecosystem Services* (PES).

10 Water Governance in Southern Africa and the Orange Basin

In the context of the river basin, negotiations between the riparian states about water in the wider region has been organised under the umbrella of the Southern African Development Community (SADC). The goal of SADC was to promote integrated regional development, "on the basis of balance, equity and mutual benefit of all states" (SADC 2007). South Africa joined SADC in 1994.

Due to the critical role of water in any nation's development, water can be seen as a determinant of prosperity. In relation to the Orange basin itself, a river basin authority has been established known as the Orange-Senqu River Commission (ORASECOM 2000), made up of the four riparian states, Lesotho, South Africa, Botswana and Namibia. ORASECOM came into existence in 2000, through multilateral agreements between Lesotho, South Africa, Namibia, and Botswana. These four riparian states in the Orange basin all benefit, to varying degrees, from the river.

Being the first whole basin commission in southern Africa, ORASECOM was a major step forward for the region, and its legislation took into account the revised SADC Water Protocol on Shared Rivercourses (SADC 2011a). Due to its higher levels of economic development, South Africa has historically had significant political power over how water from the basin is used, but Namibia and Botswana (with their important mining interests), ensure their riparian rights are formally recognised through membership of ORASECOM, for strategic reasons. Lesotho, the key water provider in the basin, is also a partner in ORASECOM.

ORASECOM is the forum in which issues such as benefit sharing can be discussed, along with other technical issues. All the riparian states participate as members of ORASECOM, but operational water management issues are implemented at the country scale by the relevant national water ministries in each country. Most of the activities of ORASECOM are of a technical nature, through the Technical Advisory Committee, but its overall goal of balanced economic development is supported at both this technical level, and also at the political level through ministerial representation. If agreement cannot be reached on technical solutions to specific issues, discussion reverts to political negotiation, and this comes under the jurisdiction of the accepted instruments of International Water Law (Sullivan and Fisher 2011).

In addition to its membership of ORASECOM, Lesotho is involved internationally in bi-lateral agreements with South Africa, through the Lesotho Highlands Water Commission, and the Lesotho Highlands Development Authority (LHDA). For Lesotho, the upstream water collector, large amounts of foreign exchange have been earned as a result of their water storage agreement with South Africa. Benefit sharing from water storage, and the potential from the payment of ecosystem service charges, have, and will continue to have, a significant impact on the financial wellbeing of the country.

One area of new legislation which arose from the extensive impact assessment studies associated with the construction of the huge Lesotho Highlands Water Scheme, concerned the recognition of the need to allocate water for the environment. It is notable, that since the establishment of this transboundary water scheme, many other countries, basins and regions have adopted a more formalised approach to water allocation for the environment.

At the national scale, each riparian state has a national Department of Water Affairs, each implementing its own water laws and water policy. At the local level, Local Government Authorities (LGAs) have the responsibility to operationalize the law. Countries are divided into water management units, and regulatory issues for each area are implemented through both the ministerial departments and the local governments who oversee planning regulations and municipal water use. Other government agencies such as agriculture ministries or irrigation departments also have some jurisdiction over water systems, and are important stakeholders with significant influence over land use, water regulations, etc. Water Utilities and Water User Associations are important groups influencing both the resource and its use. Local customary rights are also significant in some areas (Kranz et al. 2005).

At the national scale in Lesotho, building on the Water Resources Policy first approved in 1999, the 2001 Lesotho Environment Act provides the basis for much water legislation, particularly relating to water quality, including procedures for pollutant and effluent discharge licensing. In addition to giving the government authority to 'ensure rational exploitation and management of Lesotho's water resources', and responsibility to 'ensure access to potable water by all people of Lesotho' this national policy also mandates that 'all environmental aspects of water will be protected'. In addition to these 3 core principles, the policy guarantees free potable water for basic human needs, but then creates a fee system for payment for water consumption rates above that amount. At the local scale in Lesotho, agencies such as Water User Associations and LGAs are now obliged to comply with the national legislation, but they can still retain some independence through local by-laws (Chakela 1997).

In contrast with Lesotho, Namibia is a dry, desert country depending largely on groundwater. The Water Act of 1968 which embodied the common law aspects of water, and the 1962 Waterworks Act, and its 1983 Amendment, have all attempted to regulate water resources as the country has developed. The establishment in 1968 of the Water Apportionment Board, under the jurisdiction of the Department of Water Affairs, has been responsible for all water licensing and regulation. Other, later, health and environmental acts have also supplemented this water governance process (NMET 2002).

In Namibia, this decentralised policy gave the right (and responsibility) for any community to determine which solutions and service levels they wanted (for example through Water-Point Committees), subject to resource and environmental constraints. Water Utilities Corporations were given legal recognition under the 1970 Water Utilities Corporation Act, which put in place rules on their structure and function. As part of the government decentralisation process which gained ground in the 1990s, the Namibian Water Resources Management Review was carried out in 1997, to recommend a more appropriate institutional structure for the water sector, and institutional change is slowly happening, especially in recognition of the need for an allocation for an *environmental reserve*, as is the case in South Africa.

In South Africa, the National Water Resource Strategy (NWRS) provides the foundation for water management, with three main objectives:

- To achieve equitable access to water,
- To achieve sustainable use of water, and
- To achieve efficient and effective use of water for maximum social benefit.

With water now included within the Constitution of South Africa "Everyone has the right to have access to sufficient water (Section 27 (1) b), popular expectations put pressure on the government to deliver. This has been a challenge, and although DWAF has focussed on implementation, demographic growth has made water for all an almost impossibility.

Within this new legislation, water has now been shifted from private to public ownership, with water becoming a public good managed for people by the government. The government forms the law and implements policy. Water ownership is no longer embodied in riparian rights. Water management itself has been devolved to the local governments, in cooperation with the Catchment Management Agencies (CMAs). For water management purposes, the Orange basin in South Africa has been divided into 5 Catchment Management Agencies (CMAs) stretching thousands of kilometres from the border with Lesotho, down to the delta in the Lower Orange (de Coning and Sherwill 2004).

At the local level, municipalities, Water User Associations and Irrigation Committees control water use through formal Agricultural Water Management Plans, or Water Services Development Plans, all of which are overseen by the Department of Water Affairs and Forestry (DWAF). A Water Conservation and Water Demand Management Strategy has been put in place to support this process. This has enabled the development of a useful data record to support national and local policies, including those to develop Integrated Water Resources Management (IWRM) (Pegram and Mazibuko 2003).

National and water statistics have been kept to a very high standard in South Africa, so significant robust data is generally available across the country, including the South African part of the Orange basin. How such data can be used is illustrated through the measurement of Water Vulnerability, in Section 11. At the international scale, DWAF has put in place an International Liaison Directorate to comply with the now legally mandated requirement of 'achieving fair arrangements with neighbours who share our rivers'.

The fourth riparian state in the Orange basin is Botswana, a large dry landlocked country, keenly aware of its own water stress status. While having developed a National Water Master Plan in the 1990s, the country has continued to legislate on water resources and their management, and to make amendments to this legislation for over a decade. Communities are actively engaged in water development and management at the local scale, through District Councils, and for industrial scale users, alternative technologies such as desalination and water reuse are important developments. At the international scale, Botswana has been involved in building interconnecting water transfer schemes with various neighbouring countries, but in terms of its involvement in the Orange, participation in ORASECOM so far has been mainly for strategic reasons. Botswana has also participated actively through the various SADC water protocol meetings (SADC 2011a).

In addition to the various agreements within ORASECOM, other current bilateral agreements are also in place in the Orange, between South Africa and Lesotho, in the form of the Lesotho Highlands Water Scheme. The treaty to bring into existence the transboundary Lesotho Highlands Water Project was signed in 1986 and from this was formed the Lesotho Highlands Development Authority and the Joint Permanent Technical Committee. This was made up of representatives from the 2 involved riparians (South Africa and Lesotho). In 1999, this committee was reformed as the Lesotho Highlands Water Commission (Conley and van Niekerk 2000; Ashton and Turton 2004).

While this arrangement of international trade in water storage is currently mutually beneficial for the two partners, it is possible that future decisions about water allocations could be different. Should Botswana and Namibia consider gaining greater access to their perceived share of water from the Orange basin, transboundary pipelines from the Lesotho Highlands could be their approach water sharing in the future. While this of course would impact on South Africa, it would have little ground for objection, since South Africa itself has already constructed such a transboundary water transfer. Perhaps when the current agreement with South Africa comes to an end, Lesotho will be able to sell its storage potential to the highest bidder, putting South Africa's water security at risk. It is likely however that this can be avoided through reallocations to Boswana and Namibia, via multilateral arrangements on other transboundary rivers such as the Limpopo and the Zambezi, in which South Africa is a partner.

11 Core Challenges for the Orange Basin

The pressures on water resources from both demand and supply are significant in most parts of the Orange River basin. Clearly, the most significant pressures are in South Africa, where a rising population in a highly industrialised nation makes heavy demands on the water system. From a supply perspective, there is also a major challenge, since this part of the world is very much an arid and semi arid zone. As a response to supply side pressures however, water departments in the region have been very active in trying to secure an adequate level of supply, particularly in South Africa, where resources from large scale commercial interests have supported capital investments in infrastructure, or mobilisation of loans where needed.

Within the South African part of the basin, a complex analysis has been done to examine the pressures on water resources. The results demonstrated that there is clear diversity of water stress conditions, as measured by the *Water Vulnerability Index* (Sullivan 2010), and these conditions can arise from a variety of interacting driving factors. By understanding these drivers more clearly, we can be better equipped to develop appropriate and effective responses.

Based on a variety of indicators of both demand and supply, these pressures are summarised for several contrasting municipalities in Fig. 9.4.

From the data shown in Fig. 9.4, some examples may be noted for illustration. In the case of Golden Gate, the overall level of water vulnerability is relatively low, compared to other LGAs, but the pressure that does exist arises mostly from the supply side, rather than demand. This is not surprising, given the sparsely populated dry landscape which characterises it. In contrast, with its dense population and high level of industrialisation, the City of Johannesburg demonstrates high levels of both supply and demand related drivers of vulnerability. Information such as this has been plotted for all 87 municipalities of the basin, and is displayed in Fig. 9.5. A location map and list of the names of these municipalities is provided in the Appendix 1.

From Fig. 9.5 we can see that the places where water vulnerability is high (red coloured areas), compared to other areas where water related

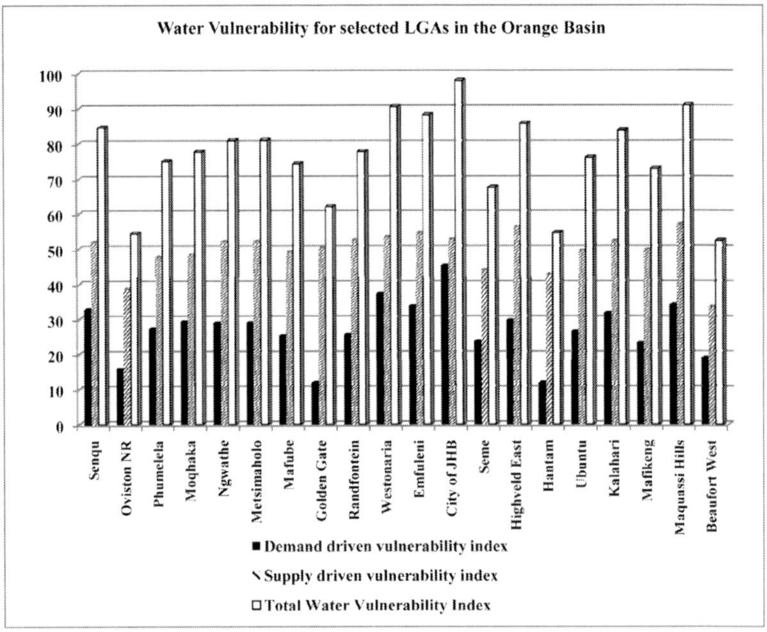

Figure 9.4. Components of the *Demand* and *Supply* side drivers of water vulnerability.

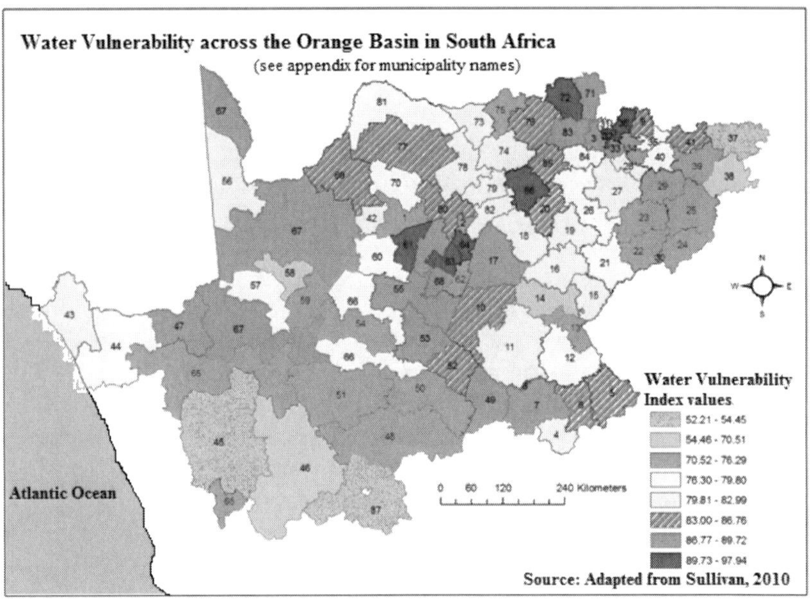

Figure 9.5. Water Vulnerability across the South African part of the Orange basin.
Source: Sullivan 2010

Color image of this figure appears in the color plate section at the end of the book.

vulnerability is less. We can then look at the detailed values contributing to the various components[4] of water vulnerability illustrated by Fig. 9.4, which can then verify the causes of that site-specific vulnerability.

12 Potential Conflicts in the Orange as a Transboundary Basin

At the core of all water management challenges is the need to share water between 3 primary uses (Turton 2003b):

- For domestic use and livelihood needs (as a social good)
- As an input into large scale agricultural and industrial production (as an economic good)
- As a crucial and essential input into the environment (as an ecological good)

There is significant potential for conflict over these different water uses, even at a household scale (Solomon and Turton 2000; Salman 2004). For example, while a man may want to use the water he has brought from the well to irrigate his vegetable garden, his wife may want to use that water to bathe her child. At the village or community level, how can water be shared between households, and who has responsibility for maintenance of the water system? At the national scale, what priorities should be given to these dimensions in a large river basin with several competing users, as in the Murray Darling Basin, in Australia (Sullivan 2013). At the international scale, how can the competing needs of the huge populations upstream in the Nile basin be eclipsed by the needs of those downstream in Egypt? How can the masses of people dependent on the ecology of the Mekong basin be ignored by the needs of the industries in Thailand or indeed China?

In the Orange basin, the current water needs of the different riparian countries are clear. This does not mean however that they will always be this way. Across the basin, in addition to water for domestic and municipal use, water is an important factor of production for resource-intensive industries, and the overall level of economic development. For example, in South Africa, over 800,000 hectares of irrigated agriculture is fed from the Orange. In addition, in the lower reaches of the basin, several towns both near the river and relatively further away, receive their water from the Orange (NES 1999). This includes water for a new rural water supply project north of the River in the Kalahari. Water is also taken from the Orange for stock watering and to support tourism. In addition to the water sharing arrangements with Lesotho outlined above, South Africa has a number of other transboundary

[4] For more details on the WVI methodology, see Sullivan 2010.

water sharing arrangements, and there is little doubt that water security is high on the agenda in South African politics (Turton 2003a).

For Lesotho, regulation of the flow of the Orange through water infrastructure is the way to maximise returns on its resources. The benefit sharing achieved through the Lesotho Highlands Water Project enables South Africa to benefit from regulated flows being available when needed, while Lesotho receives royalty payments for water storage services in the Mohale and Khatse dams (Klasen 2002). It is anticipated that these payments should be distributed within Lesotho so that standard of living for people across the whole country can be raised through better infrastructure, more jobs and better health outcomes. To date, progress in this respect has yet to be fully realised (de Jonge Schuermans et al. 2004).

In the lower part of the Orange basin, Namibia is particularly dependent on the Orange River for 2,200 hectares of irrigation, which takes about one third of all water needs taken by Namibia from the basin. In addition, water from this river is used to support two large zinc mines, some diamond mines, and small scale community needs. Future needs for Namibia may include developments associated with the Kudu gas field, and further development of the successful grape growing industry that Namibia has built up through exports to the USA. While this latter development has the potential to create up to 10,000 seasonal jobs, Namibia is currently dependent on water being released from South Africa. In the future, it is likely that a dam specific to the needs of Namibia will be built in the lower reaches of the Orange, but this would have to be done in cooperation with South Africa. This has been a long issue of contention between these two countries, and remains an example of a current conflict over water in the Orange basin (SADC 2011b; NMET 2002).

Since much of Botswana forms part of the Kalahari desert, it is highly vulnerable to water stress (Heyns 1995). Although it contributes little to the run-off of the Orange, it is connected through the Molopo-Nossob River system, and current investigations through ORASECOM are looking at the opportunity to increase water supplies to Botswana through transboundary water transfers from South Africa, and Lesotho. Transfers from Lesotho directly to Gaborone are technically possible, but at present, Botswana also gets water from other international sources, notably the Zambezi and Limpopo. As South Africa is also part of those basins, is possible in the future that some form of intra-basin water trading will take place within the region as a whole, in place of specific off-take from the Orange. Since ORASECOM is the body through which such agreements can be made, it is clearly advantageous from a geopolitical perspective that Botswana is included in the Orange River basin (Ashton and Turton 2004).

13 Future Impacts on the Basin

Like every place in the world, the Orange basin will be subject to a wide variety of both anticipated, and un-anticipated impacts of global change. These will include climate-related impacts on ambient temperature and rainfall regimes, and on the number and frequency of extreme events (droughts, floods, hail, cyclones, etc.) (Arnell et al. 2003; Joubert and Hewitson 1997; Hulme et al. 2001). On top of these expected changes[5] there will also be significant changes arising from the dynamics of rapidly expanding and evolving social systems. This means it is highly speculative to consider what may lie ahead, even if basing such estimates on a *business as usual* (BAU) scenario. BAU scenarios are often seen as being the worst option, as the impact of improved governance is anticipated in better futures (as for example in the 'policy first' scenario of UNEP's Global Environmental Outlook (UNEP 2003; 2007). Various groups including DWAF have developed future scenarios to assist with water planning, and although there may be some disagreement with some aspects of the details of these, there is a general consensus that the Orange basin is in a part of the world where both economic growth and demographic change will bring about major pressures on water resources (Vörösmarty et al. 2010). The effectiveness of management responses to these pressures can be assessed using various evaluative tools, and the promotion of more adaptive management approaches will help to build the resilience that such a basin of extremes will need for a successful future. 'Softer' approaches such as more effective use of wetlands in water treatment, distributed storage and more consultative approaches to water allocation would all contribute to this process (Mysiak et al. 2010).

14 Conclusions

More effective water management is one way to foster more equitable opportunities for development, and this can be examined more explicitly using the Water Poverty Index which shows in Fig. 9.2 that in spite of living in the '*mountain water tower*' of the region, people of Lesotho have poor access to water, and gain little economic benefit from the water of the Orange basin, compared to people in the other riparian countries. For Lesotho, there is a clear need to improve water access, both for domestic use and irrigation. In terms of efficiency of water use, as shown by the *Use* component in Fig. 9.2, efforts should be made to improve returns to water use through appropriate industrial expansion based on efficient water use systems, and improved

[5] For full details of the latest available understanding of the biophysical science underpinning climate change, and how it is likely to influence Southern Africa, see IPCC (2013).

efficiency in agricultural water use. In comparison to Lesotho, both Namibia and Botswana perform relatively well on efficiency of water use, as does South Africa, with these three latter countries earning high returns from mining, in particular. The wealth generated from this mining of natural capital is being converted in part at least to human capital, as can be seen by the better performance of Botswana and Namibia on capacity, compared to South Africa and Lesotho. While South Africa has the lowest per capita level of resources, they do actually deliver the best level of access of the four countries in the basin, with Lesotho struggling to maintain improvements in access in the face of more rapid population growth. A consequence of the urgent pressure to deliver better access in the Lesotho Highlands Scheme, has been the rapid installation of latrines. In many upland areas, these are now clearly dotted across the beautiful landscape, without consideration of the concept of *convenience* for the recipient households.

Planetary boundaries (Rockström et al. 2009) are made up of the cumulative effect of all our actions, so management of the catchment system holistically within its biophysical limits including the constraints of the hydrological cycle provides the foundation for sustainable management. As such, IWRM is the key to living within our water boundaries. In the context of a hydrological system as complex as the Orange basin, it is clear that adaptive management is the key to operational success. This enables flexibility of options to be maintained, and if implemented correctly, assures continued stakeholder involvement. The consultation process as a whole, and peoples' ownership of it, is crucial to the success of water management initiatives, from a local to a transboundary scale. Local communities and authorities have an important influence in basin management, and they must be embedded in the whole process of IWRM. Transnational meetings to foster such cooperation between Local Government Authorities within transboundary basins should be encouraged.

Meaningful water accounting and valuation are also essential for more effective water management at the basin scale. Linking water use to economic efficiency is a challenge for econometricians and national statistical agencies. In the Orange basin, this is more achievable than in other areas, due to the significant capacity in the government departments responsible for water management in the various riparian countries of the basin. The results of the analysis for the Water Vulnerability Index have shown how water vulnerability arises from an interaction of both supply and demand drivers. Understanding the overall impact of this vulnerability and these interactions can suggest what explicit measures are needed, and at what location, to build greater resilience at local government level in the water sector.

Equity is the key to development success in the Orange River basin. The countries of southern Africa are amongst the most unequal societies in the world and the basin community presently consists of some of the poorest and richest groups in society, even at a world scale. Not only is there great disparity in income distribution, but there is also great inequity in access to livelihood resources, including both water and land. Without a significant attempt to address these national and regional inequities, southern Africa as a region, and the Orange basin within it, are unlikely to have a smooth development trajectory. Better water accounting systems and schemes to support payments for ecosystem services underpin future water management success. These could, potentially, not only remedy the current levels of inefficient and inequitable water use but also bring about meaningful benefit sharing across the Orange basin, and greater prosperity and wellbeing for all.

References and Further Readings

Alexander, W. 2005. The Orange River project: from dream to reality. Civil Engineering 13(6): 4–7.

Anderson, M.D., H. Kolberg, et al. 2003. Waterbird populations at the Orange River mouth from 1980–2001: a reassessment of its Ramsar status. Ostrich 74(3-4): 159–172.

Arnell, N.W., D.A. Hudson and R.G. Jones. 2003. Climate change scenarios from a regional climate model: estimating change in runoff in southern Africa. Journal of Geophysical Research 108.

Ashton, P. and V. Ramasar. 2002. Water and HIV/AIDS: some strategic considerations in Southern Africa. Hydropolitics in the developing world: a Southern African perspective. A. R. T. R. Henwood. African Water Issues Research Institute (AWRIU): 217–235. Pretoria.

Ashton, P. and A.R. Turton. 2004. Transboundary water resource management in southern Africa: challenges, opportunities and lessons learned. Water development and cooperation.

Biggs, R., E. Bohensky, P.V. Desanker, C. Fabricius, T. Lynam, A.A. Misselhorn, C. Musvoto, M. Mutale, B. Reyers, R.J. Scholes, S. Shikongo and A.S. van Jaarsveld. 2004. Nature supporting people: the Southern African Millennium Ecosystem Assessment. Council for Scientific and Industrial Research, Pretoria, South Africa.

Bohensky, E., B. Reyers, et al. (eds.). 2004. Ecosystems Services in the Gariep Basin. Sun Press, Stellenbosch.

Chakela, Q.K. (ed.). 1997. State of the Environment in Lesotho. National Environment Secretariat (NES), Ministry of Environment, Gender and Youth Affairs, Maseru, Lesotho.

Conley, A.H. and P.H. van Niekerk. 2000. Sustainable management of international waters: the Orange River case. Water Policy Journal 2(1-2): 131–149.

de Coning, C. and T. Sherwill. 2004. An assessment of the water policy process in South Africa (1994 to 2003). Water Research Commission 1–57.

de Jonge Schuermans, A.M., J. Helbing, et al. 2004. Evaluation of Success and Failure in the International Water Management Orange River Basin, South Africa. Zurich 19.

Department of Water Affairs and Forestry. 2004. Internal strategic perspective: Orange River system. Pretoria, Department of Water Affairs and Forestry 1–99.

Diederichs, N., D. O'Regan, C.A. Sullivan, M. Fry, M. Mander and C. Haines. 2005. Orange River basin baseline assessment report. NeWater Project Report 3.8.

Earle, A., D. Malzbender, A. Turton and E. Manzungu. 2005. A Preliminary Basin Profile of the Orange/Senqu River. INWENT in cooperation with the African Water Issues Research Unit (AWIRU), University of Pretoria, South Africa.

Falkenmark, M. 1989. The massive water scarcity now threatening Africa—why isn't it being addressed. Ambio 18: 112–118.
Heyns, P. 1995. The Namibian perspective on regional collaboration in the joint development of international water resources. International Journal of Water Resources Development 11(4): 467–492.
Hirji, R., P. Johnson, P. Maro and T. Matiza-Chiuta (eds.). 2002. Defining and Mainstreaming Environmental Sustainability in Water Resources Management in Southern Africa. Maseru/Harare SADC. IUCN, SARDC, World Bank, Washington DC.
Hoover, R. 2001. Pipe Dreams: The World Bank's Failed Efforts to Restore Lives and Livelihoods of Dam-Affected People in Lesotho, International Rivers Network 1–62.
Hulme, M., R. Doherty, T. Ngara, M. New and D. Lister. 2001. African climate change: 1900–2001. Climate Research 17: 145–168.
IPCC. 2013. Working Group I (WGI) contribution to the IPCC Fifth Assessment Report. Climate Change 2013: The Physical Science Basis.
Joubert, A. and B. Hewitson. 1997. Simulating present and future climates of southern Africa using general circulation models. Progress in Physical Geography 21: 51–78.
Klasen, S. 2002. The costs and benefits of changing in-stream flow requirements (IFR) below the Phase1structures of the Lesotho Highlands Water Project. Lesotho Highlands Development Authority, Maseru, Lesotho.
Kranz, N., E. Interwies and A. Vorwerk. 2005. Case Study: Orange Basin—South Africa and Lesotho: Governance, institutions and participation. NeWater Project Deliverable 1.2.1.
Leiman, T., B. Conradie, et al. 2000. Macro economic reforms and sustainable development in Southern Africa. Midrand, WWF and Development Bank of South Africa 1–50.
McKenzie, R.S. and A.R. Craig. 2001. Evaluation of river losses from the Orange River using hydraulic modelling. Journal of Hydrology 241(1-2): 62–69.
Mysiak, J., H.J. Henrikson, C.A. Sullivan, J. Bromley and C. Pahl-Wostl. 2010. The Adaptive Water Resource Management Handbook. Earthscan, London.
National Environment Secretariat, Lesotho (NES). 1999. State of the Environment in Lesotho, Ministry of Environment, Gender and Youth Affairs, Government of Lesotho.
Nel, J., G. Mare, et al. 2004. South African national spatial biodiversity assessment 2004 Technical Report: Volume 2: River Component. Stellenbosch, Council for Scientific and Industrial Research. 1–52.
NMET, Namibia Ministry of Environment and Tourism. 2002. Namibia State of the Environment Report Executive Summaries and Indicators, Directorate of Environmental Affairs.
ORASECOM. 2000. The Establishment of the Orange Senqu River Commission Pretoria.
Pegram, G. and G. Mazibuko. 2003. Evaluation of the role of water user associations in water management in South Africa. Pretoria, Water Research Commission 81.
Rockström, J., W. Steffen, K. Noone, Å. Persson, F.S. Chapin III, E. Lambin, T.M. Lenton, M. Scheffer, C. Folke, H.J. Schellnhuber, B. Nykvist, C.A. de Wit, T. Hughes, S. van der Leeuw, H. Rodhe, S. Sörlin, P.K. Snyder, R. Costanza, U. Svedin, M. Falkenmark, L. Karlberg, R.W. Corell, V. J.Fabry, J. Hansen, B. Walker, D. Liverman et al. 2009. Planetary boundaries: exploring the safe operating space for humanity. Ecology and Society. 14(2): 32.
Rooseboom, A. 1983. Reservoir Storage Losses Due To Sedimentation.Workshop On Hydrological and Water Supply Aspects of Irrigation Research, University of Pretoria, Pretoria, South Africa.
SADC. 2007. Regional Water Policy Gabarone, 2007.
SADC. 2011a. Regional Strategic Plan 3, 2011–15, Gabarone 2011.
SADC. 2011b. Peace Dividends Report, Gabarone, 2011.
Salman, S.M.A. 2004. Shared watercourses in the Southern African development community: challenges and opportunities. Water Policy 6(1): 25–38.
Sene, K.J., D.A. Jones, et al. 1998. Rainfall and flow variations in the Lesotho Highlands. International Journal of Climatology 18(3): 329–345.
Solomon, H. and A. Turton (eds.). 2000. Water Wars: An Enduring Myth or Impending Reality. African Dialogue Monograph Series No. 2. Accord Publishers, Durban.

Sullivan, C.A. 2002. Calculating a water poverty index. World Development 30: 1195–1210.
Sullivan, C.A. 2010. Quantifying water vulnerability: a multi-dimensional approach. Stochastic Environmental Research and Risk Assessment, DOI 10.1007/s00477-010-0426-8.
Sullivan, C.A. 2013. Planning for the Murray-Darling basin: lessons from transboundary basins around the world. Stochastic Environmental Research and Risk Assessment.
Sullivan, C.A. and D.E. Fisher. 2011. Managing wetlands: integrating natural and human processes according to law. Hydrological Sciences Journal 56(8): 1640–1655.
Sullivan, C.A. and J. O'Keeffe. 2011. Water, biodiversity, and ecosystems: reducing our impact. *In*: Q. Grafton and K. Hussey (eds.). Water Resources, Planning and Management: Challenges and Solutions. Cambridge University Press, Cambridge.
Sullivan, C.A., J.R. Meigh, A.M. Giacomello, T. Fediw, P. Lawrence, M. Samad, S. Mlote, C. Hutton, J.A. Allan, R.E. Schulze, D.J.M. Dlamini, W. Cosgrove, J. Delli Priscoli, P. Gleick, I. Smout, J. Cobbing, R. Calow, C. Hunt, A. Hussain, M.C. Acreman, J. King, S. Malomo, E.L. Tate, D. O'Regan, S. Milner and I. Steyl. 2003. The water poverty index: development and application at the community scale. Natural Resources 27: 189–199.
Turton, A. 2003a. An overview of the hydropolitical dynamics of the Orange River basin. International waters in Southern Africa. M. Nakayama. Tokyo, Japan 136–163.
Turton, A.R. 2003b. The evolution of water management institutions in select Southern African International River basins. pp. 256–294. *In*: C. Tortajada, O. Unver and A.K. Biswas (eds.). Water as a Focus for Regional Development. Oxford University Press, London.
UNEP (United Nations Environment Programme). 2003, 2007 Global Environmental Outlook New York.
Vörösmarty, C.J., P.B. McIntyre, M.O. Gessner, D. Dudgeon, A. Prusevich, P. Green, S. Glidden, S.E. Bunn, C.A. Sullivan, C. Reidy Liermann and P.M. Davies. 2010. Global threats to human water security and river biodiversity. Nature 467(7315): 555–561.

Appendix 1

South African Local Government Authorities in the Orange basin

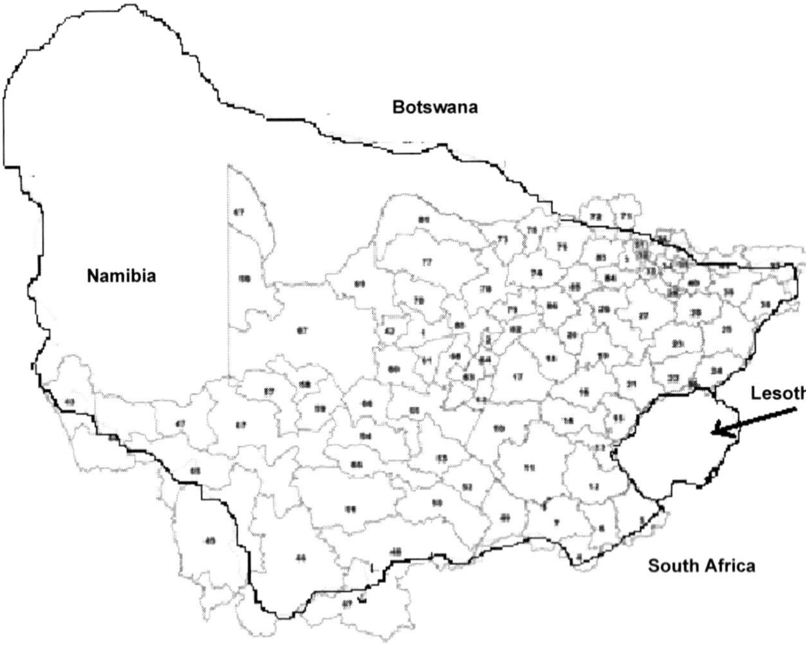

Municipality ID	Name	Municipality ID	Name	Municipality ID	Name	Municipality ID	Name	Municipality ID	Name
M 1	Ga-Segonyana	M 21	Setsoto	M 41	Highveld East	M 61	Kgatelopele	M 81	Molopo
M 2	Phokwane	M 22	Dihlabeng	M 42	Gamagara	M 62	Sol Plaatje	M 82	Lekwa-Teemane
M 3	Merafong City	M 23	Nketoana	M 43	Richtersveld	M 63	Dikgatlong	M 83	Ventersdorp
M 4	Inkwanca	M 24	Maluti o Phofung	M 44	Nama Khoi	M 64	Magareng	M 84	Potchefstroom
M 5	Senqu	M 25	Phumelela	M 45	Hantam	M 65	Namaqualand	M 85	Klerksdorp
M 6	Maletswai	M 26	Moqhaka	M 46	Karoo Hoogland	M 66	Bo Karoo	M 86	Maquassi Hills
M 7	Gariep	M 27	Ngwathe	M 47	Kh?i-Ma	M 67	Benede Oranje	M 87	Beaufort West
M 8	Oviston NR	M 28	Metsimaholo	M 48	Ubuntu	M 68	Diamondfields	Municipalities in the South African part of the Orange basin are very variable, and include desely populated areas, such as the City of Johannesburg, as well as remote desert and mountain areas.	
M 9	Ekurhuleni	M 29	Mafube	M 49	Umsombomvu	M 69	Kalahari		
M 10	Letsemeng	M 30	Golden Gate	M 50	Emthanjeni	M 70	Moshaweng		
M 11	Kopanong	M 31	Randfontein	M 51	Kareeberg	M 71	Rustenburg		
M 12	Mohokare	M 32	Westonaria	M 52	Renosterberg	M 72	Kgetlengrivier		
M 13	Naledi	M 33	Emfuleni	M 53	Thembelihle	M 73	Setla-Kgobi		
M 14	Manguang	M 34	Midvaal	M 54	Siyathemba	M 74	Tswaing		
M 15	Mantsopa	M 35	Lesedi	M 55	Siyancuma	M 75	Mafikeng		
M 16	Masilonyana	M 36	City of JHB	M 56	Mier	M 76	Ditsobotla		
M 17	Tokologo	M 37	Msukaligwa	M 57	Kai !Garib	M 77	Kagisano		
M 18	Tsewlopele	M 38	Seme	M 58	||Khara Hais	M 78	Naledi		
M 19	Matjhabeng	M 39	Lekwa	M 59	!Kheis	M 79	Mamusa		
M 20	Nala	M 40	Dipaleseng	M 60	Tsantsabane	M 80	Greater Taung		

10

Political Economy versus Comparative Advantage in the Nile Basin: Short Term Advantages or Long Term Gain?

Simon Thuo[1] and Phil Riddell[2,*]

SYNOPSIS

This chapter looks at water allocation decisions in the Nile Basin from a geopolitical perspective. All its riparians are dependent to greater or lesser extent on Nile waters in this food insecure region. But based on the fact that some riparians are net contributors and others net consumers of the Basin's water, the chapter questions the relevance of the river's traditional sub-division into the Equatorial Nile and Eastern Nile sub-basins. The chapter therefore begins by defining the basin in hydro-political terms before going on to:

- summarize the transboundary water management and allocation challenges as currently perceived by the riparian countries, before describing how a combination of climate change and population growth is going to make things very much worse

[1] Policy & Strategy advisor on water resources management; Institutional and social inclusion specialist for water development, Kenya, Rwanda and Uganda; Member, Global Water Partnership Reference Group-technical support for water, climate & development. Email: simonthuo@gmail.com

[2] Freelance International Water Policy and Management Specialist, Pays de Gex, France; currently Water Resources Management Adviser at the Ministry of Electricity, Dams, Irrigation and Water Resources, South Sudan. Email: philriddell@hotmail.com

* Corresponding author

- suggest transboundary approaches for solving the problems or reducing their impact while dealing with the need to understand: the role of trade; the need for socio-economic transformation and the difference between economic and physical efficiency
- assess current transboundary efforts, as per the Nile Basin Initiative
- compare these with the political reality
- package some conclusions and recommendations in the form of four scenarios ranging from business as usual to an ideal future

Keywords: Aswan dam, Burundi, Democratic Republic of Congo, Egypt, Equatorial lakes, Eritrea, Ethiopia, farming systems, fishery, floods, food insecurity, geopolitics, hydropower, Kenya, Lake Kyoga, Lake Victoria, Nile delta, Nile Waters agreement, riparian states, Rwanda, silt, Sudan, Tanzania, Uganda, World Food Program

1 Introduction

Most water professionals are familiar with the concepts of physical and economic scarcity of water (Molden et al. 2001). But there is also political scarcity. Examples of political scarcity of water include India's reluctance to charge poor farmers for the energy needed to extract water from their wells, many of which are also subsidized or grant financed. As a result of free access to groundwater, water tables are falling and dry season base flows in surface streams are being compromised (EPCO 2000). In the Nile Valley itself, Egypt's agricultural sector (which to all intents and purposes is entirely dependent on irrigation) contributes just over 14% to Egyptian GDP,[1] yet provides 55% of all livelihoods and employs some 30% of the workforce.[2] Because of this, and although other uses of Egypt's share of the Nile are thought likely to have greater economic benefits,[3] the perceived political cost of allocating water out of agriculture is considered unaffordable. With negligible exceptions, all of Egypt's irrigation water currently comes from Lake Nasser, stored behind Aswan Dam. Since its completion in 1964, Aswan Dam has retained some 5 billion tons of silt, compromising hydropower and leading to a loss of a 1.5 km wide slice on

[1] 2012 figures: http://www.gfmag.com/gdp-data-country-reports/280-egypt-gdp-country-report.html#axzz2fLLxcrnc.

[2] 2011 figures: http://www.ifad.org/operations/projects/regions/PN/factsheets/eg_e.pdf.

[3] It has been suggested on an informal basis by several commentators that the use of L Nasser water to preserve navigation depths in the Nile downstream may be more beneficial for instance, due to the costs incurred in maintenance dredging: but this needs to be scientifically verified.

the fertile crescent—Egypt's most valuable land, whilst there is increasing intrusion of salt water from the Mediterranean sea into the Nile delta (Abd-Elhamid and Javadi 2008). Lake Nasser has a very high stage storage relationship, which combined with its low altitude and high ambient temperatures means that its evaporation losses are very high (estimated at 10 km^3/yr[4]). Yet options for saving annual evaporation losses from Lake Nasser by storing Egypt's water in the Ethiopian highlands—where such losses would be much less due to lower evaporation rates and stage/storage relationships—are considered politically unacceptable.

But this introduces another example of political scarcity in the basin.

During the course of the Eastern Nile Irrigation and Drainage Study[5] when faced with this very opportunity, Ethiopian water managers said that any water stored in Ethiopia would be used for irrigation within Ethiopia (despite costs/m^3 of irrigation development from such dams being approximately seven times those of using the same water downstream in say, Sudan because of the lengthy contour canals that would be needed to get the water from the mountainous terrain to flatter land).[6]

When we look at water allocation decisions in the Nile Basin through a political lens, we find similarities between the financial crash of 2008 and the potential for a collapse of the Nile Basin's water economy. As with the financial sector, decisions remain predicated on short term gain, locking in first mover advantages for the country, rather than higher economic and environmentally sound long term benefits across the basin through joint or coordinated investments in shared waters (despite there being myriad institutional options to facilitate this, options that are seemingly ignored).

This chapter will begin by defining the basin in hydro-political terms before going on to:

- summarize the transboundary water management and allocation challenges as currently perceived by the riparian countries, before describing how a combination of climate change, land degradation and population growth is going to make things very much worse

---

[4] This figure is widely assumed in the literature, although recent studies carried out with increasingly sophisticated equipment tend to suggest even higher figures, even up to 16 km^3.

[5] Carried out between 2008 and 2010 by BRLi of France on behalf of the Nile Basin Initiative.

[6] Conversations during execution of the Eastern Nile Irrigation and Drainage project (see footnote 5), between the Ethiopian water managers and one of the authors when discussing a possible cascade of dams in the montane reaches of the Blue Nile.

- suggest transboundary approaches for solving the problems or reducing their impact while dealing with the need to understand: the role of trade; the need for socio-economic transformation and the difference between economic and physical efficiency
- assess current transboundary efforts, as per the Nile Basin Initiative
- compare these with the political reality
- package some conclusions and recommendations in the form of four scenarios ranging from business as usual to an ideal future

2 Defining The Nile

2.1 The Nile itself

The Nile is unique in many ways. It is for instance and by some counts, the World's longest river;[7] it has more riparian states (11) than any other major river; it is the only major river that flows South to North; it has the longest hydrological records of any river; its basin has one of the fastest growing populations in the world (see for instance Lamere 2012) and it is the world's most vulnerable river to climate change.[8] Its geo-political significance goes back at least to Roman days (and even further according to biblical accounts). For instance, Egypt's ability in the days of Cleopatra to supply the Mediterranean Basin with cereals, and by being a price maker, bring the mighty Roman empire to its knees—was entirely due to its ability to harness the Nile's annual floods and the fertility they carried down in the form of sediment from its vast hinterland. Ethiopia also used the Nile as an instrument of foreign policy when King Lalibela threatened to dam the Blue Nile in order to emasculate the Egyptian economy. One thousand years on, Ethiopia is actually implementing Lalibela's plan in the form of the Grand Ethiopian Renaissance Dam. As expected, Egypt is not happy about this—a situation that has been exacerbated by Ethiopia's recent alignment with the upstream countries within the Equatorial Lakes region—of which more is discussed below. We will also examine the political ramifications of this later.

But first it is necessary to understand a little more about where the Nile's water comes from, and where it goes to.

The layman usually thinks of the Nile in terms of a White Nile and a Blue Nile which meet in Khartoum. But until recently, the water manager usually thought in terms of two sub-basins: the Eastern Nile, the riparians of

---

[7] See for instance: http://en.wikipedia.org/wiki/List_of_rivers_by_length.

[8] This statement is based on multiple sources, including the IPCC and UNEP, reviewed during preparation of Thuo's study "The Colours of Nile", in progress at the time of writing and hence as yet unpublished.

which were Egypt, Eritrea, Ethiopia and Sudan; and the Equatorial Lakes, for which the riparians were Burundi, The Democratic Republic of the Congo (DRC), Kenya, Rwanda, Tanzania, Uganda. These of course represent a geo-political delineation. However, several factors suggest a different way of thinking. The separation of South Sudan from the North and Ethiopia's change of allegiance along with Fig. 10.1 lead to the inescapable conclusion that some riparians are net contributors of water and that the others are net users. This thinking is useful in two ways. First it delineates the Nile on a hydro-political basis, and thereby secondly, provides a useful platform on which to build a discussion of the very real conflict between political economy[9] and comparative advantage[10] in the basin as a whole. But before

COUNTRY	BASIN	POLITICAL AFFILIATION	NETT CONTRIBUTION	
Egypt	Eastern Nile		no	net consumers
North Sudan	Eastern Nile		no	
South Sudan	Eastern Nile	to be confirmed	yes	total net contribution: 84.18 km^3
Eritrea	Eastern Nile		yes	
Ethiopia	Eastern Nile	Equatorial Lakes	yes	
Uganda	Equatorial lakes		yes	
DRC	Equatorial lakes		yes	
Kenya	Equatorial lakes		yes	
Tanzania	Equatorial lakes		yes	
Rwanda	Equatorial lakes		yes	
Burundi	Equatorial lakes		yes	

Figure 10.1. Alignments, Allegiances and Contributions in the Nile Basin.

[9] Used in this context to mean "*how political forces affect the choice of economic policies, especially as to distributional conflicts*": Wikipedia.

[10] Used in this context to mean the most productive and sustainable use of the available water resources in a particular location.

proceeding to identify and assess the many problems being faced in the basin, it is necessary to note that its long-term management should not be predicated on the basin being a discrete, stand-alone unit. This is because the Nile casts a big and important shadow across the broader region and indeed beyond.

The shadow of the Nile

The Nile does not exist in isolation, with all its challenges and opportunities lying only within its defining watershed. The Greater Horn Region, which is usually taken to comprise Djibouti, Ethiopia, Eritrea, Kenya, Somalia, Sudan, South Sudan and Uganda and as such clearly includes territory which is not included in the Nile Basin itself, is one of the world's most food-insecure regions.[11]

Yet according to the same source, in the future, the impacts of climate change, as well as growing populations and declining per capita agricultural production, are expected to exacerbate the chronic food insecurity already endemic in this region.

As one of the least developed areas in Africa, there is limited capacity to respond to drought or food crises. To prevent humanitarian emergencies, the Horn of Africa needs to strengthen its ability to build long-term resilience and tackle the root causes of the region's vulnerability. The Nile Basin, with its vast undeveloped land resources, especially in the Sudan and parts of Ethiopia, has the potential to address increasing food security, poverty and environmental problems not only in its riparian regions, but also the Greater Horn as a whole. But this would require a different water allocation paradigm to that which prevails, and to which we will return in due course.

Opportunities for economic growth also lie beyond the basin in the form of agricultural exports. For instance, within the next 30 years India needs 150 million tons of cereals—a 60% increment over its present production which has already maxed out. With sea level rise threatening traditional agricultural export regions of South Asia, Africa is best placed to capture this emerging market. The Nile region's exports already include high value horticulture, fruits, tea, coffee and cotton, etc. But with expanded irrigation these could be increased, and with production secured thereby in terms of quality and quantity, more investment could be attracted into added value opportunities.

Finally and controversially, Egyptian policies also mean that Nile water itself is transferred out of the basin via the Mubarak Pump Station and the

[11] (http://na.unep.net/geas/getUNEPPageWithArticleIDScript.php?article_id=72).

Sheik Zayad Canal into the Western Desert to the New Valley Project at Toshka; and via the Al Salam Canal into West Sinai (The world's first and only inter-continental water transfer infrastructure).

Before proceeding to a discussion of the very real conflict between political economy and comparative advantage in the Nile Basin (as defined in footnotes 9 and 10) it is first necessary to i) understand the nature of the problems faced in the basin; and ii) assess their likely trajectory as climate change begins to bite and populations continue to rise before suggesting transboundary approaches to solving them: regional solutions to local problems it were.

3 Current Needs and Incoming Threats

3.1 Current needs assessment

As already stated, the Nile Basin has the most rapidly growing population in the world, projected to increase from 180 million in 2005 to 648 million by 2030.[12] In turn, this will drive increased social and economic demands; while urban growth, economic diversification and changes towards more sophisticated diets together suggest a need to intensify and increase agricultural production, which for much of the basin will mean increased withdrawals for irrigation. Local and transboundary conflict over water and pasture already exact a major toll on human life, livelihoods and security in the region; while high climatic variability (which is expected to worsen) causes chronic hunger and distress from increasing droughts and flash floods.

Although a large majority particularly of the rural population subsist off the land (mainly outside the formal cash economy); upheavals in the financial sector cause immediate and lasting damage to local economies and food security precisely because so little money is available to tide over difficult economic times. There is no money to import and distribute grain unless UN agencies such as World Food Program step in, which unfortunately again destroys nascent local markets as subsidized or free imported food products flood the local markets.

The combination of i) droughts and floods—affecting ever larger numbers—in combination with ii) increased environmental degradation as populations invade critical water catchments and marginal lands simultaneously and coupled iii) with fragile economies, could create disastrous, potentially irreversible situations that could easily lead to social and economic collapse for the entire Nile region.

[12] As for footnote 8. Other estimates are even more extreme, as for instance in Lamere 2012.

A possible way out of this imbroglio is to improve regional cooperation. Regional trade in agricultural products for instance will mobilize productive comparative advantage, thereby increasing the productivity of the natural resources involved and achieving economies of scale while creating new jobs along the market chains.

Ethiopia's potential development of "basin-level" hydro-power infrastructure at unit costs that are much lower than in surrounding countries that have less suitable topography provides real case demonstration of how to avoid the need to allocate scarce development finance to cover the higher unit costs of more local solutions elsewhere. And storing water at altitude in reservoirs with low stage/storage ratios will avoid the high evaporation losses from low altitude reservoirs with high stage/storage ratios—the example cited above, on Lake Nasser refers. In addition, in all Nile countries, transformation of the huge and increasing pool of the unemployed (particularly the youth) from a primed bomb waiting for the merest tripping point into a productive and entrepreneurial workforce, will be critical in creating economic growth in the many rural areas that have so far been left out of the development process.

In order to understand this better it is helpful to develop a typology of basin level approaches that could solve or at least ameliorate problems currently encountered or perceived at riparian state level. The comprehensive needs assessment carried out during the Conceptual Design of the Nile Basin Decision Support System identified a total of 30 water related problems being faced by the Nile Basin's riparian states. Of this total, seven have transboundary implications (see Fig. 10.2).

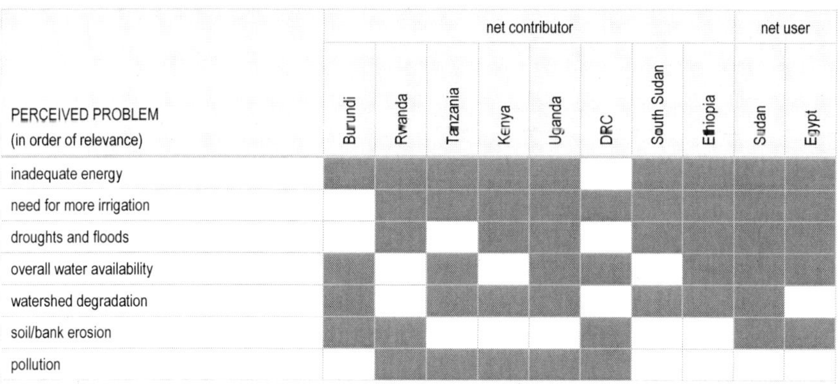

Notes: 1. *The needs assessment was carried out before South Sudan's independence, hence the allocation of problems between South Sudan and Sudan does not represent the expressed view of the countries themselves but rather represents the authors' opinions.*

2. *The needs assessment was carried out in consultation with the member countries of the Nile Basin Initiative.Eritrea had, and still has NBI observer status only, and hence was not involved in the consultation process.*

Figure 10.2. Transboundary Water Related Problems Identified by the Riparian States.

3.2 Incoming threats

The needs assessment summarized above reflects the **current** problems being faced in the basin. Under current management and allocation paradigms for instance, there is already not enough energy or water to adequately support existing livelihoods and economic growth opportunities, while water quality is falling in terms of both chemical pollution and sediment overload. But in the absence of increased inter-riparian cooperation at the political, economic and social levels, a combination of various incoming threats is going to make things a lot worse. The threats themselves, which include climate change, population growth and poor land management tend to be interlinked, so it is impossible to deal with each separately.

As already mentioned; the Nile is considered to be the world's most vulnerable river to climate change and in fact is already a very dry basin. Despite being almost equal in area to the Congo Basin, in a typical year the Nile's flow is only 5% of the Congo's. This is exacerbated by the fact that rainfall in the Nile basin—which in any case typically falls on only 20% of the basin—is 300% to 500% more variable than the rest of the world's. To make matters worse, the zone in which the Nile lies is expected to face double the world's average climate change induced temperature rise. Add to this the facts i) that the majority of the Nile's riparians feature among the world's poorest 30 countries by several reckonings—with the DRC and Burundi consistently in the poorest three; and ii) that the basin is characterized by the world's highest population growth rates—and the possibility of a catastrophic climate related impact on people, their livelihoods and their national economies becomes clear, and hence so does the possibility of armed conflict: the aroma of which is never far away in this already volatile region!

To make matters worse, Lake Victoria and the Albertine Lakes, which together feed the steady flow of the White Nile south of the Sudd Swamps, are themselves vulnerable to a combination of increased evapo-transpiration and reduced inflows. As a result of this, Lake Victoria may even dry up. If so, it would not be the first time in its brief history. According to the geological record, the lake did not even exist until around 12,000 years ago when a terrestrial uplift to the west caused streams, that hitherto had flowed into the Congo Basin, to flow northwards thereby creating the White Nile. But lateritic deposits indicate that the Lake has dried up several times since then and it could be drying out once again now: but this time with more serious consequences. Sediment resulting from land degradation in its catchment is driving up the bottom of this, and indeed other shallow lakes in the basin—such as Lake Kyoga—to extent that once dry this time, there is a very real chance that they may never fill again.

Population contributes to and is affected by changes in hydrology and flow regime. For instance, catchment degradation due to population pressure and poor land management compromises both soil cover and hydrological function. As soil cover deteriorates, farming systems become more and more extensive thereby increasing flood risk and sediment loads. These in turn reduce both the amount of water that can be managed within a catchment and the lifetime, utility or cost effectiveness of water management or storage infrastructure. Sudan for instance in addition to being prone to floods emanating from denuded hillsides upstream, spends between US$200 and $300 million annually removing sediment, largely of Ethiopian origin, from its dams and canals. This is reportedly 70% of its entire annual water sector budget; while to the North East some 70% of Eritrea's dams are already full of silt.

Problems accruing to population growth will not be limited to the Equatorial Lakes region. According to a 2006 UNEP report, the Nile basin as a whole will experience an average decrease of 75% of available water per capita by the end of the century. This would not necessarily matter if there was plenty of water to begin with. But as shown in Fig. 10.3, for several of the countries, the situation in 1990 was already challenging and by 2025 the water security of all the countries considered[13] will be vulnerable at best, with five of them actually water scarce.

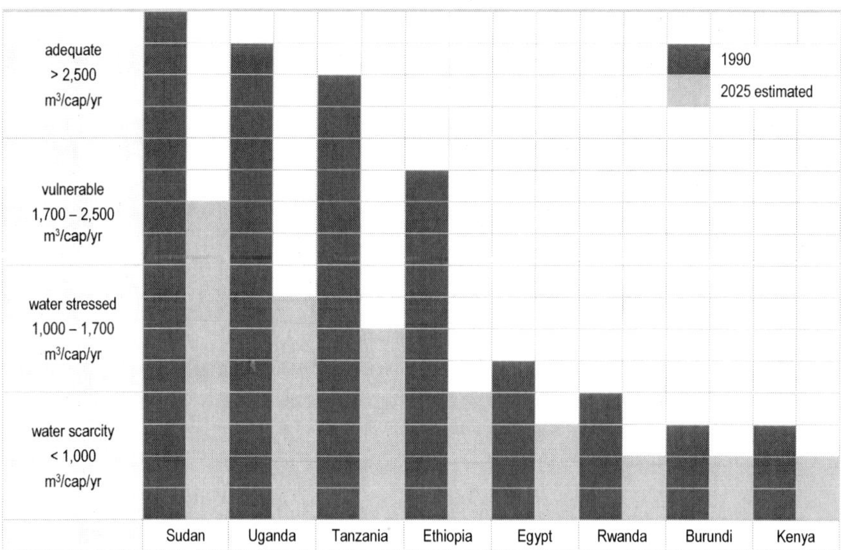

Figure 10.3. Water Security Expectations in the Nile Basin to 2025 (United Nations Economic Commission for Africa).

[13] DRC and Eritrea were not included in this analysis—but from other sources, Eritrea is expected to be stressed.

The economic damage this could cause will be extensive; but as with population, economic activity is not only affected by climate related environmental stress, it also contributes to it. The ways that it does this are not confined to the Nile Basin alone. The conversion of wetlands for housing and industrial development risks incurring changes to landscape and ecosystem services everywhere it is practiced. These services include groundwater recharge and hence the maintenance of dry season surface baseflows; flood attenuation and water purification, all of which relate to the problems identified by the Nile riparians and which could be solved, or reduced by means of a regional approach. But also, wetlands sometimes represent a low cost alternative to wastewater treatment and nutrient retention. This could be significant in the case of countries, such as those in the Nile Basin, which have many competing demands on scarce financial resources. Similarly, deforestation is a common emerging country problem—where regardless of whether it accrues to the need for timber or the need for land—it can result in large volumes of lost soil (and hence fertility), clogged wetlands and increased flooding. But civil unrest in much of the Basin (Burundi, the DRC, Rwanda and Sudan in particular) drives hundreds of thousands of people into the forests with similar results.

Box 10.1 Valuing Ecosystem Services.
A valuation exercise for the city of Kampala, Uganda, showed that the nearby Nakivubo Swamp provided an economic value of between USD 1 million and USD 1.75 million a year in wastewater purification and nutrient retention services. Researchers concluded that the services provided by the Nakivubo Swamp created a much cheaper means of treating Kampala's wastewater than the expansion and maintenance of new wastewater facilities. Moreover, public funds were simply not available to replicate the natural ecosystem services provided by the swamp. Despite these findings, policy makers have been slow to protect the area and the wetland's ability to remove nutrients and pollutants has been greatly reduced over the past decade (UNEP Green Economy Brief).

As the major water using sector and principal source of livelihoods in the basin, agriculture is especially important and hence is worth a closer and more specific look. Studies have indicated that the higher temperatures, compromised groundwater recharge, reduced water availability and desertification in the basin are likely to reduce crop yields by up to 23%. This would be serious enough given the expected population increases, but given the higher food prices that are expected to pertain for the foreseeable future, household food security will be increasingly difficult to achieve. The effects will moreover, be registered at the macro-economic level. Work done by the World Bank in Kenya for instance (Hirji 2009), confirms that rainfall variability (which is expected to increase with climate change in the basin) unsurprisingly has a major effect on agricultural GDP.

And finally with respect to incoming threats there is the matter of floods which not only threaten a wide range of economic activities, infrastructure and habitat, but also reduce the amount of water that can be managed for productive purposes. Recent downscales by ACCESS[14] of regional circulation models (RCM) warn of shorter, more intense rain seasons in the White and Blue Nile catchments. The combination of degraded watersheds; heavier silt loading and intensifying hydrological events is thus expected to increase the destructive power of future floods in the basin.

Clearly, when taken together and in the absence of transboundary mitigating measures, the incoming threats will take a huge toll on human development, economic growth and stability in the basin. Luckily there are transboundary options for both mitigating and militating these threats.

4 Regional Solutions to Local Problems

Figure 10.4 reworks Fig. 10.2 into a typology for regional cooperation: regional solutions to local problems as it were.

Taking each in turn:

1. *Water management and economically efficient allocation at basin/sub-basin level—sometimes called "basin welfare"*: Simply stated this solution is predicated on the reasonable assumption that water wasted as a result of sub-optimal use in one location, is no longer available for better use in another: thus if water could be saved instead of wasted, then more could be made available for productive use elsewhere. We can

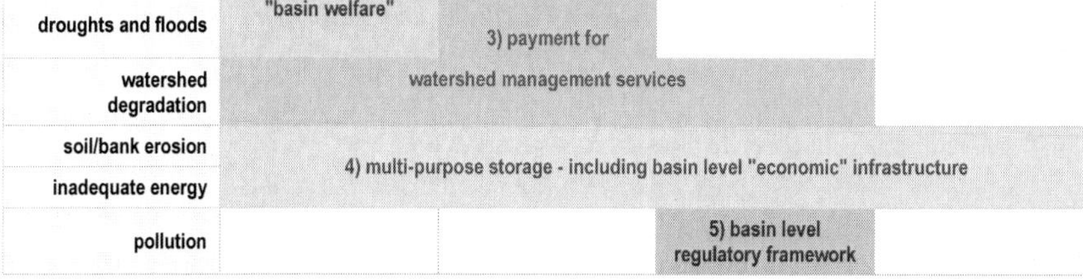

Figure 10.4. A Typology of Regional Solutions to local Water Related Problems.

[14] African Centre for Collaboration of Earth System Sciences (ACCESS) May 2013.

illustrate this with a real example from Indonesia, namely the JatiLahur basin. This is a small basin which begins approximately 100 km to the South East of Jakarta, the capital city, though which it discharges into the Java Sea. Close to the south eastern extremity of the watershed is a hydro-electric dam.

Once water has passed through the turbines there, it flows through an area of wetland rice which it supplies with irrigation water before reaching the city and hence the sea. Some time ago, the farmers found that with improved crop husbandry and irrigation water management, they could grow more rice with less water. This gave them two choices: either to increase their irrigated areas, or leave the saved water in the river. They chose the latter. The increased stream flows meant that capture fisheries increased downstream of the irrigation offtake, and extra water became available to bottling plants in Jakarta, who paid the irrigation farmers accordingly. Thus a more basin oriented allocation of water increased the economic returns to it from those accruing just to energy and rice production to include additional benefits from increased fisheries and industrial use—while benefitting also, the rice farmers pockets.

This example also illustrates the important difference between physical efficiency and economic efficiency. Physical efficiency is simply a measure of how much water a process needs as compared to how much water is withdrawn in order to supply that need. Economic efficiency on the other hand is a measure of how the value of production compares with the opportunity cost of the water consumed for the purpose of production. Important studies carried out for instance by IFPRI (Cai et al. 2001) and IWMI (Keller et al. 1996) confirm that in the absence of suitable mechanisms to allocate the saved water towards uses closer to its opportunity cost (a process sometimes known as increasing the economic mobility of water), then increases in physical efficiency can actually reduce access to water and decrease its overall productivity. Another example illustrates this very well.

The physical water use efficiency associated with the irrigation of coffee in Vietnam's Southern Highlands in the mid 1990's can fairly be said to have been high. However, all available water was allocated to a continually expanding coffee area. So much coffee was produced in fact, that global coffee prices collapsed even while the availability of water for production uses elsewhere in the basin was drastically reduced. In other words, not only was this a socio-economic catastrophe for the poor, highly indebted coffee farmers, but the economic returns on water use in the basin fell through the floor. In other words, a highly efficient economic disaster!

The transboundary relevance to the Nile Basin will be readily apparent.

2. *Regional trade based on comparative productive advantage*: This solution is somewhat related to the first; but is more finely tuned to agriculture, which as we have seen is the predominant water using sector in the basin and the most socio-economically significant. Clearly and along the same lines as solution 1, water that is allocated, for reasons of political economy—perhaps self-sufficiency—to a crop that is thirsty in comparison to local water availability, is no longer available for more efficient use in another location. However, should this water be allocated to where the crop could be grown cheaper and with less water, regional trade based on comparative advantage will increase market and other activities along the value chain and hence will diversify livelihoods and create jobs. As a result, direct dependence on fragile and/or scarce natural resources is reduced while the productivity of the natural resources and jobs per drop will increase. The validity of this solution is supported by work carried out under the FAO-Nile project which established, contrary to prevailing "expert opinion" that the Nile riparians would be much better off in terms of food security by opening up regional trade in agricultural products between themselves while simultaneously restricting imports from more developed regions.
3. *Payment for watershed management services* is likely to be a little esoteric in the context of this chapter which, having a political facet, is more concerned with governmental/public sector opportunities; whereas this approach is normally associated with the private sector. Private hydropower operators in South America and India pay farmers in their dams' catchment areas to prevent soil loss from their fields and hence sedimentation in the dams. This is considered by the power companies an acceptable and necessary cost to maintain profitability. In like manner, the Coca Cola company considered the feasibility of subsidizing improved farming systems in the upper reaches of Vietnam's Dong Nai river in order to i) improve base flows and ii) reduce flood risk—again, a commercial transaction. More radical, but more relevant here, would be the rehabilitation of selected upper catchments in the Yangtze Basin, where the work was financed not by public funds, but rather by a stock issue. This was attractive to stockholders because the rehabilitation of the watersheds was predicated on high value horticultural or arboreal produce to be sold in high grade markets such as Beijing or Shanghai. But what makes this interesting from the political perspective was a law promulgated

as part of the initiative and which required the developers to guarantee at least a 300% increase in income for the indigenes.[15]

Before proceeding to examine the fourth solution, it is worth noting, in passing, that watershed management services are not necessarily limited to sediment and flood prevention. Wetland conservation also has a potential role to play not least as regards flood attenuation, water quality, micro-climatic regulation and biodiversity. The Nile basin has at least three major wetlands: the Kyoga system in Uganda, the Mashar swamps at the border between Ethiopia and South Sudan, and the Sudd swamps. Kyoga is already suffering from industrial and urban encroachment and there is renewed talk of draining the Sudd, but economic analysis often confirms that wetlands contribute more to basin level economics when conserved rather than when developed.[16]

4. *Multi-purpose storage—including basin level "economic" infrastructure*: It will be obvious to the reader that, in theory, dams can serve many purposes: storage, amenity, power generation, fisheries, water supply, navigation depths, sediment trapping, flood attenuation, etc. Furthermore, as already noted, reservoirs in Ethiopia's narrow highland valleys would suffer from less evaporation losses than lower altitude, more extensive ones such as Lake Nasser. However, multi-purpose benefits in the Nile Basin may mean that a dam constructed to meet a need in one country, may have other benefits in another. This has two significant implications. First is that transboundary institutions will be needed to allocate costs and benefits in an equitable and mutually agreeable fashion, and institutional cooperation has political costs. Secondly, multi-purpose dams are generally sub-optimal to operate because of competing operating rules between say: power generation and irrigation. This will have to be well understood by all concerned, and again is likely to have political implications.
5. *Basin level regulatory framework*: This is also more or less self-explanatory, except to note that it is likely to be low cost in both financial and political terms. This is because the benefits will accrue "closer to home". Chemical pollution is going to cause a problem in the country of origin, even before it moves downstream into another; while large, expensive

---

[15] Unfortunately it has not been possible to find any references to cite regarding this programme. But the scheme was visited by one of the authors in 1999 and was found to have been implemented at landscape scale and to great effect.

[16] This statement is justified by a large body of literature—see for instance http://agecon2.tamu.edu/people/faculty/woodward-richard/paps/WetMetaData.PDF—including studies commissioned by Riddell during preparation of the Madhya Pradesh Integrated Water Resources Management Strategy, 2002–2004.

dams are not the only way to solve problems of sedimentation. Smaller scale structures such as so-called "sand dams" can also do the job, while creating local sources of filtered water with minimal to nil mosquito populations; low evaporation losses and which can supply domestic needs, livestock and vegetable gardens. Similarly, watershed rehabilitation for instance creates short term jobs while securing long term livelihoods in the form for instance of terraced agriculture and/or agro-forestry/alley cropping. Furthermore, it will be obvious that all of these measures increase the economic efficiency of water use in the basin.

So then, what is being done to initiate or mobilize the regional solutions to local problems?

5 Sharing the Vision

The Nile Basin Initiative (NBI) is a partnership among the Nile riparian states that "seeks to develop the river in a cooperative manner, share substantial socioeconomic benefits, and promote regional peace and security". Its objective is *"to achieve sustainable socio-economic development through the equitable utilization of, and benefit from, the common Nile Basin water resources"* and it was formed on the 22nd February 1999 from a mutual recognition among the riparian states that some sort of basin level cooperative arrangement was needed in order to tackle the many and varied challenges faced by their citizens and their economies. The NBI has a multi-layered institutional structure.

With respect to Governance, The Nile Council of Ministers of Water Affairs is at the apex: it is responsible for policy guidance and decision making; and is supported by the NBI Technical Advisory Committee. This comprises senior civil servants, one from each riparian and is intended to provide technical advice to the Council of Ministers. It meets two to three times annually. Both committees are supported administratively by the NBI Secretariat which is based at the Initiative's headquarters in Entebbe, Uganda.

The NBI consists of three main programs. The ***Shared Vision Program*** is a Basin-wide program focusing on"...building institutions, sharing data and information, providing training and creating avenues for dialogue and region-wide networks needed for joint problem-solving, collaborative development, and developing multi-sector and multi-country programs of investment to develop water resources in a sustainable way". **The *Eastern Nile Subsidiary Action Program*** (ENSAP) is intended to develop the water resources of the Eastern Nile Basin in a "sustainable and equitable way to ensure prosperity, security and peace for all its peoples". It is managed by the Eastern Nile Technical Regional Office (ENTRO) based in Addis Ababa, and its component initiatives are grouped into fast-track investments and more complex, longer-term multi-purpose investments. The ***Nile Equatorial Lakes Subsidiary Action Program*** (NELSAP), which is managed by a Kigali based coordination unit, is an investment oriented program. Its mission is to help reduce poverty, promote economic growth, and reverse environmental degradation. As such it falls under two broad program areas: Natural Resources Management and Development, and Power Development and Trade.

Other operational units are situated in Egypt (applied training); Ethiopia (water resources planning); Kenya (agriculture) and Tanzania (power trade). As such, the NBI's basin wide orientation is more than adequately confirmed by its multi-location approach: but how has it performed?

Tangible results claimed by the NBI itself can be summarized as follows:

- Launch of the first ever State of the River Nile Basin report which provides baseline data on health of the Nile Basin.
- Member countries are now borrowing to implement NBI facilitated projects; increasing their member contributions to cover the NBI's core costs and South Sudan became a sovereign member in its own right—which all confirm a general acknowledgement of the need for a basin level organization and a particular confidence in the NBI.
- Reduced preparation costs for projects contributing to economic growth, poverty reduction and sustainable river basin management.
- Regional and national flood forecast and early warning systems which are already saving lives and livelihoods in the Eastern Nile.
- Narrowing of water resources management related knowledge asymmetries among riparians through enabling access to regional data and information.
- A growing and invaluable store of data and information in the areas of water, power, agriculture has been made available through the Nile-Information System and on-line libraries.
- New access to regional data and analytic tools for investment formulation and decision making through the Decision Support System.
- The emergence of trans-boundary dimensions in national water policies is claimed to be contributing to a growing consensus and awareness among policy makers, professionals and the riparian public at large that the River Nile is a shared resource and that trans-boundary water resources management does require joint action.

The authors see no reason to question the validity of these results, and would actually go further to stress that the NBI is staffed by highly qualified, experienced and dedicated professionals. But as we will now show, real-politics constrain national willingness to make the most of the NBI's achievements and maintains a gulf between an acknowledgement that joint action is required, and joint action itself.

6 Political Realities

6.1 The Nile Waters Agreement

The Nile Waters Agreement was signed by the Governments of Egypt and the Sudan in 1959 and was based on a mutually agreed annual average flow of 84 km^3 at Aswan, and a recognition on Sudan's part that Egypt's multi-millennia tradition of irrigation granted it a prior use right.

The agreement granted Egypt 55.5 km^3, and Sudan 18.5 km^3 of the total annual flow, leaving 10 km^3, equal to some 10% of the available water, for environmental stream flows and other purposes. But as Fig. 10.1 clearly shows, almost all of this water originates outside of the two countries sharing it and in recent years this has become problematic. For instance, some 80% of the water allocated to Egypt and the Sudan originates in Ethiopia which is ignored in the 1959 Agreement. And the other 20% almost entirely originates in the Equatorial Lakes countries, also ignored in the Agreement.

We have already made reference to FAO's work confirming the need for transboundary intra-basin trade in agricultural commodities. Similarly and also in line with the philosophy behind Fig. 10.4, other studies carried out under the aegis of the NBI have confirmed the advantages of equivalent regional solutions to problems in the energy, environmental and other sectors. Conclusions such as these have begun to raise legitimate questions among the riparian countries, questions concerning the need to supersede the Nile Waters Agreement with a new and binding agreement to pursue these regional solutions.

6.2 The Nile Basin Cooperative Framework Agreement

The Nile Basin Cooperative Framework Agreement (CFA) is intended to provide such a solution; by securing each country's vital water interests while enhancing water security and productivity for all member states. Simply stated, it is intended to transform the Nile Basin Initiative into The Nile River Basin Commission and was itself negotiated over a period of several years up to 2010 under the NBI framework. It is in fact, the first multilateral treaty successfully negotiated under the auspices of the Nile Basin Initiative. Among other things it requires the Nile Commission to monitor the equitable utilization and development of the Nile Waters and enforce the proper application of the rights and duties of the signatory countries. It also requires that the Commission is established upon ratification by the legislatures of at least six riparian countries. But instead of being a much needed unifying instrument, it has proved divisive and as yet has not been signed by all of the Nile Basin countries. Signatories to date are as follows: Burundi, Ethiopia, Kenya, Rwanda, Tanzania and Uganda but at the time of writing, Ethiopia is the only country so far to have actually ratified the agreement (which it did on June 13th 2013). The DRC is expected to sign, but the two Sudans and Egypt have yet to do so.

The crux of the disagreement hinges on whether the CFA is a new all-inclusive framework among all riparian countries aimed at creating a common basis for negotiating on water development and entitlements, thus making redundant existing "unfair, colonial, non-inclusive treaties" (the view upstream riparians hold) or if the previous treaties particularly of 1959 principally between Egypt and Sudan must be explicitly recognized as still being in force.

An additional problem is that the CFA does not include fixed water shares for each riparian country, but assumes that under the CFA, apportionment informed by international norms such as Helsinki convention or the pending UN Convention on use of Non-navigable Waters; will be negotiated continually with the decision implemented under the aegis of the Nile Basin Commission.

For the countries of the Lake Victoria Basin, this is not a particularly big deal. Analyses carried out during preparation of Uganda's Irrigation Framework Master Plan suggest i) that wholesale development of remaining irrigation potential in the Lake Victoria riparians will have negligible effects on the flows north of the Sudd Swamps; and ii) that the small reductions in flow that such development would produce, will in any case be less than the savings expected to accrue to improvements in physical water use efficiency in Sudan (Riddell 2010). In terms of water allocation therefore, the Equatorial Lakes riparians are to all extent and purpose, not stakeholders in water allocations between the two Sudans, Ethiopia and Egypt. But the secession of South Sudan (which like Ethiopia is a sovereign, net contributor to the Eastern Nile sub-basin) along with Ethiopian saber rattling has changed all that and the political ground is shifting. Ethiopia's Minister of Water was reported as saying in mid-2013 for example that "The Nile is a common resource. The Nile is a gift to the population of the all Nile Basin Countries. We will not allow a single country[17] to have full control of our shared resources. Conflicts and tensions over the utilization of the Nile are not helpful and will not lead to sustainable utilization".[18]

[17] He was referring to Egypt.

[18] http://www.voanews.com/content/nile-basin-initiative-meeting-water-share/1686319.html.

As originally drafted Article 14b of the CFA requires that member countries would work together to ensure "...not to ***significantly*** affect the water security of any other Nile Basin State." Egyptian and Sudanese representatives at negotiations in April 2010 wanted however, the article to read "Not to adversely affect the water security and current uses and rights of any other Nile Basin States". But by way of rebuttal, representatives of upstream countries said they were "tired of first getting permission from Egypt before using River Nile water for any development project like irrigation, as required by a treaty signed during the colonial era between Egypt and Britain in 1929"! And South Sudan's secession from the north has not helped. In June 2013 for instance, its Minister for Irrigation and Water Resources stated that his country would sign the CFA[19] consistent with its wish develop the waters of the White Nile flowing through its territory for hydropower and irrigation and to "...replace colonial-era deals that awarded the lion's share of the Nile waters to Egypt and Sudan".[20] But stirring the political pot even more, is the fact that by late September 2013, South Sudan had still not signed because now: "Egypt is keen on developing a mechanism for cooperation between Egypt and South Sudan".[21] It is reasonable to suspect that this might be based on the possibility of draining all or part of the Sudd Swamps to free up water that could be sold to Egypt!

At this stage it is tempting to sit back and conclude that surely, Egypt and Sudan have had it their way for too long—but it is by no means as simple as that!

To all intents and purposes, all of Egypt's water is used in the irrigation sector where levels of water productivity and efficiencies are unmatched anywhere else in the entire Nile basin (Riddell 2010 *ibid*)—at least not at sector level (individual schemes elsewhere may do better). To reallocate water away from Egypt's agricultural sector would therefore reduce the economic efficiency of water use in the basin—not just because agricultural water productivity would decline, but also because water flowing to Egypt has other benefits such as navigation and power generation, etc. The positive correlation between basin level economic efficiency and basin welfare[22] has already been mentioned. Hence in terms of basin level management the real question might concern where the best place would be for Egypt to store its water, and to what extent does it/should it reward the countries actually supplying it?

---

[19] http://www.voanews.com/content/nile-basin-initiative-meeting-water-share/1686319.html.

[20] http://allafrica.com/stories/201306201064.html.

[21] http://allafrica.com/stories/201309232232.html.

[22] In terms for instance of productive water use, equitable access to water and environmental sustainability.

In terms of comparative advantage, the calculus is simple: Egypt has clear comparative advantage in terms of agricultural water productivity while Ethiopia has clear comparative advantage with respect to storage, even of Egypt's water. And were that comparative advantage mobilized, a portion of the water stored in Ethiopia could also be used for power generation (of potential benefit to the other countries), while increasing navigation depths in Sudan's share of the Blue Nile—thereby increasing basin level economic efficiency of water use further.

Similarly, senior technical officials in Sudan concede that rather than spend between $200–300 million each year dredging the Nile to remove silt, comprehensive and more sustainable investment in watershed conservation in Ethiopian highlands would only cost a fifth of that sum while creating a strong framework for deepening transboundary cooperation.

But clearly, the prevailing rhetoric throughout the basin confirms that comparative advantage is still being sacrificed on the altar of political economy—notwithstanding the enormous and invaluable body of knowledge that the NBI has generated, and the studies that it has commissioned or facilitated confirming the potential that regional approaches have with respect to accelerating the sustainable development of each member country.

This chapter therefore ends with an assessment of what this could and should mean for the future.

7 At the Cross Roads

The future is unknowable, yet policy makers need to make decisions now in order to secure beneficial outcomes—and this is the key to the Nile's future, good or bad: why should a politician risk political capital making a decision that makes them unpopular now in order to make another politician look good in the future? This dilemma takes us to the heart of political economy as we understand the term in this context, and scenarios provide a helpful way resolve the dilemma. Although they are little more than stories about the future, by suggesting alternative future paradigms, scenarios raise awareness of the likely outcomes of different courses of action.

Accordingly, and returning to our hypothesis that water is like financial capital in that it can be invested for short term gains for a few or longer term prosperity for the many, we are suggesting that the future paradigms most relevant here are:

- The ***investment philosophy***—the choices being political economy or comparative advantage; and
- The ***scale*** at which an investment is made—these choices being either country or basin.

The first scenario ***"Business as Usual"*** in Fig. 10.5 suggests what will happen if nothing changes in existing practices. Each Nile riparian will go it alone, making investments that are more likely to support political rather than socio-economic or developmental agendas. Examples might be an expensive irrigation scheme or a large dam of limited utility near to a government minister's home town. Such investments are likely to be unfeasible both financially and economically while benefits are highly localized, water is taken out of circulation and allocated significantly below its opportunity cost and the multiplier effects of economically mobile water are lost.

The second scenario ***"Keeping it Close to Home"*** in Fig. 10.5 adopts the principle of comparative advantage, but applies it only within a particular country. Benefits are more evenly shared within the country concerned, value chains emerge and jobs are created along them in diversifying livelihood niches. Investments are more likely to be financially and economically feasible, but there is a danger that downstream interests would lose out. For instance, the Grand Ethiopian Renaissance dam under construction just 45 km inside Ethiopia's border with Sudan may well represent comparative advantage in the context of Ethiopia's many water related economic constraints (low water-resilience, excessive reliance on bio-mass for energy production, etc.) and indeed in terms of the irrigation potential that could be served by the dam (if high unit feeder system costs can be tamed, which is likely because the Grand Renaissance Dam would

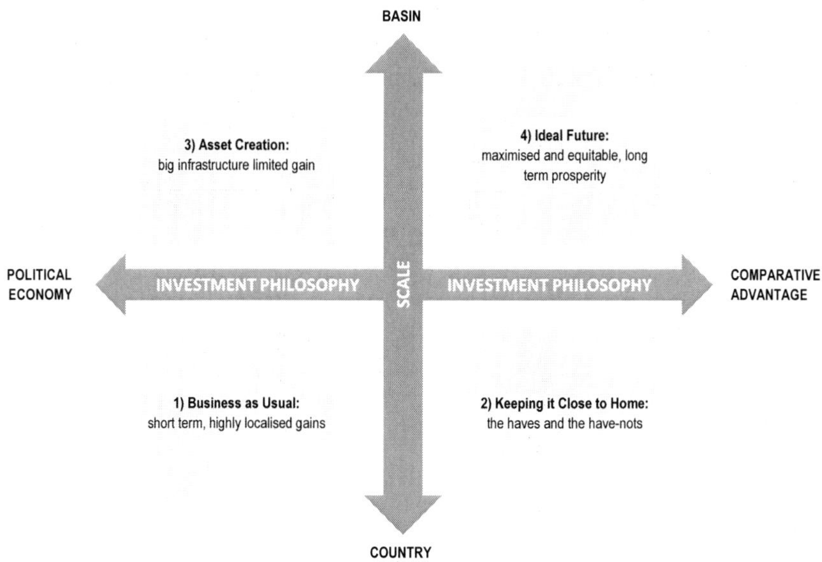

Figure 10.5. A Framework of Scenarios for use of water in the Nile Basin.

not need the expensive contour canals that dams further upstream might). But little is known about the dam's likely effect on downstream flows and hence downstream water use (which as we have seen is economically efficient in Egypt). Thus the likely outcome under this scenario would be increasing differentiation between the countries that have, and those that have not. And given the volatility of the region, this could lead to military conflict of the kind already threatened by Egypt, if Ethiopia continues on this course.

The third scenario ***"Asset Creation"*** in Fig. 10.5 recognizes the importance of basin level initiatives, but would be limited to grandiose schemes, notably large dams. These would represent political economy in terms of much needed international cooperation; but they may be very costly in environmental, social and even economic terms while locking vast amounts of water into potentially sub-optimal operating rules. Any gains could be limited in comparison to those accruable to less costly, more localized alternatives such as sand dams and mini-hydro (or alternative sources of renewable energy). Also, where cascades of such dams are involved there is the risk of manufactured scarcity (Molle 2009), and if the dams are dedicated to power generation, there is also the risk of catastrophic flooding because of operating rules that favor full supply levels whenever possible and hence reduced capacity for flood attenuation.

Finally, there is the fourth scenario ***"Ideal Future"*** in Fig. 10.5, wherein all water is allocated to the most productive and appropriate use in a given location, and benefits are shared in a way that facilitates long term peace, stability and prosperity in the Basin itself while maximizing benefits wherever else the Nile casts its shadow, not least by means of regional, employment creating trade. Infrastructure will be encountered at various scales, but all infrastructure having transboundary impact will be operated by suitable transboundary institutions and in a multi-purpose fashion to the greatest practical and technically feasible extent.

In closing, and based on their ongoing work, the authors have identified a pervasive lack of awareness at all levels with respect to the win-win benefits of a transboundary approach to water resources allocation and exploitation in the Nile Basin. Yet without the necessary levels of awareness the sanctioned discourse (Allan 2002) is likely to remain fixated on the political economy characterizing business-as-usual. But with a robust program of awareness raising, it is reasonable to expect a development trajectory towards an ideal future wherein the vested interest of some could be transformed into the best interest of everyone.

References

Abd-Elhamid, H.F. and A.A. Javadi. 2008. An Investigation into Control of Salt Water Intrusion Considering the Effects of Climate Change and Sea Level Rise. School of Engineering, Computing and Mathematics, University of Exeter, Exeter, UK.

Allan, T. 2002. The Middle East Water Question: Hydropolitics and the Global Economy. I.B. Tauris, London.

Cai, X., C. Ringler and M.W. Rosegrant. 2001. Does efficient water management matter? Physical and economic efficiency of water use in the river basin. Environment and Technology Division, International Food Policy Research Institute, Washington DC.

EPCO. 2000. Environmental status report. Environmental Planning and Coordination Organization, Madhya Pradesh.

Hirji, R. 2009. Climate Variability and Water Resources in Kenya: The Economic Cost of Inadequate Management. World Bank, Washington DC.

Keller, D., J. Keller and D. Seckler. 1996. Integrated Water Resource Systems, Theory and Policy Implications. International Water Management Institute, Colombo.

Lamere, C. 2012. Nile Basin at a Turning Point as Political Changes Roil Balance of Power and Competing Demands Proliferate. New Security Beat, Environmental Change and Security Programme, Wilson Centre, Washington DC.

Molden, D., U. Amarasinghe and I. Hussain. 2001. Water for rural development. International Water Management Institute Working Paper No. 32. Colombo.

Molle, F. 2009. The Bang Pakong River Basin Committee: Analysis and Summary of Experience. Institut de Recherche pour le Developpement, Paris and FAO, Rome.

Riddell, P. 2010. Agricultural Water Use Projections in the Nile Basin to 2030: Comparison with the Food for Thought Scenarios. FAO, Rome.

PART 3

Water Management Policy, Politics and Economics

Water, especially freshwater, is such a vital resource. Policies and projects focused on freshwater ecosystem alterations have been carried out through much of modern history, with the intensity of modifications increasing in the early to mid-1900s. Common waterway modifications, such as the construction of dams and irrigation channels, inter-basin connections and water transfers, can impact on the hydrology of freshwater systems, disconnect rivers from floodplains and wetlands, and decrease water velocity in riverine systems. This, in turn, can affect the seasonal flow and sediment transport of rivers downstream, impacting on fish migrations and changing the composition of riparian ecosystems. All of these issues require a balanced approach to their resolution. Legislation, policy formulation and the role of socio-economic forces are all part of the complex matrix that represents modern day responses to increasing demand for water and its dwindling supply relative to global population.

The five chapters examine aspects of the responses of societies concerned about ensuring a continuing supply of freshwater to service the needs of agriculture, industry, domestic use and the environment. **Du et al.,** uses the Yellow River Commission in China as a case study of how a large but mainly arid country supporting the world's largest human population has legislated to manage and allocate water from one of the world's longest rivers, **Loch et al.,** elaborate on the issues and conflicts involved in managing Australia's largest river system that services water users in five separate jurisdictions. **Xu et al.,** present an analysis of how China has tackled the management of a large inland river basin in an arid part of north-west China, and **Krutov et al.,** summarize the present situation in the Aral sea basin and examine the role of the Republic of Tajikistan in the Inter-state Aral sea commission. **Kingsford et al.,** analyze the issues in the Lake Eyre basin in central Australia.

11

Sustainable Management of Water Resources in the Yellow River Basin: Main Issues and Legal Approaches

Du Qun,[1] Zhuang Chao[2] and Simon Spooner[3,*]

SYNOPSIS

This chapter reviews and analyzes developments in the legislative and administrative arrangements in the Yellow river basin, one of China's most important river basins. The Yellow River drains a semi-arid area about as large as France but with twice the population from which the total runoff is relatively small but the sediment load is greater than any other river on earth. The river basin managers face serious challenges of a rapidly developing society demanding more water while past overexploitation and climate change mean less is becoming available. Floods are potentially disastrous, having killed millions of people in the past; droughts are severe; and industrial, urban and agricultural pollution have led to a water quality crisis and environmental degradation. The Yellow River Conservancy Commission (YRCC)

[1] Collaborative Innovation Center of Judicial Civilization, PRC, Research Institute of Environmental Law, Law School, Wuhan University, PRC.
Email: qdu@wuhan.edu.cn

[2] Research Institute of Environmental Law, Law School, Wuhan University, PRC.

[3] Atkins Water & Environment, Epsom, UK.
Email: simon.spooner@atkinsglobal.com

* Corresponding author

was created to help coordinate management across the river basin but its authority and jurisdiction are not complete, encountering conflicts with the administrative provinces of the basin and between water, environmental and other technical departments. The goal is to achieve integrated management of quantity and quality, surface and ground water, river and watershed. The question is how to bring about the necessary regulatory controls and behavior changes—by legal instruments or by policy and administrative means, by developing a special Yellow River Law or by developing national legislative and political mechanisms and implementing appropriately to the special needs of the Yellow River Basin.

Keywords: Bohai Sea, eco-civilization, floods, gauging stations, hanging river, HuangYuanKuo, Lanzhou, legislation, levees, LongMen, permits, runoff, SanMenxia, sediments, Tuo DaoGuai, water dispatch, water pricing, water resource management, XiaoLangDi, Yellow River Law

1 Introduction

The Yellow River genuinely is a river like no other. Originating in the high mountains, rain and snow melt provide a year long base flow that runs down through arid lands where the unusual Loess Plateau geography means that when the rains do come the flood washes off vast quantities of sand and silt and the river carries more sediment across the wide flat NE China plain and into the sea than any other river on earth. The challenges for people living in the Yellow River valley are the need to build and continuously maintain complex irrigation and flood protection systems and the constant threats of the river flooding and dropping massive loads of sediments that can bury whole cities and result in the river dramatically changing course and direction. Only an organized society can survive under these pressures. When managed well the Yellow River valley can be a place of plenty in the midst of harsh lands. For these reasons the Yellow River valley has long been at the center of China's political, economic, and social development. The dual nature of the river has resulted in the simultaneous use of the phrases "the cradle of Chinese civilization" and "China's sorrow" to describe the Yellow River.

The efforts to manage the river, its basin catchment, its resources and its people are the subject of this chapter and bring together the two major themes of this book, namely, *people* and *place.*

1.1 Geographic description of YRB

The Yellow River, or Huanghe, is the second longest river in China. Originating in the Bayangela Mountains in western China, the river drops a total of 4,500 m as it loops north into the Gobi Desert before turning south through the Loess Plateau and then east to its terminus in the Bohai Gulf. In total, the river flows over 5,400 km, passes through nine provinces and autonomous regions and drains an area considerably larger than the area of France (see Box 11.1).

Box 11.1. Yellow River Basin: some facts.

Drainage: 794,712 km^2

River length: 5,464 km

Irrigation area: 5 million hectares

Population: 113.68 million

Water uses: Irrigation, domestic and industrial use, electricity generation, supporting ecological health of areas of high conservation value

The Yellow River is commonly divided into three reaches Upper, Middle and Lower as indicated in Fig. 11.1.

The upper reach of the Yellow River drains over half of the total basin area and extends from the river's origin in the Bayangela Mountains to the Toudaoguai gauging station near city of Datong. While the upper reach provides a large part of the basin's surface runoff (YRCC 2002a), the contribution comes from two distinct geographic backdrops characterized by counteracting physical processes. On the Tibetan Plateau where the Yellow River begins, steep rock slopes draining onto flatter areas with deep

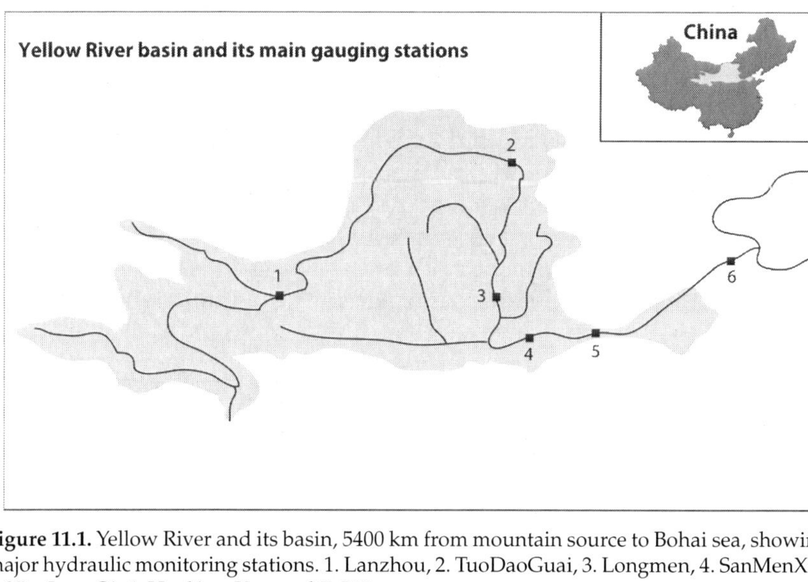

Figure 11.1. Yellow River and its basin, 5400 km from mountain source to Bohai sea, showing major hydraulic monitoring stations. 1. Lanzhou, 2. TuoDaoGuai, 3. Longmen, 4. SanMenXia, 5. XiaoLangQi, 6. HuaYuanKou and 7. Lijin.

soils high in humus, low evaporation, and high moisture retention produce high runoff coefficients. This, combined with relatively high precipitation levels, result in this western-most region of the upper reach contributing 56 percent of the entire river's total runoff above the Lanzhou gauging station (based on pre-1990s averages). As the river moves northward from there into the Ningxia/Inner Mongolian plains and the Gobi Desert, the evaporation rate increases to levels several times that of precipitation. As a result, the section from Lanzhou to Toudaoguai shows a net outflow from the main channel into the adjacent wetlands.

The middle reach, covering 46 percent of the basin area and providing an additional 43 percent of total runoff (based on pre-1990s averages), sits between the Toudaoguai and Huayuankou gauging stations. From Toudaoguai, the river begins its "great bend" to the south into and through the Loess Plateau. The middle reach of the river plays a significant role in basin water balances and availability for human use for two reasons. First, the reach includes some of the river's major tributaries, such as the Fenhe and Weihe, which contribute substantially to total flow. Second, as the river turns southward, it cuts through the Loess Plateau and its highly erodible soils. These soils contribute massive quantities of silt to the main stem and its tributaries, providing 90 percent of the river's total sediment and resulting in average sediment loads unprecedented amongst major waterways. Unpredictable and intense summer storms in the reach exacerbate the sedimentation problem and are the major cause of the Yellow River's historically devastating floods.

The lower reach of the Yellow River commences at Huayuankou and forms one of the most unique river segments in the world. Here the sediment transported from the middle reach begins to settle as the river spills onto the flat North China Plain, producing a consistently aggrading bed and a naturally meandering and unstable channel. To stabilize the channel, successive river managers have constructed levees to hold in the river. While such structures may succeed in the short term, their success depends on repeatedly raising the levee walls to keep pace with the sedimentation which elevates the level of the channel that runs within these walls. Over time, the process of levee rising has created a "suspended" river in which the channel bottom is above ground level, sometimes by as much as 10 meters. This raising of the channel above the level of the neighboring countryside has clear implications for the severity of flooding because when levees break the whole river can change course, never returning to the original channel. This also alters the meaning of the term "basin" in the Yellow River context.

With the channel above ground level, rainfall on surrounding lands cannot drain into the river nor can tributaries enter. This essentially means that the river "basin" becomes a narrow corridor no wider than the few kilometers breadth of the dyked channel. With almost no inflow, the contribution of the lower reach is limited to only 3 percent of total runoff.

From Table 11.1 it may be seen that the water resources situation in the Yellow River has been reasonably stable over recent decades with trends of increasing reliance on groundwater abstraction (which consistently exceeds the renewable resource) and in water use the shift from agricultural to urban and industrial uses, but still with irrigation consuming 70% of available water in 2011, which much lower than the 86% in 1980. Overall around 60% of the available resource is consumed each year which is well above the level of 40% which would generally be considered sustainable.

In Table 11.2 the longer term picture is of decreasing surface water resources over the period from the 1950's to the 1990's with the total resource of the Basin falling from more than 60,000 Mcm to less than 45,000 Mcm, by contrast the last 7 years of data from 2004 to 2011 show total surface runoff back up to around 55,000 Mcm (Table 11.1).

1.2 Institutional structure of the water management in the YRB

The Yellow River Conservancy Commission (YRCC) has the primary responsibility for water management in the Yellow River basin this includes flood, sediment, ice, drought, water quality, water allocation and environmental water management. This work is carried out in an integrated way, with one of the key objectives being investing in management actions that jointly improve ecological outcomes and enhance the welfare of the people who live in the Basin. YRCC follows an adaptive management model, whereby the actions are based on a continuous review of science-based investigation and monitoring, and an awareness of the changing needs of the communities that live in the Basin.

Six key Yellow River issues managed by YRCC are:

- Land degradation in the Loess Plateau
- Risk of loss of life and economic losses due to major floods
- Build up of sediment within the river system
- Shortage of water to meet all human and ecological demands
- Protection of the ecological values of the Delta
- Maintaining water quality of an acceptable standard

Table 11.1. Water resources and water use by source and by user type 1980 to 2011.

Year	Water Resources			Water Use by source				Water Use by User Type				%Resource
	Surface	Ground	Total	Surface	Ground	Other	Total	Urban	Industry	Agriculture	Envir. flow	Consumed
2011	62.09	11.85	73.94	26.85	12.9	0.69	40.44	4.8	6.55	28.14	0.96	55%
2010	56.89	11.1	67.99	26.26	12.68	0.28	39.23	4.4	6.15	27.76	0.91	58%
2009	55.17	10.52	65.69	25.67	12.71	0.19	38.57	4.33	5.69	27.87	0.68	59%
2008	45.42	10.49	55.91	25.39	12.81	0.22	38.42	3.98	6.08	27.72	0.65	69%
2007	54.21	11.33	65.54	24.91	12.95	0.25	38.11	3.99	6.15	27.45	0.52	58%
2006	45.6	10.83	56.43	25.62	13.7	0.29	39.61	3.94	6.04	29.25	0.37	70%
2005	65.73	9.9	75.63	24.48	13.32	0.35	38.15	3.83	5.56	28.41	0.36	50%
2004	51.85	10.95	62.8	23.79	13.21	0.21	37.21	4.80	6.47	25.21	0.32	59%
2002				25.17	13.73		38.90	3.43	5.38	29.33		
1993				27.89	12.85		40.74	3.32	5.35	31.27		
1980				27.4	8.44		35.84	1.58	2.74	30.82		

All figures in bcm, Source: Ministry of Water Resource (MWR). 1993, 1980; YRCC 2002b, 2004–2011

Table 11.2. Rainfall and Runoff in Yellow River Basin, 1956–2000.

Area (000 km²)		Time Period					1990s Change From Average
		1956–70	1971–80	1981–90	1991–00	Average	
Upper 368	Rain (mm)	380	374	373	360	372	–3%
	Runoff (bcm)	35	34	37	28	34	–16%
	Runoff yield (%)	25%	25%	27%	21%	24%	
Middle 362	Rain (mm)	570	515	529	456	523	–13%
	Runoff (bcm)	29	21	23	15	23	–34%
	Runoff yield (%)	14%	11%	12%	9%	12%	
Lower 22	Rain (mm)	733	689	616	614	671	–8%
	Runoff (bcm)	1.5	1.1	0.6	0.0	0.8	–100%
	Runoff yield (%)						
Basin 752	Rain (mm)	482	451	455	413	454	–9%
	Runoff (bcm)	65	56	61	43	57	–24%

Source: Zhu et al. (in English), 2003

2 Main Issues Regarding Water Resources and Environment in the Yellow River Basin

2.1 Water quantity management: Water shortage and the imbalance of water supply and demand

The growing population, rapid economic development and urbanization in the YRB are intensifying water shortage and environmental degradation. In the last decade, the coal and chemical industries in autonomous regions of Inner Mongolia, Ningxia and Xinjiang and in Shanxi Province have consumed large amounts of water resources. There were 32 coal and chemical key projects under construction and operation in the National 11th Five-Year Plan (2006–2010) and 15 new key projects are to be initiated in the National 12th Five-Year Plan (2011–2015). These coal and chemical industrial projects will together demand 11.1 bcm of water per year, namely 3.04 million m^3 per day, about the same as the water demand for Beijing in 2012 and very much more than current industrial consumption (Wen 2012). The total amount of available water resources is relatively inflexible so this additional water requirement can only come by a reduction in the water used by agriculture. The contradiction between the industry and agricultural irrigation and other use, such as urban demands, environmental conservation, power generation, ice prevention, fisheries, and shipping water gives rise to the greatest challenge of the water quantity management in YRB. There is not enough water available to meet everybody's requirements so all have to make do with less than they would like.

Actually, almost all cities in the YRB face different degrees of water shortages. Although taking into account implementation of strengthened water-saving measures (but excluding water transfer from other river basins), the gap between water resource demand and supply of the Yellow River is estimated to grow to 7.5 bcm in 2020 (Chen 2011).

In order to overcome the shortage of Yellow River water supply, irrational water use occurs, for example over extraction of the groundwater to meet short term needs. In the longer term, irrational water use leads to serious environmental harm and disaster. Statistics showed that in 2010 (YRCC 2010a), the long-term excessive mining of groundwater formed 4 cones of depression in artesian groundwater and 8 cones of depression in non-artesian groundwater covering a total of 87,777 square kilometers namely 11% of the YRB.

The south north water transfer has been conceived to address some of these shortages. The central route of the transfer will bring water to some of the cities in the east of the Yellow River Basin, such as Zhengzhou, Handan and Shijiazhuan and on to Beijing which, though actually outside of the river basin were being supplied with water from the river and will continue to be so. But the cost of this water and the limited amounts that it can supply mean that this transfer will merely off set the growing demand from these cities that would otherwise have to be drawn from the Yellow River. The western route is planned to transfer large quantities of water from the Upper Yangtze into the headwaters of the yellow River. This would have major impacts providing 10 to 15 bcm to the main river but the engineering challenges and costs involved are such that this is unlikely to be constructed in the foreseeable future. Smaller transfers through the Qingling mountains to the Wei Valley are under way and more are under construction and will bring local benefits, but no basin level solutions.

2.2 Water quality management: soil erosion and water pollution

Despite the many challenges in water quantity management, the water quality issue in YRB cannot be neglected. "The 2010 Surface Water Quality Bulletin of the Yellow River Basin" pointed out the Yellow River faced the threat of water quality deterioration and resultant water function degradation from severe water pollution; it showed that the surface water quality of 33.9% river flow of Yellow River was inferior to Class V (YRCC 2010b). At this standard the water is not fit for any purpose without treatment and there have been reports of outbreaks of cancer associated with consuming water from the river (Lu et al. 2008).

The economy of the YRB is based on coal and chemical industries with high water consumption, heavy pollution and low levels of clean production. Large amounts of untreated or ineffectively treated industrial waste water and untreated urban effluent are discharged into the river. In 2010, the effluent discharges into the Yellow River basin streams was 4.4 billion m^3, of which civil, industrial and commercial waste water were respectively 1.2, 2.9 and 0.34 billion m^3, accounting for 27%, 65% and 8% of total effluent production (YRCC 2010c). With the heavy load of effluent discharge, the urban sewage treatment rate is only 13%, the same as the average rate for the whole of China (Ministry of Environmental Protection 2005). Agricultural runoff is also a significant source of pollution, especially from large scale livestock rearing, contributing about 40% of COD and Ammonia pollution of the river. While The Ministry for Water resources is responsible for the quality of water in the river and can coordinate basin wide through YRCC, the Ministry of Environmental Protection is responsible for the control of discharges of pollutants to the river, but must exercise enforcement of standards through the provincial Environmental Protection Bureaus over which it has only partial control to influence municipal governments and industry and even less control over polluting agricultural practices.

2.3 Flood management: a hanging river

The basic characteristics of the Yellow River are described as "less water but with too much sand, and both temporal and spatial imbalances in the distribution of water and sand". The frequent cutoffs in the downstream reach of the Yellow River caused obstructions to sediment transport and deterioration in the ecological environment, e.g., biodiversity and coastal wetlands, and an aggravation of flood risks.

Historically, the flood disasters of Yellow River occurred so frequently that there once existed a saying "the Yellow River broke its riverbank twice every three years, rechanneled once every one hundred years". From this saying we understand that flood management is of a very critical issue in the YRB. Indeed the 1887 and 1931 floods of the Yellow river are noted as the most serious natural disasters in the history of the world with death tolls in the many millions.

Most water in the Yellow River originates from the upper reach above Lanzhou. However, more than 90% of sediment comes from the middle reaches as the river flows through the Loess Plateau region from Hekou Town to Tongguan. Floods carry large amounts of sediment into the downstream river. Since the downstream lacks sufficient water to transport this sediment, most of it is deposited in the river bed, making the Yellow River a hanging river above the surrounding country. This situation generates great flood control risks. As the downstream flood plain of

Yellow River stretches to the Huai River and Hai River basins, the Yellow River flood control issues, particularly in the lower reach, has become the country's most serious concern since it relates not only to the water and grain security but the very survival of people cities in the yellow river flood risk areas of the North China Plain.

Flood control dams and strengthened dykes have, in recent years, protected against the devastating floods in the Yellow river plain but the economic and human losses caused by flood disasters along the river are still huge. For example, the flood in July to August of 2012 in Gansu Province affected 29.6 million people living in 9 Cities (or autonomous prefectures) and 31 counties (districts). In this flood 2.8 million people were temporarily relocated and more than 1 million people were given emergency living assistance, 300 houses collapsed and 6,000 houses were damaged to various degrees; besides, 20.8 thousand hectares of crops were affected. In total, the direct economic losses amounted to about 620 million RMB Yuan (Tian 2012).

3 The Legal Framework of YRB Water Resource Management and Development of Key Legal and Administrative Instruments for River Basin Management

In order to confront the specific management issues in the Yellow River Basin, a legal, regulatory and institutional framework has been established through which unified water utilization dispatch and allocation, water permit and water rights transfers serve as main instruments to implement sustainable and efficient use of water resources.

3.1 Legal and regulatory framework for water utilization and management in the YRB

At present, there are five national laws related to the management and protection of water resources of YRB: *The Environment Protection Law (1989); Water Law (2002); Water Pollution Prevention and Control Law (2008); Water and Soil Conservation Law (2010);* and *Flood Control Law (1998).*

In response to water pollution issues in the Yellow River, the major implementing law is the national law of *Water Pollution Prevention and Control Law*. It was revised in 2008 and stipulated that the prevention and control of water pollution shall conduct unified planning on the basis of river basins or regions. It requires that specialized national regulations or river basin level regulations for water pollution prevention and control for Yellow River Basin be developed. The current management objectives for YRB are mainly focused on the water quantity and with less emphasis on water quality protection and pollution control. The 2011 Number 1 Policy document on Water sets out a framework for comprehensively and strictly managing water resources in accordance with the "3 red lines" of water allocation, water use efficiency and pollution load control. The current legal framework does not give full control to the river basin authority to control and allocate water resources and pollution loads but requires a balance between the River basin level and the administrative regions of provinces, counties and municipalities in the YRB.

A specific regulatory framework, consisting of rules and accompanying comprehensive provisions, has been developed for water resource utilization in YRB. The principal documents are: *Available Water Allocation Program of Yellow River (1987), Interim Measures of the Yellow River Basin for the Management of Water Resource Appraisal for Construction Projects (2003); Regulations for Yellow River Water Dispatch (2006) and Implementing Measures for the Management of Water Rights Transfer of Yellow River (Trial) (2004).*

Flood control and drought resistance are important aspects of water resources utilization and management in the Yellow River. A large number of regulatory rules have been formulated for the prevention of and coping with flood and drought. These rules include: *The Responsibility System for the Water Dispatch of the Downstream of the Yellow River (Trial) (2002), The Measures for the Cap Control Management of the Yellow River Water Taking Permits (2002), The Provisions for the Emergency Response, Investigation and Resolution of Significant Water Pollution Incidents in the Yellow River (2003), The Provisions on the Emergency Response and Disposal of Water Dispatch Incidents (2008),* and *The Preparatory Program for the Resistance of Drought in the Yellow River Basin (2008).*

3.2 The implementation of main legal instruments for the water resource management in YRB

3.2.1 Allocation plan for available water of YRB

Since water is the scarcest resource for economic development in YRB region, the core social and economic issue concerning water resource is how to allocate the limited available water of Yellow River among various jurisdictional areas, or further how to coordinate the interest parties in equal and fair use of water of Yellow River. Until around 1987, all parties competed in taking water from the Yellow River and in excessive extraction of ground water for their own regional development with no concern for the consequences for impact on flow in downstream reach. This aggressive taking of ground water resulted in ground water levels in Taiyuan and Xi'an and other cities declining rapidly and brought about serious geological and environmental problems such as water depression cones. Therefore, the development of a unified water resources allocation plan for water taking from Yellow River, together with a cap control indicator system, was of great necessity and became the first priority task to implement river basin management.

The first unified *Available Water Allocation Program of Yellow River* was thus initiated by Yellow River Conservancy Commission (YRCC) and approved by State Council. It was known as *The 87 Program* as it took effect in 1987. *The 87 Program* was a milestone for YRCC tasking them to actually implement integrated water allocation schemes in the YRB. Nationwide, *The 87 Program* was the first integrated river basin water allocation plan for national key rivers and lakes administrated by specialized river basin management institutions under the Ministry of Water Resources. The aim of *The 87 Program* was to balance water supply and demand and to solve contradictions and disputes in water use among various jurisdictions in YRB (see Li and Squires, this volume).

The 87 Program allocated initial water rights to the provinces and autonomous regions. *The 87 Program* indentified that the natural runoff of the Yellow River was 58 bcm, of which 37 bcm was defined as available to be allocated to 9 provinces and autonomous regions in the YRB and the adjacent Hebei Province and Tianjin Municipality (the latter two jurisdictions are adjacent to YRB and suffered from water shortage at that time). The remaining 21 bcm was planned for sand transportation and other ecological water uses. *The 87 Program* served as the fundamental basis for the water resource allocation in the YRB provinces (and autonomous regions) as well as to implement other instruments for basin water resources management, e.g., water permit and trading system. However, *The 87 Program* only provided macro water allocation indicators under average natural flow conditions. It could not give specific guidance on management of Yellow River during years of drought. Additionally, the Program only provided a cap control indicator of annual water allocation for each province (or autonomous region) and lacked seasonal or other sensitive water allocation indicators (Peng and Hu 2006).

While implementation of *The 87 Program* had strengthened the water management of YRB and alleviated some contradictions between supply and demand of water resources, it did not solve water sharing issues in the Yellow River nor prevent competition for Yellow River water by concerned parties. Considering the conditions of Yellow River water supply and demand had changed, and in particular increasing concern for water required for ecological and environmental functions of Yellow River, the National Planning Committee and Ministry of Water Resource in 1998 formed and implemented a new unified available water allocation plan—*the Program of Yellow River on Annual Allocation of Available Water and Quantity Dispatch of Mainstream Water* (hereafter called *The 1998 Program*).

The 1998 Program brings significant progress in comparison to *The 87 Program*. First, it reduced the available water taken from the Yellow River to 31 bcm, 6 bcm less than that in *The 87 Program*. Secondly, it distinguished water allocation plans for three scenarios: high, normal and dry years; and in each scenario water allocation control indicators were prepared and respective guaranteed rates of water allocation were provided for the whole YRB to comply with. Third, the implementation of The 1998 Program was strengthened through a new institution—the Water Resources Management Bureau (WRMB), which was established in the YRCC. In March 1999, WRMB issued its first water transfer mandate, which historically kicked off unified water allocation for the entire YRB. The implementation of *The 1998 Program* has had very positive impacts. It has eased the tight contradiction between water supply and demand in YRB and has successfully prevented the cutoff (cease-to-flow) in the downstream reaches of Yellow River since the 12th August of 1999. *The 1998 Program* and its implementation set up an important administrative model or practice for comprehensive river basin management, in which the provinces, autonomous regions and municipalities within the river basin are being required to initiate and execute their local water planning and programs within the context of the central governments' or river basin commissions' overall planning and programs.

The practices of *The 1998 Program* and WRMB were formally recognized by the revision of the 1988 *Water Law* in 2002. From then on the *2002 Water Law* has provided the legal basis for river basin management institutions (e.g., YRCC) and authorizes them with the power for unified and comprehensive water resources management and river basin water quantity dispatch.

3.2.2 Integrated water quantity dispatch system of the Yellow River

The integrated water quantity dispatch system of the Yellow River is another effective legal instrument for integrated water resource utilization and management for YRB. It also has close links with the available water allocation program of YRB. The Yellow River water allocation program, approved by the State Council, is the foundation of the Yellow River water dispatch, in another word, water dispatch has to abide by the water allocation program.[1] The integrated water dispatch system also defines the quantitative water rights for jurisdictional areas, which further constitutes the foundation for water rights transfer or trade among interested jurisdictional parties.

[1] See the Article 3 and 10 of *the Regulations of the State Council on Water Regulation of the Yellow River.*

The integrated water quantity dispatch system of the Yellow River began in March 1999. It was prepared as a formal legal instrument in 2002 under the revised *Water Law*. In 2006, the State Council issued the *Regulations for Yellow River Water Dispatch*, providing more specific rules for the implementation and execution of Yellow River water dispatch.[2]

The *Regulations for Yellow River Water Dispatch (2006)* sets out the principles of water dispatch, including the cap control, section flow control, hierarchical administration and responsibilities system. It authorizes the YRCC to have the administration power for water dispatch, supervision and of surveillance (Chao and Wang 2011). This regulation has the following stipulations:

(i) The period of Yellow River water dispatch is from 1st July to 30th June of the following year.

(ii) The scope of Yellow River water includes the main stem and tributaries of the upper and middle reaches of the river and the main stem of the lower reaches.

(iii) The water dispatch principles are to comply with the river basin planning and long-term water supply planning, to give full consideration of water resources conditions of YRB, including the water resource situation, status and projected future trends of water supply and demand balance.

(iv) The objectives of water dispatch and allocation are to meet the needs of residents living in urban and rural areas, to make reasonable arrangements for water utilization among agriculture, industry, the ecological environment, and ensure water flow is maintained to the downstream sections of the Yellow River.

(v) The adjustment of the water allocation plan shall comply with procedures to consult with the 11 provinces in YRB; the adjustment proposal shall be reviewed by the State Planning and Development Committee and finally be approved by the State Council. Later *the Regulating Rules for Yellow River Water Dispatch (Trial)* were enacted in 2007 to provide detailed regulatory requirements for annual total water control and water dispatch in the trans-jurisdictional tributaries in the non-flood season and other matters.

(vi) The responsibility system of the leadership of the local people's government and the YRCC and water dispatch affiliated institutions or competent departments are applied as a political and legal instrument

[2] The *Regulations for Yellow River Water Dispatch* were formulated on the basis of *Water Law 2002*. It further built up the *Implementing Measures of the State Council for the Water Taking Permits* and *the Instructive Opinions of the Ministry of Water Resources on Water Right Transfer Pilot Work of the Yellow River Main Stream in Inner Mongolia and Ningxia* and in order to realize optimal allocation and efficient utilization of water resources of Yellow River and regulate the water rights transaction in YRB.

to ensure the effective implementation of integrated water dispatch. The YRCC is responsible for the supervision and inspection of the Yellow River water dispatch and implementation. The water administrative department at local levels and the executive institution responsible for water dispatch within the YRCC are in charge of the execution, inspection and supervision of water dispatch plans in authorized jurisdictional areas of the Yellow River. Specifically, the YRCC is responsible for releasing the tributary water dispatch quantity quota and for comprehensive supervision and inspection; the responsible local authorities are accordingly to ensure water taking from the respective tributaries is in compliance with regional water allocation quota controls and cross-sectional flow indicator controls.

(vii) The emergency response mechanism for water dispatch incidents has also been established. The Provisions on the Emergency Response and Disposal of Water Dispatch Incidents (2008) stipulate procedures for prediction, monitoring and reporting of the Yellow River Water emergencies and incidents, and define the reward and punishment measures for water dispatch emergency management.

3.2.3 Water taking permit system of the Yellow River

The water taking permit system is one of the basic legal instruments to manage water resources. The above-mentioned available water allocation and water quantity dispatch systems could be seen as the tools to carry out water allocation at various regional and provincial levels in the first instance, whereas the water taking permit system deals with the secondary instance of water allocation, in which substantial users will have rights to take and use water from Yellow River through the legal procedures and under given legal conditions. The YRCC recognizes that the stringent implementation of the water quantity dispatch system and water taking permit system is the most effective approach to prevent river flow cutoff in the downstream reaches of the Yellow River except under extreme drought conditions.

There are a series of ministerial provisions to regulate water taking permits. They are: *the Provisions on Review and Approval Procedures for Water Taking Permit (1994), the Measures for the Supervision of Water Taking Permission (1996) and the Provisions on Water Quality Management in Water Taking Permit (1995 and 1997)*. These policy documents were integrated into a formal legal formulation in 2004 by the *Regulation of State Council for Water Permit Management*. This regulation formally specifies the administrative requirements for water permits including application and acceptance, review and decision, and reiterates the roles of river basin management institutions that are affiliated to the Ministry of Water Resources with regard to their responsibilities for water taking permits. In 2006, when the

Regulations of State Council on Yellow River Water Dispatch were launched, the *Regulations of State Council on Water Taking Permit and Management of Collecting Water Resource Fees* were promulgated and implemented, providing a better legal basis to examine and approve water permits to construction projects in the Yellow River Basin. According to the above mentioned regulations, the YRCC exercises the administrative power for water permits in the Yellow River Basin.

Implementing the system of water taking permits in YRB has strengthened the total control of water taking, curbed the rapid growth of water demand, and allowed balancing water allocation between regions. In addition, the water taking permit system has also allowed water rights to be transferred in order to optimize water resource management of the Yellow River. For this purpose, the YRCC formulated a special decree, *the Implementing Rules for the Management of Water Taking Permit of the Yellow River* in 2009, dealing with the application, approval and supervision of water taking permits in YRB. In particular, it further clarified and strengthened the supervision power of the YRCC in water taking permit in the YRB.

Economists suggest pricing and the price system is the means to effectively curb demand in a market economy; a reasonable water price system can orient water demand (Cai and Yang 2010). In order to reflect this principle, levying small fees on water resources in YRB commenced in the 1980s. However, the price of water from the Yellow River is far too low to exert economic control on demand. The existing water pricing system mainly reflects only the operational cost of water supply companies, without any cost of water resource management or any price expression of the scarcity nature of water resources of Yellow River. In this sense, we can say that the real water price mechanism of Yellow River has yet to be established.

3.2.4 Pilot practice of water rights transfer

A water right transfer pilot in the YRB arose from a sincere concern over poor water efficiency, unreasonable water structure in the YRB and the reality of the water consumption of some provinces and regions of the YRB. The allocated quotas are regularly exceeded and defined indicators concerning river flows are violated. The introduction of a water right transfer system is expected to help ease the existing problems in the function of marketing mechanism in the Yellow River. In 2003, the YRC initiated the water right transfer pilot which allowed a province or autonomous region in the YRB to transfer saved water rights on the premise that overall it continued to consume water in compliance with the water quota and indicators defined by the State Council.

Usually, the agriculture sector consumes the majority of water resource and has large space to improve water efficiency by application of better water saving technologies, equipment and infrastructure, but with very narrow profit margins on produce has few financial resources to invest in water saving. On the other hand, the industrial sector has an ever-increasing water demand and the cost of water only makes up only a small portion of their final product cost. Therefore the water rights transfer system was designed to allow an industrial water user who has insufficient water rights allocated to them to be able to start production to invest in water saving in agriculture and pay farmers who then officially transfer a portion of their water access rights to the industry. From then on, as the farmers are using less water to produce the same crops they save money by having to pay less in water resources fees. This water rights transfer is a unique solution which effectively places a value on water rights for industry as the local cost of investing to save that amount of water. This is quite different to water rights trading in the free market sense. Though an excellent, pragmatic and simple solution, strong regulation and monitoring are still required to ensure that first there is a clearly established and documented baseline situation and then, once a transfer is agreed, agricultural users really do reduce their water use rather than simply producing more and that industrial users promptly pay for the investments required and also do not exceed the rights that have been transferred to them.

The Implementing Measures for the Management of Water Rights Transfer of Yellow River (Trial) (2004) issued by the YRCC regulated the water rights transfer in the pilot projects and state that all pilot projects of water rights transfer in the Yellow River Basin should adhere to three basic principles. First, water resource cap control must be obeyed. The transferred water rights must be part of a water quota and observe the indicators in accordance with the integrated available water allocation program approved by the State Council. Second, the water rights in transfer must be clarified and be able to be accounted. The pilot province or autonomous region should take care to prepare the implementing programs of water taking permit and assign water taking quota or targets to substantial water users within the cap control of regional available water. In a word, the initial water rights authorized by the procedure of water taking permit must be clear. Third, a command-control approach and marketing mechanism should be jointly applied in the water right transfer process. At present, almost all steps of water right transfer have been conducted in command-control approach, except for the pricing or compensation of transferred water rights which involves limited civil negotiation and marketing elements.

The pilot practice of water right transfer first took place in two autonomous regions of Ningxia and Inner Mongolia, guided by the *Instructive Opinions of the Ministry of Water Resources on Water Right Transfer*

Pilot Work in Ningxia and Inner Mongolia (2004). The water administrative departments of the autonomous regional governments in conjunction with the development and planning committees formulated overall plans for the water rights transfers which were approved by the YRCC. The YRCC identified 5 water right transfer pilots in Inner Mongolia and Ningxia. The two water right transfer pilot projects in Ningxia were water-saving projects in Qingtongxia Hedong Irrigation Area and in—Huinong Canal Irrigation Area, Hexi Corridor. The water saved from these two projects was transferred to the third phase expansion project of Ningxia Dam and Power Plant and the Ningdong Maliantai Power Plant[3] (Hu and Chen 2004). Compensation of transferred water resources or water rights is a key aspect of the process of water rights transfer. It ought to reflect the value of water resource in a market-oriented base. The water right transfer price in the pilot was calculated as the product of total cost of water rights transfer per unit of water quantity, the period of water rights transfer and the transferred water quantity. Nevertheless the actual total cost of water rights transfer should include—in theory but have not yet covered in reality—water right conversion costs and reasonable profits, in which the cost of engineering construction and operation, the amount of ecological compensation, the water right conversion period and other factors to be taken into account.

Up to the end of 2008, the YRCC had approved 526 water rights transfer projects transferring a total of 228 million cubic meters water. The total investment of the related water-saving projects was 1.226 billion Yuan (Wang and Tian 2010). In Ningxia, after the implementation of water right conversion, the total water use diminished from 7.8 bcm in 2005 down to 7.76 bcm in 2006 and 7.36 bcm in 2011. Apparently all of this reduction could be attributed to agricultural water use or irrigation as the first 3 YRCC-approved pilot projects in Ningxia has been completed. The total investment in agricultural water saving in these three pilot projects have accounted for 151 million Yuan, building up the annual water transfer capacity of 0.54 million cubic meters and saving water 30 million cubic meters annually (Wen 2012).

[3] The total investment of water-saving projects of the third phase expansion project of the dam and power plant is 49.327 million yuan, with the capacity of saving 18 million cubic meters water every year; the total investment of water-saving projects of Ningdong Maliantai Power Plant is 57.6 million yuan, with the capacity of saving 21.5 million cubic meters water every year. These water right transfer projects not only change the traditional model of the agricultural water-saving that always relies on governments' investment, but also broadens the investment channels and fund raising sources for agricultural water-saving irrigation construction.

3.2.5 The system of water function zones for water pollutants control

The system of water function zones is the main instrument of *Water Law (2002)* of People's Republic of China to protect water bodies from pollutant discharges.[4] Since 2000, in accordance with the overall plan of the Ministry of Water Resources, the YRCC has defined the water function zones for the main Yellow River. The provinces and autonomous regions of YRB also accomplished water function zoning for their own jurisdictional areas. The Yellow River Basin is divided into 488 first level water function zones and 465 secondary level function zones (YRCC 2008). In 2003, the Ministry of Water Resources issued the *Measures on the Management of Water Function Zones*. Its Article 2 set out that water bodies should be managed in defined water function zones based on their capacity to carry pollutants. The management plans for water function zones are used to guide rational planning of water resource utilization and development. Different water function zones have restrictive or protective measures for development and utilization activities. That is to say, the total amount of pollutants allowed in each type of water function zones should be audited according to water quality requirements of the water functional zone and the natural purification capacity of water bodies. Based on the water function zoning, a series of pollution control measures, for example, the environmental impact assessment, pollutant discharge controls, water quality monitoring and information reporting would be executed effectively. In 2005, the Water Resources Protection Bureau of the Yellow River Basin officially released *"No. 1 Water Resource Quality Bulletin of Water Functional Zones of the Yellow River"*. This publication has helped the relevant governmental authorities as well as the public understand the status of water quality in water functional zones. Thus it can be considered an important initial step for public participation in the river basin management of Yellow River by beginning with the public's access to the basic information.

In the *Water Function Zoning of National Key Rivers and Lakes* that was ratified by the State Council in 2011, 171 water function zones of the Yellow River basin were listed as the first level water function zones nationwide and 234 as the secondary level water function zones. This means more than 50% of water function zones of the YRB including all mainstream and main

[4] The third clause of the article 32 of the *Water Law* stipulates: The administrative department for water resources under the People's Government at or above the county level or the river basin authority shall, on the basis of the water quality required by a water function zone and the natural purification capacity of the water bodies of the zone, check and define the pollution-receiving capacity of the water areas there and make proposals to the administrative department for environmental protection on limitation of the total amount of pollution discharged to the said areas.

important branches of the Yellow River are of national significance. From 2004 to 2011 auditing work with regard to respective pollutant carrying capacity was completed for all water function zones in YRB by the Water Resources Protection Bureau of the Yellow River Basin (YRB Water resource Protection Bureau 2012).

In 2011 Water Resources Protection Bureau of the Yellow River issued *the No. 4 Water Resource Quality Bulletin of Water Functional Zones of the Yellow River*. It showed that 58 water function zones of the Yellow River Basin, accounting for 45.0% of the length of all water function zones in the YRB, were meeting the national standards of water resource quality, whereas 71 water function zones, comprising 55% of the total, failed.

The high non-compliance rate shows the restrictive and regulative measures for pollutant discharge to water bodies have been not well implemented. Water function zones are in nature a technical tool that relies on strict implementation of legal pollutant discharge measures, for example, allocating pollutants discharge permits to specific social bodies, monitoring pollutant discharge, coordinating supervision and execution of water law between water resource authorities and environmental protection authorities, etc. With the water resource quality targets set for each water function zone and effective implementation of pollutant discharge control measures, the pollutant discharge control for the Yellow River would be effective. Since 2007, the YRCC has carried out four joint inspections for water law enforcement taking account water function zoning in order to correct polluting enterprises' illegal acts and has been remarkably effective in water law compliance. According to the 2011 Yellow River Bulletin basin-wide assessment of 188 water function zones showed 87 were compliant, a rate of 46.3%.

3.3 Recent national policy developments and reform

In January, 2011, the State Council issued their No. 1 policy document on Water Resources Reform and Development.[5] This sets out a 10-year program of reform and investment in water resources and flood protection infrastructure and well as water resources pricing reform and institutional reform. Since 2011 implementation directives and regulations have been in preparation and issued periodically[6] and the principles of the No. 1 Water policy document have been incorporated into

[5] *The Decision of the Central Committee of CPC and the State Council on Accelerating the Water Conservancy Reform and Development*, issued on 31 December 2010 see: http://www.gov.cn/jrzg/2011-01/29/content_1795245.htm.

[6] For example 2012 Number 3 Policy document of the state council on strict water resources management see: http://www.gov.cn/zwgk/2012-02/16/content_2067664.htm.

the *12th Five-Year Plan* at national, provincial and local levels and become a key part of delivering the 'Eco-Civilization and Beautiful China' policy of the 18th Party Congress. The No. 1 policy document also introduces financial reform in the water sector. The details of the reform of the water resources fees in a program to 2015 are set out in supporting documents in 2012 and 2013[7] which bring in sharp rises in fees and set differential rates for surface and ground water to encourage preservation of over-exploited ground water and much higher fees in water-scarce are as of Northern China (e.g., Beijing surface and groundwater fees will rise to CNY1.6 and 4.0 per cubic metre respectively, while Inner Mongolia Fee to rise to 0.5 RMB/m^3 for surface water and 2 RMB/m^3 for groundwater). For many years, Cities have had progressive water tariffs rising in the order; domestic; institutions; services industry; production industry; and tourism, with domestic water generally at 25 percent of the highest tariff. Agricultural water use, outside the urban water networks, is at much lower tariffs. Even these increased rates will still be below the level to exert strong market economic influence but they will significantly increase the value of the water rights.

The major policy and institutional reform introduced in the No. 1 policy document is the introduction of river basin management based on the 'three red lines' which are:

- The first red line for water allocation sets water quantity objectives in rivers, lakes and groundwater. It requires the total-quantity control of water abstraction.
- The second red line for water use efficiency sets objectives for water use efficiency. This will accelerate the development of national standards regarding water-use quotas for high water consumption industries and the service industry.
- The third red line for pollution load management sets maximum permissible total pollution loads for Water Functional Zones, which cover the catchments, reaches of rivers and lakes that must meet specific water quality standards.

For each of these red lines targets and indicators are set at the level of defined sections of Rivers (water functional zones). These also match with

[7] In June 2012 the National Development and Reform Commission (NDRC), with Ministry of Water Resources (MWR) and Ministry of Housing and Urban-Rural Development (MOHURD) as co-signatories, issued Planning Notice 1618 on implementation of progressive water tariffs by end of 2015. See http://www.ndrc.gov. cn/zcfb/zcfbtz/2012tz/t20120621_486982.htm. In January 2013 the NDRC, with MWR as co-signatory, issued Decision 29 on bulk water prices to be set according to water scarcity in different regions and with groundwater assigned a higher value than surface water. See http://www.ndrc. gov.cn/zcfb/zcfbtz/2013tz/t20130114_522933.htm.

targets at local, regional, river basin and national levels. Achievement of the targets is incorporated into the official cadre performance assessment process of the relevant Communist Party of China (CPC) organization department. It makes individual officials personally responsible for achieving the outcomes in their annual letters of responsibility through quantified key performance indicators. This is a significantly different control mechanism which is available in China and has more influence on the actions of key actors than general legal or administrative incentive systems. Working out exactly how to frame these indicators will be key to successful implementation of national water resources and river basin management policy. The most important indicators in the official Cadre assessment have been GDP growth, population control, Social stability and employment but environmental factors have recently been introduced with much higher priority than in the past.

For water quality aspects the Ministry of Environmental Protection (MEP) also has its National Pollution Control program focused on its role in controlling and reducing pollution discharges from industry and from municipal wastewater to meet the river water quality objectives. The Ministry of Housing and Urban Rural Development (MOHURD) sets the plans for Urban Water supply and wastewater investment.

Each of these ministries has departments in the provincial and municipal governments which take technical direction from the central ministry, but financial and administrative direction from the local government. This can lead to conflicts of interest between the objectives for achieving national environmental improvements and local economic and social management objectives.

4 The Main Challenges for Sustainable and Integrated Water Resource Management of YRB

It is widely recognized that integrated river basin management is an essential component of sustainable management in YRB. The integrated water management in YRB could be interpreted by the function and operation of the YRCC and a series of legal instruments as discussed in the previous section. Although there have been many good practices in this regard, fully implementing integrated management is still a challenge in YRB.

4.1 The integrated institutional system for water resource management

The institutional mode for the water resource management in the Yellow River Basin lies in two independent but interacting systems, one is the regional administrative system where provinces or autonomous regions in YRB play roles for water resource management within particular jurisdictional regions; the other is the river basin management system, where YRCC as the river basin management institution has the role for the integrated water resource management for the entire river basin. YRCC is administrated by the Ministry of Water Resources like all other institutions for national key rivers or lakes. In the power hierarchy, provincial governments are at the same level of administrative power as the Ministry of Water Resources, in that they are each ultimately responsible to the State Council; in this hierarchy the YRCC is at a lower level than provincial governments in the YRB. Therefore the common understanding of the most serious challenge in institutional integrated management of YRB is the difficulty to achieve institutional coordination and cooperation in water resources management among the provincial (and lower level) administrative authorities for water resources (Xu 2013).

In general, regional administrative authorities for water resources have more substantive and executive powers than YRCC, who have mainly macro-planning roles with limited executive functions. The main responsibilities of YRCC are: water allocation, water quality protection, soil and water conservation and flood protection. If the component responsibilities of YRCC in water resource management are not clear or are conflicting, the integrated water management for YRB could fail. The frequent YR flow cutoff cases in 1980's are an example.

The revision of *Water Law* in 2002 intended to provide a clear institutional status for river basins management institutions and in particular for national key rivers like the Yellow River. The law mainly the Articles 12 and 15 of *Water Law* 2002 state that the combined system of the regional management and river basin management is the general institutional arrangement to be applied in water resources management, in which the river basin management had a higher legal status than regional water resources management. According to the *Water Law 2002*, the main tools for the river basin management include the preparation of river basin planning, approval of the water pollutant carrying capacity, water quality monitoring, setting limits on sewage outfalls, formulating water allocation and water regulation plans and the implementing defined integrated water regulatory tools for example issuing water taking permits. By law, these integrated management tools serve as the foundations for the regional water resources management in the YRB provinces.

Although the river basin institution's administration is criticized, the integration of these issues is quite technical and there is a lack of substantial intervention into the implementation process. For example, in the management concerning planning, the YRCC is hardly involved in the planning procedures for the large reservoirs or water diversion projects located along the main stream of Yellow River. The provincial regional authorities and concerned ministerial departments play dominant roles and YRCC lacks a supervision mechanism for such planning. Similarly, in the water permit system, due to a lack of binding administrative tools for violation and punishment, the YRCC faces difficulties in carrying out effective supervision and management over the water extraction by provinces and autonomous regions, especially where the water-taking occurs in tributaries of the Yellow River. Therefore it is common practice for the provinces to circumvent the strict Yellow River abstraction control procedures by taking their waters from the tributaries before they enter the main river stem.[8]

With regard to water quality management and pollution load control, YRCC Water Resources Protection Bureau (WRPB) has responsibility for calculating the carrying capacity and pollution load allocations for each water functional zone but they only have authority to issue control quotas, permits and conduct monitoring in the river stream and at the point where any sewer, ditch or tributary enters the main river or YRCC controlled main tributary. There is no linked permitting system connecting water taking and water discharging and it is not possible for YRCC to get directly involved in controlling the pollution discharges from industry because that is the responsibility of the local environmental protection bureaus. The WRPB is nominally a joint agency with the Ministry of Environmental Protection, but in practice this link is very weak and does not afford integrated management of polluting discharges. Significant reform of the Discharge permitting regime is required to put in place effective mechanisms for coordinated management.

Moreover, as an institution created for YRB water management, YRCC has competence only upon the matters specified by *Water Law 2002* or those granted by the Ministry of Water Resources, and cannot have the power to manage other issues affecting the Yellow River without specific authorization.

4.2 Integrated marketing mechanism for water right transfer

Article 7 of Water Law 2002 elaborates the principle of "users pay for water resources", so does Article 48 of this law for the water resource fee with regard to water use permit and water taking. The State Council and the Ministry of Water Resources provide specific rules on how to collect water resource fee. All these rules constitute a basic legal basis for

[8] Finance and Economics Net. "Struggle over Yellow River water: water rights transfer and marketization process hits a snag," 28 July 2011, available at http://www.caijing.com.cn/2011-07-28/110791023.html.

the market oriented water resource management. However, the current water resource management method in YRB is still heavily reliant on the administrative means rather than the market mechanisms. The economic and social conditions are not realistic for the market to reallocate water resource. As water resources are defined as wholly state-owned property, and water resources are of the most essential resources for overall regional developments in the dryland or semi-dryland areas like YRB, governmental command-control is the main mechanism for water management. For example, the current water price is not market defined, and reflects only the engineering cost of water resources whilst water resources' natural capital nature and ecological function services are neglected. The low water price may be considered necessary given the needs of rural households and their economic status, however, this low price also encourages extensive water use in irrigation and leads to low efficiency of water use. The low water price has been one of the reasons to promote water right transfer.

The pilot practice of water right transfer in YRB discussed above is an innovative policy experiment tied with water rights trading or market based transactions but its nature has not yet that of a market-based institution. In the pilot areas, the allocation of primary water right, the transfer of the water right to the new location and the price of water rights are all driven and controlled by the government. The water right holders have no liberty to make a decision nor act as market players. The involved governments are the water right supplier, water market regulator, water market agent and water pricing decision maker. Therefore, there still exists quite a distance between water right transfer and water right trade although trade is the future direction of transfer.

4.3 The efforts and dilemma of the integrated special law—"Yellow River Law"

It has been the ongoing topic for a decade to initiate an integrated special law for YRB, the *Yellow River Law*. The intent of this law is to not only follow the general principles of the *Water Law*, but also reflect the characteristics of the Yellow River so that the healthy and orderly management of the Yellow River could be carried out. As early as in 2001, the Asian Development Bank (ADB) provided China with a technical assistance (TA) to explore the feasibility of formulation of the Yellow River Law. One output of this ADB TA was to provide with an expert draft of Yellow River Law. In 2009, the YRCC seriously launched the legislative work plan of The Yellow River Law, but this plan did not result in proposed legislation. The Yellow River Law has still faces dilemmas and now has few advocates.

The opposing views to creating a special Yellow River Law assert that the specific legislation of the Yellow River could result in inconformity with existing basic legal principles and institutional system of the river basin management. The existing laws and regulations of the State Council dealing with water pollution control and water resources utilization should be sufficient for YRB water resource management. The problems of YRB water resource management are not a lack of legislation but ineffective implementation of the existing relevant laws and regulations for Yellow River management (Zhou and Wang 2008).

However, the supporting views for a special law for YRB are: (i) The current water law system is difficult to adapt to special needs of the Yellow River; *Water Law 2002* and other water related laws were quite inadequate to implement integrated river basin management (Zhang and Wang 2008; Li 2011; Wang and Zhong 2011). (ii) The long history and rich experience of YRCC excising river basin management can provide valuable lessons for formulating the Yellow River Law (Li and Wang 2005). (iii) From the example of foreign legal practice of river basin management, for example, the Tennessee Valley Authority Act (United States), Colorado River Administration Law (United States), Murray-Darling River Agreement (Australia) and Waikato River Basin Administration Law (New Zealand), a specialized river basin law could consolidate and facilitate the legal status of the river basin institutions and more successfully integrate management. (iv) In 2011, the formulation and taking effect of *Administrative Regulations for Taihu Lake Basin* created a comprehensive legislative precedence in the field of national key rivers or lakes in China.

Furthermore, supporters of the "Yellow River Law" put forward the key elements for the proposed law. They were: (i) the intensified legal status and competence of YRCC and the clearer institutional system for integrated river basin management; (ii) the integrated management principles for water quality and water quantity to ensure the river flow of water and soil conservation, flood control and healthy water environment; (iii) the improved legal mechanism for water allocation and transfer, water right transaction and compensation; and (iv) the intensified administrative supervision power and executive procedures for YRCC to carry out substantial river basin management.

The dilemma of the Yellow River Law demonstrates that legislation is the systematic action of the society and restricted by many social, economic and political factors. Indeed, whether the formulation of Yellow River Law is put into practice or not still requires exploration and discussion, and the establishment of effective integrated management in the Yellow River basin in order to deal with water allocation and water pollution, and water and soil conservation will be the consensus of the interested parties in the society.

The final goals may be achieved through a specific Yellow River law or as looks more likely now through the further refinement of national laws, regulations and policy directives.

4.4 The role of stakeholder participation in the integrated river basin management

For water resources and some aspects of soil and water quality management the River Basin Commissions have been established in the major river basins (Yangtze, Yellow, Pearl, Songliao, Hai and Huai). These have trans-provincial jurisdiction and are departments of the central water ministry, but have limited direct influence on the provincial and local administrations. This creates a complex framework in which to attempt to implement the widely lauded goals of "Integrated River Basin Management" or IRBM.

Integrated River Basin Management can be defined as a "process of coordinating conservation, management and development of water, land and related resources across sectors within a given river basin, in order to maximize the economic and social benefits derived from water resources in an equitable manner while preserving and, where necessary, restoring freshwater ecosystems" (GWP 2000).

In the Chinese context there are cultural and political differences which mean that the objectives and the means of achieving the objectives will be different compared to Western countries. The fragmented nature of the institutions managing the resources extends to the Principle Laws which are drafted by sectoral ministries and then coordinated to a limited degree by the State council before being passed as directives to be implemented in each Provinces' legal systems. These laws tend to be general in their wording, have limited mechanisms for enforcement and may conflict, overlap, or have gaps with Laws drafted by different Ministries. Also the ultimate objectives of equitable distribution of resources and preservation or restoration of natural ecosystems may have very different priorities in China where the natural environment is regarded as being a resource to serve the needs of society and is rarely accorded intrinsic value and where equitable distribution means equitable between those persons in the positions of Authority responsible for securing the resources for the needs of their administrative area and its people.

Introduction of stakeholder participation in the development of River basin management planning as has been attempted in Europe through the Water Framework Directive would be a very large leap for China. It would generally be seen as upsetting to the current systems and balance of power and so not necessarily desirable. None the less there have been many pilots of stakeholder involvement in river basin planning—such as by the World bank in the Tarim Basin,[9] and also in the Yellow River basin through the EU-China River Basin Management Program.[10] The experience from these pilots on stakeholder participation in China may be reviewed in the materials produced by the EU-China RBMP.[11]

This chapter has looked at how the legal framework for the management of water resources in the Yellow River Basin has evolved and how pilots of water rights trading in the yellow River basin are allowing the Authorities to better manage scarce resources by facilitating stakeholders to enter into direct exchanges of investment and expertise to resolve conflicts in the equitable access to water. The stakeholder participation in the integrated river basin management is generally more difficult to implement at a macro scale such as at Yellow River Basin, and more practical to conduct at a smaller geographical scale for example in the water saving program in Shiyang river basin (Du 2012).

The experiences in the Yellow River can contribute towards the national goal, as set out in the 2011 No. 1 policy document, of China evolving into a water saving society. This requires the participation of every citizen to achieve but will only come into being with effective and adaptive leadership from the water management authorities supported by appropriate laws regulations and incentive mechanisms.

5 Conclusions

The Yellow river is an exceptional river, despite only moderate flows from its vast but arid catchment it carries the greatest sediment load of any river on earth. This makes it especially challenging to manage and potentially deadly, representing the world's greatest risk of loss of life to a natural disaster should a major flooding incident occur. Managing this river and the development of huge and complex irrigated agriculture systems has been at the heart of the historical development of Chinese society.

9 The World Bank invested in Tarim River Basin management projects and took efforts to push stakeholders, in particular the farmers or water users, to be involved in water use planning and decision-making process. See http://www.tahe.gov.cn/e/action/ShowInfo.php?classid=65&id=1146. Subsequently, the irrigation committee of Tarim River was established. It included the Chair of the Water Use Association of Farmers (WUAF). However observed from the newsletters report of the Tarim River Basin official website, almost no activities of these institutions especially WUAF were covered. See http://www.tahe.gov.cn/zzjj/.

10 See IRBM Action Framework in a Selected Pilot Basin in the Yellow River Basin, Phase 1 and 2Key Issues and Draft Action Framework Report, EU-china RBMP December 2010, www.cewp.org.

11 See RBMP Resource centre at www.cewp.org or direct athttps://www.dropbox.com/sh/a7oxglsbnymvdh2/4Gd-q6GuHH/10%20RBMP%20Knowledge%20Centre/01%20publications.

The rapid urbanization and industrial development of the river basin is placing huge pressures on the management water resources which needs to make a transition from an agricultural to an urban and industrial economic model. This requires the reallocation of resources and the need to implement effective water quality protection. New river basin management principles are required to steer development to be harmonious with the river and build an ecological civilization.

This requires integrated solutions for water quantity and quality, between the main stem and the tributaries, between surface and groundwater and between the river and its catchment for soil conservation. The establishment of the Yellow River Conservancy Commission as a trans-jurisdictional administrative body has gone some way to achieving this but responsibilities for different aspects of the river system still remain outside of YRCC and better mechanisms for cooperation between authorities are required to achieve integrated river basin management.

Moves over the last decade or so to produce a Yellow River law to integrate functions across the basin were moving towards this but ultimately the solution in the Yellow River needs to be part of a national solution that is consistent and replicated in other major river basins. Therefore in recent times the consensus has moved away from preparing a separate law specifically for the Yellow River.

The 2011 Number 1 policy document for water sector reform and subsequent supporting regulations goes some way to achieve this. The "3 red lines" principle produces a framework for water quantity management, water savings incentives and water quality protection and this is set in the context of a greatly increased 10 year water investment program. More work is still required to define the mechanisms for implementation such as enhanced water abstraction and discharge permitting systems, comprehensive monitoring and data management, standards and objectives based on environmental outcomes and the economic and market based management of water. This policy document also sets out the uniquely Chinese approach to implementation through the national system of key performance indicators for senior officials assessment that now includes environmental and water resources outcomes and standards compliance amongst the factors determining their pay and career progression.

Water is a vital part of the economy and needs to be properly valued in the economic system. Increasing water resource tariffs can help to raise the value of water and focus attention on water saving. However this needs to be balanced with social programs to protect the financial interests of poor farmers and poorer water users.

The direct water rights trading approaches pioneered in the Yellow River basin are a significant step forward in allowing the negotiation of a unique value of water abstraction rights based on the local circumstances relating the need and willingness to pay of the buyer and the costs and consequences to the seller of achieving the water use efficiencies necessary for them to maintain their production with less water.

The critical challenges in the Yellow River Basin still remain as how to implement these systems through the levels of administration and stakeholders participation across the river basin both vertically and horizontally and to codify this in the systems of Policy directives, Laws, regulations and standards at national and local levels to make YRB a better living place for the people.

Appendix Lists of the main Laws, regulations and important policies that this paper reviews

National Laws of the National People's Congress

Water Law (2002)
Water Pollution Prevention and Control Law (2008)
Water and Soil Conservation Law (2010)
Flood Control Law (1998)
Environment Protection Law (1989)

Regulations and Decisions of the State Council

The Available Water Allocation Program of Yellow River (1987)
The Program of Yellow River on Annual Allocation of Available Water and Quantity Dispatch of Mainstream Water (1998)
Interim Measures of the Yellow River Basin for the Management of Water Resource Appraisal for Construction Projects (2003)
Implementing Measures for the Management of Water Rights Transfer of Yellow River (Trial) (2004)
The Regulation of State Council for Water Permit Management (2004)
The Regulations of State Council on Water Taking Permit and Management of Collecting Water Resource Fees (2006)
The Regulations of State Council for Yellow River Water Dispatch (2006)
The Regulating Rules for Yellow River Water Dispatch (Trial 2007)
Water Function Zoning of National Key Rivers and Lakes (2011)
The 2011 Number 1 Decision of the Central Committee of CPC and the State Council on Accelerating the Water Conservancy Reform and Development (2011)

Policy Documents of Ministries or YRCC

The Provisions on Review and Approval Procedures for Water Taking Permit (1994)
The Measures for the Supervision of Water Taking Permission (1996)

The Provisions on Water Quality Management in Water Taking Permit (1995 and 1997)
The Responsibility System for the Water Dispatch of the Downstream of the Yellow River (Trial) (2002)
The Measures for the Cap Control Management of the Yellow River Water Taking Permits (2002)
The Measures on the Management of Water Function Zones (2003)
The Provisions for the Emergency Response, Investigation and Resolution of Significant Water Pollution Incidents in the Yellow River (2003)
The Instructive Opinions of the Ministry of Water Resources on Water Right Transfer Pilot Work in Ningxia and Inner Mongolia (2004)
The No. 1 Water Resource Quality Bulletin of Water Functional Zones of the Yellow River (2005)
The Provisions on the Emergency Response and Disposal of Water Dispatch Incidents (2008)
The Preparatory Program for the Resistance of Drought in the Yellow River Basin (2008)
The Implementing Rules for the Management of Water Taking Permit of the Yellow River (2009)
The No. 4 Water Resource Quality Bulletin of Water Functional Zones of the Yellow River (2011)

Acknowledgments

Authors acknowledge the two editors Victor Squires and Hugh Milner for their insights and scholarly comments to this chapter.

References

Cai, Ximing and Yi-Chen Yang. 2010. Water Allocation Management in the Yellow River Basin: Potential for Water Trading? Yellow River Basin: Living with Water Scarcity. International Food Policy Research Institute (IFPRI), Washington, DC, p. 689.
Chao, Gen-fang and Guo-yong Wang. 2011. The Research on Legal and Institutional Construction of Basin Management. China Water Hydraulic Power Press, Zhengzhou.
Chen, Xiao-jiang. 2011. The speech at the 2011 commander's meeting for the Yellow River flood control and drought relief, website: http://www.hwswj.gov.cn/swjcms/html/1/481/2011-5-10/46855.html, access time: May 12, 2012.
Du, Qun. 2012. Ecosystem Conservation Law in China: Legal Study on Integrated Ecosystem Management and Ecological Compensation, Higher Education Press, Beijing, p. 209.
GWP (Global Water Partnership). 2000. GWP Technical Advisory Committee Background Papers No. 4.
Hu, Yu-rong and Qi Chen. 2004. The practice and cognition of the Yellow River water right transfer. Journal of China Water Resources 15: 45–48.
Li, Jing-xin and Chun-li Wang. 2005. 32 Representatives of the People's Congress propose a bill of Yellow River Law to save "mother river". Journal of People's Yellow River 7: 22.
Li, Xiaoqiang. 2011. Accelerating legislation process for the Yellow River. The Journal of China Water Resources 20: 27–29.

Li, Yuhong and V.R. Squires. 2014. Integrated water resources management of inland river basins in the Hexi Corridor, Gansu, China with special reference to the Shule River Basin. pp. 82–100. *In*: V.R. Squires, H.M. Milner and K.A. Daniell (eds.). River Basin Management in the Twenty-first Century: Understanding People and Place. CRC Press (this book).
Lu, Wen-Qing, Shao-Hua Xie, Wen-Shan Zhou, Shao-Hui Zhang and Ai-LinLiu. 2008. Water pollution and health impact in China: a mini-review. Open Environmental Sciences 2: 1–5.
Ministry of Water Resource (MWR). 1993, 1980 Bulletins of Water.
Ministry of Environmental Protection. 2005. Investigation Report of the water pollution in Yellow River Basin: "mother river" has become "three evils", see: http://www.zhb.gov.cn/hjyw/200506/t20050628_68082.html, access time: September 30, 2012.
Peng, Xiang and He-ping Hu. 2006. Evaluation on allocation system of available water of the Yellow River. Journal of Yellow River 4: 41–43.
Tian, Yue. 2012. The recent flood disasters in Gansu province killed 5 people and caused direct economic losses of 630 million RMB, see: http://lzcb.gansudaily.com.cn/system/2012/08/01/012634772.shtml, access time: August 23, 2012.
Wang, Yahua and Fuqiang Tian. 2010. Evaluation and prospect on the pilot program of water right transfer in the Yellow River Basin. The Journal of China Water Resources 1: 21–25.
Wang, Yining and Yuxiu Zhong. 2011. Thinking on the formulation of the "Yellow River Law". The Journal of China Water Resources 2: 1–3.
Wen, Jing. 2012. All pending projects depend on Yellow River water and the high consumption of water becomes the biggest concern of coal chemical industry, website: http://news.xinhuanet.com/energy/2012-09/18/c_123728021.htm, access time: September 18, 2012.
Wen, Yu-jiao. 2012. The century journey for China's environmental protection—water flows upwards: the implementation of the most stringent water resources management system. China Environmental Newspaper, July 10, 2012: 6.
Xu, Lin. 2013. Problems and countermeasures of water administrative enforcement in the upper and middle reaches of the Yellow River Basin. The Journal of People's Yellow River 7: 15–19.
YRB Water resource Protection Bureau. 2012. Important water function zones of the YRB take account of 50% or more in the all water function of YRB see: http://www.yrwr.com.cn/html/1/8/2012-2-17/2752.html, access time: June 12, 2012.
YRCC (Yellow River Conservancy Commission). 2002a. Yellow River Basin Planning. YRCC website: http:// www.yrcc.gov.cn/, access time: March 15, 2002 (In Chinese).
YRCC. 2002b. 2004–2011, Bulletin of Yellow River Basin. website: http://www.yellowriver.gov.cn/other/hhgb/, access time: October 20, 2012.
YRCC. 2008. Water resource management and dispatch of the Yellow River. The Yellow River Water Conservancy Press, 1st Edition.
YRCC. 2010a. The 2010 Bulletin of Yellow River Basin, website: http://www.yellowriver.gov.cn/other/hhgb/, access time: September 10, 2012.
YRCC. 2010b. The 2010 Surface Water Quality Bulletin of the Yellow River Basin, website: http://www.yellowriver.gov.cn/other/hhgb/, access time: September 18, 2012.
YRCC. 2010c. The 2010 Bulletin of the Yellow River Water Resources, website: http://www.yellowriver.gov.cn/other/hhgb/, access time: September 18, 2012.
Zhang, Jian-wei and Yan-ling Wang. 2008. Some legal issues on drafting the Yellow River Law. pp. 470–474. *In*: The Proceedings of the 2008 National Symposium of Environmental and Resources Law: The Legal Studies on Sustainable Utilization of Water Resources and Water Environment Protection. The Research Institute of Environmental Law, Wuhan University.
Zhou, Guidang and Zhanjiang Wang. 2008. A legal thinking about the reasons and control of the pollution in the Yellow River. The Journal of Qinghai Social Sciences 1: 108–110.
Zhu Zhongping, Mark Giordano, Cai Ximing and David Molden. 2003. Yellow River Basin Water Accounting, website http://www.iwmi.cgiar.org/assessment/files_new/publications/Workshop%20Papers/IYRF_2003_Zhu.pdf, access time: December 13, 2013.

12

People versus Place in Australia's Murray-Darling Basin: Balancing Economic, Social Ecosystem and Community Outcomes

Adam Loch,[1,*] Sarah Wheeler[2] and David Adamson[3]

SYNOPSIS

Public policy requirements to satisfy triple-bottom-line (i.e., economic, social and environmental) objectives, particularly in regard to the sustainable development of resources have impacted on river basin management. This chapter uses the Murray-Darling Basin (MDB) in eastern Australia as a basis for discussion as it provides useful insights into how early river basin management in the MDB emphasized *people* issues. There has also been a gradual recognition of the relevance of *place* in the MDB and the implications of this are discussed. This chapter also examines the resultant trade-off requirements by considering dimensions of people and place in the context of the MDB.

[1] Early Career Development Fellow, Center for Regulation and Market Analysis, School of Commerce, The University of South Australia, GPO Box 2471, Adelaide, SA 5001 Australia.
Email: adam.loch@unisa.edu.au

[2] Senior Research Fellow, Center for Regulation and Market Analysis, School of Commerce, The University of South Australia, Adelaide Australia.

[3] Senior Research Fellow, Risk and Sustainability Management Group, School of Economics, Colin Clark Building, The University of Queensland, St Lucia QLD 4072.

* Corresponding author

Keywords: trade-off, management objectives, policy, legislation, triple bottom line, hydrology, irrigation, wetlands, Ramsar, salinity, Community Risk Profile, governance, drought, water markets, water trading, Murray-Darling Basin, economics, social impacts, environmental planning

1 Introduction

River Basin management (RBM) can be characterized by sets of trade-off decisions, where the potential (or very real) benefits from one choice are relinquished in favor of another choice perceived to be more desirable or beneficial. The incidence of RBM varying decisions have been given further weight in recent decades by public policy requirements to satisfy triple-bottom-line (i.e., economic, social and environmental) objectives, particularly in regard to the sustainable development of resources (Hacking and Guthrie 2008). The list of management objectives can be expanded to include community outcomes as an important management requirement. This chapter examines these trade-off requirements by considering dimensions of people and place through a discussion of RBM outcomes in Australia's Murray-Darling Basin (MDB). Specifically, we characterize *people* as the economic, social and community aspects of the MDB, while *place* is comprised of environmental attributes. Using the MDB as a basis for discussion provides useful insights into how early RBM in the MDB emphasized *people* issues; the gradual recognition of *place* relevance in the MDB; examples of the transitional RBM approaches adopted, and their effectiveness. The more recent MDB Plan approach now seeks to entrench sustainable RBM across *people* and *place*; and despite acknowledging that Basin managers often fail to learn from one another (Campbell et al. 2013), we still believe that lessons might be learned by RBM practitioners internationally.

2 A Snapshot of the MDB

The MDB is a complex system. It covers 14% of Australia's landmass, or more than 1 million square kilometers (Fig. 12.1).

Its rain fed and irrigated agriculture is responsible for 39% of the nation's diverse agricultural output including cereal grain, citrus, apple and wine-grape crops as well as livestock such as cattle, sheep and pigs (Crase et al. 2012). The hydrology of the Basin is complex, comprising 23 (often interconnected) river valleys into which rainfall brings an average of 530,618 gigalitres (GL; one billion litres). Of that rainfall, 94% evaporates from the Basin, 2% drains into groundwater aquifers and a mere 4% is converted

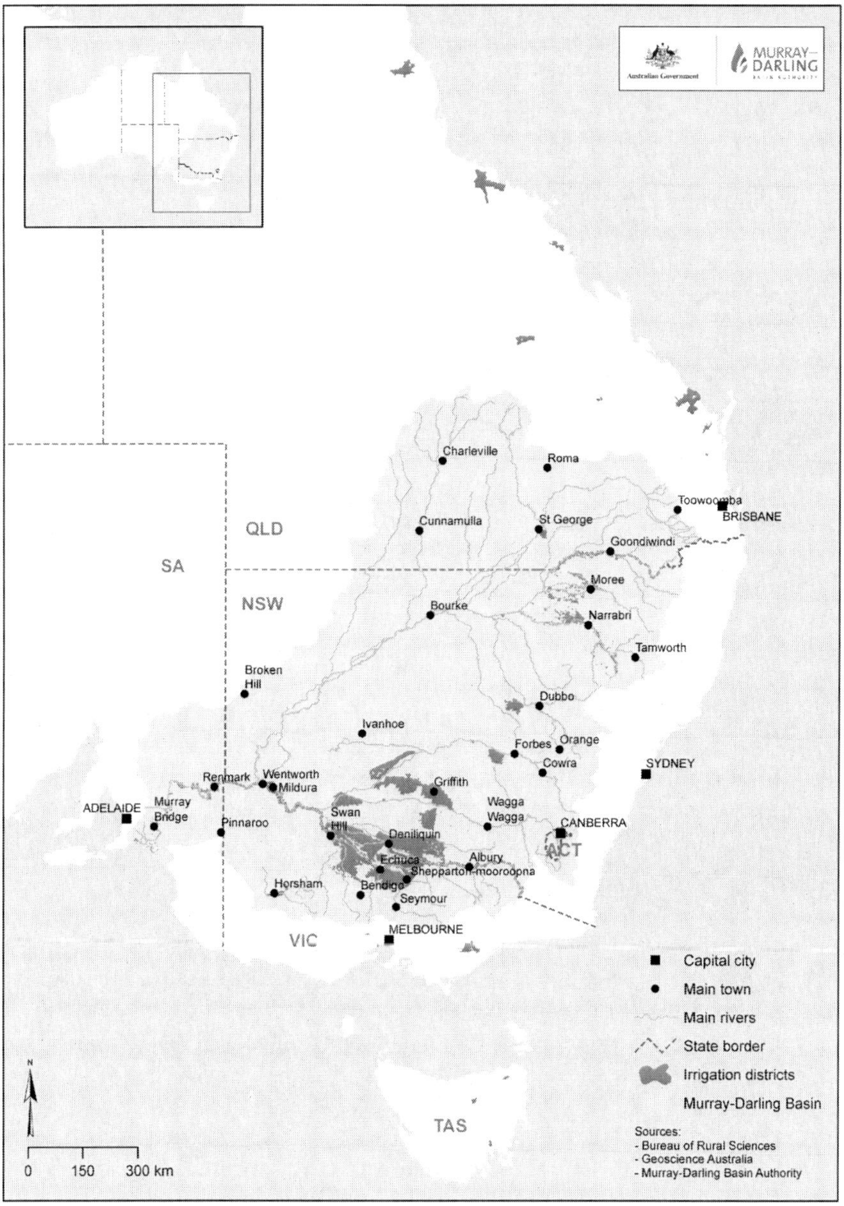

Figure 12.1. The MDB is a complex system. It covers 14% of Australia's landmass, or more than 1 million square kilometres. Source: (Murray-Darling Basin Authority).

into useable runoff. Much of that runoff is captured in an extensive scheme of storage and delivery structures, particularly in the southern part of the Basin, which support 65% of the nation's total irrigated land (MDBA 2009).

The Basin's rain fed and irrigated agricultural production base, together with a range of other important economic activities (e.g., manufacturing which represented AUD$10.75 billion in 1992 (MDBC 2006)), underpin many urban and rural communities. These range from small townships of less than 1,000 people to major population centers, such as Canberra, the nation's capital with a population of around 370,000 as at June 2011. In total, approximately 11% of Australia's population resides in the MDB.

Environmentally, the MDB has 16 Ramsar-listed wetlands supporting 95 flora and fauna species that are dependent on aquatic inundation for their existence. There are an additional 30,000 wetlands throughout the Basin that contribute significantly to species habitat and biodiversity maintenance (MDBA 2010a). These wetlands are supported by and occasionally connected to the river systems; one of which, the River Murray, enjoys iconic status among many Australians for its amenity and recreational values. As such, the Basin has deep levels of symbolism among many in Australia's society, including significant cultural heritage values and land and water management interests by indigenous populations as the MDB is home to 34 major indigenous groups (Ross 2009). Thus Basin management can generate considerable debate (Campbell et al. 2013). Sustainable management of the Basin entails judicious governance arrangements to manage alternative choices between economic, social, community, and environmental outcomes. However, previous Basin management tended to emphasize economic and social dimensions to the detriment of environmental needs. The issue is complicated as environmental needs are also driven by social and economic dimensions. Elements of the Basin's history remain influential in contemporary RBM for the MDB, resulting in difficult trade-offs.

2.1 Historical RBM arrangement

In the late 19th century Australia's focus on the MDB was largely riparian water rights for stock and domestic use and navigation rights allowing river trade. State governments provided a fragmented basis for administration of these rights, as well as an early basis for disagreements on how to best use or share the available resources. Prior to Federation in 1901, formal structures and policies to provide coordinated management of MDB water resources were either rudimentary or non-existent.[1] Alfred Deakin played a notable role in the initial examination of irrigation in the United States, which was instrumental in Australia's avoidance of seniority of rights approaches, together with an emphasis on state water laws and ownership of water resources via the Constitution. The Constitution ensured the states

[1] Federation was the process by which the original six British colonies that made up Australia combined into nationhood.

'reasonable use' of water assets while the National Government retained the powers to ensure free trade between states.

Early irrigation communities were established around this time, and expanded across the southern MDB. These early irrigation schemes were affected by a series of dry years in the 1890s that, together with the economic crash related to the end of the gold-mining boom, saw the collapse of the wool industry. Extreme MDB droughts in 1901–02 (known as the Federation Drought), 1912 and again in 1914 dramatically reduced the level of navigation, irrigation and riparian water uses. State governments played a large role in attempts to avoid similar affects in future by facilitating expansion by supporting soldier settler schemes after the 2nd World War, constructing water storage and supply schemes, and implementing a range of commodity subsidization schemes to help establish fledgling irrigated industries.[2] Despite some concerns about the economic viability of such schemes, very few limits were applied to the development of irrigation schemes in the MDB, and irrigated agriculture began to play an integral part in state economies (Davidson 1969).

Political differences between the states toward water quality, availability and use, as well as adverse political bargaining power for upstream states, necessitated a stronger governance model for MDB water management and sharing arrangements (Craik and Cleaver 2008). In 1914 (ratified in 1915) a River Murray Waters Agreement was established between the states. This agreement initiated a period of investment in water flow management and storage capacity that resulted in weirs, locks and dams constructed throughout the southern MDB to counter flow and supply variability. State governments increased their spending on infrastructure to store and supply water for rural townships and irrigation projects, under the firm belief that secure water supply for livestock and food production were national priorities. The national government also began to play a role in these projects under the guise of nation building, and irrigation farming itself achieved iconic status among much of the Australian population (Davidson 1969). Thus, between the 1940s and 1980s there was a tenfold increase in storage capacity across the MDB, such that total water use increased by 65% of the available resource (NWC 2011e).

2.2 People-centric development priorities

A shared investment in infrastructure by the states was, however, where many similarities ended. Each state tended to focus on different agricultural

[2] Soldier settlement schemes were an incentive for men to take an active role in the Australian armed services by rewarding returned soldiers with a parcel of land in rural areas, from which they could carve out a new living after the war.

commodity production that, while not problematic, certainly complicated the sharing and management of water resources across the Basin. New South Wales (NSW) focused on annual crops such as cotton and rice that were more appropriate for lower reliability water supply arrangements in that state, and maximizing cropping yields which provided incentives to use all available water. Victoria used water to grow pasture or engage in perennial horticulture which motivated irrigator's requirements for stable and secure supply. Likewise, South Australia (SA) was dominated by horticulture, and a secure annual water supply for permanent crops and for Adelaide was their paramount concern (Crase 2008).[3] In accordance with these requirements, states issued different types of water rights, adopted different (more conservative) means of estimating water availability, and instituted different policies about how water could be used and transferred.

Although periodic drought events occurred again during this 'developmentalism' phase (Crase 2011), the focus of MDB water management was largely people-centric with emphasis on economic gains through social nation building projects that maximized the use of (seemingly) abundant resources. Some states began to realize the limits to water use, and introduced volumetric allocations. In the late 1960s an awareness of over-allocation emerged, and states (starting with SA) commenced placing moratoriums on the issue of new water entitlements. NSW introduced volumetric water allocations for the River Murray in 1975, followed by an extension to the Gwydir and Macquarie Valleys in 1978 based on a recognized need to limit further water entitlement issue. In SA, similar action led to a general 10% reduction in the total volume of water entitlement rights (Bjornlund and O'Callaghan 2003). By the mid-1980s there was wider state and national government recognition that water resources in the MDB had been over-allocated to economic and social users. Over-allocation presents an issue when the full volume of water that is able to be extracted at a point in time exceeds an environmentally sustainable level of extraction (Young 2012). Examples of the effects of such over-allocation included reduced environmental values in the Macquarie Marshes and Gwydir Wetlands in NSW, river bank collapses along the Barwon and Darling Rivers, as well as rising toxic blue-green algal blooms and increased salinity levels in many MDB waterways (ARMCANZ 1996). This recognition heralded a shift in MDB water management emphasis toward a focus on place issues more generally in Australia.

---

[3] The River Murray at Morgan is a primary back-up source of water supply for the City of Adelaide when local reservoirs run low. At current demand levels up to 200 GL can be extracted for urban use during such periods.

3 Understanding 'place' in the MDB

As recognition of the importance of environmentally sustainable water management took hold around the world, water managers in the MDB began to evaluate and establish ecosystem and community water needs, and to assess their governance arrangements in relation to these needs. Although there are many examples of early MDB initiatives to understand place in the Basin and address sustainable water management (e.g., an expansion of the River Murray Water Agreement (1982) to include the management of environmental issues; intergovernmental meetings to establish broad policy agenda and competition payment frameworks (COAG 1992; Industry Commission 1992); working group reports suggesting ways to achieve that agenda (Working Group on Water Resource Policy 1994); and the setting of national ecosystem water provision requirements (ARMCANZ 1996)), a pivotal stage in these discussions can be found in the setting of a defined limit on further water extraction.

The catalyst for this limit (known as the Cap) was a 1995 audit of river flow-regimes in the MDB, which concluded that median annual flow-to-sea levels were 27% of pre-development; and that severe drought-like flow patterns now occurred in 60% of years as compared to 5% of years under natural conditions (MDBC 2004). As any further growth in water extractions would exacerbate this problem, the Cap was made effective from July 1 1997, limiting total development of MDB water resources to the level reached in 1994–95 (Bain et al. 1996). As such, although development could occur elsewhere around the Basin, water would now have to be reallocated from existing uses to facilitate such development (Bjornlund 2005). These expected limits to further extraction were widely discussed before their introduction, and existing water users had opportunity to strategically adjust before its implementation. Consequently, many irrigators sought to augment their access to supplementary water and/or take advantage of gaps in licence moratorium restrictions, leading to an 8% increase in water extractions between 1984 and 1994 (NWC 2011e). Much of this additional extractive capacity was found in the activation and/or trade of previously unused (sleeper) or under-used (dozer) water entitlements (Crase et al. 2009). This activity placed an additional burden on an already stretched system, but the limit to further MDB water extraction meant that assessments of place water requirements could be undertaken; namely environmental and community water requirements as discussed further below.

3.1 Environmental issues

Since the introduction of the Cap, significant progress toward understanding environmental water requirements in the MDB has been achieved. There is now a much greater recognition, if not always understanding, of the relevant ecosystem requirements and their interconnected nature. However, as a result of this improved ecosystem knowledge, the complexity of the role that river basin managers play in the MDB has grown significantly. Further, there is now substantial social and political expectation to deliver complex environmental outcomes via RBM arrangements, which factor heavily in any national discussion of the MDB. The complexity of RBM issues in the MDB can be highlighted in the following sections.

3.1.1 Surface water

The River Murray has a highly complex surface water flow regime, and in each sub-catchment flows are influenced by different climatic factors. In addition, rainfall and temperature patterns vary significantly between the northern and southern Basin zones (Loch et al. 2013). MDB surface water flows also vary in terms of the space they occupy and the times at which they occur (Howard 2008). For instance, high flows trigger breeding among fish and allow them to move up or down the river in search of suitable habitats, and they keep estuaries open. Low flows in summer sustain base habitats and maintain fish refuges. By contrast, spring floods aid the regeneration of wetlands and floodplains, while replenishing river channels (MDBA 2010a). Infrastructure projects implemented prior to the 1990s have dramatically altered surface flow away from its natural pattern to meet consumptive needs. For example, in 2008 river storage infrastructure captured up to 103% of mean annual run-off and provided around 81% of total extracted water toward human uses (Howard 2008). Consequently, storage of water for irrigation and community use has effectively negated the natural pulse system with one of uniformity (Knights and Milner 1987) (Fig. 12.2).

The gradual move toward 'working river-systems' in the MDB has also resulted in lower volumes of available water for wetlands and ephemeral floodplain areas that support the breeding patterns of wildlife and water-ecology species. There have been changes to MDB vegetation levels and a general decline in riverbank forests as a consequence of surface flow pattern changes (Howard 2008), as well as substantial impacts on native fauna species during periods of water scarcity (MDBA 2010a). Changes to surface flows have also impacted on other water issues (such as groundwater, salinity levels and return flows) due to their interactive nature across the Basin.

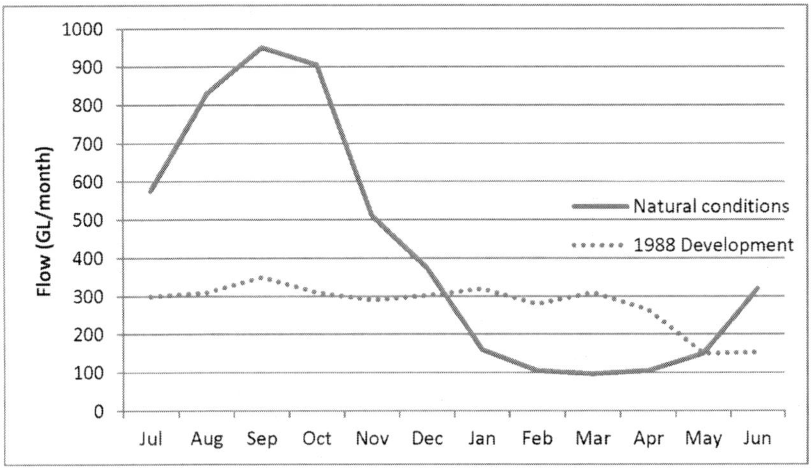

Figure 12.2. Effect of water regulation on Murray River flow. Source: (MDBC 2001).

3.1.2 Groundwater

The maintenance of Basin ecosystem health is not merely a function of the volume of surface water flow. Groundwater comprises about 17% of accessible water in Australia, accounts for over 30% of overall consumption; and is increasingly being tapped because of greater water demand (Goesch et al. 2007). About 6,000 GL of groundwater is extracted nationally in Australia, primarily for irrigation, but this can increase by over 70% in some catchments during extended dry conditions (Richardson et al. 2011). Groundwater is more difficult to manage than surface water because it is less visible, recharge is more difficult to measure than stream inflow, and use levels can be poorly monitored. It is only in recent years that large scale metering of groundwater has occurred in the MDB.

Similar to surface water issues, groundwater use can impact on groundwater dependent ecosystems, water quality, land subsidence, reduced dilution and assimilation of contaminants and sea water intrusion in coastal aquifers (Goesch et al. 2007). As of 2012, annual groundwater extractions from the Basin were estimated to be 1,744 GL per annum (NWC 2013), and in many places extraction exceeds recharge capacity with poor long-term outcomes for groundwater levels (Cosier et al. 2012a). If current patterns are continued, aquifer recharge through rainfall will not match extraction. Recognizing that we have reached/exceeded sustainable levels of groundwater use—when limited signals of that threshold are available—thus seems highly improbable, with attendant consequences for the long-term health of the MDB. Over-extraction of groundwater may also lead to declining agricultural output. This may occur due to adding water to crops beyond the optimal level of watering, the creation of soil water-logging from excessive irrigation, increasing well depth raising the costs of extraction, and higher soil salinity levels (Athukorala and Wilson 2012).

3.1.3 Salinity

Salinity became a concern in the MDB during the 1980s and 1990s, largely due to raised water tables caused by irrigation practices along the Murray River (Adamson et al. 2007) and the clearing of deep-rooted vegetation for farming or grazing purposes. The irrigation areas of NSW and Victoria were particularly affected (Rendell 2011) although the contributions of saline groundwater inflow to the Murray River in South Australia were also a major concern. Due to the importance of this issue, a variety of initiatives were established in the MDB to deal with the problems, including: a 1989 interstate agreement to address salinity and drainage issues; salt interception and removal schemes (MDBMC 2008); salinity credit trading schemes (Connor et al. 2008); and improved modelling of surface-groundwater interaction for environmental protection (Beverly et al. 2011). As such, salinity has possibly become one issue of place that is quite well understood and managed in the MDB. However, in part this management success is due to the fact that the Millennium drought (2001–02 to 2009–10) reduced water supply to such low levels that much of the historic salinity problem began to recede. With a return to more normal water supply and irrigation application arrangements at present, it is possible that salinity management will again feature as an issue of place for MDB rain fed and irrigated producers.

3.1.4 Return flows

Water use practices have variable consequences for the return of unused agricultural water in the form of surface-water runoff or ground-water recharge. Return flows can have environmental consequences by influencing river hydrology. Improvements in water use efficiency have seen less water returned to river systems. The response in most settings has been to reduce allocations in proportion to the efficiency gains achieved (Young 2010). Between 1993–94 and 2009–10, drought, water use efficiency gains, changing on-farm practices and improvements in drainage collection have seen return flows decrease (Loch et al. 2012). If climate change reduces future water availability in the MDB, further improvements in water use efficiency may lead to even greater detrimental return flow impacts (Adamson and Loch, in press).

These environmental knowledge gains since the 1980s have provided a greater sense of place for the MDB, particularly in regards to its value as an ecosystem and the importance of its sustainable management in future. It is now important to return our attention to people issues in the MDB.

4 Understanding Community (e.g., 'People') Issues in the MDB

Communities in the MDB have also been viewed as vulnerable to shifting priorities around water sharing and use. Therefore, community has been an increasingly targeted domain to improve knowledge about likely impacts of water reallocation (whether through policy or climate change effects) in order to formulate effective and acceptable policy.

4.1 Continuing reform focus

It must be recognized that community apprehension in the MDB toward water reallocation sits within a longer term context of almost constant reform over the past four decades. The Australian Productivity Commission (2005) acknowledged there was a perception that some rural communities had been particularly hard hit by the consequences of the National Competition Policy (NCP) agenda to reform water and other industry sectors. Further, farmers and farming communities were viewed as having faced a series of long-term stresses, including: pressure to find economies of scale due to a steadily declining terms of trade; continuing decline in agriculture's importance and employment, despite productivity growth; ongoing population and services transfer from smaller communities to larger ones; increasing pressure on family farms and the loss of farming land to non-agriculture uses; an ageing farming workforce; and continuing water reform pressure to reallocate existing resources (EBC et al. 2011; Miller 2011).

Population, economic diversity, dependence on irrigation and location are the criteria that most influence how well rural communities will respond to reduced water availability (EBC et al. 2011). Empirically, it has been suggested that community vulnerability is linked to the: irrigation intensity and incidence; the percentage of the workforce employed in agriculture; the ratio of employment in agriculture to employment in other sectors; the number of households dependent on employment in agriculture; and the ratio of employment in agriculture to employment in downstream processing employment (ABARE-BRS 2010). The amount spent on irrigation per capita is another important influence. Communities most likely to be adversely affected have relatively high levels of per capita spending on irrigation. Table 12.1 demonstrates the way the intersection of size and dependence on irrigation creates differing risk profiles for communities.

Table 12.1. Community Risk Profile.

Category One: Small, dependent centres	**Category Three: Large, dependent centres**
These communities: • are somewhat geographically isolated • have contracting economies and population loss in the 20–44 year age bracket • lost taxable incomes, locally based businesses and discretionary income • find it difficult to recruit and retain staff in key service sectors • see environmental water reallocation as likely to intensify these trends.	These communities: • are vibrant communities, but likely to become much less buoyant if they face a future with reduced water • lost young people during the drought • also frequently acquired a population of people of low socio-economic status seeking cheaper housing • have less discretionary spending and makes the retention of key services much more difficult.
Category Two: Small, diversified centres	**Category Four: Large, diversified centres**
These communities: • combine irrigation with other sectors, such as tourism, providing an income stream in addition to irrigation and irrigation service industries • are somewhat quarantined from the adverse impacts of declining water availability • have stable population bases • ongoing viability is closely related to secure water for viticulture and sufficient river flows to provide tourism opportunities • face lower threats to gaining and maintaining services than for those in category one.	These communities: • are regional centres reasonably insulated from the impacts of declining water availability • experienced significant gains in population, frequently absorbing the population drain of smaller centres • have also experienced a growth in their service sector • are generally speaking more resilient and possess considerable capacity for adaptation to reduced future water availability.

Source: (EBC et al. 2011)

4.1.1 Importance of irrigation in the MDB

Much of the work examining community vulnerability to water reallocation has now come to understand how irrigation, as a once iconic industry sector in Australia, still features heavily in the sense of place for rural townships; especially in the MDB. Irrigated agriculture uses much-less land, delivers greater flow-on employment and economic gains than rain-fed farming, and is an important employer in all regions either directly and indirectly through food and fiber processing. Across MDB regions, at least 75% of total farm operating expenditure takes place in the regional economy—typically within 50 kilometers of the farm gate—and almost all irrigated production is processed in the same region or a nearby region (Miller 2011). A reduction in economic water availability that decreases the extent of irrigation in a local area, forcing a switch to rain fed farming or other less water-intensive agronomic mechanisms, will adversely affect the local economy (Marsden Jacobs Associates 2010).

While reforms promoting deregulation and competition generate a net benefit for rural Australia, they often have adverse impacts for some rural communities, particularly small ones (Productivity Commission 2005). While rural communities can be assisted in the efforts to adjust to structural change requirements, and have been in the MDB, much Government assistance targets individual farmers or particular agricultural sectors, rather than assisting rural communities to reorient their economies. Overall, rural communities with diversified economies respond much more effectively to policy reforms (e.g., NCP reforms) that have the potential to create significant structural change (RMCG 2013; EBC et al. 2011). In the MDB specifically, it has been argued that in order to aid communities potentially affected by reduced water access, governments or river basin managers should provide: investment funds for future planning for and by communities; technical advice about developing new locally based business opportunities; support structures for rural communities via national, state and local government collaboration; and facilities to redirect money earmarked for irrigation infrastructure to developing community based development (Miller 2011).

4.1.2 Drought and location impacts on vulnerability

Such analysis aside, water reallocation has been greeted with near universal resistance by rural communities. They fear further pressure on family farms, leading to fewer farms (Barr 2009). It is also feared that these developments would fray the social fabric of rural communities and that property prices would also decline, trapping some people in decaying communities (Fenton 2007; Edwards et al. 2008). The Millennium drought provided something of a natural experiment about what happens to rural communities facing reduced water access. That drought period resulted in declining productivity, especially in the dairy, rice and cotton sectors, the closure or mothballing of cotton and rice processing plants, with community wide impacts, impaired financial viability of many individual farm enterprises and an increased need for subsidy or drought support payments by smaller farms. In addition, rural communities did contend with the closure of local support services and contraction of key community institutions (e.g., local secondary school) to larger rural townships, falls in residential property prices, a consequent change in the local population/ social mix, and increasing salinity, bank collapses and decreasing tourism levels particularly below Lock One in SA (EBC et al. 2011).[4] However, it is

[4] GPS coordinates: 139.6084117, –34.3554489.

important to note that there are long-term rural structural and other world influences that played a large part in this rural decline, and it is very difficult to separate out each influence from one another.

Finally, there is a clear north-south divide between the Basin's irrigation communities. In the southern MDB regions (excluding the SA Murray River below Lock One) and Murrumbidgee irrigation districts (NSW), irrigated agriculture ranges from 79% to 92% of total agricultural production; whereas in the northern MDB regions it accounts for 15% to 52% of total production. This suggests that southern MDB communities are likely to be more affected by the introduction of water reallocation arrangements (Marsden Jacobs Associates 2010). In particular, the SA Murray River below Lock One, Riverland (SA), and Sunraysia (northern Victoria/southern NSW) districts may be particularly at risk due to reductions in water availability (Marsden Jacob Associates 2010), and it is likely that increased environmental allocations (e.g., flows to the Lower Lakes at the Murray River mouth) will have to be sourced from southern areas. Related rural communities are placed at further risk because they are already suffering considerable socio-economic disadvantage, which is a key indicator of a community's capacity to respond to adversity (Marsden Jacobs Associates 2010).

This greater understanding of place issues in the MDB highlighted a requirement to achieve better policy outcomes and associated RBM coordination. The analysis of the first round economic impacts by Mallawaarachchi et al. (2010) where minimal economic loss on a partial equilibrium analysis were expected, have been confirmed at the second round level (or general equilibrium analysis) where minimal economic impact after compensation is evident throughout the economy (e.g., Dixon et al. 2011). Armed with this knowledge, MDB planners and river basin managers have sought to manage economic, social, ecosystem and community objectives through a process of water reallocation within sustainable level of extraction constraints. At the same time the Millennium drought changed the perceptions of water entitlement reliability. This not only created short-run impacts on water markets expressed as high prices, but also forced industry restructure and adjustment to cope with changes in water supply and cost functions. This pressure also created negative private and emotional response, sometimes with unfortunate tragic loss. The next section outlines the range of initiatives undertaken in the MDB to first establish that sustainable level of extraction and then manage water reallocation toward achieving it.

5 Steps Toward Coordinated MDB Governance

Table 12.2 summarizes the evolution of coordinated governance arrangements in the MDB from the establishment of the River Murray

Table 12.2. Evolving coordinated governance initiatives.

Evolution Governance Mechanisms for MDB Management	
Year	**Initiative**
1915	River Murray Waters Agreement; water sharing NSW & Victoria; entitlement for SA
1917	River Murray Commission established to implement 1915 agreement
1982	River Murray Waters Agreement amended to allow management of environmental problems
1985	Murray-Darling Basin Ministerial Council established to provide integrated planning and management of water, land and environmental resources
1987	Murray-Darling Basin Agreement
1989	First Interstate Environmental Agreement to address salinity and drainage
1992	Murray-Darling Commission established (replacing River Murray Commission)
1996	Commonwealth National Heritage Trust established in response to crisis in water quality
1997	The Cap arrangements put into place, if not fully agreed to/signed by all states
2002	The Living Murray Initiative; identify projects to improve river health and establish icon sites in the Basin
2004	National Water Initiative; improved management of water and increased compatibility between states to facilitate expanded water trading
2007	Water Act; established Murray-Darling Basin Authority to create MDB Plan
2008	Water Amendment Act; referred some state powers to the Commonwealth
2010	Water for the Future Initiative: 10 year plan to balance water needs of farmers, communities and the environment
2010	Guide to the Proposed Murray-Darling Basin Plan released
2012	Murray-Darling Basin Plan passes through Federal Parliament into law
2014	All states ratified the Murray-Darling Basin Plan

Sources: (NWC 2011d; Loch et al. 2013; State Library South Australia 2013)

Waters Agreement (1915) to the Basin Plan in 2012. Since we have discussed the earlier initiatives previously, we will continue our discussion below with the 1992 establishment of the Murray-Darling Basin Commission (MDBC) which was put in place by the national government to extend RBM arrangements to the whole of the Basin.[5]

Coordinated management through the MDBC was aimed at allowing for proactive Basin administration to safeguard its environmental health, and that of its associated communities, while ensuring its economic productivity. It highlighted a need to recognize and respond to the needs and issues of different jurisdictions involved in RBM, and to find mechanisms

[5] Prior to the 1992 changes the River Murray Agreement was limited to the Murray River (below Hume Dam) and the Darling River below Menindee. There was no power relating to land issues. Each State was solely responsible for water management issues in river systems and their tributaries.

that would allow resolution of their differences. While it was essential for limiting further entitlement approvals and the introduction of water trade, a critical factor working against the prioritization of environmental water requirements (e.g., ARMCANZ 1996) to deliver reliable ecological flows to (at least) key environmental sites in the MDB (e.g., state water sharing plans) was the strengthening of consumptive property rights following the introduction of the Cap in 1997 (Loch et al. 2011). Since the Cap limits to extraction were predominantly comprised of consumptive (e.g., irrigation) water rights, there was no scope to derive environmental flows without jeopardizing reliable and secure supply to irrigators/rural societies. This raised issues of how much water the environment actually needed, as much as where it would eventually be found. A further issue was the requirement to define water entitlements in comparable terms to meet the requirements of the National Competition Policy, such that inter-State water trading could be made possible. While individual state governments sought to meet their obligations under COAG agreements to initiate water sharing plans, practical and physical water reallocation toward place issues stalled.

In response, the MDBC developed a new initiative aimed at recovering physical water for place needs: The Living Murray Initiative (TLM) (MDBMC-2002). Targeting six 'icon' environmental sites, the TLM initiative sought to recover 500 GL of water over five years (2004–2009) through a mix of water efficiency infrastructure projects and purchases of water rights from willing sellers. The original budget for this initiative was AUD$500 million, and recovered water by participating state and national governments was to be applied to the icon sites to improve environmental health. A 2006 review increased the TLM budget by a further AUD$500 million over five more years (from 2006–07), but actual recovered water was limited in these early years by reduced water market activity and slow infrastructure saving contributions. By 2007, however, extended impacts of the Millennium drought spurred national government action to accelerate recovery outcomes.

5.1 Drought initiative outcomes

As irrigators, rural communities and the environment entered their sixth year of drought condition, and as some state water sharing plans were suspended outright (NWC 2009), the then Prime Minister John Howard released a *National Plan for Water Security* (Howard 2007) to address what was viewed as a water resource crisis in the MDB. The plan allocated AUD$10 billion over 10 years to increase water right purchases from willing irrigators and invest in irrigation infrastructure efficiency projects, such that water could be returned to the environment. A national *Water Act* (2007) was also passed into law, with the requirement that a new Murray-Darling

Basin Authority (MDBA) be established from the MDBC to draft and oversee a comprehensive RBM plan for the entire MDB. This plan would make greater use of the existing water markets to recover environmental water, and build on TLM projects to generate water savings in irrigation districts that could contribute further to ecosystem flows. Although a change of government occurred later that same year, the aims of this plan remained largely intact under a new *Water for the Future* program, but the available funding grew substantially to AUD$12.9 billion (Wong 2008). Water markets were still viewed as playing a key role in meeting the objectives for water recovery, particularly from economic users to generate national/social welfare. That is, there was little political will to reallocate water to the environment without adequately compensating the consumptive users of water, and markets offered a (then) politically-palatable means of achieving that outcome.

5.2 Water markets in the MDB

This stage of the MDB water reform process could be viewed as retaining a balance between people (i.e., economic focus on water market and rural social welfare improvements via infrastructure projects) and place (i.e., environmental flow provision) issues. Like other global settings, Australia has adopted market-based instruments as a means of reallocating water between various users. Effective markets manage rising demand for water among competing users and ensure that it is used efficiently for desired ends, while still promoting environmental, economic and social sustainability (Bjornlund et al. 2013; Grafton and Jiang 2011). In Australia two common types of water trade products are: a) *Water entitlements (or permanent water)* that provide a long-term access to a share of the total consumptive pool of water resource(s) with a reliability of supply factor (low, general or high), and b) *Water allocations (or temporary water)* which is the volume of water that water entitlement holders receive during a given water year, dependent on the available water in storages, expected inflows, system losses, demand expectations, delivery capacity and other factors (NWC 2011b).

5.2.1 Effective water market properties

An important determinant of how effectively markets can deliver the benefits discussed above is the design of markets, their associated institutions and the regulation to which they are subject. Loch et al. (2013) note that while markets have an important role to play in allocating water within and between rural, urban and environment interests, market failure is possible and markets therefore need careful design and regulation. Strong connections between water markets and water policy are necessary for effective RBM.

The MDB Cap highlights this point. Although the Cap resulted in stronger property rights for consumptive users, as well as increased water right values for existing holders and incentives to trade where transfers had been limited previously (Loch et al. 2011), its implementation resulted in significant short-term increases to total extraction from the Basin, and reduced water availability for allocation to environmental flows that reduced water for the environment. Secure property rights are crucial for encouraging investment in water use and productive outputs (Bjornlund and O'Callaghan 2003), as well as promoting effective reallocation of water within market structures between users (Loch et al. 2012). However, in the MDB the range of different rights and trade processes have resulted in a fragmented and at times cumbersome market platform lacking a nationally recognized set of entitlements and in some cases (e.g., stock and domestic) not having any transferable water rights (NWC 2011a).

5.2.2 Water market benefits in the MDB

As the Millennium drought period continued, water markets played an important role in helping people cope with water scarcity impacts. For instance, annual crop producers in NSW reduced their water use by 36% during the drought, with much of this water being transferred to perennial crop producers in SA and Victoria (NWC 2012). As such, interstate water transfers grew from 19% to 65% as a response to differing state demands for water resources (Loch et al. 2013) and over 60% of NSW's water allocation exports went to SA; most in 2008/09 following particularly difficult allocation conditions in 2007–08. The total volume of water in the allocation market increased from 8% to 30% of available water, as seasonal allocations began to diminish dramatically (NWC 2011b).

The benefits of these transfers can be seen in crop production figures between 2005–06 and 2008–09. At the height of the drought, CGE modelling suggests that the gross value of irrigated agricultural production (GVIAP) dropped by 29% from AUD$5.5 to $4.3 billion as water availability fell by 53% (NWC 2012).[6] However, intra-regional and inter-regional water trading

[6] Estimates of the GVIAP are useful for informing government and industry decision-makers. Typically, such estimates are often used to demonstrate the contribution of a particular commodity, region or industry to economic output. While estimates of GVIAP provide important information about irrigated activity in a region or industry, it is important to recognize some limitations. For example, GVIAP cannot be used to compare the net benefits from using irrigation water in alternative uses (for example, in producing rice or vegetables). In this case the relevant information would be the net benefit (revenue less all costs) of using an additional megalitre of irrigation water in alternative uses (NWC 2012).

reduced MDB job losses by ensuring that productivity was maintained at higher than expected levels without water trade capacity. It is further estimated that between 2006–07 and 2010–11 MDB agricultural production was AUD$845 million higher than it would have been without interregional water trading (NWC 2012). Overall, production in the southern MDB was AUD$4.3 billion higher than it would have been without water trading and on farm reallocation of water between irrigation activities. Moreover, the benefits of water trading did not stop at the farm gate.

Water trading increased Australia's gross domestic product by AUD$220 million in 2008–09, with attendant benefits for the MDB states. Between 2006–07 and 2010–11 NSW water trade increased the net value of irrigated agriculture by AUD$136 million, while Victoria and SA benefitted by AUD$885 million and AUD$103 million respectively. It also helped avoid what might have been substantial income losses: respectively NSW (AUD-$760 million), Victoria (AUD-$2,256 million) and SA (AUD-$419 million) (NWC 2012). Figure 12.3 and Figure 12.4 reveal the extent of declines in GVIAP for the Basin and changes in GVIAP for particular commodities, respectively (Ashton et al. 2011).

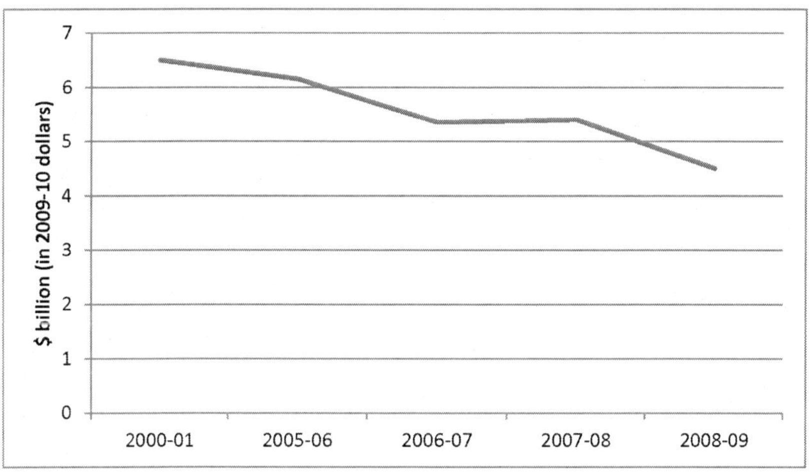

Figure 12.3. Gross Value of Irrigated Agricultural Products in the MDB, 2000–2009. Source: (Ashton et al. 2011).

5.2.3 Other benefits of MDB water trade

Water markets have also performed relatively well in reallocating water to the environment; although Wheeler et al. (2013) caution that a distinction between acquisition and management of environmental water needs to be

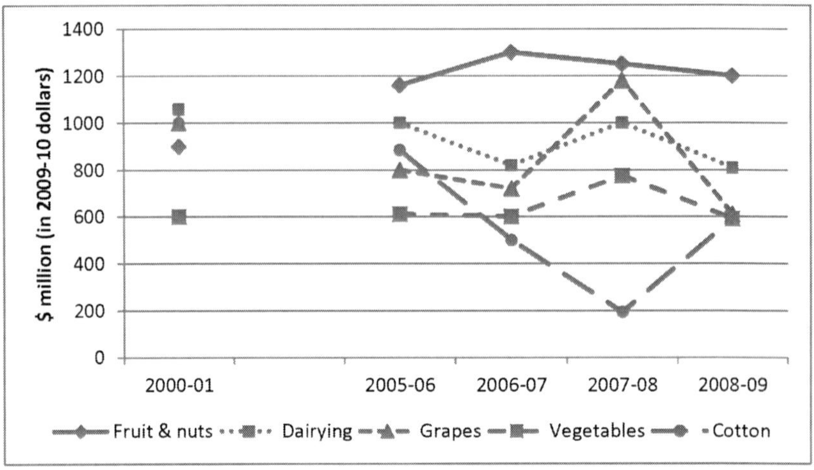

Figure 12.4. Selected Gross Value of irrigated production in the MDB, 2000–2009. Source: (Ashton et al. 2011).

fully tested. That said, Grafton (2011) argues water markets are having a positive effect on the MDB environment. This is largely due to water moving downstream during the drought, reducing summer flow stress and creating no change to winter flow patterns—admittedly patterns that have now been altered for over a century. While negative impacts may occur where water trade results in a detrimental change to the volume, location and/or timing of water use, environmental benefits due to trading are likely to be greater in drier rather than wetter periods (NWC 2012). Specifically, environmental concerns around water trade focus on several issues:

Salinity: The salinity levels of groundwater are substantially higher in the downstream sections of the MDB. If trade continues to shift large volumes of water downstream, higher on-farm water use increases the recharge of groundwater and seepage into rivers, likely leading to higher in-stream salinity levels, which in turn may need to be assessed and addressed by further salinity off-setting works or measures. Water trading may, however, redress this impact by transferring water to farms with high value products. They are likely to use better practices, making the release of salt into water courses less likely (Loch et al. 2013). Irrigators who sell water are more likely to be marginal. If they exit the industry, by selling their water, the socio-economic impact may therefore be less (Wheeler et al. 2012). A side benefit of drought is that with less water moving through the soil, salt remains trapped, providing misleading indications of water quality.

Transmission losses: Surface water evaporation, seepage from river-channel bottoms, river banks leakage or overbank losses during high-flow events can cause transmission losses. Water trade and regulatory arrangements that provide flows ranging from low in-stream flows to high overbank flooding could improve ecological outcomes for the MDB (Loch et al. 2011), but changes in the location of water extraction from downstream trade generate higher transmission losses because water is transported greater distances. Effects of trading on surface evaporation or seepage are expected to be negligible because the surface area and size of the water delivery route remain unchanged. However, increased levels of downstream water movement since 2007–8 could increase the potential losses from leakage (NWC 2012).

Upstream/downstream impacts: Water trade can result in changes in location and time of water use, altered hydrology and ecological assets via changes to the flow regime, and direct impacts on return flows and water quality (NWC 2012). Because rice and dairy producers are located upstream and horticulture downstream, water typically moves downstream during periods of drought, with changes to flow patterns. Further, efficiency improvements may have had negative outcomes for downstream producers where 'excess' water from inefficiencies would no longer travel downstream to support irrigation and environmental flows (Adamson and Loch, in press).

Carry-over: Prior to the implementation of carry-over provisions, many users either used all their water or sold their excess in the allocation market leading to changes in the location and timing of water use (NWC 2011c). This tended to increase water use in some areas, frequently with negative environmental outcomes. Irrigators in SA (in 2008–09) also suggested that without carry-over provisions, they would typically allow some unused water to flow downstream each season. This aided environmental flow and dilution (Loch et al. 2013).

Exchange rate and tagged trade: MDB water transfer exchange rates were initiated in 1998 to try and limit the environmental and supply security consequences of water reallocation due to trade (Bjornlund et al. 2013). In all downstream transfers an exchange rate of 1.0 applies. In upstream transfers, an exchange rate of 0.9 applies to counteract reduced supply security in the Darling River and Lake Victoria (DEWHA 2009).[7] The application of exchange rates on water trades during the pilot program was designed to yield environmental benefits because gains from transfers are applied to the riverine environment. Therefore, third-party impacts (e.g., environmental flow reductions) potentially associated with exchange rate water trade are much less likely to eventuate under tagged trade arrangements (Loch et al. 2013).

[7] GPS Coordinates: 141.2841298E, –33.9763408N.

Ultimately therefore, water trade in the MDB appears to have helped both people and place adapt to water reform requirements. However, when a draft MDB Plan was released in 2010 by the MDBA the focus for stakeholders changed dramatically under what was perceived to be an over-emphasis on the reallocation of Basin water toward place (environmental) issues.

6 A Failure to Balance People and Place

Without much public consultation, but consistent with the intent of the *Water Act (2007)*, the MDBA released its Guide to the MDB Plan in October 2010. A comprehensive document, the Guide provided detailed information about the thinking behind ecosystem targets, key ecological indicator and assessment sites across the Basin, and importantly a range of scenarios for water recovery and ecological outcomes that had been modeled. Although recovery of up to 7,000 GL of water in the Basin had been assessed (with substantial environmental gains), the preferred recovery target figure range was between 3,000 and 4,000 GL (MDBA 2010a). This was based on extensive ecological modeling of outcomes at the identified key sites (MDBA 2010b), and assessments of impacts on rural communities throughout the Basin (MDBA 2011). Environmentalists greeted the plan with some enthusiasm because of its ostensible commitment to restoring Basin health (Kildea and Williams 2011). Rural communities, however, did not react well to the Guide and perceived it to have serious problems with its socio-economic analysis; heavily skewed toward environmental priorities and outcomes; and poorly discussed/consulted within stakeholder groups prior to its launch (Crase 2011).

The Guide fallout resulted in the resignation of the MDBA Chair, Michael Taylor; the seeking of legal opinions regarding the intention of the *Water Act (2007)*, and a Parliamentary Inquiry into the future of the MDB Plan (Australian Parliament 2011). Further, the science underlying the modeling for communities in particular was questioned by the people who effectively owned the plan (Crase et al. 2011; Miller 2011). Consequently, river basin managers were forced to drastically revise the MDB Planning process, their understanding of how to address economic, social, environmental, and community trade-offs in the Basin, and their approaches to water acquisition/management. In general, public policy surrounding the MDB Plan: i) shifted from general to targeted and strategic water market buyback approaches (i.e., purchases of whole districts and/or parcels strategically aligned to environmental sites of interest); and ii) emphasized more public investment in infrastructure efficiency and environmental works projects to deliver improved flows and rural welfare benefits (Australian Parliament 2011). The Plan was also largely rewritten.

7 Transitioning Back to People and Place

Following the appointment of a new MDBA Chair, extensive public consultation throughout the Basin and the recommendations of the Parliamentary Inquiry the MDB Plan finally passed through national parliament into law in November 2012 (MDBA 2012). The Plan set the sustainable diversion (extraction) limit (SDL) target at 2,750 GL, which was to be achieved by 2019. Its passage was not without controversy though. The NSW Government opposed the Plan in its original form (NSW Office of Water 2012) and the Queensland government was highly critical of potential negative impacts on agriculture and agricultural communities. Similarly, while the SA government was broadly supportive (apart from making 71 recommendations for improvement) they also argued strongly that 2,750 GL per annum were not adequate to safeguard the health of the Basin's lower reaches (SA DEWNR 2012). In contrast, the Victorian government argued that 2,750 GL per annum placed too onerous a burden on irrigators and communities (Hawker Britton 2012). Finally, independent groups such as the Wentworth Group of concerned scientists argued that the draft plan did not provide sufficient information about the volume of water required to maintain a healthy ecosystem, did not sufficiently factor the impacts of climate change into the Basin's future and did not model the impact of increasing groundwater extraction. It also asserted that the proposed reduction to water market purchases and focus on infrastructure improvements had not been properly evaluated (Cosier et al. 2012b).

Accurately predicting the likely impact of the Plan on agricultural output and its associated economic consequences is crucial, while the implications for rural communities will also need careful assessment. Several analyses have been conducted about the impacts of a 3,000 GL/annum reduction to water availability for productive uses, and estimate a 0.2–0.7% reduction in GDP (e.g., Mallawaarachchi et al. 2010). Greater dependence on rain fed farming and the impacts of water trade, however, mitigate much of this reduction. Areas with more marked dependence on irrigation may suffer more significant impacts. Most regions would experience short term reductions mitigated by unrestricted intra-regional trade, the removal of existing cap barriers to out-of-district water trade and targeted purchases within strategic irrigation districts. Further, Mallawaarachchi et al. (2010) also found that reductions in water for irrigators to ensure environmental flows would precipitate a reduction in irrigation net returns of between 16% and 20% compared to a 2007–08 baseline. In dollar terms, this would represent a decline in economic surplus of between AUD$371 and $405 million per annum. However, the changes in water use and productive value would vary markedly by crop type and region. The second round impacts barely registered.

These post-Plan release shifts in the focus on public policy and information needs reflect a transition once more toward balanced attention to issues of people and place in the MDB. In support of this position, Miller (2011) argues that the MDB Plan would likely adopt a reactive and minimalist response to the potential impact on rural communities (place), instead focusing on making good the losses that might accrue through the implementation of the plan. This approach would reduce proactive approaches that aid communities to reorient and diversify their economies, in effect retaining the status quo in relation to rural communities. Improving rural irrigation infrastructure, for example, is regarded as a lifeline because it is perceived to keep water within local communities by promoting more efficient use. However, it is suggested this does little to actually aid the requisite transformation of MDB communities (Miller 2011; Crase 2011). The Plan is an attempt to promote holistic management of the Basin. Managing the Basin as a single entity, rather than its management by individual states, has the potential to facilitate recognition of social, economic and environmental factors. Though, it should be remembered that management of the Basin remains an inherently political process, and it would be naive to assume that technical information alone will be sufficient to arbitrate disputes between key stakeholders; notably those related to the economic, social, community and environmental aspects on the Basin.

8 Managing the MDB into the 21st Century

We believe that future integrated RBM in the MDB must consider a wide range of factors. In particular, the advent of climate change will prove to be particularly challenging in the future, as not only will river managers have to plan for the 'median' or most likely to occur scenario of a reduction in river inflows, they will also have to develop plans for scenarios where river flows are decreased substantially. For irrigators themselves, they also will have to plan for adjustment to future water scarcity. Government policy should focus on the risk of future water shortages, and how planning for water shortages increases profitability and farm viability, rather than using the term 'climate change' to irrigators (Loch et al. 2013). There needs to be more market-based instrument policy development to increase farmers' adaptive capacity to climate change. The efficiency and effectiveness of on and off-farm irrigation infrastructure investment need careful reconsideration (e.g., because of return-flow issues, increased future energy costs, increased future water charges). The increasing over-extraction of groundwater and its use as a substitute for surface water allocations will also gain increasing prominence over the next few decades. There will be an increasing need for markets to offer new products that optimize flexibility in water use for both irrigators and the environment. Such developments include, trade in allocations for environmental flows (Wheeler et al. 2013), counter-cyclical trade between irrigation and environmental water holders (Kirby et al. 2006), option contracts in both rural and urban markets (Heaney et al. 2004; Leroux and Crase 2010) and urban markets (Loch et al. 2013).

Finally, government will need to continue attempting to balance the benefits and costs attached to integrated RBM in the MDB. There will be continuing struggle between the needs of people (in terms of economic and social dimensions) and the needs of place (in terms of the environment). Such a struggle is made even more complicated by the fact that the needs of place overlap with economic, social and hydrological dimensions that span across six political boundaries (four State Governments, one Territory Government and the national Government). Nevertheless, Australia has embarked upon the largest buy-back of environmental water in the world, giving large precedence to 'place' issues, whilst at the same time recognizing the fact that 'people' are also just as important. The importance of people issues are shown in the billion dollar programs that have been developed, and reinforcing legislation that irrigators are the owners of water, and that any reallocation of water from people to place is based on voluntary transfers only.

Acknowledgements

We are grateful for helpful comments received on this paper by the book's editors, and for the research assistance of Jane Edwards. This research was in part funded by a National Climate Change Adaptation Research Facility grant, SD11-116.

References

ABARE-BRS. 2010. Indicators of community vulnerability and adaptive capacity across the Murray-Darling Basin: a focus on irrigation in agriculture. ABARE-BRS Client Report 43099. ABARE-BRS, Canberra.

Adamson, D. and A. Loch (in press). Possible negative feedbacks from 'gold-plating' irrigation infrastructure. Agricultural Water Management. http://dx.doi.org/10.1016/j.agwat.2013.09.022.

Adamson, D., T. Mallawaarachchi and J. Quiggin. 2007. Water use and salinity in the Murray-Darling Basin: a state-contingent model. Australian Journal of Agricultural and Resource Economics 51: 263–281.

ARMCANZ. 1996. National principles for the provision of water for ecosystems. Occasional paper SWR No. 3, July. Dept. of the Environment, Sport and Territories, Canberra.

Ashton, D., M. Oliver and T. Formosa. 2011. Overview of recent changes in irrigated agriculture in the Murray-Darling basin: 2006/07 to 2008/09. Waterlines Report Series No. 64. National Water Commission, Canberra.

Athukorala, W. and C. Wilson. 2012. Groundwater overuse and farm-level technical inefficiency: evidence from Sri Lanka. Hydrogeology Journal 20: 893–905.

Australian Parliament. 2011. Of drought and flooding rains: inquiry into the impact of the guide to the Murray-Darling basin plan. House of Representatives Standing Committee on Regional Australia, Canberra.

Bain, D., A. Close, H. Milner and A. Hee. 1996. Striking the Balance—Managing the Water Resources of the Murray Darling Basin beyond the Year 2000. Hydrology and Water Resources Symposium. Hobart, Australia: Institution of Engineers, Australia.

Barr, N. 2009. House on the hill: the transformation of Australia's farming communities. Land & Water Australia, Canberra.

Beverly, C., A. Roberts, M. Hocking, D. Pannell and P. Dyson. 2011. Using linked surface–groundwater catchment modelling to assess protection options for environmental assets threatened by dryland salinity in southern-eastern Australia. Journal of Hydrology 410: 13–30.

Bjornlund, H. 2005. Irrigators and the new policy paradigm—an Australian case study. Water Policy 7: 581–595.

Bjornlund, H. and B. O'Callaghan. 2003. Property implications of the separation of land and water rights. Ninth Annual Pacific-Rim Real Estate Society Conference. Queensland, Brisbane, 19–22 January.

Bjornlund, H., S. Wheeler and P. Rossini. 2013. Water markets and their environmental, social and economic impact in Australia. pp. 68–93. *In:* J. Maestu (ed.). Water Trading and Global Water Scarcity: International Perspectives. RFF Press/Taylor and Francis/Routledge.

Campbell, I., B. Hart and C. Barlow. 2013. Integrated management in large river basins: 12 lessons from the Mekong and Murray-Darling Rivers. River Systems 20: 231–247.

COAG. 1992. Council of Australian Governments communiqué. 7th December 1992.

Connor, J.D., J. Ward, C. Clifton, W. Proctor and D. Hatton Macdonald. 2008. Designing, testing and implementing a trial dryland salinity credit trade scheme. Ecological Economics 67: 574–588.

Cosier, P., R. Davis, T. Flannery, R. Harding, L. Hughes, D. Karoly, H. Possingham, R. Purves, D. Saunders, T. Bruce, J. Williams and M. Young. 2012a. Analysis of groundwater in the 2011 draft Murray-Darling Basin Plan. Wentworth Group of Concerned Scientists, Canberra.

Cosier, P., R. Davis, T. Flannery, R. Harding, L. Hughes, D. Karoly, H. Possingham, R. Purves, D. Saunders, B. Thom and J. Williams. 2012b. Statement on the 2011 draft Murray-Darling Basin Plan. Wentworth Group of Concerned Scientists, Sydney.

Craik, W. and J. Cleaver. 2008. Modern agriculture under stress—lessons from the Murray-Darling. Murray-Darling Basin Commission, Canberra.

Crase, L. 2008. An introduction to Australian water policy. pp. 1–16. *In*: L. Crase (ed.). Water Policy in Australia: The Impact of Change and Uncertainty. Resources for the Future, Washington DC.

Crase, L. 2011. The fallout to the guide to the proposed basin plan. Australian Journal of Public Administration 70: 84–93.

Crase, L., S. O'Keefe and B. Dollery. 2009. Water buy-back in Australia: political, technical and allocative challenges. *In*: 53rd Annual Australian Agricultural and Resource Economics Society Conference. Cairns, Australia, 11–13 February.

Crase, L., S. O'Keefe and B. Dollery. 2011. Some observations about the reactionary rhetoric circumscribing the guide to the Murray-Darling Basin Plan. Economic Papers 30: 195–207.

Crase, L., S. O'Keefe and Y. Kinoshita. 2012. Enhancing agrienvironmental outcomes: market-based approaches to water in Australia's Murray-Darling Basin. Water Resources Research 48: W09536.

Davidson, B. 1969. Australia Wet or Dry? The Physical and Economic Limits to Expansion of Irrigation. Melbourne University Press, Carlton, Victoria.

DEWHA. 2009. NWI Water trading study: Final report Appendix A. Department of Environment, Water, Heritage and Arts, Canberra.

Dixon, P., M. Rimmer and G. Wittwer. 2011. Saving the southern Murray-Darling basin: the economic effects of a buyback of irrigation water. Economic Record 87: 153–168.

EBC, RMCG, Marsden Jacob Associates, Econsearch, G. McLeod, T. Cummins, G. Roth and D. Cornish. 2011. Community impacts of the guide to the proposed Murray-Darling Basin plan: Report to the Murray-Darling Basin Authority. Murray-Darling Basin Authority, Canberra.

Edwards, J., B. Cheers and H. Bjornlund. 2008. Social, economic and community impacts of water markets in Australia's Murray-Darling Basin region. International Journal of Interdisciplinary Social Sciences 2: 1–10.

Fenton, M. 2007. A survey of beliefs about permanent water trading and community involvement in NRM in the Loddon-Campaspe irrigation region of northern Victoria. North Central Catchment Management Authority and Department of Primary Industries Victoria, Huntly.

Goesch, T., S. Hone and P. Gooday. 2007. Groundwater management—issues affecting the efficient allocation of groundwater. Agricultural Commodities 14: 201–211.

Grafton, Q. 2011. Economic costs and benefits of the proposed basin plan. pp. 245–262. *In*: D. Connell and Q. Grafton (eds.). Basin Futures: Water Reform in the Murray-Darling Basin. Australian National University E-press, Canberra.

Grafton, Q. and Q. Jiang. 2011. Economic effects of water recovery on irrigated agriculture in the Murray-Darling Basin. Australian Journal of Agricultural and Resource Economics 55: 1–13.

Hacking, T. and P. Guthrie. 2008. A framework for clarifying the meaning of Triple bottom-line, integrated, and sustainability assessment. Environmental Impact Assessment Review 28: 73–89.

Hawker Britton. 2012. The Murray-Darling Basin Plan. Hawker Britton Group Pty Ltd, Adelaide.

Heaney, A., S. Beare and A. Hafi. 2004. Trading with the environment: using water options to meet environmental demands. Australian Commodities 11: 606–610.

Howard, J. 2007. The National Plan for Water Security. Prime Minister & Cabinet, Canberra.

Howard, J.L. 2008. The future of the Murray River: amenity re-considered? Geographical Research 46: 291–302.

Industry Commission. 1992. Water Resources & Waste Water Disposal. Australian Industry Commission, Canberra.

Kildea, P. and G. Williams. 2011. The Water Act and the Murray-Darling Basin Plan. Public Law Review 22.

Kirby, M., M. Qureshi, M. Mainuddin and B. Dyack. 2006. Catchment behaviour and counter-cyclical water trade: an integrated model. Natural Resource Modeling 19: 483–510.

Knights, P. and H. Milner. 1987. Natural Streamflow Patterns: A Surrogate for Environmental Requirements. Australian Resources Council: Specialist Workshop on Instream Flow Management. Sydney, Australia.

Leroux, A. and L. Crase. 2010. Advancing water trade: a preliminary investigation of urban irrigation options contracts in the Ovens Basin, Victoria, Australia. Economic Papers 29: 251–266.

Loch, A., H. Bjornlund and R. McIver. 2011. Achieving targeted environmental flows: an evaluation of alternative allocation and trading models under scarce supply—lessons from the Australian reform process. Environment & Planning C: Government & Policy 29: 745–760.

Loch, A., H. Bjornlund, S. Wheeler and J. Connor. 2012. Allocation trade in Australia: a qualitative understanding of irrigator motives and behaviour. Australian Journal of Agricultural and Resource Economics 56: 42–60.

Loch, A., S. Wheeler, H. Bjornlund, S. Beecham, J. Edwards, A. Zuo and M. Shanahan. 2013. The role of water markets in climate change adaptation. Griffith University, Gold Coast: National Climate Change Adaptation Research Facility (NCCARF).

Mallawaarachchi, T., D. Adamson, S. Chambers and P. Schrobback. 2010. Economic Analysis of Diversion Operations for the Murray-Darling Basin Plan: Returns to Irrigation under Reduced Water Availability. Business School Commercial, The University of Queensland, Brisbane.

Marsden Jacobs Associates. 2010. Delivering the Basin Plan: economic & social profiles of Murray-Darling Basin communities. Appendix 3A: A community survey technical report for the Murray-Darling Basin Authority. MDBA, Canberra.

MDBA. 2009. About the Basin—Quick facts [Online]. MDBA, Canberra. Available: http://www.mdba.gov.au/water/about_basin [Accessed 14 July 2009].

MDBA. 2010a. Guide to the Proposed Basin Plan: Overview. Murray-Darling Basin Authority, Canberra.

MDBA. 2010b. Guide to the Proposed Basin Plan: Technical Background Volume 2. Murray-Darling Basin Authority, Canberra.

MDBA. 2011. Guide to the Proposed Basin Plan Technical Background Part III: Appendix C Irrigation District Community Profiles. Murray-Darling Basin Authority, Canberra.

MDBA. 2012. Proposed Basin Plan—A Revised Draft. Murray-Darling Basin Authority, Canberra.

MDBC. 2001. Rivers as Ecological Systems: The Murray-Darling Basin. Murray-Darling Basin Commission, Canberra.

MDBC. 2004. Water audit monitoring Report 2002/03: report of the Murray-Darling Basin commission on the cap on diversions. Murray-Darling Basin Commission, Canberra.

MDBC. 2006. Basin statistics [Online]. Murray Darling Basin Commission, Canberra. Available: http://www.mdbc.gov.au [Accessed 22 July 2013].

MDBMC. 2002. The Living Murray: A discussion paper on restoring the health of the River Murray. Murray-Darling Basin Ministerial Council, Canberra.

MDBMC. 2008. Report of the independent audit group for Salinity 2006–07. Murray-Darling Basin Ministerial Council, Murray-Darling Basin Commission, Canberra.

Miller, C. 2011. The future of the basin: thriving or dying communities? pp. 193–216. *In*: D. Connell and Q. Grafton (eds.). Basin Futures: Water Reform in the Murray-Darling Basin. Australian National University E-press, Canberra.

NSW Office of Water. 2012. NSW Response to the MDBA Basin Plan—Community Information. NSW Department of Primary Industries—Office of Water, Sydney.

NWC. 2009. Australian water markets report 2008–09. National Water Commission, Canberra.

NWC. 2011a. Australian water markets report: 2010–2011. National Water Commission, Canberra.

NWC. 2011b. Australian water markets: trends and drivers, 2007–08 to 2009–10. National Water Commission, Canberra.

NWC. 2011c. Australian water markets: trends and drivers, 2007–08 to 2010–11. National Water Commission, Canberra.

NWC. 2011d. Strengthening Australia's water markets. National Water Commission, Canberra.

NWC. 2011e. Water markets in Australia: a short history. National Water Commission, Canberra.

NWC. 2012. Impacts of water trading in the southern Murray-Darling Basin between 2006–07 and 2010–11. National Water Commission, Canberra.

NWC. 2013. Australian water markets: trends and drivers 2007–08 to 2011–12. National Water Commission, Canberra.

Productivity Commission. 2005. Review of national competition policy reforms. Productivity Commission, Canberra.

Rendell, R. 2011. Regional imperatives for change. pp. 30–47. *In*: J. Langford, J. Briscoe and N. Taylor (eds.). The Australian Water Project Volume One, Crisis and Opportunity: Lessons of Australian Water Reform. Committee for Economic Development of Australia, Melbourne.

Richardson, S., R. Evans and G. Harrington. 2011. Connecting science and engagement: setting groundwater extraction limits using a stakeholder led decision-making process. pp. 351–366. *In*: D. Connell and Q. Grafton (eds.). Basin Futures: Water Reform in the Murray-Darling Basin. ANU Press, Canberra.
RMCG. 2013. Predicting socioeconomic impacts of the Murray-Darling Basin plan on Victoria: Final Report. RMCG Consultants, Bendigo, Victoria.
Ross, S. 2009. Response to preliminary assessment report: a submission to the Natural Resources Commission of NSW Riverina Bioregion Regional Forest Assessment River Red Gums and Woodland Forests. Murray Lower Darling Rivers' Indigenous Nations, Melbourne.
SA DEWNR. 2012. South Australia's response to the draft Basin Plan. South Australian Department of Environment, Water and Natural Resources, Adelaide.
State Library South Australia. 2013. River Murray Timeline [Online]. State Library, Adelaide, South Australia. Available: http://www.slsa.sa.gov.au/murray/timeline.htm [Accessed 23 July 2013].
Wheeler, S., A. Zuo, H. Bjornlund and C. Lane-Miller. 2012. Selling the farm silver? Understanding water sales to the Australian government. Environmental and Resource Economics 52: 133–154.
Wheeler, S., D. Garrick, A. Loch and H. Bjornlund. 2013. Evaluating water market products to acquire water for the environment in Australia. Land Use Policy 30: 427–436.
Wong, P. 2008. Water for the Future—Speech to the 4th Annual Australian Water Summit. Conference & Exhibition Center, Sydney.
Working Group on Water Resource Policy. 1994. The first report of the working group on water resource policy to COAG. Working Group on Water Resource Policy, Canberra.
Young, M. 2010. Managing environmental water. pp. 51–60. *In:* J. Bennett, R. Kingsford, R. Norris and M. Young (eds.). Making Decisions about Environmental Water Allocations. Australian Farm Institute, Surrey Hills, Australia.
Young, M. 2012. Trading into and out of trouble: Australian water allocation and trading experience. Environmental Institute Working Paper. University of Adelaide, Adelaide.

13

The Role of Water Users Associations in Integrated Water Resource Management of Zhangye City in Heihe River Basin, China

Hui Xu,[1] Lichun Sui,[1] Yuhong Li[2] and Dawei Zhang[3,*]

SYNOPSIS

Lately governments at all levels within Gansu have placed high priority on the sustainable management of inland basins. Water User Associations (WUAs) are user-based and participatory ways to manage water resources. The objectives of WUAs are to improve water delivery, increase crop production and provide the farmers with the chance to be involved in the process of irrigation management. In this chapter we examine the processes involved in setting up WUAs in the middle reaches of the HeiHe River (a major inland river in the Hexi corridor of NW China) and summarize experience to date. A series of recommendations on the development of WUAs is offered.

Keywords: Artificial oasis, Endoheic, Gauging of flows, governance groundwater, inland river, Irrigation, Qilian Mountains, redlines, reform, Shiyang river basin, Shule river basin, terminal lakes, Volumetric cap water delivery, water price, water trading, WUAs, Yellow river basin commission, Zhangye

[1] School of Economics, Lanzhou University, Lanzhou, 730000, China.
[2] Department of Water Resources Gansu Province, Lanzhou, 730000, China.
[3] State Key Laboratory of Grassland and Agro-Ecosystems, School of Life Science, Lanzhou University, Lanzhou, 730000, China.
* Corresponding author: zhdawei@lzu.edu.cn

1 Background

There are about 18 inland river basins in western China many of these are in the Hexi Corridor in Gansu Province (Li and Squires 2008). The Heihe River, Shiyang River and Shule River are three typical inland river basins of Hexi Corridor in Gansu Province (see Fig. 13.1).

Inland or internal drainage systems (also called Endoheic) are closed hydrologic systems. Their surface waters drain to inland terminal locations (Fig. 13.2) where the water evaporates or seeps into the ground, having no access to discharge into the sea (see Kingsford et al., this volume).

Lately governments at all levels within Gansu have placed high priority on the sustainable management of inland basins. In the decades beginning 1990 several national government funded river basin rehabilitation projects were launched in Gansu Province, north-western China to make a quick response to the serious water shortage problems currently experienced in inland river basins. There are many dimensions to the underlying causes of problems in dealing with water shortages including: lack of knowledge and technology, inadequate policies and regulations, little policy coordination, inconsistencies between government policy and actions within a river basin, and lack of awareness of measures that could most effectively redress the

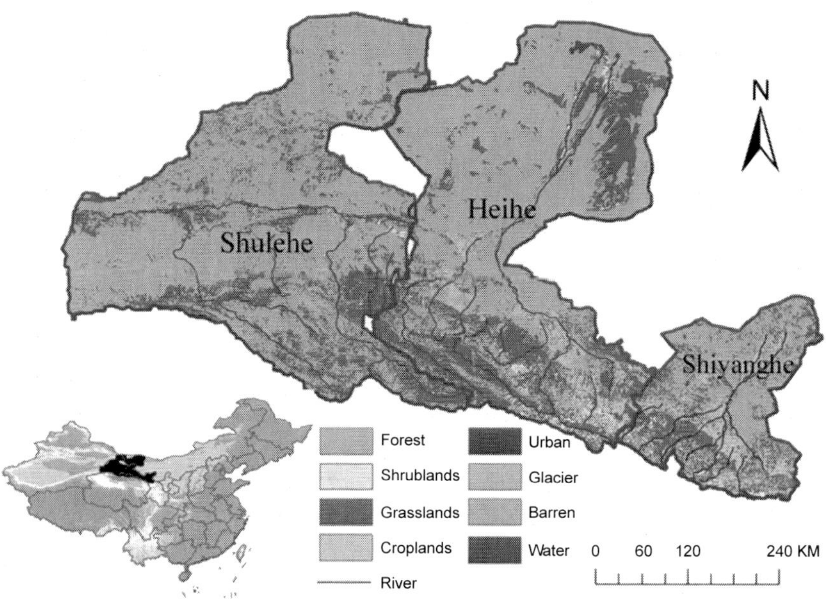

Figure 13.1. Map of three inland river basins of Hexi Corridor in Gansu Province.

Color image of this figure appears in the color plate section at the end of the book.

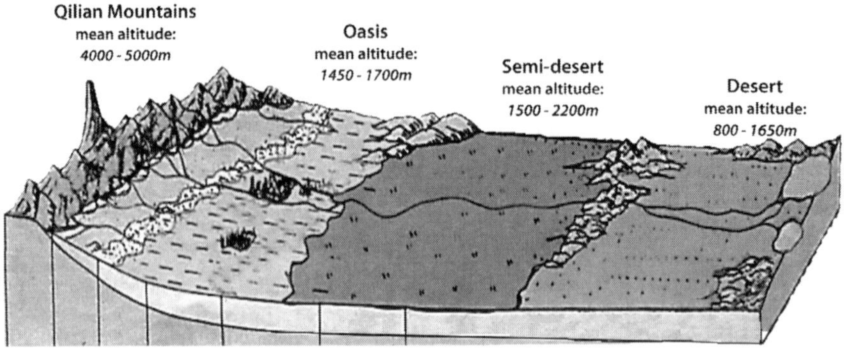

Figure 13.2. A block diagram of Heihe River. The Zhangye Oasis is in the middle reaches.

problems. If not addressed, the problems in river basins are exacerbated by economic development, migration, changing land use and climate change.

This chapter will focus on the Heihe River Basin (see Li and Squires, this volume, for discussion on the Shule and Li et al. (2013) for discussion on the Shiyang).

2 Study Area

The Heihe River Basin is the second largest inland river in the arid area of Northwest China. It has an area of about 143,000 km^2 and its main stream, with a length of 928 km, originates in the Qilian Mountains and flows through the Hexi Corridor of Gansu Province and enters into the western part of the Inner Mongolia Plateau (see Fig. 13.2).

Zhangye City,[1] located in the middle reaches of the Heihe River Basin, has an administrative area of 40,874 km^2 and has a population of 1.2046 million, with the natural growth rate of 5.23%, including a rural population of 771,200 (64.0% of the total). It has a typical continental arid climate, with the annual average temperature of 7.6°C, annual rainfall ranging from 100 to 300 mm, and annual potential evaporation capacity reaching 2,300 mm. According to the Ministry of Water Resources (MWR 2004) Zhangye City is seriously short of water, even though it uses almost all the water of Heihe River. The available water resources volume of the city is only 2.75 million m^3 (bcm), with surface water sources capacity of 2.46 bcm and groundwater sources capacity of 0.115 bcm making the per capita total volume water resources in the basin of 2,304 m^3. However, about 0.95~1 bcm of the total water annually flows through the Zhengyixia Gorge where the Heihe River exits from the Qilian Mountainsthus providing a potential

[1] In China a city is a large administrative areas, often encompassing several counties and having a population of several million.

per capita volume of about 1,250 m^3 that is far lower than the national average (Han 2011). Growing demands on water, particularly in northern China, are putting more and more pressure on Chinese ability to produce its own food. The Zhangye City, one of the major grain producing regions in Gansu Province, is highly dependent on irrigation water.

Only about 50% of the farmland is irrigated and many arable lands have been abandoned due to water scarcity. The area to which water is supplied has reduced to 205,203 ha in Zhangye City (Chen et al. 2005) while the size of the irrigated area in northern China as a whole has rapidly increased since the 1980s, from 45 million hectares (Mha) in 1979–81 to 60 Mha in 2010 (World Bank 2006; NBS 2011; Zhang et al. 2012). In 2012, agriculture accounted for approximately 90% of all water use and almost all the water in the Heihe River that reaches Zhangye is extracted for irrigation use. As a result, too little water flows into Juyanhai Lake; the lake dried out in 1992, turning an area of 200 km^2 around the lake into a desert (MWR 2004; Zhang et al. 2009).

Water saving measures should be taken to achieve greater efficiency of water use per area of crop irrigated, less water per unit of crop production (Ml/tonne), and promote water resource management. This chapter is focused on water allocation for irrigation, its management and the role of Water User Associations (WUAs).

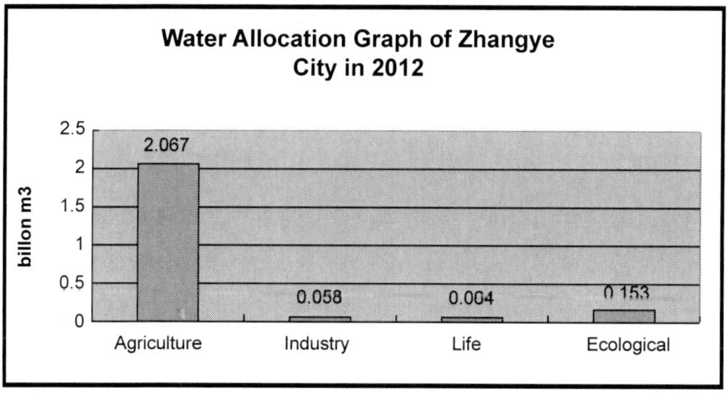

Figure 13.3. Water allocation in Zhangye in 2012. Ecological flows have increased recently as a result of the decision to allocate water to refill the terminal lake. Source: (Water Bureau of Zhangye City in 2013).

3 History of Water Resources Management

The Household Responsibility System which originated in the late 1970's, and which turned the responsibility of crop production over to farmer households, is the key source of the most important transformation in rural China. In this peaceful revolutionary process, farmers made good use of governmental reform policies and social capital to achieve their goals of autonomy in production and marketing, though they did not ask for land ownership as they did in the past. Luckily this historical reform enabled farmers themselves to leave the collective farming system which severely restrained farming productivity and flexibility. Under Chinese law, the ownership of land and water resources belongs to the State and the collectives. Farmers are entitled to use land under a long-term contracting system. Except for the other communal properties, such as offices of village committees, schools and irrigation facilities at the terminal level, most of the collectively-owned farmland was redistributed to individual households by means of a land-use contracting system. Family heads once again became the masters of their own houses, in control of farming and allocating labour within their families. This de-collectivization increased farmers' physical capital and financial capital, and liberated them from the rigid control of the state bureaucracy. Water resources management at that time was mainly done through collective ownership arrangements. The village leaders, representing the village council, were responsible for water allocation, canal operation and maintenance and water fee collection (Huang et al. 2009). This traditional water resource management system is similar to the system that governed most of China's rural water resources during the people's commune system period that came into effect in 1949.

It can be seen that the role of local people in irrigation water management was neglected in China for some time. As China's economy grows, problems and issues in natural resource management have become apparent and acute in recent years. Given China's rural water use situation and pressure on the natural resource base, corresponding governmental policy and management system reforms that are aimed at promoting participatory irrigation management and development of WUAs have been welcomed by rural communities in recent years. Establishment of water rights and water markets, promotion of participatory irrigation management, development of WUAs and construction of a water-saving society are the key ingredients in China's current water management reform.

Since the 1990s two major types of water management reforms can be observed in northern China, namely user-based, participatory management through water users associations (WUAs) and contracting out of irrigation canal management to individuals. It was estimated that by 2004 more than one-quarter of the villages in northern China had replaced traditional water resources management by either WUAs or contracting (Huang et al. 2009). Their study further shows that water availability, length and complexity of the canal system and reform-promoting policies of local governments are the main drivers of water management reforms (see Li and Squires, this volume). In subsequent research comparing the performance of the other management systems, Huang et al. (2010) find that WUAs perform much better than traditional water resources management systems in terms of maintenance expenditures, timeliness of water delivery and rates of water fee collection. Management systems for water distribution infrastructure based on contracting to individuals also perform better than traditional systems, although not as well as WUA-based systems.

3.1 The issues of the traditional irrigation system management

With the rapid development of social economy and increasing population, water resource scarcity of Heihe River Basin has become a serious problem. The middle and lower reaches, which account for 93% of the whole basin area, produce little surface runoff. The current water resources supply and demand are in serious imbalance made worse by the inefficient traditional irrigation system. Rural water sector was facing an unexpected challenge. Some serious problems have been revealed with its irrigation systems, such as low irrigation system capacity, wasteful irrigation practice, slow and irregular water delivery, low design and construction standards of irrigation infrastructure, poor operation and management of irrigation facilities due to rigid policy and management systems, as well as severe shortage of funds for restoring and maintaining both new and old irrigation systems in many irrigation districts.

3.1.1 Chaotic irrigation behavior

There are no detailed plans to distribute water resource appropriately in the irrigation area. When irrigation water is delivered at the peak of growing season, everyone scrambles for it as much as they can, but after this period of time, few people express interest in it. Some villagers in the Ganzhou district of Zhangye City describe a situation where people from the upper basin can use water resource whenever they like, however, lots of water-deficient areas in the lower basin cannot have enough water, so chaotic irrigation behaviors often occur because of lack of supervision and poor management.

3.1.2 Water fee is difficult to calculate and imposition of an arbitrary charge is common

The aim of establishing scientific water pricing policy is to ensure water delivery to proceed smoothly, improve rational disposition of water resources and save water. Traditionally water pricing policy was formulated by the administrative department acting alone, so that it cannot adapt to current situation or promote water resource saving and protection. Water fee is one of the most serious problems in the traditional irrigation system. As a general rule, the initial water fee is based on the number in a household (per capita) and the area of arable land (per unit area of land). This gives rise to conflicts so that the distribution of water resource and other management work of irrigation areas are difficult to carry out smoothly. In addition, arbitrary and double charges are serious because of the lack of a specialist organization to oversee the operation.

3.1.3 There is a serious waste of water in irrigation areas

The phenomenon of "the communal-watering system" was very common in the old days, the use of water resource was split by each water-user evenly (regardless of need) and there was no mechanism to improve users' water-saving awareness. So, the phenomenon of wasting water became more and more serious. In the past years, government puts huge investment into water projects. For example, in 2000, the State Council invested 35 billion Yuan[2] on the national water project (Liu et al. 2008). In 2012, the practical investment on all kinds of water conservancy project in Gansu Province had reached to 6.98 billion yuan (4.8 billion yuan from National Government). It was estimated that the total investment would be 7.5 billion yuan (5 billion yuan from the National Government) (Song 2012).

3.1.4 Water infrastructure lacks maintenance and the aging engineering construction is in bad condition

In some traditional irrigation areas, engineering facilities are in a stage of aging and disrepair. Water infrastructure lacks routine renovation and maintenance. The main reasons are that the main responsibilities are not clear and the system lags behind with the building and maintenance.

4 Water Users Association (WUA)

To deal with water shortage problems, the MWR initiated a pilot project in 2002 named "Building a Water-saving Society in Zhangye City". The project was done through government investments in water-saving irrigation systems by installation of meters for measuring water use and through establishing a Water Use Rights (WUR) system with tradable water quotas. Since its inception, agricultural production capacity, the efficiency of water utilization, municipal and industrial wastewater treatment and so on have made a great progress according to government sources. By 2006, about 790

[2] At that time $USD1=8 yuan.

WUAs had been established based on natural villages[3] or hydrological units of the irrigation system and the total number of managed lateral canals is 20,560; the total length is 37,126 km and the irrigation area is 210,300 ha in Zhangye City (Water Bureau of Zhangye City 2007).

4.1 The concept of a Water User Association

Ideally, a WUA is a group of water users, such as irrigators, who pool their financial, technical, material, and human resources for the operation and maintenance of a water system. The objectives of WUA are to improve water delivery, increase crop production and provide the farmers with the chance to be involved in the process of irrigation management. A WUA usually elects leaders, handles disputes internally, collects fees, and implements maintenance. Membership of WUA provides benefits. Membership is open to irrigators who may apply to join but membership also includes government representatives and other non-irrigators.

There have been pilot projects to establish WUAs in the neighboring Shiyang River Basin (Zheng et al. 2012) and there has been opportunity to learn from this experience.

The WUAs in Zhangye City were established by the local government and water administrative department taking administrative villages[4] as a basic areal unit. In general, a WUA consists of the water-user group, the membership representative conference, the executive commission and the board of supervisors.

4.2 The water-user group

The *water-user group* is formed on the basis of villagers' group or hydrological boundary. It is at a level lower than a WUA. The water-user group includes some who are also members of a WUA. The water-user group plays an important role as a bridge and tie between the WUA and water-users.

4.3 The member representative conference

The number of member representatives is determined based on the irrigation area or number of the households. In principle, one formal

[3] A 'natural village' is a group of households who are resident on a site and whose fields are nearby.

[4] In China an 'Administrative village' is also an administrative area of several km^2 that is smaller than a Township that may contain several administrative villages and hundreds of natural villages.

representative may be elected for every 100–300 mu[5] in the village or every 10–15 households whichever is the lesser. The member representatives are elected by ballot from water-users but other people (e.g., local government) may be elected as well. These one or two representatives are people who must have enough support and standing in each water-user group. In general, the membership representative conference of the WUA is held once a year and insists on the principle "to deal with the issues one by one". It is the top decision-making organization of a WUA.

4.4 The executive commission

The members of the *executive commission* are elected by the member representatives and there is one chairman who is responsible for the overall work and 1 to 2 vice-chairmen whose duty is to oversee day to day running of the commission. The tenure of the executive commission is three years. The executive commission is the main administrative department that is responsible for routine work, including contacting with Water Supply Organization, drafting internal regulations and systems, signing up water supply-use contract, drawing up and carrying out the irrigation plan, managing water right trade, and collecting water fee, etc.

4.5 The board of supervisors

In order to ensure good operation and transparent financial affairs of the WUA, there needs to be a special board of supervisors to monitor commission's routine work and financial expense of the association. Normally, the board of supervisors consists of local government, member representatives of WUA, irrigation management organizations and local farmers.

5 The Operation of WUA

The government helped the fledgling WUAs to draft its specific articles of association. Main contents of the article include: name; nature; objectives; the structure and duty of the organization; the rights and obligations of the members and the executive commission; supervision arrangements. After a WUA is established formally, the operation of a WUA focuses on mainly three respects: (i) Irrigation management. The WUA make a detailed plan for irrigation water use on the basis of local water-users' demand and actual water supply capacity. The WUAs need to ensure that irrigation distribution

[5] 15 mu=1 ha.

systems are running well. (ii) Infrastructure management. Normally, WUA need to maintain and protect irrigation canals and engineering facilities regularly within its own sphere. (iii) Finance management. In order to regulate the behavior of financial and improve efficiency in fund utilization, associations need to establish a strict accounting system and internal control, public financial revenues and accept democratic supervision.

So, as a major and concrete part of the Participatory Irrigation Management (PIM), it is widely believed that WUA could promote the development of efficiency and equity in water resources management at local level through providing appropriate incentives to water users and more opportunities to be involved in the process of irrigation management and improve the accountability of the decision-making process.

5.1 The relationship between government and WUA

In the past, the traditional management of water resources was mainly done through collective ownership arrangements. During that period, government and water resource administration management organizations played the most important role in water resources allocation. Naturally, they became the main body who could draft related policy, fix water price, and dictate the regulatory framework and so on.

However, since WUAs were founded, government should not be the leader, but should play a role as a partner. Bear in mind though that WUAs are promoted by the government, especially MWR. The MWR initiated a pilot project called "Building a Water-saving Society in Zhangye City" in 2002. After that, WUAs were gradually set up across the whole of Zhangye City. The government, through MWR, not only leads and develops WUA in the policy aspect, but gives detailed guidance on the specific operation. In preparation for WUA establishment, the government set up at water management system, project management system, financial management system and drafted basic articles of association.

In most cases, the institutional structure of self-managed irrigation district consists of the Water Supply Organization (WSO) and WUA. WSO is responsible for the unified operation and management in the canal system and the whole irrigation and drainage district. WUA is responsible for the operation and management of the lateral canal and branch canals. The basic model is "WSO+WUA+water-users". But in fact, the relationship between WUA and local water management department is still ill-defined despite several pilot projects in cooperation with international donors.

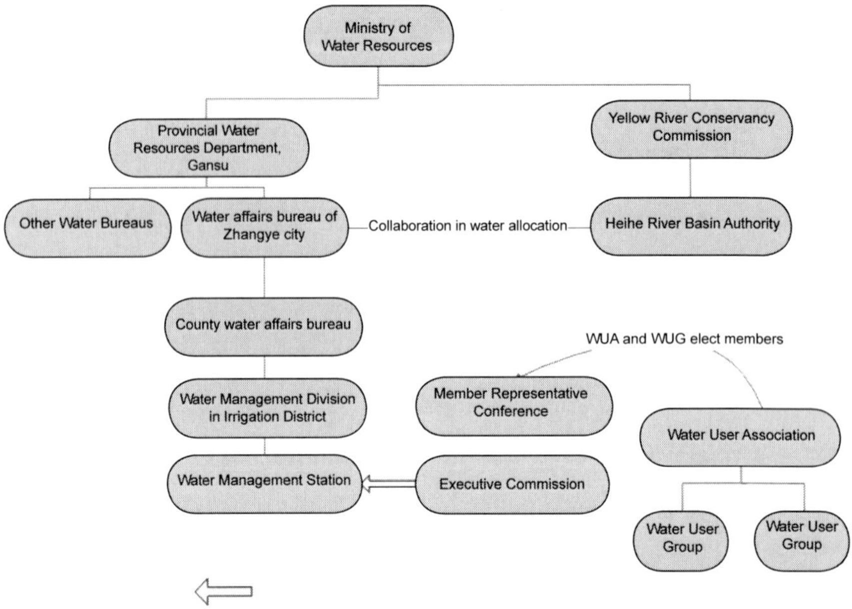

Figure 13.4. Organogram of water resource management of surface water resources in Heihe river basin.

5.2 Yellow River Conservancy Commission (YRC)

The YRC, an agency of MWR, is responsible for integrated management and supervision over water resources throughout the entire Yellow River basin but also has authority in the inland river basins of the Hexi corridor (see Du et al., this volume; Li and Squires, this volume). In order to solve the water issues, the Heihe River Basin authority was established in 2000 as a branch of the Yellow River Conservancy Commission. The role of the authority is to realize the integrated water management of Heihe River; draft water allocation plans; be responsible for construction and management of water conservancy; and solve disputes over water uses among different administrative regions.

5.3 The water supply organization

WSO actually includes all water supply management institutions, such as water affairs bureau at municipality and county level and water management station at irrigation district level. Water Bureau of Zhangye city manages 6 county water affairs bureau. Water management station in irrigation district is affiliated to county water affairs bureau. Water management station is the primary level of the county water affairs bureau and is responsible for actual operation of the irrigation channel system at main and secondary levels to match with the agreed irrigation operations plan. Operation of tertiary level channels is done by the appointed water user group members.

6 Barriers that Hinder the Development of WUAs

6.1 The problems existing in the formation of WUAs

(i) In the early days after WUAs were founded, the basic publicity and training were inadequate, villagers lacked enough enthusiasm to participate actively in this organization; (ii) In order to pursue completion rates of assignment, some irrigation areas blindly established a number of WUAs without any verification so that relevant regulations and procedures (including formal registration) cannot meet the original standard; (iii) According to formal requirement, each newly established WUA should register in the local civil department, then it can become a legitimate organization. But some WUAs were founded without registration in some places because of traditional notions or high registration fee. In some places, registry department charges a simple registration fee, which is mainly used for daily expense, such as paying for staff wages; (iv) Due to poor educational level and inexperience of members of village organizations and water-users, the operation and management of WUAs was not going well.

6.2 The problems existing in the operation and management of WUA

(i) Some WUAs are not separated from local village committee completely, so that the operation and management of WUAs cannot fully realize democratization; (ii) Irrigation engineering works are controlled by a WUA but owned by the government. Because WUAs have the right (and responsibility) to manage and use irrigation engineering but the ownership belongs to others (departments within the government) there is resistance to raising funds to support the routine work and maintenance of irrigation infrastructure; (iii) Lack of financial support may be the biggest barrier in the operation and management of WUAs. Shortage of funds makes it hard to provide for routine spending and payment for staff so that farmers' enthusiasm declines gradually. The WUA is a nonprofit organization so it needs some funds to maintain daily expense but the only revenue comes from the water fee that is collected from water users and used for buying water from the government.

7 Recommendations on the Development of WUAs

WUAs should play a significant role in solving water scarcity and promoting water use efficiency among the households in Zhangye City. But in order to make them a success, the administrators believe that it is necessary for government to take effective measures to develop and improve the current operation and management of WUAs. Some measures that should be taken include: (i) Building long-term compensatory mechanisms (such as: setting up a special fund, credit policy, international loan). This is seen as a primary task for sustainable development of WUAs. When sufficient funds are provided the organization can maximize its function and members become more active participants in WUAs. (ii) Increasing publicity about WUAs and promote awareness raising among water-users. Lots of local villagers still retain traditional views and they lack knowledge about WUAs, so it is essential to publicize some relevant information and make water users understand the advantages of new irrigation systems. (iii) Relationship between irrigation management department and WUAs need to be clearly defined through better regulations and rules. Only in this way can WUAs avoid unnecessary conflict with some government departments. Keeping good relationships is conducive to the further development of participatory irrigation management. (iv) Establishing a monitoring and evaluation system. As an independent legal entity the WUAs need to accept strict supervision from the board of supervisors, village committees and water-users. During the whole processing of WUAs, the board of supervisors need to plan the specific and regular monitoring system to make incentives for the further operation of WUA.

Generally speaking, it turns out that establishing WUAs in Irrigation districts (ID) can not only ensure water supply and improve water saving consciousness, but also it could reduce water use disputes. However, the WUAs in Zhangye City are still in an early stage and there are really many problems that are difficult to solve. The government and irrigation management department should take efforts to promote reform to create and develop a more relaxed external environment for WUAs.

7.1 Pre-requisites and steps that are still required for water trading to begin

Water trading has been suggested as a mechanism to improve productivity and water use efficiency.

(i) While water rights have been defined at broad levels, including usage type and irrigation district, etc. they haven't always been pushed down to the level of water user associations (WUA), say 100 households, using about 300,000 m^3 water, which might be the realistic level for trading in Hei He Basin. The water rights that have been defined are based on long-term average flows, which is recognized as being a problem. It is said that in dry years all the rights for different uses are reduced proportionally, but the relationship between the enduring, unvarying entitlement volume and the volume that can be used in the current season is unclear. Given the volume that can actually be used may go up or down, there needs to be very clear, stable rules for assessing the available water and making current-seasonal allocations. And there needs to be a register of current allocations, as well as of enduring rights.

(ii) Because reservoir capacity is less than half annual usage volumes, it is likely that having current-seasonal allocations won't work in the inland basins in Northwest China. Possibly allocations could be based on predicted inflows. It is crucial that the way of specifying what people can take in the current year, suits the hydrology and pattern of demands. If there are to be allocations that are tradable as clear-cut, unencumbered rights to water in the current year, then once they have been given out they should never be reduced: the allocations may have been sold, and cutting back allocations that people have paid money for would make for a messy market.

(iii) Governments must be responsible for protecting the health of ecosystems in the inland basins, and it is important to set aside enough water for this **before trading starts**. It would also be wise to have an ultimate power to review rights, e.g., every 15 years, in case the environment needs more water. The allocation plans have, in each basin, identified an amount of "water reserve", as well as an ecosystem right. This reserve is for "emergencies", but that makes it vulnerable to being misused by other users. It might be better for the reserve to be earmarked for the environment. In any case, right systems need to be embedded in law, not subject to policy changes or power plays.

(iv) Gauging of flows in the main and branch canals is in place, but accurate metering at the level of tertiary canals (WUAs) and minor canals (households or groups of households) is not often done in the inland basins. Whereas for groundwater, good meters are being installed to prevent serious overuse in the inland basins, surface water is still being distributed by rotating the times that manually-operated gates are open. This reflects the way resources are shared at this level, instead of there being individual rights. This is an area for reform. As water rights are more sharply defined and their value becomes clearer, some legal sanctions will probably be appropriate. But when people are poor and are simply taking more than they are allowed, the cut-off system (pumps in tube-wells are operated via IC card) being used for groundwater makes sense.

(v) It is quite understandable that there is anxiety about speculation and the risk of a monopoly in water trading. It would be very reasonable to start off only allowing trade between farmers, or only have limited trade to specified other uses. It will be necessary to loosen the tie between land and water to some extent, e.g., someone's land shouldn't be confiscated if they sell their water.

(vi) The process described above for getting approval to use water, namely submissions up through various layers, getting a water permit, etc. seems to be quite complicated.[6] Before each irrigation season each household has to report to its water user group and WUA about its water use plan including size of farm plots to be irrigated, type of crops to be irrigated, time for irrigation, etc. Once WUA gets all the irrigation requests from its members the WUA will compile the requests and make its own irrigation plan to be submitted to the Water Management Station that is in charge of water distribution in the Irrigation District. Being supervised by the Irrigation District Management Division, after consulting with township water resource conservancy office, the Water Management Stations will review all WUAs' irrigation plan and formulate overall water distribution plan for the whole irrigation district. Once the Irrigation District Management Division gives approval to the finalized water distribution plan for each water management station it cannot be altered. For successful water trading to occur it is necessary to be clear about what is being traded. Presumably it is the underlying right, represented in the transparency card,[7] or the current season's water that is based on this right. The water permit may be more like a delivery entitlement. Some "unbundling", at least conceptually and administratively, is in order. Decisions will be needed about whether water can be traded across reservoirs, between rivers, etc., and these should be set out in trading rules that everyone is aware of. The bigger the trading zones, the deeper the market (David Lewis, pers. comm.).

7.2 Mechanics of water trading

(i) Water trades will need to be approved and recorded by an appropriate government or municipal body, probably the body that keeps a

[6] Each household within a WUA must buy a water permit from the water management station in a process that is very complicated and inflexible. Once approved the WUA must keep water use within the approved water distribution plan and rotation.

[7] The "transparency card" is issued to each household by township Water Resource Conservancy Office. This is to ensure that water allocations are legitimate. The name of the householders, the area of irrigated land, the nominated volume of water on the permit, water price, water tariff and total allocation for the year are all recorded.

register of entitlements. If there are separate trades of current-seasonal allocations and these are managed by a more local body, approvals for these could be from that body. An agreement to trade water entitlement is basically a matter between the seller and the buyer, i.e., it is a private contract. A temporary trade comes into effect as soon as it is approved by authorized agency, the contract needs to require the buyer to hand over the money, less any deposit. A permanent trade is more serious, and it needs approval first, it is settled with the payment next, and then is recorded. While in each case arrangements could be left to the individuals, it may be helpful for a government body to provide some guidance, e.g., model contracts.

(ii) Buyers and sellers finding each other can again be left to individuals to work out, or people may choose to get help from a business that specialize in match-making, i.e., a water broker. A broker will also assist with contracts, like a real estate agent. It may be that in China the government body that approves trade, does match-making too. For example, it keeps a list of offers from sellers, and when a buyer accepts an offer, the trade can forthwith be approved and recorded by the same body. However, there is some risk involved; indeed there is risk even in a separate government body offering a match-making service. There must be no suspicion that certain buyers are favored, no secret commissions, etc. (David Lewis, pers. comm.).

In summary, it must be clear that each trade has to be approved by the water authority. Some trades are arranged privately but often brokers are used to find matches. The benefits of trade are huge, e.g., giving choice to users, reallocating water to its best use, promoting efficiency. In regulated systems trade is possible over wide areas. Water metering is essential for trade. Stealing water should be punished, e.g., heavy fines or several months in jail for a first offence. Unbundling the land right and water right from delivery rights facilitates trading. Above all, for trading clear entitlements are needed, with clear stable rules about how much water goes to each entitlement. Extra water for the environment must come from buying entitlements or from investment in water savings.

8 The *"Three Red Lines"* Policy Might be the Key to Solve Water Problems in the Hexi Corridor

The *"three red lines"* policy formulated by the Ministry of Water Resources China is regarded as the most stringent water policy in China's history. The National Government requires that the national water sector should implement the strictest water resources management system, endeavor to promote water management from supply-based to demand-oriented, from random and over-exploitation to rational and orderly development, from intensive utilization to highly-efficient utilization, from post accident remedy to precaution. One of the objectives is that by 2020, a complete water resources management and supervision system shall be set up to ensure rational water allocation, water-conserving society shall be established to realize a marked increase of water utilization efficiency and benefit, water quality of drinking water sources and water ecological conditions of key regions shall be greatly improved, over-exploitation of groundwater shall be effectively controlled, and water supply capacity for socio-economic development should be greatly enhanced.

In order to implement the strictest water management system, the core is to define the "three red lines" or safeguard lines, namely the "red line" of water development and utilization, used to control water utilization quantity, the "red line" of water use efficiency to curb waste of water, and the "red line" of pollutant carrying capacity in water function zones to strictly control pollutant discharges. In order to achieve the objectives, China's water sector should carry out the following key tasks.

(i) **Water allocation should be rationalized**, focusing on the control of total water quantity. China will promote the formulation of water allocation plans in river basins and regions and establish an index system guiding water abstraction licensing to cover all river basins, provinces, municipalities and counties. China will strictly manage water allocation and utilization, and strengthen integrated water resources allocation in order to satisfy the needs of domestic, industrial and ecological water use.

(ii) A **water conservation society should be developed**, aiming at improving water use efficiency and effectiveness. China will set up an evaluation system assessing local and industrial water use efficiency, water consumption quota for unit products, promote technology of water saving products and strengthen technology upgrading, intensify water conservation. The government will promote reform of the water rights and pricing system.

(iii) **Water resources protection should be reinforced**, based on water functional zoning. The country will improve its rapid response mechanism for water pollution accidents, and strictly monitor pollutant discharge outlets. China will optimize water allocation and maintain the health of rivers, lakes and ecosystems, strictly manage groundwater exploitation and utilization, control groundwater development quantity, and issue bans and quotas for groundwater exploitation in over-abstraction areas so as to rehabilitate the environment and ecosystem.

(iv) **Water management system reform should be deepened**, concentrating on integrated water resources management. China will improve the water management system that features river basin management and administrative management, intensify integrated water resources planning, allocation and regulation in river basins, strengthen integrated water affairs management in administrative regions, and promote integrated rural and urban water affairs management.

(v) **Water resources management should be improved**, supported by science and technology advancement. The government will speed up construction of water monitoring and control systems, and set up a monitoring and management platform which meets the demands of total quantity control of water utilization, water function zoning management and water resources protection, so as to promote information technology application and modernization of water resources management.

As the 12th Five-year plan of China stipulated, it is urgent to carry out the strictest water resources management system, strengthen the controls over the total quantity and quota management of water resources, and build an innovative water saving society. According to the 12th Five year plan of Zhangye City speeding up the pace of irrigation water-saving transformation, renovating water conservancy infrastructures, and forming innovative participatory irrigation management to promote efficiency of agricultural water use are prioritized. Therefore, water management reform in Zhangye City is more likely to be successful in improving water use efficiency and possibly even farm income levels, if those characteristics that boost the development and operation of WUA are taken into account. Local authorities should take notice of the experience gained in neighboring river basins such as Shiyang (Aarnoudse et al. 2012) and Shule (Li and Squires, this volume) and the work of the international donors such as DIFID, ADB and AusAid in approaches that integrate water use from both surface and groundwater sources through locally-based WUAs.

Guidelines for better functioning of WUAs arise from several studies in NW China, especially from experience gained in the Shiyang Basin (Zheng et al. 2012). WET (2012) describe how water allocation to WUAs could be improved according to the principles of fairness, efficiency and environmental sustainability. They also describe how the water rights of WUAs could be volumetrically defined and capped through the issue of Group Water Entitlements (GWEs) at the point at which WUAs pay for bulk deliveries. Below this point, farmers would continue to pay for water on an area basis, as delivery and monitoring infrastructure in irrigation districts (IDs) in China, is not in place to monitor individual entitlements at the household level.

A volumetric cap on the water rights of WUAs needs to fully consider existing patterns of water use within and between WUAs, and the experience of farmers, WUA representatives and supervisory staff who administer the present systems. Hence it is proposed (WET 2012) that rights allocation follows existing practice by linking land and water rights. In other words, rights assigned would be directly linked to the (existing) irrigated areas of each WUA, and could not be negotiated upwards by a WUA seeking to expand its irrigated area or plant more water intensive crops, for example. Hence one objective of defining and enforcing WUA-based GWEs would be to end the "requirements approach" to water use planning that currently prevails so that, in future, *water savings* rather than *additional supply* would be used to maintain or increase farm production and farmer incomes.

Too little is known about what farmers want but studies in the Shiyang basin and elsewhere in NW China have been helpful in capturing their priorities and in pinpointing failings in the present system of water allocation and supply (Box 13.1). There is plenty of room for improvement and as Zhangye City gains experience in administrating WUAs and as actual water users get to have more influence in the running of the WUAs we can expect to see a shift from current rigidly controlled top-down arrangements to something closer to how WUAs are meant to work.

Box 13.1. Key issues to be resolved in the development and extension of WUA.

Criticisms of the present situation on WUAs

- Little participation by farmers in WUA or even information about their role
- WUA not used directly for water resource management, no impact on water use
- Inconsistently enforced
- Ineffective, unrepresentative
- Standard 'solutions' applied little impact on productivity of water
- No power to stop increasing water being diverted to industry
- *Impact* of WUAs not known—no way to assess it

What do the farmers want from WUA?

- Reliable delivery of water in accordance with an agreed allocation
- Clear decision-making, taking account of their interests
- Technical assistance to cope with declining allocation
- Compensation for loss of traditional rights
- Management of water quality and wider water environment issues
- Reliable delivery of water in accordance with an agreed allocation

Box 13.1. contd....

Box 13.1. contd.

What is needed to overcome these issues: an integrated approach to water management

- Rational and agreed water use plan—based on knowledge of the resource and uses, and how they are linked
- Financially sustainable management system
- Established and protected water rights
- Compliance with and enforcement of rules
- Participation in design to ensure acceptability and implementation
- Complementary measures to optimize water use and productivity
- Monitoring, auditing and adapting the plan

In conclusion, the government has been seeking alternative ways to decentralize its traditional water management system, and WUAs are considered as an effective approach for this purpose in order to solve the problem of contradiction between water supply and demand, increase water use efficiency and promote fairness in water resource allocation. It is essential to popularize WUAs across Zhangye City because since 2005 WUAs have played a role in innovative irrigation management mode and encourage farmers to participate in on-farm water management.

Acknowledgements

We thank Dr Victor Squires and Mr Hugh Milner for their valuable suggestions on the chapter. We are grateful to Peter Waldheim for Figure 13.2. Thanks to Heihe River Basin Authority, Water Bureau of Zhangye City for the documents provided. We thank Dr. Qun Du for giving the chance to write the chapter. This work was supported by the Chinese National Natural Science Foundation (grant no. 91125020 to D.Z. and grant no. 71373109 to H.X.).

References and Further Readings

Aarnoudse, E., B. Bluemling, P. Wester and W. Qu. 2012. How WUAs facilitate direct groundwater regulation. A case study of Minqin County, China. Water Policy Research Highlight 23: 1–5.

Chen, Y., D. Zhang, Y. Sun, X. Liu, N. Wang and H.H.G. Savenije. 2005. Water demand management: a case study of the Heihe River Basin in China. Physics and Chemistry of the Earth 30: 408–419.

Du, Q., C. Zhuang and S. Spooner. 2014. Sustainable management of water resources in the Yellow River Basin: Pressure, Issues and Countermeasures. pp. 243–275. *In*: V.R. Squires, H.M. Milner and K.A. Daniell (eds.). River Basin Management in the Twenty-first Century: Understanding People and Place. CRC Press, Boca Raton.

Han, C. 2011. Impact of participatory irrigation management of Water Users' Association on Farmer's water-using behavior in Zhangye irrigation District. China Population, Resources and Environment 7: 126–131.
Huang, Q., S. Rozelle, J. Wang and J. Huang. 2009. Water management institutional reform: are presentative look at northern China. Agricultural Water Management 96: 215–225.
Huang, Q., J. Wang, K.W. Easter and S. Rozelle. 2010. Empirical assessment of water management institutions in northern China. Agricultural Water Management 98: 361–369.
Lewis, D. (pers. comm.). Water Rights and Trading in Victoria. Slides used for visit to Gansu province, China, November 2011.
Lewis, D. (pers. comm.). Visits to Two Stressed River Basins in Northwest China to Share Experience of Water Trading, November 2011.
Li, F., G. Zhu and C. Gao. 2013. Shiyang River ecosystem problems and countermeasures. Agricultural Sciences 4(2): 72–78.
Li, Y. and V. Squires. 2014. Integrated Water Resources Management of inland river basins in the Hexi Corridor, Gansu, China with special reference to the Shule River Basin. pp. 82–100. *In*: V.R. Squires, H.M. Milner and K.A. Daniell (eds.). River Basin Management in the Twenty-first Century: Understanding People and Place. CRC Press (this book).
Liu, J., R. Meinzen-Dick, K. Qian, L. Zhang and L. Jiang. 2008. The impacts of Water Users' Associations on household production in central China. China Economic Quarterly 7(2): 465–479.
Liu, J., R. Meinzen-Dick, K. Qian, L. Zhang and L. Jiang. 2010. The impact of irrigation management transfer on household production in central China. China Economic Quarterly -9(2.
NBS. 2011. China Statistical Yearbook 2011. National Bureau of Statistics of China, Beijing.
Song, A. 2012. http://www.gs.xinhuanet.com/news/2012-10/01/c_113265304.htm.
WET. 2007. Water Entitlements and Trading Project (WET Phase 2) Final Report December 2007 [in English and Chinese] (Beijing: Ministry of Water Resources, People's Republic of China and Canberra: Department of the Environment, Water, Heritage and the Arts, Australian Government.
World Bank 2006. China: Water Quality Management–Policy and Institutional Considerations. World Bank, Washington, D.C.
Zhang, J., F. Zhang, L. Zhang and W. Wang. 2009. Transaction costs in water markets in the Heihe River Basin in Northwest China. International Journal of Water Resources Development 25(1): 95–105.
Zhang, L., N. Heerink, L. Dries and Q. Futian. 2012. Water Users Associations and Agricultural Water Use Efficiency in Northern China. The Agricultural & Applied Economics Association's 2012 AAEA Annual Meeting, Seattle, Washington, August 12–14, 2012.
Zhangye City. 2012. The record of water conservancy in Zhangye city http://www.zhangye.gov.cnhttp://www.cenews.com.cn/xwzx/cysc/sk/201304/t20130429_740712.htm.
Zheng, H., Z. Wang, R. Calow and Y. Wei. 2012. Water rights allocation management and trading in an irrigation district: a case study of northwestern China. pp. 67–78. *In*: Kumar, Manish (ed.). Problems, Perspectives and Challenges of Agricultural Water. InTech Publishers, Delhi http://cdn.intechopen.com/pdfs/31497/InTech-.

14

Republic of Tajikistan: Its Role in the Management of Water Resources in the Aral Sea Basin

Anatoly Krutov,[1,*] Sulton Rahimov[2] and Anvar Kamolidinov[3]

SYNOPSIS

The chapter discusses the challenges of water management faced by the Aral Sea basin countries of Central Asia, as well as the role and position of the Republic of Tajikistan in addressing these challenges.

Since independence, problems have appeared in the management of the water storage and control infrastructure. Most facilities were built by the Soviets to serve the needs of the entire Aral Sea basin. In many cases, infrastructure located in one country is for the benefit of other countries. The division of maintenance and operational responsibility for trans-boundary water infrastructure is often in dispute. There are serious problems with seasonal water coordination. Upstream states

[1] Principal Research Fellow at the State Oceanographic Institute, Moscow, Russian Federation, 2, Pobedy Street, Building 1, Apartment 62, Reutov, Moscow Region, Russia, 143966.
Email: ankrutov@yahoo.com

[2] First Deputy Minister of Energy and Water Resources, Republic of Tajikistan, 5/1 Shamsi Street, Dushanbe Republic Tajikistan, 734054.
Email: r.sulton@gmail.com

[3] Senior Research Fellow, Tajikistan Branch of the Scientific Information Centre of Interstate Coordination Water Commission (SIC ICWC), 16, Behzod Street, Apartment 1, Dushanbe, Tajikistan, 734013.
Email: anvarkamol@gmail.com

* Corresponding author

would like to use water resources for their own winter power generation needs which is in conflict with the downstream states' summer irrigation needs.

With the contraction of economies in the region after independence, governments cut maintenance budgets. A lack of funds for operation and maintenance has resulted in a gradual decline in the condition of the irrigation and drainage systems.

Massive expansion of irrigation developments, associated with the diversion of river water created serious environmental problems in the Aral Sea Basin. In addition to that, new challenges such as climate change impact to water resources and increasing water demand because of population growth and economic development make the existing situation more acute.

The recent efforts in this direction give hope for progress toward positive outcomes. Proposals by the Republic of Tajikistan are based on an integrated approach to water management issues with due regard to natural, historical, and geopolitical conditions. This approach should help meet the challenges and prevent inter-state conflicts.

Keywords: Afghanistan, Amu Darya, Aral Sea basin, Central Asia, China, conflict resolution, energy, glaciers, groundwater, hydropower, irrigation, Kazakhstan, Kyrgyzstan, lakes, Pamir, regional cooperation, Syr Darya, Turkmenistan, Uzbekistan

1 Introduction

The Republic of Tajikistan is a landlocked mountainous country in Central Asia. Mountains account for more than 93% of the country's territory, with mountain peaks uplifting to 3000–7000 m. Nearly half of the country's territory is situated at elevations of over 3000 meters above sea level. Tajikistan is home to the highest mountain peaks of the Pamir and Alay.

Tajikistan borders Kyrgyzstan in the north and northeast, Uzbekistan in the north and west, Afghanistan in the south, and China in the east. The total area of the country is 142,600 km^2.

Tajikistan has continental climate with hot summers and mild winters. In the lowlands, the July average temperatures are above 30°C, while in the mountains average temperatures range from 6 to 16°C. The frost-free period lasts about 250 days. Winters are warm and short, but penetration of air masses from the north can plummet temperatures to minus 15°C and lower. In the southern areas the average January temperature does not drop below –2°C. In the mountains the January temperatures are in the range of –10–+12°C. In the Pamir climate changes from arid to polar depending on the elevation. In the Eastern Pamir, winter temperatures can be as low as –40°C.

Figure 14.1. Major River Basins in The Republic of Tajikistan.

Color image of this figure appears in the color plate section at the end of the book.

The average annual precipitation for Tajikistan is about 690 mm. It varies from less than 100 mm in the southeast to 2,400 mm in the central part of the country.

2 Water Resources

Tajikistan's water resources are formed by the melting of glaciers and snowfields, which account for over 13 km^3 or 25–30% of the annual runoff and some 50% during the vegetative period. Glaciers and snowfields cover about 8% of the Tajikistan territory (UNEP 2006). In Tajikistan, there are some 14 thousand glaciers with a total area of 11,146 km^2. The total volume of all the glaciers is 845 km^3.

Occupying around 20% of Aral Sea basin (350,000 km^2), the mountainous area, primarily within the Amu Darya and Syr Darya river basins, feeds about 90% of surface runoff. On average, annually, 116 km^3 of river flow is formed in this mountainous area. Out of this volume of 116 km^3, 64 km^3 of water is formed on the territory of Tajikistan, i.e., more than 55% of annual flow of the Aral Sea—from Amu Darya basin 62.9 km^3 and 1.1 km^3 from Syr Darya basin. Most of the glaciers are located in the Obihingou, Gount, and Mouxou River basins.

Table 14.1. Water flow to the Aral Sea Basin.

Country	Amu Darya		Syr Darya		Total	
	km³/y	%	km³/y	%	km³/y	%
Kazakhstan	-	-	4.50	12.12	4.50	12.12
Kyrgyzstan	1.90	2.42	27.40	73.77	29.30	25.30
Tajikistan	62.90	80.17	1.10	2.96	64.00	55.4
Turkmenistan (with Iran)	2.78	3.54	-	-	2.78	3.54
Uzbekistan	4.70	5.99	4.14	11.15	8.84	7.6
Afghanistan	6.18	7.88	-	-	6.18	5.4
Total	**78.46**	**100.0**	**37.14**	**100.0**	**115.60**	**100.0**

Source: Main Provisions of the Water Strategy of the Aral Sea Basin, Ministry of Water Resources and Land Reclamation 1996

3 Rivers

In Tajikistan, there are 947 rivers longer than 10 km. The most important rivers are the Panj, the Vakhsh, the Zarafshan, and the Kafirnigan. The Amu Darya (the largest river in Central Asia) is formed by the confluence of the Panj and the Vakhsh on the Afghanistan/Tajikistan border. In the north of the country, a number of small rivers are tributaries of the Syr Darya River, which is formed by the confluence of the Naryn and the Kara Darya in the Fergana Valley. The highest water discharge in the Tajikistan rivers is observed in June–August, the period of maximum snowmelt.

Table 14.2. Main Rivers of the Aral Sea Basin.

River	Length km	Catchment area km²	Annual flow km³	Main tributaries
Amu Darya	2,540	309,000	78.46	Panj, Vakhsh, Kafirnigan, Surkhan Darya, Sherobod, Kunduz
Panj	921	114,000	32.0	Pamir, Vakhan Darya
Vakhsh	786	39,100	20.0	Surkhob, Obi-Hingov
Kafirnigan	387	11,600	5.3	Sardai-Meyona, Sorbo
Zarafshan	877	17,700	5.1	Fan Darya, Kshtout, Magian
Syr Darya	3,019	219,000	37.14	Naryn, Karadarya, Chirchik, Akhangaran
Naryn	807	59,900	15.14	Kashkasu, Maytara
Karadarya	180	30,100	3.78	Tar, Kara Kuldzha

Source: Main Provisions of the Water Strategy of the Aral Sea Basin, Ministry of Water Resources and Land Reclamation 1996

4 Lakes

In Tajikistan, there are about 1,300 lakes with a total water area of 705 km^2. Most of the lakes (73%) are located in the Pamir-Alai Mountains at an elevation of 3,500–5,000 m above sea level. The largest lakes such as Karakul, Sarez, Jashilkul, and Iskanderkul formed as a result of earthquakes and rock falls. The Tajikistan lakes hold the total of about 46.3 km^3 of water, including over 20 km^3 of fresh water. Their total water area is 705 km^2 (MLRWR 2006).

5 Water Reservoirs

The country has 10 reservoirs, including 3 reservoirs in the Syr Darya Basin, and 7 in the Amu Darya. The reservoirs hold the total of 16 km^3 of water, with the water area of about 690 km^2. The largest water reservoirs include: the Nurek Reservoir, with a volume of 10.5 km^3 of water located on the Vakhsh River; the Kayrakum Reservoirs (4.16 km^3) on the Syr Darya River. Tajikistan enjoys enormous hydropower generation potential, estimated to be some 527,000 million kW/h/year (UNEP 2006).

6 Groundwater

Underground waters, as used in the national economy, are mainly confined to the quaternary alluvium of the major river valleys (the Syr Darya River, the Kafirnigan, the Vakhsh, the Kyzylsou, the Yakhsou) and intermountain basins. In Tajikistan, 60% of groundwater is involved in the formation of the Syr Darya flow, while 20% of the Amu Darya flow is derived from underground water.

The country enjoys an estimated 18.7 km^3 of renewable groundwater resources including 3.0 km^3 of confirmed reserves. It has more than 20 thermal and mineral water resources are found in more than 200 locations. Groundwater is abstracted to satisfy the needs of the public water supply, irrigation and industry. The yield rate is not more than 6,500,000 m^3/day. Of these, 46.2% of underground water is used for drinking purposes, 44.8% is used for irrigation, and about 9% is used in industry (GovRT a 2001).

7 Drainage and Return Water

Most of the Tajikistan territory is piedmont and mountain area which results in significant surface runoff and high levels of drainage from irrigated areas. The surface irrigation (furrow) technology and rocky soils also contribute to high return flows of about 4.0–4.5 km^3 per year. Drainage water from irrigated land makes up 3.5–4.0 km^3, and the remaining 0.50 km^3 is from municipal and industrial wastewater. The main part of the return flow flows back to rivers and only about 10% is used for irrigation.

Since there are not many saline lands in the country, the return drainage waters have insignificant salinity levels and do not have any substantial effect on the salinity of the Amu Darya river flow as measured by salinity levels in the Kerky River Station in Turkmenistan, where mean annual salinity is no more than 0.7 g/l.

8 Water Regulation

Water regulation in Tajikistan is based on the principles of prioritizing the use of water for drinking and household needs, payment of services for the delivery of water to consumers, while taking into account economic and ecological interests.

According to the Constitution of the Republic of Tajikistan, water, like other natural resources, is the exclusive property of the state, and the state guarantees their effective use in the interests of the people.

The Constitution and the Water Code serve as a basis for the water legislation. In addition there are also other laws such as the *Law on Drinking Water and Drinking Water Supply, Law on Water Users Associations*, the *Law on Environment Protection*. In specific cases, the water sector is also regulated by many other laws and regulations.

9 Water Resources Management

The current system is governed by the Government of the Republic of Tajikistan, local governments and authorities, as well as designated state agencies mandated to regulate the use and protection of water resources:

- The Ministry of Energy and Water Resources of the Republic of Tajikistan (MEWR RT) is a national-level authority for water policy and regulation;
- Agency on Land Reclamation and Irrigation under the Government of the Republic of Tajikistan is a national level authority for operation and maintenance of irrigation and drainage systems;
- Committee for Environmental Protection under the Government of the Republic of Tajikistan is responsible for state control over the use and protection of water;
- The Geological Authority is responsible for groundwater exploration and use;
- State Supervision of Safety in Industry and Mining Service under Government of the Republic of Tajikistan is responsible for the control

of the rational use of healing waters, chemical, thermal and industrial groundwater and therapeutic mud;

- The Anti-monopoly Service under the Government of the Republic of Tajikistan regulates tariffs associated with the use of water and other resources.

The water resources management functions are divided among different ministries and departments. Water infrastructure is management by the central and regional governments. Under this structure, there is considerable overlap between the functions of regulation, governance and service delivery within the sector.

10 Use of Water Resources

The main consumers of water in Tajikistan include agriculture, industry, municipalities, fisheries, recreational, and natural systems. Along with them hydropower is also one of the most important water user in Tajikistan providing about 98% of electricity in the country.

The actual annual water diversion of Tajikistan for the period from 1985 to 2008 was about 10.0–14.5 km^3 or 17–20% of the total renewable water resources of the country. Water intake for the needs of public utilities accounted for some 400 million m^3/year, i.e., no more than 3.0% of the total water intake in the country.

Agriculture is the main water consumer. For example, in 2012 agriculture received 7.84 km^3 of water from all the water sources. Industry does not use that much water nowadays (not more than 0.3% of the total intake).

Each year, the fishery sectors uses about 100 million m^3 of water diverted from surface water sources (LML 2012).

Taking into account the fact that flows from Tajikistan contribute more than 55% of water resources of the Aral Sea Basin, the country plays an important role in the management of this very sensitive issue in the region.

11 The Aral Sea Basin

Located in the heart of Central Asia, the Aral Sea Basin covers an area of 2.2 million km^2 and is the home base of more than 60 million people. The basin comprises the drainage area of two large rivers, the Amu Darya and the Syr Darya and the area around the Aral Sea.

These two rivers begin in the Pamirs and Tien Shan mountain systems and run approximately 2,500 km through the mountainous upstream countries of Afghanistan, Tajikistan and Kyrgyz Republic as well as through the downstream countries of Turkmenistan, Uzbekistan and Kazakhstan.

The Amu Darya in the south is a large river with a flow of approximately 80 km^3/year. The Syr Darya in the north is about half this size. Both rivers end in the Aral Sea, via deltas located in north-western Uzbekistan (Amu Darya) and in south western Kazakhstan (Syr Darya).

Climate and soil conditions in the Aral Sea Basin are very favourable for different agricultural crops, especially cotton and rice. Intensive cultivation of cotton and rice began here with the arrival of Czarist Russia in the late 19th century. The production of cotton accelerated in the 1920s after the Russian revolution. Under the Soviet Virgin Lands Campaign of the 1950s, large-scale irrigation projects began, and vast tracts of the Central Asian desert were ploughed, planted, mainly with cotton, and watered from one of the two rivers. From 1950 to 1990, 3.2 million hectares of new land came under cultivation and the total area of irrigated land increased to 8.0 million hectares. The result was a drop in water volume entering the Aral Sea and an increase in river salinity.

Until the 1960s the Aral Sea was the world's fourth largest inland body of water. The inflow into the Aral Sea from Amu Darya and Syr Darya was about 60 km^3 annually, precipitation is about 6 km^3 and groundwater is about 5 km^3. Evaporation from the surface of the Aral Sea was about 63 km^3 annually. With these inflows and outflows the level of Sea was stable between 50 to 53 m. Massive irrigation diverted large volumes of water annually from the two main rivers and has led to a substantial decrease in the volume of water reaching the Aral Sea. In 1990 the Aral Sea split into two parts—northern small sea and southern large sea. The water salinity of the northern sea is gradually decreasing as inflows from the Syr Darya River dilute the salt water. Fish have been reintroduced.

The water level in the southern body of the Sea, however, continues to drop and the salinity to increase. In 1997 salinity in the southern Aral Sea was around 40 g/l (sea water is 35 g/l), a fourfold increase causing many of the fish and other water species to die out. Today the southern Aral Sea is biologically almost dead and has shrunk by approximately 80% in volume and 90% in area. The shrinking of the Aral Sea has resulted in the exposure of sediments containing high levels of pesticides and herbicides, and the former wet delta environment is threatened by desertification with severe effects on human and animal life and on biodiversity. The dry polluted sediments are sometimes mobilized in dust storms. Mismanagement in irrigation and drainage, a lack of funds for operation and maintenance, and insufficient investments for rehabilitation of irrigation and drainage infrastructure have resulted in about one third of the irrigated land becoming saline, and crop yields being reduced by about one third. Excessive use of water for irrigation has generated large volumes of highly saline drainage water, some of which is disposed of in desert sinks to minimise return flows to the rivers. These effects have had a major adverse socio-economic impact on the people living in the Amu and Syr Darya Deltas, and their livelihood and quality of life has been severely affected.

Local drinking water has become too saline to drink and is often polluted by pit latrines. As a result, there is a high incidence of diarrheal disease and other health problems in the region. Life expectancies in the districts near the sea is nine years less than those of surrounding nations.

In the areas around the Aral Sea, there have been severe environmental impacts. Large sections of the Amu Darya and Syr Darya delta wetlands dried up as the river waters were diverted, and one of the most important migratory bird feeding grounds in Asia is now less than one fifth its former size. The climate directly around the Aral Sea has become more continental, with greater annual fluctuations in temperature and humidity.

The approximately 2.7 million people living adjacent to the Aral Sea have become economically impoverished as fishing stopped and huge tracts of agricultural lands were degraded with salt.

Irrigated agriculture is still the backbone of agricultural production in all of the Aral Sea Basin countries. The total area of irrigated land has not changed significantly since independence from the Soviet Union, with reductions in some States being offset by expansion in others. Major reasons are the differences between the States in reactions of the agricultural sector to privatization efforts; public sector investment policies; and economic capacity to maintain the existing irrigation and drainage infrastructure.

Although the total area under cotton has declined since 1990, it remains one of the most important crops and accounts for the majority of export earnings. Since 1990 a significant shift in cropping patterns has occurred in favour of wheat, basically as a response to government policies to achieve self-sufficiency in that commodity. The area planted to wheat has more than doubled, primarily at the expense of fodder crops and cotton, leading to declining soil fertility and a sharp reduction in animal production. The area devoted to rice has reduced substantially in the lower reaches of the Amu Darya and Syr Darya rivers.

A clear and formally accepted strategy on management of the Aral Sea is not yet available. The dream of restoring the Sea to its former level is still alive. This is expressed in the name of the key basin institution, the International Fund for Saving the Aral Sea, and in the school songs in the delta areas. Experts and practitioners, however, know that restoration of the Sea is impossible and they act accordingly. No variant of full restoration currently appears economically or environmentally feasible.

The current policy framework has few incentives for the efficient use of resources. Government regulation and ownership of agricultural land provides no incentive for the producers to improve or protect the land. Policy changes and the creation of incentives to conserve irrigation water have potentially big benefits from both reduced water use and decreased salt mobilization.

Water infrastructure maintenance is also a problem. As economies in the region contracted after independence, governments cut maintenance budgets, resulting in poor maintenance of most of the water infrastructure. Moreover, many irrigation projects lack operating or suitable drainage systems, which has led to salinization and waterlogging of fields. Many irrigation projects have directed saline runoff to the rivers, resulting in water quality problems downstream. Some projects avoid draining runoff into the rivers by directing it into the desert and creating hundreds of small, saline lakes, which often mobilize additional salt.

Since independence, problems have appeared in the management of the water storage and control infrastructure. Most facilities were built by the Soviets to serve the needs of the entire Aral Sea basin. In many cases, infrastructure located in one country is for the benefit of other countries. The division of maintenance and operational responsibility for trans-boundary water infrastructure is often in dispute. Further, there are problems with seasonal water coordination. Upstream states would like to use water resources for their own winter power generation needs which is in conflict with the downstream states' summer irrigation needs. Therefore, a strategy that defines a mutually beneficial system of rational water use is required.

12 Water Policy of Tajikistan

Tajikistan fully recognizes the importance of water resources management on the regional level and builds its water policy on base of this understanding.

The main directions of the national water policy of Tajikistan are formulated in the *Concept for the Rational Use and Protection of Water Resources in the Republic of Tajikistan* adopted by the Government of the Republic of Tajikistan (GovRT a 2001). Management of the water sector is now based on a combination of administrative, territorial, and basin principles.

The water resources management reform was launched with the adoption of the Decree of the President of the Republic of Tajikistan on April 8, 1996 N 460 *On the Introduction of Fees for the Supply of Water to Customers from the State-Owned Irrigation Systems*. It was further developed throughout the Decree of the President of the Republic of Tajikistan, September 16, 2008 N 541, which approved the *Concept for the Improvement of the Public Administration in the Country*, and the *Statement of the President of the Republic of Tajikistan to the National Parliament*, April 15, 2009, which stated the need for the exploration and introduction of new principles of water management on the basis of basin and integrated management with the appropriate division of political functions and water delivery functions.

The Government of the Republic of Tajikistan is currently considering options for the reform of water sector and opportunities of applying the basin principles of water management along with the establishment of a designated entity to implement the national water policy.

13 Regional Water Management Policy of Tajikistan

The international and regional water policy of Tajikistan is based on the principles of equality and justice, avoidance of substantial damage to the littoral countries, mutual respect, cooperation and mutual benefit. At various levels, Tajikistan calls for the sharing of its rich water and energy resources.

Tajikistan implements its regional water policy through its current water management mechanism on the basis of the *Almaty Agreement* (Agreement a 1992), the *Nukus Declaration of Central Asia countries,* 1995 (Declaration 1995) and other international and regional agreements the country joined.

14 Regional Water Allocation

Tajikistan's position regarding water allocation in the Aral Sea basin is based on the *Concept for Integrated Use and Protection of Water Resources* developed in the last century. These documents set forth water intake allocations directly from the Amu Darya and Syr Darya channels. According to these documents, annual water diversions from the Amu Darya and Syr Darya channels amounted to 11.31 km^3, or 13.4% of the total water intake by all Central Asia countries.

Table 14.3. Water Diversions from the Amu Darya and Syr Darya rivers.

Country	**Amu Darya**		**Syr Darya**		**Total**	
	km^3/y	**%**	**km^3/y**	**%**	**km^3/y**	**%**
Kazakhstan	-	-	10.01	44.12	10.01	11.9
Kyrgyzstan	0.40	0.60	0.39	1.72	0.79	0.9
Tajikistan	9.50	15.40	1.81	7.98	11.31	13.4
Turkmenistan	22.0	35.80	-	-	22.0	26.1
Uzbekistan	29.60	48.20	10.48	46.19	40.08	47.6
Total	**61.50**	**100.0**	**22.69**	**100.0**	**84.19**	**100.0**

Source: Updated Concept for Integrated Use and Protection of Water Resources of the Amu Darya (1987), Updated Concept for Integrated Use and Protection of Water Resources of the Syr Darya (1984)

It is hard to consider such water allocation as a fair one because, having the largest volume of emerging water resources, Tajikistan and Kyrgyzstan have the lowest share of total water use.

Despite such a substantial impairment of interests, Tajikistan and Kyrgyzstan, even after independence, based on principles of mutual respect and good neighborliness, agreed to accept these schemes as the basis for the utilization of water resources.

It was affirmed by the *Agreement of 1992* and *Nukus Declaration of 1995* (Obligations, Part I. The commitment to the principles of sustainable development: "We agree that the Central Asian states recognize previously signed and existing agreements, contracts and other legal acts regulating relationships between them on water resources in the Aral Sea basin, and take them to a steady performance"). However, one must consider that it was proposed to develop a new strategy for water allocation in the region, which, unfortunately, has not yet been implemented. Several attempts in this direction were not successful; for various reasons.

15 Problems and Prospects for Regional Cooperation

Interstate water disputes between the countries of Central Asia were started by subsequent independent planning by these countries after the collapse of the USSR.

The problem of the regional water resources distribution emerged from the fact that the countries of the region view cooperation in the light of their own policies for energy and water resources self-sufficiency. The policy of non-cooperation to ensure minimal dependence on other countries is costly to all the countries concerned. Effective management of transboundary water resources requires more intensive regional cooperation.

In this context, Tajikistan has a clear understanding that no progress is possible without proper cooperation. The country is fully committed to a dialogue and cooperation on the basis of fundamental principles such as respect for sovereignty, equality and mutually beneficial cooperation, good neighborly relations, and good implementation of international agreements.

This is strengthened by practical actions of the country at the regional and global levels. Tajikistan is one of the most proactive members of the international community in promoting water issues globally. At the initiative of Tajikistan, the UN General Assembly declared the year of 2003 as the International Year of Freshwater and the years of 2005–2015 as the International Decade for Action Water for Life.

To attract the attention of the world community to the development of cooperation in transboundary river basins, which are home to more than 40% of the world's population, the UN General Assembly declared the year of 2013 as the International Year of Water Cooperation. This was also initiated by the Republic of Tajikistan.

16 Tajikistan Vision for Addressing Water and Energy Issues in the Region

The Republic of Tajikistan has repeatedly expressed its willingness to cooperate with all the stakeholders at the regional level. The country firmly believes that the water resources use must be addressed together with the energy sector issues. In this regard, Tajikistan assumes that unlike irrigation, whose history dates back several millennia, hydropower, and in general the entire energy sector in Central Asia began to develop only in the 20th century.

As well as in the irrigation sector, the energy sector in the region got its boost from the 1930's to 1990's. In those years, an entirely new sector was born in the region—electricity generation. In the mid 1990's, the total installed capacity of all electric power plants in the region was 41 million kW.

During the period of the USSR, the infrastructure of the energy sector in the region was established on the principles of mutual integration. Immediately after the collapse of the Soviet Union, integration strengthened in an environment of market economy. The operating Joint Energy System comprised hydropower schemes (HPS) of Tajikistan and Kyrgyzstan, as well as thermal power plants in Kazakhstan, Turkmenistan and Uzbekistan. Until 2009, the use of such a framework allowed Tajikistan to maintain a high level of integration with other countries in the former Soviet Union, and above all with the republics of Central Asia.

The recent years have seen a dramatic change of the situation. Tajikistan adopted energy security concept as one of the main priorities of its economic development strategy. Under this strategy, the country in 2006 initiated construction of the South-North 500 kV transmission line, which was commissioned in December 2009. This line now connects two isolated energy systems of Tajikistan's north and south delivering electricity directly to the northern Sogdiysky Oblast, bypassing Uzbekistan.

But at the same time, the previously operational energy exchange system between Tajikistan and Uzbekistan stopped. And Tajikistan is no longer in a position to import electricity from other countries, and the volume of its export-import operations plummeted.

The previously effective system of integrated use and management of water and energy resources broke down. Tajikistan was forced to rely solely on its own hydropower resources. Thus the water flow regulation regimes have been changed to generate more electricity in winter time, when the demands to power are high. This has aggravated the pre-existing competition between irrigated agriculture and hydropower.

Nowadays, in the Aral Sea Basin there is an escalating conflict between the interests of the irrigation sector, developed mainly in the downstream countries (Kazakhstan, Turkmenistan, and Uzbekistan) and hydropower sector of mostly upstream countries (Tajikistan and Kyrgyzstan). This problem is of inter-state importance. Both of these sectors require different modes of water flow regulation by reservoirs. To generate electricity, it will be necessary to accumulate water in the summer and use it in the winter, in the coldest, energy-deficient period. While, the irrigation sector, on the contrary, wants to accumulate water in the winter and use it in the summer, during the growing season.

Water resources are the main source of energy in Tajikistan, with the hydropower potential estimated at 527,000 million kWh (of which 317,000 million kWh are cost-effective). Nowadays, this potential is used only at 3–4% (Petrov 2009). Tajikistan assumes that the use of hydropower potential can contribute to addressing the challenges of water management in the region. The most important priority is to ensure water security and guaranteed water delivery for irrigation in all the countries of Central Asia in dry years by constructing water reservoirs to regulate the flow of rivers in the seasonal and long-term context.

17 The Role of Regional Organizations

Historically, the Aral Sea Basin had sustainable water management, represented by traditional practices of managing small independent irrigation systems, with their own local water diversions, unlined canals and a primitive system of water distribution, i.e., oasis irrigation systems (Spoor and Krutov 2003). The water management and distribution policy was adopted and implemented at local levels. Changes taking place in the traditional methods of water management can be traced from the period of the conquest of Central Asia by the imperial Russia. Expansion of irrigated areas, mainly for the production of cotton and rice, began in the late 19th century, when it was decided to replace the traditional methods of management and irrigation of small areas with a large farms irrigation system.

After the coup of 1917, the Bolshevik leadership continued this policy by initiating the implementation of large-scale irrigation projects to ensure the so-called national cotton security and provide the national light industry with raw materials for processing. The process of expanding irrigated areas accelerated in the 1920s after the Russian Civil War and the strengthening of the Soviet power in the region. In the late 1930's, the system irrigated 2.5 million hectares, and by 1950 the irrigated areas grew to 4.7 million hectares and the water from the two largest water systems irrigated vast desert areas.

Since then, the local population lost control over water resources. Planning, management and control passed to the central government. Central Asia was divided into management zones on the basis of administrative principles that ignored the hydrological boundaries. The collectivization brought about changes of its own by bringing together water users regardless of whether they were located downstream or upstream. The new management system was established on strictly vertical structures including at the national, regional, district and local levels with weak horizontal links between agencies. It should be noted here that with a central system controlling production and resource limitations, this appeared to be the most appropriate method of management.

Understanding the limitations of the system, and, in particular, the loss of its operational efficiency in the last years prior the collapse of the Soviet Union, the government initiated the establishment of basin organizations, which would be responsible for the management of water resources at the basin level. Basin Water Organizations (BWOs) were established in many basins and some of them are still operational. The main objective of BWOs was to allocate water resources between the republics and manage the main head diversions from the Amu Darya and Syr Darya.

The collapse of the Soviet Union and the transition from a centrally planned to a market economy caused the institutional and other problems water resources management. The newly independent countries had to reinvent almost overnight a cooperation framework between the countries with different levels of economic development and with different interests in terms of water use. At the same time, these countries were guided by the Soviet era regulatory framework that ignored the requirements of environmental protection and sustainable development for the sake of short-term economic benefits, and production of large volumes of products without regard to its quality. The legacy of that system of management, pricing and lack of interest in outcomes was not conducive to the rational use of scarce water resources. In addition, the budget problems that rose during the transition period, led to delays in the payment for services provided and hence insufficient coverage of operational costs and expenses to maintain irrigation infrastructure, not to mention its modernization (Spoor and Krutov 2003).

The countries of Central Asia were very quick to respond to the changes and the need to establish a new framework for the allocation of water resources. Ministers of Water Resources of the newly independent states declared on September 12, 1992, in a joint statement that the joint management of water resources will be based on the principles of equality and mutual benefit. The Interstate Agreement of 18 February 1992 established the Interstate Commission for Water Coordination (ICWC) at the level of Ministers of Water Resources of the Central Asia countries to coordinate annual water allocations and modes of reservoirs operation in the basins of the Amu Darya and the Syr Darya. According to the Agreement, consensus decisions of ICWC were binding on all the five countries, while BWOs were mandated to ensure implementation of these decisions (Agreement a 1992).

In 1992–1994, donors helped establish new interstate organizations such as the International Fund for Saving the Aral Sea (IFAS) and the Interstate Commission on the Aral Sea (ICAS) with the Executive Committee. It was assumed that the main IFAS objective would be tapping and pooling of funds, while EC IFAS would be responsible for the implementation of Aral Sea Basin Plan (ASBP). In addition, the countries established the Interstate Commission on Sustainable Development (ICSD), whose main focus was issues related to environmental protection in the region. This cumbersome structure was a compromise and perhaps the best solution for the time, allowing alignment of the countries interests.

The meeting of the Heads of the States-Founders of IFAS in February 1997 adopted fundamental decisions for bringing IFAS and ICAS into one organization and their restructuring, as well as forming the Executive Committee, providing for rotation of its chairperson, the location of the Committee and the size of contributions to the IFAS. Since then, neither the IFAS Charter nor the EC IFAS Terms of Reference underwent any significant changes (Agreement b 1999).

During the entire period the IFAS operation, since the date of its establishment in 1997 up to the present time, the Executive Committee, following the chairperson, was correspondingly located in Almaty (1993–1997), Tashkent (1997–1999), Ashgabat (1999–2002), Dushanbe (2002–2008), Almaty (2008–2013) and since August 2013, the Executive Committee of IFAS has been moved to Tashkent.

18 Analysis of the Situation and Trends

In the long-term perspective, the water use problem in the Central Asian region could become more acute due to high population growth. According to some estimates, population growth in Central Asia may increase the demand for water in the coming twenty years by 40%. This could catalyze inter-state conflicts.

The economic damage from this and other unresolved problems will be borne by all the countries in the region. Thus, according to experts of UNDP, the unresolved problems of water resource exploitation in Central Asia lead to annual losses of about 1,700 million dollars directly related to poor water management (UNDP 2006). The paradox is that in case of a rational use of water in Central Asia, there will be enough water to meet all the needs. However, this is not currently possible due to excessive water losses at all levels. The inefficient use of water resources and high water losses are directly associated with the outdated system of agriculture resulting in water consumption per unit of production three, sometimes ten times higher than the world's average. Another reason is lack of funds for the rehabilitation and modernization of the irrigation infrastructure. According to experts, rational water use and modernization of the irrigation infrastructure will save up to half a year flow of the transboundary rivers in the region (UNDP 2006). This can only be achieved by joint efforts. To solve this problem, the countries of Central Asia should be proactive in the regional water resources cooperation.

In this regard, the only possible approach is settlement of disputes through negotiations to reach mutually beneficial agreements. And there is a clear need for integrated water resource management that will help optimize operating modes water reservoirs and power stations with due regard to both national and regional interests.

The need for sustainable and safe water use in Central Asia calls for the development of new forms of international cooperation on the basis of integrated water resources management. At the same time, climate change mitigation and adaptation measures should be incorporated into the development plans, programs and policies at both national and regional levels.

The growing diversity can make it very difficult to develop common integration conditions that will be willingly accepted by all members of the grouping. And the costs of negotiation and coordination will also grow. In other words, the heterogeneity reduces the attractiveness of integration and diversity may turn even efficient integration groupings into an unstable system.

On the other hand, the effectiveness of the institutional environment and management, mutual trade, safety and joint infrastructure are factors that determine the stability of an integration alliance. At the same time, increased efficiency of the integration grouping increases its sustainability. Relatively more developed institutions that provide the best opportunities for aggregation of preferences and interaction between different groups, are capable to meet the challenges of diversity to a greater extent, and to capitalize on its creative energy.

However, even an adverse situation is not viewed as a basis for assuming that this or that formation is doomed. On the contrary, having gone through the phase of instability (for example, due to political will), such an institution may have a positive impact on its member countries, which will automatically contribute to a sustainable configuration.

Leveraging this through international and regional factors can quickly bring good results. This could be achieved by developing and proposing to the Central Asia states and potential donors an action plan (roadmap) that will help achieve significant progress in creating a favorable environment of trust and cooperation in the region. This, in turn, will help achieve the objectives—as set forth by the Heads of State—of addressing the emerging environmental and socio-economic challenges in the Aral Sea Basin (Vinogradov and Langford 2011).

The existing IFAS-based mechanism for regional cooperation comprised of three poorly cooperating (among themselves) intergovernmental organizations requires a lot of improvements. This poorly oiled mechanism of cooperation hinders the development of an optimal policy with regard to the solution of the Aral Sea Basin problems. It prevents a coordinated and effective implementation of such a policy. Neither does it give a clear answer to the questions related to the implementation of the strategies and programs.

In private conversations, national experts, representatives of regional organizations and donors expressed a range of views on the extent and depth of possible changes in the organizational structure of cooperation. On one side of the spectrum of opinions some recommend improvements to the exchange of information and streamlining coordination between the existing regional organizations. On the other side, fundamentally new joint regional organization and structures of cooperation are proposed. Between these extremes, suggestions were made to improve interaction between the three existing operational regional cooperation structures (IFAS—EC IFAS, ICWC and ICSD) by changing some functions and mandates.

It should be borne in mind that IFAS currently has a limited role. Despite the inherent potential incorporated during its initiation it remains no more than a promising platform for regional dialogue to overcome the social and environmental crisis in the Aral Sea basin. In response to this situation, the Heads of States in their statement of 28 April 2009 (Statement 2009) set forth the goal to develop proposals to improve the regional cooperation mechanism in the framework of IFAS. Thus, IFAS was tasked to focus, on refining and delineating responsibilities within the existing structure, clarifying mandates and establishing reporting procedures and interaction between the IFAS structural units, and on improving coordination and linkages with international organizations and donors.

The key issues facing the regional institutions were and remain issues associated with its place in the regional cooperation. Answers to these questions are important, because they not only affect the interests of the IFAS founders, but also the interests of potential donors who often have their own quite strict policies concerning the preparation and implementation of the donor-financed projects.

Over the past decade, there have been several attempts to study and understand the nature of the problems faced by the countries of Central Asia and suggest solutions. A large number of investment projects were implemented resulting in significant practical results. These projects could have delivered much better outcomes and the progress would have been faster, provided the member countries used the same approach towards solving common problems. There are other contributing issues such as lack of necessary interstate agreements, lack of funding, lack of national and regional water management strategies and the lack of a change management plan that could establish priorities and ensure effectiveness of the projects.

In the management of transboundary water resources in Central Asia a clash of interests is inevitable but it should be seen as a positive factor for regional development. At the same time, when planning activities, one must bear in mind that in the past, there were some gaps that need to be taken into account in the planning and implementation of projects.

19 Conclusion

Lack of a regional approach to the management and use of water resources in the Aral Sea basin will inevitably lead to conflicts. Sustainable use of water resources can only be achieved if all the parties concerned have agreed upon their actions, which should be based on the principles of integrated water management.

Introduction of IWRM principles into international practice means acceptance of the principle of equity in the use of water resources by all water users, including environmental components, irrigation, energy, industry, fisheries and domestic sector. The main priority should be given to the satisfaction of the water and sanitation needs of communities.

Local water and energy users' involvement in commenting on national and regional water policy and strategy is very limited. The matters under consideration at the national level are determined by politicians and advised by technocrats and scientists. Water User Associations in most of the cases were established under the Government's control and financed by donors. Energy sector is under the Government's control. Private enterprises contribution to energy production and distribution is negligible. Water and energy users are not well organized to have a voice.

One way to resolve international conflicts should be to find common approaches to regional integration, including integration at the economic level, to search for and find ways to reconcile the interests. All conflicts must be resolved through negotiations with due regard to the interests of all parties.

The main priority of water policy in Central Asia should be striving for overarching cooperation. Only overarching cooperation and integration can be the basis for addressing all the interstate water issues.

A well-developed regional cooperation is one of the key aspects of the solution of water and energy problems. Unfortunately this is not true of the Central Asian region. Although after gaining their independence, the countries of the region were able to ensure stable operation of the water sector under the difficult conditions of transition to a market economy by establishing regional institutions and signing several important regional agreements in the region, there is still a great potential for cooperation in the water and energy sectors, which can be of great benefit to all the countries concerned.

One approach addressing the conflict between irrigation and hydropower in a radical way is to develop a new water resources use concept with due regard to the needs of all water users including natural features rather than to limit any of these areas or to submit one to another.

The main problem, which becomes a kind of barrier to comprehensive water and energy cooperation in Central Asia, is national policies and interests of the countries in the region. Countries view their regional cooperation through the prism of a water and energy self-sufficiency policy. The policy of non-cooperation to ensure minimal dependence on other countries is costly to all the countries concerned. Effective management of transboundary water resources requires more intensive regional cooperation.

In this context, Tajikistan has a clear understanding that no progress is possible without proper cooperation. The country is fully committed to a dialogue and cooperation on the basis of fundamental principles such as respect for sovereignty, equality and mutually beneficial cooperation, good neighborly relations, and good implementation of international agreements.

Acknowledgement

The authors wish to thank Ms Anna Coote for her assistance in editing this chapter.

References

Agreement between the Republic of Kazakhstan, the Republic of Kyrgyzstan, Uzbekistan, Tajikistan and Turkmenistan "On Cooperation in Joint Management, Use and Protection of Transboundary Water Resources" (Agreement a). 1992. http://www.icwc-aral.uz/statute1_ru.htm.

Agreement about the Status of IFAS and its Organizations (Agreement b). 1999.

Concept for Integrated Use and Protection of Water Resources of the Amu Darya (Concept). 1987.

Government of the Republic of Tajikistan (GovRT a). 2001. The Concept of Rational Use and Protection of Water Resources in the Republic of Tajikistan (2001), N 551. http://cow-leech.ru/docs/index-4977.html.

Joint Statement of the Heads of the Founders of the International Fund for Saving the Aral Sea (Statement). 2009. http://ec-ifas.org.

Landell Mills Ltd (LML). Strategic Plan for Reform of the Water Sector in the Republic of Tajikistan (2012), http://www.unece.org/fileadmin/DAM/env/water/npd.

Ministry of Land Reclamation and Water Resource of the Republic of Tajikistan (MLRWR). 2006. Water Sector Strategy of the Republic of Tajikistan. http://www.undp.tj/files/reports/waterstrategy_rus.pdf.

Nukus Declaration of the Central Asian Countries and International Organizations on Sustainable Development in the Aral Sea Basin (Declaration). 1995.

Petrov, G. 2009. The Problems of Water and Energy Resources of Transboundary Rivers in Central Asia and Their Solutions; Optima, Dushanbe, Tajikistan.

Spoor, M. and A. Krutov. 2003. The "Power of Water" in a divided Central Asia. *In*: Amineh. Central Asia in Global Politics. Conflict, Security and Development, Volume 2. Brill Publishers, Leiden and Boston, Numbers 3–4, 2003, pp. 593–614.

UNDP. 2006. Report on Human Development in Central Asia.

UNEP. 2006. Action Plan for Integrated Water Resources Management and Water Efficiency. National Report of the Republic of Tajikistan.

Vinogradov, S. and V.P.E. Langford. 2011. Managing transboundary water resources in the Aral Sea Basin: in search of a solution. Int. J. Global Environmental Issues. 1(3/4): 345–362.

15

Lake Eyre Basin—Challenges for Managing the World's Most Variable River System

Richard T. Kingsford,[1,*] Justin F. Costelloe[2] and Fran Sheldon[3]

SYNOPSIS

The Lake Eyre Basin (LEB) is in the most arid region on the Australian land mass. There are two major river systems and many small rivers across the basin and all are unregulated with intermittent to temporary flow regimes. These rivers are ranked among the most variable rivers in the world. Many of the environmental and socioeconomic values and processes in the LEB are supported by flows in these rivers. Our aim in this chapter was to identify the key characteristics of the basin. We discuss the current decision-making processes and opportunities to protect this magnificent river basin in central Australia, one of the last unregulated large river basins in the world.

Keywords: Australia, aquatic life, birds, indigenous people, culture, recreation, mound springs, Great artesian Basin, Mining, flow regimes, floods, Simpson Desert, Kati Thanda, Georgina-Diamantina River, Cooper Creek, 'boom' and 'bust' ecology, fish, anastomosing channels, monsoon, refugia, benthic productivity, invasive species, livestock, pastoralism

[1] Australian Centre for Ecosystem Science, University of NSW, SYDNEY 2052, Australia. Email: richard.kingsford@unsw.edu.au
[2] University of Melbourne, Victoria, Australia.
[3] Australian Rivers Institute, Griffith University, Brisbane, Australia. Email: f.sheldon@griffith.edu.au
* Corresponding author

1 Introduction

The Lake Eyre Basin occupies about one sixth of the Australian continent (1.2 million km^2) and includes parts of three states and one territory: New South Wales, Queensland, South Australia and the Northern Territory (Fig. 15.1). It is in the most arid region on the land mass where temperatures can exceed 50°C and annual rainfall is less than 250 mm in the central part of the basin. Potential evaporation always exceeds annual rainfall; the driest part around (Lake Eyre, only averages about 125 mm a year but the areal potential evaporation is more than an order of magnitude greater (~2,500 mm). Much of its productivity and environmental value is dependent on rainfall and how this drives flows in the basin's major rivers. The basin is endorheic with its rivers carrying periodic large floods (Kotwicki and Isdale 1991) which inundate a range of different wetlands, including floodplains; some large floods or sequences of floods eventually reach Kati Thanda (Lake Eyre).

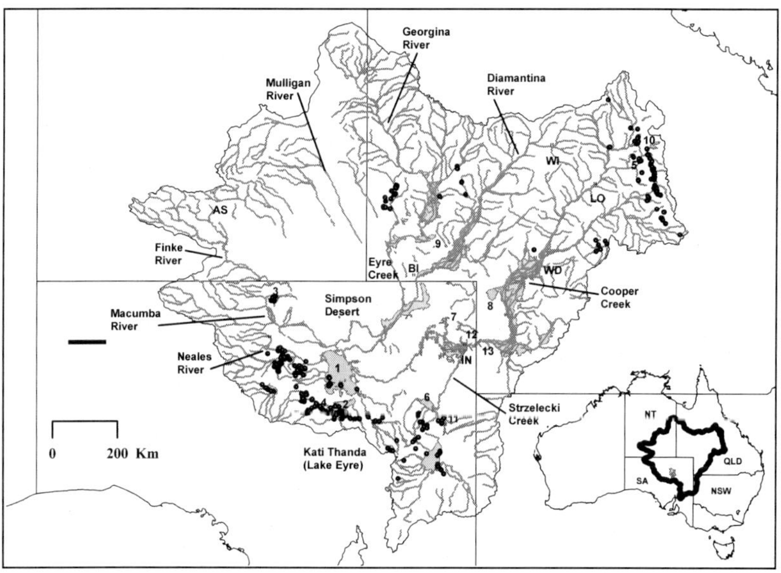

Figure 15.1. Location of the Lake Eyre Basin in central Australia (inset), including the three states (NSW-New South Wales, QLD-Queensland, SA-South Australia) and the Northern Territory (NT), showing its major rivers and wetlands in arid Australia. Numbers mark locations of the gauges (Cullyamurra (12), Nappa Merie (13)) used to show flow regime of Cooper Creek and major wetlands: Lake Eyre North-Kata Thanda (1); Lake Eyre South-Kata Thanda (2); Dalhousie Springs (3), Wabma Kadarbu springs (4); Aramac Springs (5); Lake Blanche (6), Coongie Lakes (7), Lake Yamma Yamma (8); Lake Machattie (9); Lake Galilee (10); Lake Callabonna (11). Major towns are shown (Alice Springs (AS), Winton (WI), Longreach (LO), Windorah (WD), Innamincka (IN) and Birdsville (BI)).

Color image of this figure appears in the color plate section at the end of the book.

There are two major river systems and many small rivers across the basin and all are unregulated with intermittent to temporary flow regimes. These rivers and their dependent floodplains flow through a variety of different landforms, including uplands and deserts (e.g., Simpson Desert). The two large rivers, which eventually flow to Kati Thanda (Lake Eyre), comprise the Georgina-Diamantina River catchment and the Cooper Creek catchments. They provide most of the water reaching Lake Eyre-Kati Thanda (Kotwicki and Isdale 1991). These rivers are ranked among the most variable rivers in the world (Puckridge et al. 1998; McMahon et al. 2008a). Many of the environmental and socioeconomic values and processes in the basin are supported by flows in these rivers. Aquatic organisms have life histories that reflect the highly variable 'boom' and 'bust' ecology of these ecosystems (Kingsford et al. 1999; Puckridge et al. 2000; Boulton et al. 2006; Brock et al. 2006; Bunn et al. 2006a). Human communities are also highly reliant on the rivers (Measham and Brake 2009).

There is a relatively sparse population (0.5 person km^2) with only about 60,000 people (Measham and Brake 2009). Most people live in the major towns in the basin but even these do not have large populations, including the largest of Alice Springs (~25,000 people) and Longreach (~4,300 people), with many of the other small towns with less than a thousand inhabitants (e.g., Birdsville, Innamincka, Winton) (see Fig. 15.1 for locations). The Lake Eyre Basin (LEB) also has a relatively high aboriginal population, concentrated in the Northern Territory and South Australia where 40–90% of the resident population lives (Measham and Brake 2009). Most inhabitants in the Lake Eyre Basin are supported by the pastoral industry, mining, tourism or government.

Our aim in this chapter was to identify the key characteristics of the basin that support the most variable river system in the world (Puckridge et al. 1998), LEB's dependent aquatic assets and also the communities which depend on it. There is considerable discussion about the long-term future of the river basin with increasing pressures of water resource development and potential impacts from mining, particularly coal seam gas and shale gas exploration and development. We discuss the current decision-making processes and opportunities to protect this magnificent river basin in central Australia, one of the last unregulated large river basins in the world.

2 The Rivers and Groundwater and their Dependent Ecosystems

Apart from rainfall, there are two main sources of water in the Lake Eyre Basin: groundwater from the Great Artesian Basin (GAB) and stream flow in the rivers and creeks of the Lake Eyre Basin. These supply a wide range of different inland aquatic ecosystems, ranging from GAB-fed springs that could have ages in the order of >100,000 years (Prescott and Habermehl

2008), to ephemeral floodplains and large lake systems, including salt and freshwater lakes (Table 15.1), supplied by episodic flow in rivers. The source, distribution, extent, frequency and duration of water in the landscapes define the types of aquatic ecosystems (Table 15.1).

The Great Artesian Basin, the largest freshwater aquifer in the world, covering 1.7 million km^2 extends north of the Lake Eyre Basin but much of this underground water is also within the Lake Eyre Basin. There is an estimated 149,000 ML of water which is replenished annually by rainfall on the Great Dividing Range along the eastern margin of the GAB; water flows underground slowly towards the southwest, taking about two million years (Smerdon et al. 2012) and this also provides dependable water to remarkable groundwater ecosystems. They include the relatively small ecosystems with cup-shaped large mounds in South Australia, formed over thousands of years as the artesian water from the Great Artesian Basin discharges and deposits sediments and salt (e.g., Wabma Kadarbu Mound Springs, Blanche Cup, The Bubbler). The most extensive spring group is the Dalhousie Springs (see Fig. 15.1) with more than 60 different springs extending over more 50,000 ha (http://www.environment.gov.au/heritage/places/national/witjira-dalhousie-springs/) but there are also many other smaller spring groups, including Aramac springs in the northeast of the Lake Eyre Basin. These springs contain a significant number of endemic plant and animal species and have extraordinary conservation value (Fensham et al. 2011; Murphy et al. 2013). An extremely rare fish, the red-finned blue-eye *Scatuiginichthys vermeilipinnis*, is found in a small isolated spring system in the northwestern part of the Lake Eyre Basin (Kerezsy and Fensham 2013). This patchwork of permanent groundwater ecosystems contrasts with the large flood dependent ecosystems supplied by river flows.

The rivers that flow north to south in the Lake Eyre Basin are one the most important drivers of its ecology and the lives of people who depend on the basin. The largest systems are the Cooper Creek catchment and the Georgina-Diamantina River catchment. Cooper Creek receives water from the Thomson and Barcoo Rivers (Fig. 15.1).

The other large catchment to the west has the Georgina River flowing into Eyre Creek where it can be joined by the Mulligan River before flowing through the Simpson Desert to join with the Diamantina River during large flood events about every 4–5 years. The Diamantina is the most regular contributor to Lake Eyre and flows into the lake every second year on average, but floods that inundate most of the surface area of Lake Eyre have a return period of about every eight years (Kotwicki and Isdale 1991). Cooper Creek also has a major distributary system, Strzelecki Creek, which branches off to the south and flows to lakes just north of the Flinders Ranges (e.g., Lake Blanche). There are also four rivers which flow towards Lake Eyre from the west. Two of these, the Neales River and Macumba River,

Table 15.1. Description of different types of freshwater ecosystems in the Lake Eyre Basin, their distribution, dependency and major threats to their sustainability.

Ecosystem	Description	Distribution	Dependence	Threats
Groundwater ecosystems (springs, mound springs, boggomosses)	Range from mound springs (area) to large springs such as Dalhousie	Mainly occur on the edges of the Great Artesian Basin (Habermehl 1980)	Adequate flow rates from the Great Artesian Basin	Groundwater depletion from mining
Waterholes	Deep channelized sections of the river, Can form on primary and floodplain channels, the latter usually around points of flow constriction (e.g., large sand dunes)	Throughout larger rivers but deepest waterholes are restricted to the major rivers (Cooper and Diamantina-Georgina)	Unrestricted upstream flow paths and water levels not severely affected	Water resource extraction from river flow and standing waterholes, floodplain infrastructure affecting flow paths, pollution, overfishing
Channels	Vast array of channel types, from rocky and sandy single channels to complex, mud dominated anastomosing channels. Channel types can vary between reaches and within reaches	Sandy channels are mostly restricted to the northwestern rivers (Finke, Macumba), cobbled channels to the streams draining the Flinders Ranges. Mud-dominated, anastomosing channels are characteristic of the middle reaches of the Cooper and Diamantina-Georgina	Unrestricted upstream flow paths	Water resource extraction from river flow, floodplain infrastructure affecting flow paths
Floodplains	Wide floodplains for a given catchment size are characteristic because of low gradients. Complex, anastomosing floodplains of the major rivers are defining features	See above	Unrestricted upstream flow paths	Water resource extraction from river flow, floodplain infrastructure affecting flow paths, burning or other destruction of riparian and floodplain perennial vegetation

Freshwater lakes	Relatively rare but freshwater lakes occur in the middle to lower reaches, particularly where dunes interact with flow paths to form broad basins	Lower Cooper (Coongie Lakes, Lake Hope) and lower Georgina (Machattie, Mipia) are most prominent examples	Unrestricted upstream flow paths	Water resource extraction from river flow and standing waterholes, floodplain infrastructure affecting flow paths, pollution, overfishing
Salt lakes	Generally form as small closed basins within the Lake Eyre Basin but occasionally occur connected to rivers, particularly in lower reaches where unconfined groundwater is very shallow	Include Lake Eyre-Kathi-Thanda, Lake Galilee, different lakes in the Coongie Lake system on Cooper Creek and Lake Callabonna on the Strzelecki Creek	Unrestricted flow paths	Infrastructure development

experience flows every year but only contribute inflow to Lake Eyre every two years on average. The other two western catchments, the Finke River and Todd River, have not been observed flowing into Lake Eyre and large floods currently terminate in the Simpson Desert (Fig. 15.1).

The Lake Eyre Basin rivers produce five different broad habitats: waterholes, ephemeral channels, floodplains, freshwater lakes and salt lakes (Table 15.1). As well, there are many small claypans, swamps and interdunal areas isolated from these main river systems which fill when there is rare high local rainfall. These different ecosystems are spread throughout the Lake Eyre Basin, although they are more concentrated in the lower regions of the major rivers where there are extensive floodplains. The floodplains of the Cooper Creek and the Georgina-Diamantina are the largest, sometimes extending for over 80 km across the direction of the river. As these rivers flow towards Kati Thanda (Lake Eyre), floods inundate these floodplains, flowing down many different small anastomosing channels and overflowing out of waterholes. In some reaches, the rivers also flow into large freshwater lake systems, including Coongie Lakes, Lake Yamma and Lake Machattie (Fig. 15.1). These lakes will often hold water for up to three to four years after filling (Kingsford et al. 1999). There are also salt lakes, including the largest Kati Thanda (Lake Eyre), covering about 9,500 km^2 and Lake Galilee in the northeast of the catchment.

The defining characteristic of the flow regime of the major Lake Eyre Basin rivers is their extremely high interannual variability (Fig. 15.2).

Figure 15.2. Anastomosing channels are a feature of Lake Eyre basin river systems. Photo R.T. Kingsford.

Even compared to other large arid zone rivers, the Lake Eyre Basin rivers have about double the coefficient of variation (McMahon et al. 2008a). This is largely explained by the northern parts of the Lake Eyre Basin occurring close to the southern margin of the North Australian Monsoon and so flows are primarily driven by summer seasonal rainfall (Allan 1985). In years with enhanced monsoonal activity (i.e., mainly during La Niña episodes), large tropical rainfall events push south and generate very large floods contrasting years with low monsoonal activity which typically have small rainfall events (Allan 1985; Kotwicki and Isdale 1991; Kotwicki and Allan 1998; Puckridge et al. 2000; Costelloe et al. 2006; McMahon et al. 2008b). As a result, the large rivers are intermittent, although flowing nearly every year over their upper and middle reaches with seasonal summer rainfall. Rainfall varies from about 600 mm in the northeast to less than 150 mm in the south (Measham and Brake 2009; McMahon et al. 2008b).

The rivers receive decreasing tributary input as they flow into increasingly arid areas and, with high loss rates from evaporation, infiltration and ponding, the mean annual flow decreases in the lower reaches (Knighton and Nanson 1994; McMahon et al. 2008a). Important wetlands occur in the low reaches of the rivers which small annual flows may only just reach (e.g., the Ramsar listed Coongie Lakes wetlands on the lower Cooper and the Goyder Lagoon floodplain on the Diamantina-Georgina). Downstream of these wetlands, the frequency of flow decreases and permanent fresh waterholes are generally absent.

The geomorphic setting also exerts a major influence on the flow regime of the major rivers. The northern and eastern headwaters begin at low elevations (300–400 m) and with the large size of the basin, this results in rivers with extremely low gradients, particularly in the mid to low reaches of the rivers. Consequently, the rivers have developed extremely broad floodplains with spectacularly complex, anastomosing flow pathways (Knighton and Nanson 1994). Large floods inundate 10,000s km^2, producing inundated floodplain widths of 40–80 km, whereas small annual flows generate orders of magnitude less inundated area (Fig. 15.2).

During periods of no flow, fluvial pools ('waterholes') are sporadically distributed across deep parts of the channel system. In the reaches that receive annual flow, many of these waterholes are permanent and sustain aquatic biota between flow events. Most of the deep waterholes have maximum depths of 4–6 m after flow ceases (McMahon et al. 2008a), but some extend to 12–25 m.

The small catchments of the Lake Eyre Basin, to the west of the Lake Eyre, generally lie within the more arid central parts of the basin and so have more temporary flow regimes. Streamflow events are modestly summer dominant but flows are relatively common outside of summer when frontal systems push up from the south. There has been little hydrological monitoring of these small catchments but they typically experience one to two flow events per year. Waterholes are generally shallow, commonly <4 m maximum depth upon flow cessation, and so the catchments are more vulnerable to drought periods drying out the fluvial aquatic refugia.

3 Environmental Values

The Lake Eyre Basin epitomises 'boom' and 'bust' ecology for freshwater dependent organisms, where matching flood and dry disturbances are punctuated by small flows (Kingsford et al. 1999; Bunn et al. 2006a). As floodwaters recede, waterholes within the complex channel network become the critical resource in the increasingly dry landscape and provide critical refugia for many aquatic dependent organisms (Sheldon et al. 2010). Throughout the often extensive dry periods, these waterholes provide refuge for obligate vertebrates (Kingsford et al. 2006) and freshwater invertebrates (Boulton et al. 2006; Carini et al. 2006) sustained primarily by algal productivity which is most productive around the waterhole littoral edge (Bunn et al. 2003; 2006b; Fellows et al. 2007). This high benthic productivity is primarily governed by light availability and, even under the naturally high levels of turbidity found in these waterholes, benthic algae are the dominant source of carbon at the base of the food web (Burford et al. 2008), although disturbance to the waterhole edge may limit benthic production with consequences for secondary biomass (Fellows et al. 2009).

During dry periods, waterholes within the channel network become aquatic islands in an otherwise terrestrial landscape, however during floods they become extensively connected. Animals show differential dispersal between the waterholes, even during extensive flooding. For some taxa, waterhole refugia have high connectivity, reflected in the similarity of their invertebrate communities (Carini et al. 2006) while for others there is little dispersal along the channel networks and populations within waterholes are effectively isolated (Baker et al. 2004). At a community level the spatial differences in invertebrate communities between waterholes are significantly less than temporal differences at one waterhole between different phases of a flood 'cycle' (Marshall et al. 2006). Algal communities in waterholes also include phytoplankton, diatoms and cyanobacteria (McGregor et al. 2006), increasing in diversity and abundance following flooding which brings nutrient influxes (Costelloe et al. 2005b). These waterholes are also among the few permanent habitats, apart from large lakes, for fish species during dry periods with the composition of fish communities in waterholes reflecting previous flooding patterns, waterhole structure and connectivity between waterholes (Arthington et al. 2005; 2010). Some fish species continue to breed in isolated waterholes, providing advantages where flood regimes are highly unpredictable, allowing fish communities to capitalise on flooding and extensive habitats (Kerezsy et al. 2011). Periodic flooding is also fundamental for maintaining fish populations in the long term, as many fish species primarily survive on low value resources while 'trapped' in waterholes during the dry phases and capitalise on high value food resources when floodplains are inundated (Balcombe et al. 2005). Waterbird communities can build up on waterholes but predominantly comprise the fish-eating birds (cormorants, Australian pelicans).

Waterholes need to hold water through long dry periods (Fig. 15.3) to sustain their biota and this is highly dependent on the depth of the waterhole, its connectivity to groundwater and frequency of replenishment floods (Hamilton et al. 2005; Costelloe et al. 2007). For 15 waterholes on Cooper Creek, evaporative losses reduced volume to 10%, after 6–23 months without replenishment (Hamilton et al. 2005).

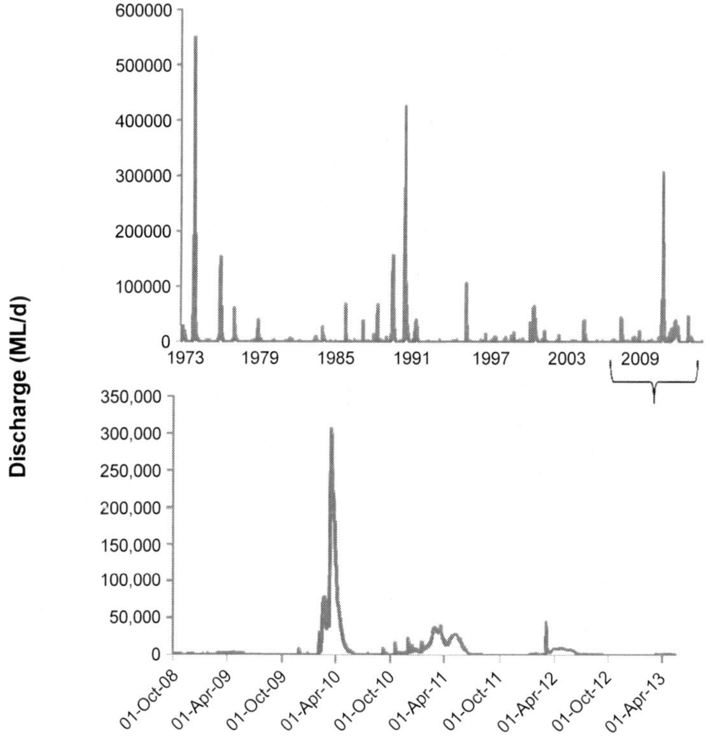

Figure 15.3. Flow record for the lower Cooper Creek illustrating the very high interannual variability in discharge characteristic of the LEB rivers. Top panel shows the full period of record of daily flow (ML (106l) per day, 1973–2013) while the bottom panel shows the most recent flood. Data were from the combined Cullyamurra and Nappa Merrie gauging station records (near Innamincka, see Figure 1) and are courtesy of the South Australian Department of Environment, Water and Natural Resources and the Queensland Department of Natural Resources and Mines.

Small flows in the river can sustain waterholes especially through dry periods (Bunn et al. 2006a), but the longest period of record with no streamflow in one of the major rivers was about 21 months (Bunn et al. 2006a). Some distributary reaches in the low parts of the Lake Eyre Basin (e.g., Strzelecki Creek, Eyre Creek) may not flow for decades or more. As water evaporates without flood inputs, the water becomes increasingly saline, increasing towards the lower end of the catchment (Costelloe et al. 2005a). During dry periods, salinity limits transpiration rates and much of the vegetation on the floodplain ceases to grow and or becomes dormant, including seeds in the soil (Brock et al. 2006; Costelloe et al. 2008). Some riparian trees draw on groundwater resources during dry periods, even though these may have higher salinity than surface water (Costelloe et al. 2008).

During large flood events, large quantities of water flow down the river channels, quickly filling up waterholes before inundating adjacent floodplains and lakes (Table 15.1). The extent of inundation depends on the quantity of the rainfall delivered to a river and the preceding antecedent conditions (Leigh et al. 2010) but this inevitably triggers a 'boom' event in biological productivity which extends from primary productivity (Brock et al. 2006; Bunn et al. 2006b; Costelloe et al. 2005b), secondary productivity (Boulton et al. 2006; Shiel et al. 2006) through to the top of the food web (Kingsford et al. 2006; 2010). Floods distribute widely, connecting all types of habitats across the landscape and allowing movement of freshwater organisms. Floods are critically important to different plant communities, structuring their form and life history (Capon 2003). There are different gradients created, depending on the size and frequency of the flooding, increasing in tolerance of aridity the further away from the channels (Brock et al. 2006). The timing and frequency of flooding were important predictors of the abundance and composition of seedling germination from seedbanks on Cooper Creek floodplain (Capon and Brock 2006; Capon 2007). Fish production within the rivers is particularly sensitive to flood magnitude (Puckridge et al. 2000; Balcombe et al. 2007), with most of the fish in Cooper Creek (11 of 12 species) showing a defined breeding response to flooding, with larvae observed on the floodplain, following widespread flooding which was a catalyst for significant movements and increased growth (Balcombe et al. 2007). Diets of fish also shift to wider range of items, predominantly aquatic algae and diatoms (Balcombe et al. 2005). The high variability of flooding possibly favours native fish species over the two alien species in the system because of the relative advantage of native species to capitalise during floods and survive during dry periods (Costelloe et al. 2010). Floods are also a key determinant of turtle population and behaviour, stimulating movements and breeding (Kingsford et al. 2006). Floods attract large numbers of waterbirds which move in with

Figure 15.4. Waterbirds breed on the Lake Eyre system, often flying in from more than 1000 km away. Photo R.T. Kingsford.

the widespread inundation of habitats (Roshier et al. 2001a) and remain as long as there is flooding (Kingsford and Porter 1993; Kingsford et al. 1999). Waterbirds are capable of long movements from other catchments (Roshier et al. 2001b; 2002; 2006; 2008; Kingsford et al. 2010) although the mechanisms by which they find habitat remains poorly known (Kingsford et al. 2010). Different species seem to adopt life history strategies, reflecting the abundance of their prey (Kingsford et al. 2010). For example, populations of Australian pelicans *Pelecanus conspicillatus* build up and breed when there are abundant food resources in large lakes (e.g., Lake Yamma Yamma in the Cooper Creek Catchment). When floods are sufficiently large to inundate Lake Eyre-Kati Thanda, invertebrate communities become established (Bayly 1976) providing abundant food resources for fish (Ruello 1976) and a high diversity and abundance of waterbird species, many of which breed (Waterman and Read 1992; Kingsford and Porter 1993). During dry periods, all species concentrate on large lake systems, such as Coongie Lakes, which retain water for long periods between floods. They will also move to habitats in other river basins in Australia.

4 Human Pressures

Water is critical to the distribution of people and their activities in the Lake Eyre Basin. Aboriginal people relied on the productivity of the rivers and the perennial springs to sustain their communities throughout the deserts of Australia (Smith 2013). They respond to the same "boom" and "bust" in flooding and resources, concentrating during wet times and dispersing during dry times (Smith 2013). During the dry periods, Aboriginal people also utilized water found in animals (e.g., aestivating frogs) and plants and the different perennial wells scattered throughout the deserts (Hercus and Clarke 1986; Gibbs 2009). Pastoralism is widespread and highly dependent on the permanent water source provided by the Great Artesian Basin and the bores that abstracted this water for livestock and landholders (Gibbs 2009).

About 36% of people in the Lake Eyre Basin are employed in agriculture, predominantly grazing (Measham and Brake 2009) but there are also many tourists, particularly visiting the flood spectacles. In 2001, more than two million people visited key areas in the Lake Eyre Basin, making tourism a major industry (Schmiechen 2004). There are also large mining developments, particularly oil and gas, gold, copper, zinc and uranium deposits where employees "fly in and fly out". In addition, there has long been pressure to develop the water resources of the Lake Eyre Basin for irrigation (Kingsford et al. 1998).

The rivers of the Lake Eyre Basin are currently unregulated and experience only minor water resource use and this lack of regulation and development is unusual for a basin of this size and underpins its good ecological condition. However, the basin is experiencing constantly changing pressures on land use and water resource demands, particularly from expanding energy industries, involving established conventional oil and gas production, and developing unconventional gas fields (shale gas, coal seam gas) and geothermal energy. Currently the energy industries are concentrated in the Cooper Creek catchment and metalliferous mines occur in the headwaters of the Georgina-Diamantina catchment and close to the southern margin of the Lake Eyre Basin. The major pressures from these industries are infrastructure development affecting flow paths, possible increased use of surface water, demand on groundwater and pollution from waste water produced.

Development of water resources in the Queensland part of the Lake Eyre Basin was recently announced by its Government. Since legislation prevents new water licences being issued, this development would involve the activation of irrigation licences which have never been used, and would be effected by allowing the trade and breaking up of these licences to be used further up the river. Potential irrigators are currently not allowed to develop dams to store the water alongside the river and so would need to pump directly on to their irrigated crops. This limits some of the development but this policy can easily be changed in the future. Such development has occurred in other rivers in Australia (Kingsford 2000), particularly in response to the need for annual returns on investment in such highly variable rivers. The very high temporal variability of flow in the Lake Eyre Basin means that the rivers are poorly suited to water resource developments that require a consistent, minimum level of extraction each year (Walker et al. 1997). The most serious concern is the impact that withdrawal of water from small to medium floods would have on the longevity of different waterholes and the variability of flows which is so important for the integrity of the river and its dependent ecosystems (Kingsford et al. 1998; Hamilton et al. 2005; Sheldon et al. 2010; Arthington and Balcombe 2011). These small to moderate floods are often vital precursors to the larger floods that can penetrate through the system to the lower lakes and wetlands (Leigh et al. 2010). In addition, establishment of irrigation in the catchment may encourage developers to harvest water that flows over the floodplain for irrigation further reducing flows downstream. Irrigation development can also introduce chemicals into a system which is currently largely not affected by pollution.

These impacts may be exacerbated by projected effects of climate change, although these are currently highly uncertain. Forecasted changes in rainfall patterns range from increases to decreases across the 5th to 95th percentile uncertainty range (http://www.climatechangeinaustralia.com.au). Increased tropical ocean temperatures could increase the frequency of large rainfall events but more extended drought periods are also possible. Undoubtedly, increased temperatures will place thermal stress on aquatic habitats and likely lead to increased rates of evaporative losses from disconnected waterholes, reducing their resilience as aquatic habitats.

The Queensland Government also signalled relaxation of another environmental policy decision by potentially allowing the mining and exploration of conventional and unconventional gas to occur on the floodplains of the Lake Eyre Basin rivers in Queensland (Fig. 15.1). This could not occur within or a protected zone, yet to be defined on the floodplain. It is not clear how much of the 1:100 year flood frequency zone and more frequently flooded areas will be protected by this proclamation but there is considerable concern that protection will only apply to frequently flooded areas. This allows developments to be largely unregulated outside this zone. Such developments in the form of infrastructure (roads, levees) may significantly affect flow patterns across the floodplain and impact on downstream ecosystems and dependent human communities. Even within the protected zone, it is not clear what forms of regulation will exist and whether these will adequately protect flow patterns. The floodplain areas and their anastomosing channels are particularly vulnerable to changes in infrastructure that will impact on flow patterns. Further, there may be pollutants generated from mining developments. In an adjacent catchment, a tailings dam at Lady Annie Mine near Mt Isa burst with a heavy storm and spilled pollutants into the nearby creek system in 2009, significantly lowering its pH and causing widespread loss of biodiversity for up to 52 kilometres downstream, resulting in half a million dollar fine for owners (http://www.miningaustralia.com.au/news/lady-annie-mine-fined-for-spill). For extraction of gas, there is often a requirement to use water, mixed with sand and chemicals, to fracture underlying rocks so that the gas can be collected, resulting in waste water which remains polluted. It is not clear how this process will be managed in the Lake Eyre Basin rivers. There are large areas of the Lake Eyre Basin which grow organic beef, with no chemicals, and they predominantly rely on the natural flows of the rivers on floodplains to generate the vegetation growth for their livestock. This significant industry could be affected.

Invasive species threaten the Lake Eyre Basin rivers and their endemic communities. These include various weed species (e.g., prickly acacia *Acacia nilotica*) and vertebrates. In particular, exotic fish species are among the most serious of invasives. Mosquito fish *Gambusia holbrooki* threaten the entire species of one of the artesian springs (Kerezsy and Fensham 2013) and are a widespread threat to spring communities (Fensham et al. 2011). Goldfish *Carassius auratus* occur throughout the Lake Eyre Basin and seem to be increasing over time in abundance (Costelloe et al. 2010). Redclaw crayfish *Cherax quadricarinatus* from tropical rivers have also recently invaded the Cooper Creek system, possibly through active promotion as an aquaculture species in the 1990s or by fisherman. The species may be a serious competitor of the endemic blue-claw crayfish *Cherax destructor*. The latest invader is the cane toad *Rhinella marina* which can affect terrestrial and aquatic organisms. It is not clear how much this species will be able to tolerate the extreme dry periods.

The effects of livestock on the floodplain vegetation of the Lake Eyre Basin has yet to be investigated in detail but there is currently limited evidence of long-term impacts on vegetation diversity and structure (Brock et al. 2006). The practice of removing or considerably reducing livestock numbers during dry periods would assist in reducing such impacts. The most obvious long-term change to the Lake Eyre Basin rivers and its catchments is the loss of medium-sized mammals with little evidence of major changes for other features of the system (Silcock et al. 2013).

The Lake Eyre Basin is increasing in popularity as a tourist destination, particularly during flood periods when large numbers of people visit the inundated Lake Eyre. This has certainly focused more attention of the environmental values of the area given that most visitors come for the wild nature of the landscape and its biodiversity. There are increasing challenges with this influx on infrastructure (e.g., roads) and the waste generated. Many people camp around waterholes where they usually build a camp fire. This has become an increasing problem given the number of visitors to some sites and the impact on dead wood around waterholes which provide habitat for invertebrates, small mammals and reptiles. Around some waterholes, which have already lost much of their woody debris, live branches are also cut from floodplain trees. Given the slow growth of many floodplain trees, this has become an increasingly serious issue in need of management.

5 Government and Community Management—A Partnership

The Lake Eyre Basin is managed primarily by the relevant States and the Northern Territory (Fig. 15.1). Under the Australian Constitution, land and water management was the primary responsibility of the State and Territory jurisdictions in the Australian Federation. National Parks, water and vegetation management plans all have separate legislation, departments and Ministers within each of the jurisdictions. Policies largely reflect the prevailing views of the government in power. In spite of this, there is a Lake Eyre Basin Agreement (http://www.environment.gov.au/water/publications/environmental/rivers/lake-eyre/pubs/agreement.pdf) between South Australia, Queensland and the Australian Government. It now also includes the Northern Territory. The goal of the agreement is to maintain the natural variability of the surface water and related natural resources. Separate legislation exists in each of the participating governments to enable the legislation. This institutional partnership arose primarily through two concerns. Local landholders were aggrieved that there was a strong recommendation to nominate the Lake Eyre Basin for World Heritage Listing in 1993 (Reid 1994). This was soon followed by a proposal to develop irrigated cotton on Cooper Creek at Windorah which received widespread condemnation for its potential impact on ecosystems and downstream communities (Walker et al. 1997; Kingsford et al. 1998; Gibbs 2009). The management of water continues to be challenging with a predominantly Eurocentric development focus, reflected in water legislation which for a long period did not adequately define large floodplains as part of the river (Gibbs 2009). This has continued to plague regulators attempting to constrain water resource development and ensure long-term connectivity of rivers throughout their catchment.

The Lake Eyre Basin Agreement is implemented through different advisory bodies to a Ministerial Forum (http://www.lebmf.gov.au/agreement/index.html) of participating governments, which meets to discuss issues of interest, recognising that the rivers are substantially spanning jurisdictional borders. There is a Senior Officers Group (jurisdictional departmental officers), Community Advisory Committee (consisting of all major stakeholders including mining, grazing, tourism and Aboriginal communities) and a Scientific Advisory Panel (technical expertise). Each can advise the Ministerial Forum separately and all meet at least twice a year to further adaptive management of the Lake Eyre Basin Rivers. This process is supported by a small secretariat team within the environment agency of the Australian Government. There are also independent natural resource management organisations within each of the jurisdictions, the South Australian Arid Lands NRM (Natural Resource Management) Board (SAAL); Desert Channels Queensland (DCQ); Northern Territory (NT) NRM Board; and the Western Catchment Management Authority in NSW. These largely focus on programs within their jurisdictions although there is still some collaboration. There is a biennial conference held somewhere in the Lake Eyre Basin, allowing for the communication of new research and other programs. In particular, there is a strong focus on assessment of the condition of the rivers and establishment of strategic adaptive management processes. A key challenge will be to detect changes in indicators, given the significant variability. This may require analyses to focus on particular bands of variability in an indicator (e.g., fish diversity) during particular floods, rather than the entire flow regime.

The Lake Eyre Basin Rivers Assessment aims to provide the information on the long-term condition of the environmental assets of the rivers of the Lake Eyre Basin (http://www.environment.gov.au/water/publications/environmental/rivers/lake-eyre/assessment.html). This is to maintain their condition among the last unaltered dryland rivers in the world. With increasing knowledge and data, it is anticipated that problems may be avoided in terms of long term impact. More recently this rivers assessment is embedded within a Strategic Adaptive Management framework (Kingsford et al. 2012) where there is a focus on setting a vision and goals that guide management and collection of data relevant to issues being managed. Despite this, there remains a strong focus on physical variability which may not adequately include social, cultural and ontological aspects of water (Gibbs 2006).

This Lake Eyre Basin Agreement process produces considerable involvement of communities and governments throughout the Lake Eyre Basin and represents a powerful engagement mechanism. For example, there is an Aboriginal member from each jurisdiction on the Community Advisory Committee. There have also been separate Aboriginal Forums (Birdsville, Tibooburra), driven by the Community Advisory Committee, held where there is much more involvement of Aboriginal communities in the future management and collaboration of the basin. These have continued to reiterate the fundamental importance of the river and its variability in defining the culture and ongoing involvement of Aboriginal communities in the Lake Eyre Basin. In particular, the last forum called for strong protection of the rivers from human pressures with a declaration for wild rivers (http://www.lebmf.gov.au/publications/pubs/aboriginal-forum-event-summary-2011.pdf).

There is also a Great Artesian Basin Agreement and these major Agreements should be complementary.

Ultimately, the major weakness of the process is that it is dependent on jurisdictional government policy direction. This was particularly well demonstrated by current policies for the rivers of the Lake Eyre Basin in Queensland after a change of government in 2012. The previous government (Labor) was pursuing a policy agenda to define the rivers of the Lake Eyre Basin as Wild Rivers under separate legislation which would make all resource development legislation subservient. The LNP (Liberal National Party) government took power in 2012 and is in the process of revoking this legislation and establishing development on the rivers of the Lake Eyre Basin in Queensland, despite the Lake Eyre Basin Agreement. To do this, it established an advisory process with a Western Rivers Advisory Panel (WRAP), which included members of the Community Advisory Committee, Scientific Advisory Panel, Aborigines and other stakeholders. There were no representatives of South Australia or the environment movement despite these being represented on a previous advisory process for the nomination of the Lake Eyre Basin rivers as wild rivers. The final report recommended activation of current irrigation licences by some members; this was not a consensus decision (http://www.dnrm.qld.gov.au/data/assets/pdf_file/0015/82500/wrap-report.pdf). These two outcomes, Wild Rivers nomination and development of rivers, are diametrically opposed and represent a major weakness in the current government and community partnership processes which affect the long-term sustainability of the river and its communities.

6 Future of the Lake Eyre Basin Rivers

The Lake Eyre Basin rivers are among the most spectacular in the world and largely remain in their natural state, particularly in relation to flow regimes. Up to now, no significant water resource development has occurred and the impacts of livestock grazing over more than a century have not clearly degraded the overall diversity and structure of flood dependent vegetation communities. Invasive species continue to affect biodiversity and livestock production. There is currently a considerable push for conventional and unconventional mining of gas resources which may significantly impact on the integrity of the entire Lake Eyre Basin, particularly if there are impacts which affect water flows and quality. It is not clear whether the current agreement in place will be able to ensure the protection of this unique river system and it will depend on communities of the Lake Eyre Basin convincing elected governments of the best path for sustainability of this unique desert river, the most variable in the world.

Acknowledgements

We thank Victor Squires and Hugh Milner for their perseverance in ensuring that this contribution could be completed.

References

Allan, R.J. 1985. The Australian summer monsoon, teleconnections, and flooding in the Lake Eyre Basin. South Australian Geographic Papers (2). Royal Geographic Society of Australasia, South Australian Branch.

Arthington, A.H. and S.R. Balcombe. 2011. Extreme flow variability and the 'boom and bust' ecology of fish in arid-zone floodplain rivers: a case history with implications for environmental flows, conservation and management. Ecohydrology 4: 708–720.

Arthington, A.H., S.R. Balcombe, G.A. Wilson, M.C. Thoms and J. Marshall. 2005. Spatial and temporal variation in fish-assemblage structure in isolated waterholes during the 2001 dry season of an arid-zone floodplain river, Cooper Creek, Australia. Marine and Freshwater Research 56: 25–35.

Arthington, A.H., J.D. Olden, S.R. Balcombe and M.C. Thoms. 2010. Multi-scale environmental factors explain fish losses and refuge quality in drying waterholes of Cooper Creek, an Australian arid-zone river. Marine and Freshwater Research 61: 842–856.

Baker, A.M., J.M. Hughes, J.C. Dean and S.E. Bunn. 2004. Mitochondrial DNA reveals phylogenetic structuring and cryptic diversity in Australian freshwater macroinvertebrate assemblages. Marine and Freshwater Research 55: 629–640.

Balcombe, S.R., S.E. Bunn, F.J. McKenzie-Smith and P.M. Davies. 2005. Variability of fish diets between dry and flood periods in an and zone floodplain river. Journal of Fish Biology 67: 1552–1567.

Balcombe, S.R., S.E. Bunn, A.H. Arthington, J.H. Fawcett, F.J. McKenzie-Smith and A. Wright. 2007. Fish larvae, growth and biomass relationships in an Australian arid zone river: links between floodplains and waterholes. Freshwater Biology 52: 2385–2398.

Bayly, I.A.E. 1976. Plankton of Lake Eyre. Australian Journal of Marine and Freshwater Research 27: 661–665.

Boulton, A.J., F. Sheldon and K.M. Jenkins. 2006. Natural disturbance and aquatic invertebrates in desert rivers. pp. 133–153. *In*: R.T. Kingsford (ed.). Ecology of Desert Rivers. Cambridge University Press, Cambridge.

Brock, M.A., S.J. Capon and J.L. Porter. 2006. Disturbance of plant communities dependent on desert rivers. pp. 100–132. *In*: R.T. Kingsford (ed.). Ecology of Desert Rivers. Cambridge University Press, Cambridge.

Bunn, S.E., P.M. Davies and M. Winning. 2003. Sources of organic carbon supporting the food web of an arid zone floodplain river. Freshwater Biology 48: 619–635.

Bunn, S.E., M.C. Thoms, S.K. Hamilton and S.J. Capon. 2006a. Flow variability in dryland rivers: Boom, bust and the bits in between. River Research and Applications 22: 179–186.

Bunn, S.E., S.R. Balcombe, P.M. Davies, C.S. Fellows and F.J. McKenzie-Smith. 2006b. Aquatic productivity and food webs in desert river ecosystems. pp. 76–99. *In*: R.T. Kingsford (ed.). Ecology of Desert Rivers. Cambridge University Press, Cambridge.

Burford, M., A. Cook, C.S. Fellows, S.R. Balcombe and S.E. Bunn. 2008. Sources of carbon fuelling production in an arid floodplain river. Marine and Freshwater Research 59: 224–234.

Capon, S.J. 2003. Plant community responses to wetting and drying in a large arid floodplain. River Research and Applications 19: 509–520.

Capon, S.J. 2007. Effects of flooding on seedling emergence from the soil seed bank of a large desert floodplain. Wetlands 27: 904–914.

Capon, S.J. and M.A. Brock. 2006. Flooding, soil seed bank dynamics and vegetation resilience of a hydrologically variable desert floodplain. Freshwater Biology 51: 206–223.

Carini, G., J.M. Hughes and S.E. Bunn. 2006. The role of waterholes as 'refugia' in sustaining genetic diversity and variation of two freshwater species in dryland river systems (Western Queensland, Australia). Freshwater Biology 51: 1434–1446.

Costelloe, J.F., R.B. Grayson, T.A. McMahon and R.M. Argent. 2005a. Spatial and temporal variability of water salinity in an ephemeral, arid-zone river, central Australia. Hydrological Processes 19: 3147–3166.

Costelloe, J.F., J. Powling, J.R.W. Reid, R.J. Shiel and P. Hudson. 2005b. Algal diversity and assemblages in arid zone rivers of the Lake Eyrie Basin, Australia. River Research and Applications 21: 337–349.

Costelloe, J.F., R.B. Grayson and T.A. McMahon. 2006. Modelling streamflow in a large anastomosing river of the and zone, Diamantina River, Australia. Journal of Hydrology 323: 138–153.

Costelloe, J.F., A. Shields, R.B. Grayson and T.A. McMahon. 2007. Determining loss characteristics of arid zone river waterbodies. River Research and Applications 23: 715–731.

Costelloe, J.F., E. Payne, I.E. Woodrow, E.C. Irvine, A.W. Western and F.W. Leaney. 2008. Water sources accessed by arid zone riparian trees in highly saline environments, Australia. Oecologia 156: 43–52.

Costelloe, J.F., J.R.W. Reid, J.C. Pritchard, J.T. Puckridge, V.E. Bailey and P.J. Hudson. 2010. Are alien fish disadvantaged by extremely variable flow regimes in arid-zone rivers? Marine and Freshwater Research 61: 857–863.

Fellows, C.S., M.L. Wos, P.C. Pollard and S.E. Bunn. 2007. Ecosystem metabolism in a dryland river waterhole. Marine and Freshwater Research 58: 250–262.

Fellows, C.S., S.E. Bunn, F. Sheldon and N.J. Beard. 2009. Benthic metabolism in two turbid dryland rivers. Freshwater Biology 54: 236–253.

Fensham, R.J., J.L. Silcock, A. Kerezsy and W. Ponder. 2011. Four desert waters: setting arid zone wetland conservation priorities through understanding patterns of endemism. Biological Conservation 144: 2459–2467.

Gibbs, L.M. 2006. Valuing water: variability and the Lake Eyre Basin, Central Australia. Australian Geographer 37: 73–85.

Gibbs, L.M. 2009. Just add water: colonisation, water governance, and the Australian inland. Environment and Planning A 41: 2964–2983.

Habermehl, M.A. 1980. The Great Artesian Basin, Australia. BMR Journal of Australian Geology and Geophysics 5: 9–38.

Hamilton, S.K., S.E. Bunn, M.C. Thoms and J.C. Marshall. 2005. Persistence of aquatic refugia between flow pulses in a dryland river system (Cooper Creek, Australia). Limnology and Oceanography 50: 743–754.

Hercus, L. and P. Clarke. 1986. Nine Simpson Desert wells. Archaeology in Oceania 21: 51–62.

Kerezsy, A. and R. Fensham. 2013. Conservation of the endangered red-finned blue-eye, *Scaturiginichthys vermeilipinnis*, and control of alien eastern gambusia, Gambusia holbrooki, in a spring wetland complex. Marine and Freshwater Research 64: 851–863.

Kerezsy, A., S.R. Balcombe, A.H. Arthington and S.E. Bunn. 2011. Continuous recruitment underpins fish persistence in the arid rivers of far-western Queensland, Australia. Marine and Freshwater Research 62: 1178–1190.

Kingsford, R.T. 2000. Review: ecological impacts of dams, water diversions and river management on floodplain wetlands in Australia. Austral Ecology 25: 109–127.

Kingsford, R.T. and J.L. Porter. 1993. Waterbirds of Lake Eyre, Australia. Biological Conservation 65: 141–151.

Kingsford, R.T., A.J. Boulton and J.T. Puckridge. 1998. Challenges in managing dryland rivers crossing political boundaries: lessons from Cooper Creek and the Paroo River, central Australia. Aquatic Conservation-Marine and Freshwater Ecosystems 8: 361–378.

Kingsford, R.T., A.L. Curtin and J. Porter. 1999. Water flows on Cooper Creek in arid Australia determine 'boom' and 'bust' periods for waterbirds. Biological Conservation 88: 231–248.

Kingsford, R.T., A. Georges and P.J. Unmack. 2006. Vertebrates of desert rivers: meeting the challenges of temporal and spatial unpredictability. pp. 154–200. *In*: R.T. Kingsford (ed.). Ecology of Desert Rivers. Cambridge University Press, Cambridge.

Kingsford, R.T., D.A. Roshier and J.L. Porter. 2010. Australian waterbirds—time and space travellers in dynamic desert landscapes. Marine and Freshwater Research 61: 875–884.

Kingsford, R.T., H.C. Biggs and S.R. Pollard. 2012. Strategic adaptive management in freshwater protected areas and their rivers. Biological Conservation 144: 1194–1203.

Knighton, A.D. and G.C. Nanson. 1994. Flow transmission along an arid zone anastomosing river, Cooper Creek, Australia. Hydrological Process 8: 137–154.

Kotwicki, V. and P. Isdale. 1991. Hydrology of Lake Eyre, Australia - El-Nino link. Palaeogeography Palaeoclimatology Palaeoecology 84: 87–98.

Kotwicki, V. and R. Allan. 1998. La Nina de Australia—contemporary and palaeo-hydrology of Lake Eyre. Palaeogeography Palaeoclimatology Palaeoecology 144: 265–280.

Leigh, C., F. Sheldon, R.T. Kingsford and A.H. Arthington. 2010. Sequential floods drive 'booms' and wetland persistence in dryland rivers: a synthesis. Marine and Freshwater Research 61: 896–908.

Marshall, J.C., F. Sheldon, M. Thoms and S. Choy. 2006. The macroinvertebrate fauna of an Australian dryland river: spatial and temporal patterns and environmental relationships. Marine and Freshwater Research 57: 61–74.

McGregor, G.B., J.C. Marshall and M.C. Thoms. 2006. Spatial and temporal variation in algal-assemblage structure in isolated dryland river waterholes, Cooper Creek and Warrego River, Australia. Marine and Freshwater Research 57: 453–466.

McMahon, T.A., R.E. Murphy, M.C. Peel, J.F. Costelloe and F.H.S. Chiew. 2008a. Understanding the surface hydrology of the Lake Eyre Basin: Part 2—Streamflow. Journal of Arid Environments 72: 1869–1886.

McMahon, T.A., R.E. Murphy, M.C. Peel, J.F. Costelloe and F.H.S. Chiew. 2008b. Understanding the surface hydrology of the Lake Eyre Basin: Part 1—Rainfall. Journal of Arid Environments 72: 1853–1886.

Measham, T.G. and L. Brake (eds.). 2009. People, communities and economies of the Lake Eyre Basin DKCRC Research Report 45, Desert Knowledge Cooperative Research Centre, Alice Springs.

Murphy, N.P., M.F. Breed, M.T. Guzik, S.J.B. Cooper and A.D. Austin. 2013. Trapped in desert springs: phylogeography of Australian desert spring snails. Journal of Biogeography 39: 1573–1582.

Prescott, J.R. and M.A. Habermehl. 2008. Luminescence dating of spring mound deposits in the southwestern Great Artesian Basin, northern South Australia. Australian Journal of Earth Sciences 55: 167–181.

Puckridge, J.T., F. Sheldon, K.F. Walker and A.J. Boulton. 1998. Flow variability and the ecology of large rivers. Marine and Freshwater Research 49: 55–72.

Puckridge, J.T., K.F. Walker and J.F. Costelloe. 2000. Hydrological persistence and the ecology of dryland rivers. Regulated Rivers-Research & Management 16: 385–402.

Reid, J. 1994. The Prime Minister's pre-election promise of world heritage listing for the Lake Eyre basin: flight of fancy? Rangelands 16: 273–297.

Roshier, D.A., P.H. Whetton, R.J. Allan and A.I. Robertson. 2001a. Distribution and persistence of temporary wetland habitats in arid Australia in relation to climate. Austral Ecology 26: 371–384.

Roshier, D.A., A.I. Robertson, R.T. Kingsford and D.G. Green. 2001b. Continental-scale interactions with temporary resources may explain the paradox of large populations of desert waterbirds in Australia. Landscape Ecology 16: 547–556.

Roshier, D.A., A.I. Robertson and R.T. Kingsford. 2002. Responses of waterbirds to flooding in an arid region of Australia and implications for conservation. Biological Conservation 106: 399–411.

Roshier, D.A., N.I. Klomp and M. Asmus. 2006. Movements of a nomadic waterfowl, Grey Teal Anas gracilis, across inland Australia—results from satellite telemetry spanning fifteen months. Ardea 94: 461–475.

Roshier, D.A., M. Asmus and M. Klaassen. 2008. What drives long-distance movements in the nomadic Grey Teal Anas gracilis in Australia? Ibis 150: 474–484.

Ruello, N.V. 1976. Observations on some fish kills in Lake Eyre. Australian Journal of Freshwater and Marine Research 27: 667–672.

Schmiechen, J. 2004. Lake Eyre Basin Heritage Tourism Future Directions. Lake Eyre Basin Coordinating Group (http://www.lakeeyrebasin.org.au/archive/media/future_directions.pdf accessed 11/10/13).

Sheldon, F., S.E. Bunn, J.M. Hughes, A.H. Arthington, S.R. Balcombe and C.S. Fellows. 2010. Ecological roles and threats to aquatic refugia in arid landscapes: dryland river waterholes. Marine and Freshwater Research 61: 885–895.

Shiel, R.J., J.F. Costelloe, J.R.W. Reid, P. Hudson and J. Powling. 2006. Zooplankton diversity and assemblages in arid zone rivers of the Lake Eyre Basin, Australia. Marine and Freshwater Research 57: 49–60.

Silcock, J.L., T.P. Piddocke and R.J. Fensham. 2013. Illuminating the dawn of pastoralism: evaluating the record of European explorers to inform landscape change. Biological Conservation 159: 321–331.

Smerdon, B.D., T.R. Ransley, B.M. Radke and J.R. Kellet. 2012. Water resource assessment for the Great Artesian Basin. A report to the Australian Government from the CSIRO Great Artesian Basin Water Resource Assessment. CSIRO Water for a Healthy Country Flagship, Australia, Canberra.

Smith, M. 2013. The Archaeology of Australia's Deserts. Cambridge University Press, Cambridge.

Walker, K.F., J. Puckridge and S. Blanch. 1997. Irrigation development on Cooper Creek, central Australia: prospects for a regulated economy in a boom-and-bust ecology. Aquatic Conservation: Marine and Freshwater Ecosystems 7: 63–73.

Waterman, M. and J. Read. 1992. Breeding success of the Australian Pelican (Pelecanus conspicillatus) on Lake Eyre South in 1990. Corella 16: 123–126.

PART 4

People and Place

The successful implementation of river basin management, integrated or not, depends on how the local stakeholders (urban, rural, industrial, environmental, etc.) behave in relation to water. It depends on their perceptions of the role of the water. Place is paramount because inevitably those upstream will have different priorities from those downstream in a river basin. Agriculture, collectively the world's biggest user of freshwater, lays claim to vast quantities of water to produce food for the worlds' burgeoning population. Increasing awareness of impending water shortages (at crisis point in some countries or regions) and the concern about ecosystems invariably lead to a clash of opinions over water allocation priorities. Some of these issues are dealt with in this Part of 5 chapters. *Squires* looks at the pivotal role of people and the importance of place. The national water policies in Nepal are outlined and assessed by *Pradhan et al.*, while *Wenger* looks at the consequences of devastating floods, often exacerbated by human interference with rivers, and the lessons drawn from experiences on four continents. *Plant et al.*, then investigate the importance of information and communication arrangements for people working together across the Thau water territory in France. Finally *Daniell, Milner and Squires* provide an overview of a number of key issues raised in this book.

16

People and Place as Determinants of Success or Failure to Meet Management Objectives in River Basins

Victor R. Squires

SYNOPSIS

Resource management is defined as the planned manipulation of resource-producing natural systems in such a way as to optimize their overall long-term productivity. Implementation of successful management interventions depends on both people (policy makers, land users, and other stake-holders) and obviously on place. Site specificity is so important. **Conservation** (implying wise use) is really the management of *human use* of the biosphere so that it may yield the greatest sustainable benefit to present generations while maintaining its potential to meet the needs and aspirations of future generations. **Development** is the modification of the biosphere and the application of human, financial, living and non-living resources to satisfy human needs and improve the quality of human life. There are various countervailing forces at play as communities and national strive to develop at a pace that satisfies their economic goals, and communities within river basins do their best to adhere to such conservation and human development goals as are the subject of various international and national conventions. Responses by land users, and the communities in which they are embedded, to these often conflicting forces, are examined.

Visiting Professor, Gansu Agricultural University, Lanzhou, China and International Consultant, Adelaide, Australia.

Keywords: Conservation, development, sustainability, community of common concern, Waterside Reserve Program, riparian, riverain forests, sediment management, Landcare, governance, watersheds, ecosystem services, recreation, hydroelectric, navigation, ecology, floodplains, climate change, mediated modeling, common pool resources, IWRM, place-based assessment, one-size-fits-all solution, cost-benefit analysis

1 Understanding People and Place

Human communities are geographically based on territories and landscapes with specific cultural connotations and connections. Within river basins, especially large ones, it is at the level of the local *watershed* (called *catchments* in some chapters) that action is taken and management is directed. In developed market economies, communities are reflected in the voluntary clustering of land managers and the formal institutions of a form of local government. Many communities, especially in non-developed market economies, are in transition as global changes interact with community-level social structure and dynamics, and the influence of the nation-state grows or diminishes. In both market economies and those in transition, cultural aspects are reflected in methods of governance. A common problem with effective watershed governance (Ffolliott and Brooks, this volume) in river basins occurs where the geographical dimensions of watershed problems are not reflected in the jurisdictional boundaries of territorial authorities. Where the watershed is fragmented and jurisdictions cross ridges and watershed boundaries, an ad hoc collection of fragmented watersheds is the outcome and governance is seriously impeded.

Most river basins encompass what are called 'Common-pool resources'. These have two features: first, they are shared resources whose use by one person makes them less available for use by another; second, it is typically difficult to limit the public's access to them (through laws or physical barriers). Many, but not all, ecosystem services can be categorized as common-pool resources (Table 16.1). Consider, for instance, the clean water provided by an intact watershed, the pollination provided by a community of bees, or the carbon sequestration provided by a healthy forest. These are public goods, but individual use can degrade a watershed or strip a forest, compromising these benefits for all.

As we look to develop institutions to better manage ecosystem services, and ensure their resilience over time, we can benefit from the lessons learned in the management of common-pool resources.

One of the most important lessons to be learned from IWRM (Chapters 1–4, this volume) is that no one solution is appropriate in all circumstances. While certain hydrological, ecological and social principles can guide us

Table 16.1. The benefits people obtain from ecosystem services (Source: UNEP 2006).

Provisioning *Goods produced or provided by ecosystems*	**Regulating** *Benefits obtained from regulation of ecosystem processes*	**Cultural** *Non-material benefits obtained from ecosystems*
- Food - Fuelwood - Timber - Fiber - Biochemicals/medicines - Genetic resources - Pollination	- Climate regulation - Flood regulation - Water purification - Carbon sequestration - Pest and disease regulation	- Spiritual - Inspirational - Aesthetic - Educational - Recreational
Supporting Services necessary for production of other services • Soil formation and conservation • Nutrient cycling • Primary production *Supporting biodiversity conservation*		

in understanding how watersheds work and how humans interact with them, these principles will never tell us all we need to know about every basin—this understanding must come from a place-based assessment of the specific physical, ecological, climatic, societal, and economic factors shaping that particular place and people.

While there is no one-size-fits-all solution to managing ecosystem services, much can be learned by sharing our approaches, strategies, and experiences. Telling the stories of both successes and failed efforts can provide a rich source of data with which to generate new solutions. Case studies (such as those presented in this book) provide an opportunity to acknowledge both the qualities that make particular places and efforts unique as well as the elements that may be generalized across contexts. Just as data collection and assessment are necessary at multiple scales, so too we must document the successful (and unsuccessful) approaches to ensure that learning can be shared among key actors. General principles can guide us in determining the appropriate institutions and organizations needed to manage a complex natural system, but we must also take into account the cultural, social, and economic attributes of the community at hand.

While acknowledging the complexity of human and natural systems is a critical first step, we also need to move toward a deeper understanding of these complex systems and their interactions. The approach taken in medical diagnoses may offer a useful analogy (Ludwig and Bastin 2008). While guided by a broad understanding of the subsystems that makeup the human body, physicians must also seek to understand a specific patient's condition to decide what treatment is most appropriate for that patient. By developing a similar corpus of knowledge around land and water management in different physical and cultural contexts, we can better analyze the specific attributes of particular cases. We can assess how different factors, such as geography and community structures, influence different resource management schemes across different resource sectors, from forestry to fisheries to land management. This will lead to a broader understanding of linked ecosystem and human health. Our bodies require regular checkups to ensure that they are functioning well. In a similar fashion, our ecosystems also need periodic checkups so we can gather sufficient information about the resources we seek to manage and whether that management is headed in the right direction. Given the scale of the system and limitations on gathering data, it is important to develop metrics to assess whether management efforts are creating desired outcomes. In this case the general direction of the trajectory is more important than precision—we don't need to measure everything to the "nth decimal point" in order to assess whether we are moving in the right direction.

2 Action and Integration at all Scales

Identifying the appropriate scale or scales on which to act is an important step toward successful management of ecosystem services. For example, it is necessary to act on a local scale when managing a river basin that provides drinking water for a community. This will allow for meaningful participation by the community and ensure that the specific characteristics of that watershed are considered. In contrast, to effectively address the challenges of climate change, action is needed at all levels—from household-level energy conservation to international agreements to generate improved outcomes at the global level.

Even when action is primarily at the local level, actions at other scales may protect against parochial self-interest and encourage shared learning across space and time. The systems that support the provision of ecosystem services range from a neighborhood, to a watershed, to a country, to a continent, to the globe. Managing them effectively therefore requires a set of nested institutions that can encompass these multiple scales and interact and communicate with each other to create cross-scale management (Cash et al. 2006).

We need to develop more integrated institutional frameworks and find ways to enhance effective participation across institutions. Our current approach is fragmented: the ecological systems that we are interested in are governed by different agencies that don't always communicate, with diverse and sometimes conflicting regulatory structures and overlapping jurisdictions. The so called silo effect (Fig. 16.1). See also Mitchell (this volume) for a discussion of the silo effect in the Canadian context.

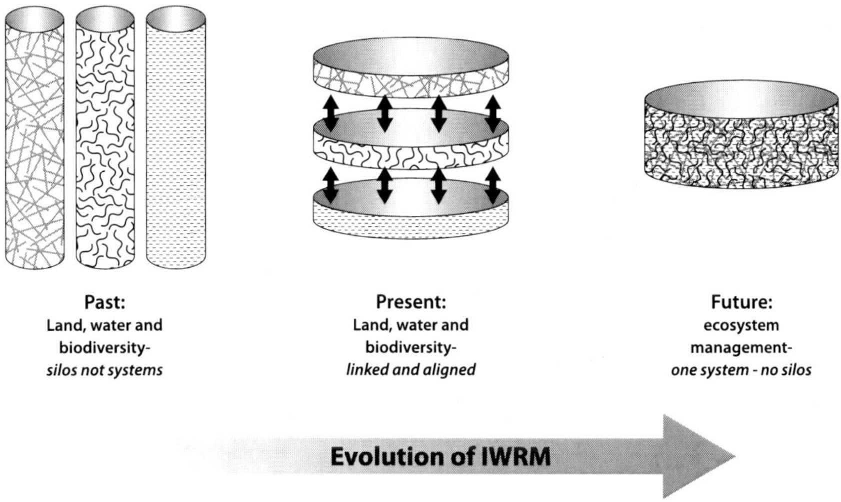

Figure 16.1. As IWRM evolves there is a shift from a silo mentality where agencies are vertically integrated but lack cross agency cooperation to a situation where ecosystem management is done in a holistic manner.

2.1 Design for resilience

Change happens, actors and circumstances change, and the social values that inform a community's priorities can shift. Increasingly, climatic and demographic changes pose fundamental and potentially seismic shifts in many communities. Not surprisingly, our strategies, frameworks, and institutions must also evolve. Though it goes against our nature, humans need to "anticipate obsolescence" of our own creations, building in mechanisms to periodically evaluate and ensure that lessons learned are understood, that plans are adapted accordingly, that trajectories toward goals have not become antiquated, and that societal goals have not changed. Nothing can or should last forever.

Recognizing that adaptation must occur and building in periodic review and renegotiation of agreements are essential elements of any long-term strategy. In addition to contributing to the resilience of a system by ensuring ongoing responsiveness to emerging issues, this approach also recognizes the need for civic engagement and active democratic participation.

3 Community Participation—A Vital Element

Participation is the key when designing management institutions for common-pool resources, the participation of relevant stakeholders is vital. And the first step to a successful cooperative management approach is communication. Possessing multiple ways to communicate about the resources or services being managed—to allow "movement beyond anonymity"—has proven to be an essential element in successful co-management of shared resources. This is closely aligned to what Ffolliott and Brooks (this volume) call locally-led initiatives.

The commitment and participation of the community is essential if the aims of IWRM are to be achieved. The physical size of most river basins and the extent and nature of environmental problems, including land and water degradation, makes it an impossible task for government alone. It is only with cooperative involvement of the basin managers and the land and water users that these resources can be both productively and sustainably managed. The community can help to improve the management of a river basin's natural resources by:

- Identifying local natural resource management issues
- Enlisting government support and funds to complement their own resources for their activities
- Developing and implementing practical management plans for their localities
- Promoting the adoption of improved management practices, e.g., sediment management
- Communicating their ideas and concerns for the management of natural resources at the local, regional and basin-wide levels

The community, with its distinctive institutional and ecological position relative to the problems of river basin management at the local level, is a crucial and largely under-appreciated influence. Community-level processes serve to amplify or attenuate forces of development that emerge from differing scales, and community-level interventions have historically proven to both slow and accelerate environmental change. Community-level perceptions provide important alternative views, though ones that may differ among stakeholders (parties interested in the outcomes). Stakeholders at the community level detect change, explain its causes, project trends, judge consequences, decide how to respond, and implement decisions. This decision-making process is mediated through psychological processes that may be far from comprehensive or rational. In many communities, voting is seen as the primary mechanism for participation in democratic systems. But the use of participatory geographic information system (GIS) technology, mediated modeling, town meetings, or other approaches can also help participants develop shared learning about a resource and shape a unique place-based approach (Daniell 2011; Loch et al., this volume).

3.1 Mediated modeling: a tool for involving stakeholders

Mediated modeling (van den Belt 2004), also known as participatory modeling or under a range of other appellations (Voinov and Bousquet 2010), is the process of involving stakeholders (parties interested in or affected by the decisions the model addresses) as active participants in all stages of the modeling process, from initial problem scoping to model development, implementation, and use. Mediated modeling can build mutual understanding, solicit input from a broad range of stakeholder groups, and maintain a substantive dialogue between members of these groups. Mediated modeling and consensus building are also essential components of adaptive management.

One product of mediated modeling is the creation of a general dynamic model that integrates ecological and economic processes in local ecosystems. Benefits and costs of different management scenarios are addressed by estimating values for harvested products, tourism, water yield, and biodiversity. Costs included direct management costs, such as sediment management, clearing invasive plants, and indirect costs, such as personnel and infrastructure.

Information is important, but problems are solved by people. In recent years, much research effort has been devoted to developing ways to put a value on ecosystem services to help ensure that they are given adequate weight in decision making and resource management efforts. While some aspects of these services are easier to assign a monetary value to than others, a variety of approaches has been developed to better capture ecosystem services' value to society (Costanza 2012).

While incorporating such values into cost-benefit analyses will help make these services more visible to policymakers, such analyses do not, in and of themselves, determine how such resources will be managed. A cost-benefit analysis cannot serve as a neutral arbiter of value, because even the best, most comprehensive cost-benefit analysis will come down to a social decision about a community's shared ideals.

Similarly, while a mediated modeling exercise can help people come to some agreement about managing their resource systems, there may still be conflicting beliefs and values regarding the use of these resources. By making all of the values and ethical considerations that influence decision making more explicit, we can better design institutions to mediate when values conflict. This information will help communities balance trade-offs between managing for the delivery of ecosystem services and other types of outcomes that may be of value to society.

Once again, ensuring that there are multiple channels for the participation and deliberation of affected stakeholders will be important to increasing the range of values represented and making sure that the priorities assigned through any management decisions are vetted as comprehensively as possible. Often enough facts are uncertain, decisions are urgent, stakes are high, and people's values matter, so detailed, time-consuming scientific research is not an option. Instead a trans-disciplinary synthesis across the natural and social sciences, relying heavily on local knowledge, anecdotal evidence, the values of affected communities, and scientific studies conducted elsewhere can be used (Costanza and Ruth 1998).

Policymakers often gravitate toward one-size-fits-all solutions and static institutions. However, when it comes to the complexity of managing human-natural systems, a more adaptive approach is required. A key to success is understanding each unique place and the people who depend on its ecosystem services. Clear communication and documentation of past success and failures at all scales not only will engage necessary participation but will help determine which successful strategies can be translated across cases and which are not translatable.

Cost-benefit analyses are helpful but limited in describing how social values will ultimately result in a particular decision. Just as natural systems evolve, so must our strategies and institutions. Like a durable good that reaches the end of its life and is then recycled, so too must we periodically evaluate and anticipate obsolescence of our management structures and tactics. As we explore ways to ensure that we are maintaining the integrity of the natural systems on which we depend, these concepts can guide us in managing natural systems in a more integrated and resilient manner. Under the right conditions, local communities can manage shared resources sustainably and successfully (Mitchell, this volume. Ffolliott and Brooks, this volume). In practice, the boundaries of human communities do not usually coincide with those of environmental systems, including watersheds, drainages, soil types, or vegetation complexes. This can cause conflict, for example, where local government boundaries do not coincide with watersheds, so that incentives for an integrated management plan (say) for erosion control or control of invasive plants are reduced.

4 Conservation as a Competing Use in Natural Resource Management

Many people support the notion of conservation of renewable natural resources and recognize the contribution to the flow of ecosystem services they provide. Most realize that natural resource development is an essential activity distinct from resource conservation but in fact they must go hand in hand. They are both part and parcel of the overall business of resource management. Conservation and development are fundamentally linked by their dependence on living resources. Both conservation and sustainable development require an attitude of stewardship, especially towards those plants, animals and microorganisms and the nonliving resources on which they depend, that could be destroyed if only short-term human interests are pursued. This is especially important when devising management plans for river basins where many competing demands vie for water allocations. Allocation of water for conservation purposes (e.g., ecological flows) as a legitimate form of land and resource use must compete with resource development for agriculture, industry and urban needs. It is a sad fact, but true, that many of the proposals for resource conservation are competitors for water in a situation where water is already overcommitted (Li and Squires, this volume; Riddell and Thuo, this volume; Sullivan, this volume) and in many jurisdictions are unlikely to receive the water allocation that is necessary to ensure the survival of some important ecosystems (Du et al., this volume).

There is increasing community and scientific concern about the conservation of the remaining ecosystems and this can only be achieved through the allocation of substantial volumes of water that would otherwise go to irrigation, or other forms of industrial use and to the burgeoning urban population for servicing critical human needs. At the same time in many regions there is increasing community interest in recreational use of the waterways, and this represents potentially one of the biggest sources of income for many regions. Recreation and tourism compete with other uses of a basin's resources, particularly water resources. Recreation on lakes, man-made lakes and reservoirs, is affected very much by the operating policy that is used in the reservoir system. Recreational fishing is another multi-million industry as is recreational navigation. Pleasure boats, holiday paddle steamers and the like are increasing in number and from the people who use those craft we are going to experience pressure for water management that facilitates their navigation, recreational and tourism needs. Some of these needs conflict with needs for drought mitigation, hydropower production and irrigation needs.

5 The *Community of Common Concern* (CCC) Concept

The term CCC has been coined to describe communities that are working together to identify and manage the issues or problems that face their particular locality. The individuals and organizations that comprise a CCC share a common concern over the condition and management of natural resources in their neighborhood/subregion. CCC will form the building blocks of basin-wide community involvement and will initiate the coordination and cooperation necessary to achieve the objectives of IWRM.

There are many models possible for CCCs; however, in general they can comprise:

- Individuals from within the community; these individuals may be elected by the community as its representatives.
- Representatives from organizations, both private and government, with a special interest in the management of a locality's natural resources.
- A coordinator responsible for the provision of administrative support for the CCCs and the coordination of the activities of relevant government agencies; the coordinator will generally be an official from the lead agency that is responsible for the provision of adequate resources for, and management of, the CCC.

The fundamental components of CCCs are that:

- The process is led by the CCC and supported by agency and specialists; a CCC has direct access to relevant ministers (or Departmental Secretaries) to resolve issues and problems that may arise;
- The process coordinates the activities of government agencies, communities and individuals;
- An 'action' (management) plan be prepared;
- Whilst the action plan focuses on the major '80%' issue, it also considers and integrates the impact on all resource sectors; the proposal is broadly supported by the community that is potentially affected;
- The proposed outcomes are achievable and sustainable.

It is anticipated that the community will develop ownership of the idea and become committed to its implementation by its being responsible for the planning process in this way. This should provide ongoing incentive, both for the initial enthusiasts as well as the slow starters in the community.

It is essential that actions plans are developed from a sound understanding of a locality's resources and the processes that contribute to their management. To ensure that there are adequate assessments of the economic, environmental and social effects of these activities, agency and other specialists will be responsible for the undertaking of the relevant investigations to support the CCC. The final action plan could include:

- A review of the problem(s)
- An assessment of the economic, environmental and social effects of the plan
- A description of the tasks to be undertaken and how they are to be implemented
- Preferred measures and the implementation targets
- Cost-sharing arrangements

- Incentives and sanctions to ensure adoption of the plan
- Responsibilities for implementing and reviewing of the plan

The detail of the plan and the supporting investigations, however, will need to be tailored to suit the magnitude of the problem(s) and the locality being addressed (see Ffolliott and Brooks, this volume).

5.1 Landcare—a viable option for resource management

On a more local scale the principles of CCC have been applied in Landcare—a community-based approach that has played a major role in raising awareness, influencing farming and land management practices and delivering environmental outcomes. The evolution of Landcare began in Australia in 1986 when landholder groups initiated community-based activities to protect and regenerate the country's land resources. Since then, various government programs have been developed and implemented to support local actions. This innovation is of interest to many other countries to address environmental challenges for sustainable land use. It has been long expanded from Australia to New Zealand and the Philippines, and continues to spread out in countries such as the United States, South Africa, Uganda and Tanzania among others. In changing landscapes and lifescapes, Landcare has become an important key driver in promoting sustainable management of environment and natural resources. It is now active across 6000 Australian landscapes and in many others in 20 countries worldwide. It is seen as a way to develop social capital in rural communities (Soebel et al. 2001; Cramb 2007). Because it involves multi-sector stakeholders, Landcare has come to mean different things to different people. The following are some of Landcare's different meanings:

- **As a MOVEMENT ...** Landcare is a movement of autonomous farmer-led organizations supported by local governments with backstopping from technical service providers that share knowledge about sustainable and profitable agriculture on sloping lands (in particular) while conserving the environment and natural resources.
- **As an APPROACH ...** Landcare is an extension approach / method that rapidly and inexpensively disseminates sustainable land management (e.g., agroforestry practices) among thousands of farmers based on the farmers' innate interest in learning and sharing knowledge about new technologies that earn more money and conserve natural resources. This embodies three basic cornerstones: appropriate technologies, partnership building and institution building.
- **As a BODY OF KNOWLEDGE ...** Landcare is a set of appropriate land management practices to care for the land in an economic and productive way.

- **As an ETHIC, a PHILOSOPHY ...** Landcare is an ethic, a philosophy that enables individuals and communities to approach agriculture in a mature and nurturing way.

Landcare is an amazing grass roots movement that harnesses individuals and groups under the ethic of caring for the land. It had its genesis in initiatives to improve agricultural productivity through sustainable land management. The movement has grown from this to a broader focus on sustainable management of all of Australia's natural resource assets and now encompasses individuals and groups across the whole landscape from coastal to urban and remote areas of Australia. Caring for the land captures a range of activities such as soil conservation, management of erosion and salinity, sustainable farm practices, restoration of native habitats, revegetation, control of weeds and pests and the development of local natural resource management skills and knowledge. It is also an approach to watershed (catchment) management (Curtis and Lockwood 2000).

The success of the Landcare model is due in part to its bottom-up philosophy. A Landcare group usually starts when community members with common objectives connect over their observations of a local environmental issue. For example, erosion of sand dunes due to mismanaged beach access or weeds affecting agricultural productivity. Groups set their own agenda, undertake work as often as they like and choose their own project sites. Groups may apply for funding from a variety of different sources to support their work including local, state, federal government and *Landcare Australia* (Box 16.1).

Box 16.1. The Landcare approach in Australia.

The Australian Government recognizes the important role Landcare and collective community action plays in managing Australia's environment and natural resources sustainably. The Australian Framework for Landcare identifies that the Landcare approach comprises:

- a philosophy, influencing the way people live in the landscape while caring for the land
- local community action putting the philosophy into practice.
- a range of knowledge generation, sharing and support mechanisms including groups, networks (from district to national levels), facilitators and coordinators, government and non-government programs and partnerships.
- Managing natural resource issues across Australia is now recognised as an intergenerational process and remains critical to the resilience of people, their communities and our natural resource base. It underpins critical issues for the future such as greenhouse gas emissions, climate change variability, water quality and availability, and food and fiber security for growing populations in Australia and around the world.

Generally, small group committees oversee operations, apply for project funding and organize communal activities like community workshops or tree planting. Most groups have one to six formal meetings annually. They may run discussion sessions, and short trips to other Landcare groups and other activities to gain and share knowledge. Some larger groups may have a paid coordinator providing part-time assistance, arranging meetings and activities and providing management guidance. Funds to pay these salaries mostly come from government. Increasingly, Landcare groups amalgamate into Landcare networks managed by community boards that take a more regional approach to environmental issues and coordinate activities to achieve watershed-wide outcomes. Networks are now a major community link to all levels of government and industry for financial support and information.

6 Floodplains: Balancing Nature and Human Needs

For countless years, the interaction between climatic conditions and water flow have forged the valleys and their floodplains (see Wenger, this volume for definition of flood plains). The landscape that we have inherited is a complex mosaic of geological activity, weather and weathering, and human innovation. Rivers and their floodplains are amongst the most fragile and threatened ecosystems in the world and the stresses associated with local environments through anthropogenic activities have a long history because of their importance as sources of water for agriculture as well as transportation routes when roads were few and also as sources of power when the Industrial Revolution was in its infancy.

Although the river and its flood plain are inseparable parts of one ecological system there is a boundary between the two. The floodplain of a river is delimited by the zone within the reach of the floods, when the river overflows its channel to inundate the surrounding areas, including wetlands and riverine forests. This is the river edge environment, a fringing zone often marked by reeds, bulrushes and other aquatic and semi aquatic plants. It is constrained by the physical characteristics of the bank material, the strong currents and shifting sediments of the open channel (see below). Another constraint, especially in the lower reaches, is high turbidity, which limits penetration of light for photosynthesis. In keeping with the general concept of an ecological boundary the river edge boundary is uniquely defined by the nature and strength of it interactions with adjacent zones.

Public riparian lands are a valuable community asset. The wise use and allocation of these lands is therefore of significance to the wider community. In particular these lands offer unique opportunities for the satisfaction of broad community expectations in the areas of recreation and nature conservation. The formal conservation parks and reserves will

Figure 16.2. In this aerial view of the Murray river near Renmark, South Australia (34.17S and 140.76E) we see the multiple land uses that the river creates. Urban, rural and conservation areas (riverine forest and wetlands) are all represented [Photo Department of Environment, Water and Natural Resources, South Australia, used with permission].

Color image of this figure appears in the color plate section at the end of the book.

never include all areas of conservation significance. The costs involved with acquisition as well as ongoing management of reserves will tend to restrict expansion of the reserves system to conservation of only the most significant areas. Nevertheless there is a need for conservation management of 'less significant' areas, to complement the formal reserve system. Management such as that outlined in Box 16.2 are seen as central to the development and maintenance of public riparian land and water resources, providing for the growth of the tourist and recreational industries while protecting key riparian, nature conservation and cultural heritage attributes. Public benefits are maximized with management undertaken in an efficient, locally responsive and dynamic fashion.

The way river and stream ecosystems, including the river edge, are currently used and managed is causing increasingly severe ecological deterioration, particularly through factors such as hydrological energy production and dams, intensive agriculture, deforestation of upstream areas of river basins, intensive use of agrochemicals, pollutants from industries in the river vicinity, urbanization of river corridors, drainage,

Box 16.2. Waterside Reserve Program in New South Wales, Australia.

The development of key river corridors for agriculture, grazing, forestry, transport and urban uses following European settlement of Australia in the late 18th century has left small and isolated pockets of natural habitat. Many of these will not be incorporated into the formal reserve system, but justify some form of conservation management. To this end, the Waterside Reserve Program was established to identify, protect and manage the small isolated areas of publically-owned land along river corridors that are of significance for nature and cultural heritage conservation. Land proposed for use as waterside reserve includes virtually all the lands in substantially natural condition and/or required for riparian public access and recreation within a multiple use context. The program involves:

- Identifying suitable areas to form a 'reserve network'
- Establishing an administrative structure to manage the areas
- Gaining public acceptance and enthusiasm for the program
- Gaining public assistance to manage the areas
- Provision for multiple land use management to off-set competing demands for the land
- Ensuring that management of areas is consistent with management (IWRM) objectives

river regulation and flood defence schemes. Developments and all other anthropogenic activities associated with riverine localities may alter the ecology of rivers, which will affect the productivity and stability of these fragile but dynamic systems. And these stresses further continue as extensive commercial navigation of many rivers, which has resulted in much artificial improvement of natural channels, including increasing the depth of the channels to permit passage of larger vessels.

6.1 Floods and flooding

Seasonal and periodic floods are necessary for maintaining the integrity of riparian habitats, floodplain ecosystems and watershed aquifers. Flood surges also provide the energy for distributing water to higher elevated wetlands (Wenger, this volume). Flooding provides multiple benefits for flora and fauna. Seasonal flood surges trigger breeding cycles among wildlife communities (Kingsford et al., this volume); seeds are dispersed in flood debris and deposited on floodplains to germinate. The ecological functions of floodplains and riparian ecotones identify them as critical environmental management zones with overriding importance. Floodplains allow floods to be accommodated beneficially without causing serious hazards or environmental damage. Rather, the water is slowed down and spread widely to where flood waters can be stored in subterranean suites of aquifers. Through seasonal flooding sequences and rainfall cycles, floodplain aquifers are continually recharged and refreshed so they can reliably discharge fresh clean water fit for food production and human consumption. In situations such that described for the Shiyang River basin in north-west China (Li and Squires, this volume) failure to allow recharge of ground waters led to widespread death of riparian vegetation, including trees that were over a century old, and ultimately to ecological migration by the poverty-stricken farmers when land became unusable. Of course these comments about floods and flooding relate to rivers in the more humid regions. Floods in drier regions take on different characteristics.

Box 16.3. Floods in arid and semi arid regions

Characteristics of precipitation in arid and semi-arid zones

Precipitation is rare and often there are high intensity ratio fall events that generate high run-off from soil surfaces that are sparsely vegetated. Serious droughts occur frequently and these may further deplete the vegetative cover. The frequency-magnitude distribution of floods shows that the ratio of magnitudes of, for example, 100 yr floods and annual floods is greater in arid than in humid environments.

- Smaller frequency of occurrence of rainstorms
- Lower magnitude of frequent storms
- Great yearly variation in storms
- Short duration and high intensity of rare storms

Characteristics of flash floods in arid and semi-arid zones

The characteristics of flash floods are their short duration, small areal extent, high flood peaks and rapid flows, and potential to cause heavy loss of life and property.

- Suddenness of occurrence
- Randomness of areal distribution
- Complex geomorphic formation reasons

Strategies for flash flood prevention

<u>Engineering measures</u>

- Building storage reservoirs
- River basin management
- Building flood protection dykes
- Clearing obstacles in the water course
- Building flood diversion storage and creating flood detention areas

Box 16.3. contd....

Box 16.3. contd.

<u>*Non-engineering measures*</u>

- Establishment of unified command system and disaster management system
- Establishment of monitoring systems for forecasting and warning
- Adequate preparedness for flood disaster management
- Setting up flood disaster evaluation and policy-making
- Flood insurance schemes
- Rapid action for disaster alleviation

6.1.1 Bankfull flow as effective discharge

The flow that fills the active channel and begins to spread onto the flood plain represents the break between channel processes and flood-plain processes. This flow is assumed to equal the dominant or effective discharge, which is defined as the discharge that transports the largest fraction of the sediment load during a specific period of time. Although extreme fluvial events can heavily erode and enlarge stream channels, steeply incising them, more modest flow regimes transport the greatest quantity of sediment material over time because modest flows occur with higher frequency. The effective discharge incorporates the principle that the channel-forming discharge is a function of both the magnitude of the event and the frequency of occurrence. For stable alluvial streams the effective discharge has been shown to be highly correlated with bankfull discharge. Bankfull discharge is the streamflow that is associated with a momentary maximum flow; it is found to govern channel shape and form. Bankfull stage is identified from geomorphic features in the field. The effective or bankfull discharge represents the single flow that is responsible for transporting the most sediment during a period of time in which channel maintenance is the most effective, resulting in the average morphological characteristics of the channel. However, a range of flows on either side of the effective discharge also carry a significant amount of the total annual sediment load. Management of sediment, how it is handled and where it is moved to is of vital importance.

7 Sediment Management in Relation to People and Place

Rivers and streams carry sediment in their flows. Sediment like water, flows downstream and supports both shorelines and habitats at the end of the line. This sediment can be in a variety of vertical locations within the flow, depending on the balance between the upwards speed on the particle (drag and lift forces), and the settling speed of the particle. In some cases, depending on the velocity, sediment will be transported downstream entirely as suspended load. In other cases it will move along the water bed as bed load by rolling, sliding, and saltating (jumping up into the flow, being transported a short distance then settling again). It may also move as a wash load. There is generally a range of different particle sizes in the flow. It is common for material of different sizes to move through all areas of the flow for a given stream condition.

Sediment in streams can adversely affect water quality and aquatic life and habitats, and it may present a hazard to downstream infrastructure and residents. Suspended fine sediments reduce the clarity of the water and possibly also biologic productivity, but the suspended sediments can also carry nutrients that may increase biological productivity. Coarse suspended sediments can bury stream habitats and can diminish the overall abundance and diversity of species.

An understanding of sediment-transport characteristics is important for applications that include flow regime change, channel-restoration efforts, and predictions of the effects of land-use changes. In order to establish river equilibrium for river-corridor protection and restoration, knowledge of the relation between flow and sediment transport and an assessment of the ability of the existing channel to transport sediment is critical. The measurement of sediment load and particle size moving through a reach at various flows allows the development of sediment-discharge curves (curves showing the relation between stream flow and sediment discharge for a specific sediment-sampling site) (ASCE 2008). Once the conditions required for sediment transport are understood, a determination of the channel dimension, pattern, and profile required for sufficient transport of the expected sediment supply can be made.

Hydrologic, geologic, geographic, and biologic factors affect the sediment load of a stream. Development (e.g., a dam on a river will interrupt sediment transport). Season, snowmelt, duration and intensity of rainstorms, watershed use, vegetation cover, watershed field slope, soil types, and human and animal activities further determine the amount of sediment entering and transported by the stream (ASCE 2008). Seasons can have a significant effect on the natural level of sediment induction into a stream. During the winter, the ground may be frozen, and precipitation may be in the form of snow. The absence of the impact of raindrops to loosen the soil and frozen ground holding the soil together are factors in transporting less sediment. During the summer, the impact of raindrops from high-intensity storms results in high concentrations and transport loads of sediment. Suspended sediments can bury stream habitats and can diminish the overall abundance and diversity of species.

Areas along the river that are used for agricultural purpose can further contribute to sediment loads. The fields maybe bare in winter and spring (or in summer fallow in the southern hemisphere) but, as the crops emerge, plant growth protects the soil from erosion. In these cases, sediment concentrations and loads for a given discharge may be low in the winter (frozen ground) and in summer (rooted crops) but high in the spring (before planting and growth of crops) and in fall (after harvest). In the winter-rainfall areas of the southern hemisphere, run off may be highest in winter and spring and sediment loads can be high from bare fallow following high intensity summer storms.

7.1 Sediment management—a major problem in many river basins

Sediment management where the principal focus is on the water typically highlights three issues:

1. Source management—addressing the type and source of sediment
2. Transport of sediment—addressing the systems that transport sediment
3. Deposition of sediment—addressing the location where sediment deposits

Management actions are tailored depending on the location they occur (Du et al., this volume) and the whether the management concerns involve a non-built environment (rivers, streams, creeks and flood plains) or a built environment (water control structures, flood levees, dams).

7.1.1 Source management of sediment

Sediments can come from anywhere and be just about anything. Organic and inorganic material alike can become the bits of matter tiny enough to allow it to be picked up and carried along with a moving fluid. Sediment management is an essential for integrated water management as the presence or absence of sediment will have significant impacts on water and its beneficial uses. Organic sediments are mostly debris from trees, plants, grasses, and animals and fish and their waste products. Inorganic sediments are divided into two main groups, these being coarse-grained sediments and fine-grained sediments. Coarse-grained sediments are boulders, cobbles, gravel, and sand, while fine-grained sediments are silts and clays. A further important distinction of the sediments is whether they are clean sediments or contaminated sediments, as this greatly affects the manner in which they can be used as beneficial material or must be isolated from their surrounding environment. Throughout this chapter the use of the term *sediment* will mean clean sediment, and if the sediment is contaminated the term *contaminated sediment* will be used.

Understanding the cumulative impacts of all past, present, and proposed human activities in watershed is important in predicting the impacts of sediment on surface waters. Sediment management is critical for the entire watershed, beginning with the headwaters and continuing into the coastal shores (except for inland river basins (Li and Squires, this volume; Kingsford et al., this volume; Krutov et al., this volume). One of the normal functions of waterways is to produce sediment, flush it through the system with some settling occurring in low areas (areas now typically used for farming or urbanization, see below) and some moving to the sea or terminal lakes to create shoreline replenishment but this important function has been interfered with.

Much sediment is the result of the naturally erosive mountains. Alluvial fans are found where sediment loads are high, for example, in arid and semiarid mountain environments, wet and mechanically weak mountains, and environments that are near glaciers. Alluvial fans develop where streams or debris flows gather speed in narrow passages then emerge into areas with greatly larger channel widths. A number of factors contribute to the severity including the degree of steep grades to flatter grades. Debris and water spill out in a fan shape depositing sediment and other debris on its way. The channels on these fans range from decimeters to several meters deep with the speed of the flows moving boulders sometimes taller than a house. These conditions are found at mountain fronts, in intermountain basins, and at valley junctions. Historic flood deposits of sediment into floodplains are the source of much of the richest farmland.

All too often these alluvial plains are the places where the majority of people (residents/businesses) settled. These inhabitants were commonly impacted by frequently fluctuating watercourse alignments (Wenger, this volume) resulting from high amounts of sediment deposition on what are really flood plains. Because they wanted more stable river or stream alignments local governments tried to assist land users, starting with agricultural development, this led to alterations in the alluvial areas' surface and groundwater hydrology. This situation often led to the construction of dams, debris basins, channels and spreading grounds. Facilities were constructed to serve agricultural and urban areas. Most of the agricultural areas later became urbanized. Farms and residential subdivisions essentially planted themselves in the very sediment disposal areas that Mother Nature had set up. Impacts from historic and current road building and land management practices continue to contribute to existing problems in many areas.

Landslides are the major producer of sediment in some regions. Road construction and poor timber harvesting techniques in the 1950s and 1960's resulted in an astronomical increase in sediment in some areas of China, California, Australia (and elsewhere). Additional system alterations also occurred as dams and channels were built for both water supply and flood protection. More and more structures changed what had been the natural hydrology, which then altered whatever system stability for sediments may have existed.

Source management occurs to prevent soil loss and adverse sediment flows from land use activities that may, without proper management, cause erosion and excessive sediment movement. Routine source management activities prevent or mitigate excessive sediment introduced into waterways due to recreational use, roads and trails, grazing, farming, forestry and construction. Excessive flows affecting erosion and sedimentation may also result from land based events such as extreme fire incidents, high water volumes, wind, and other factors. Farmers, transportation, planners, and recreation professionals are all aware that soil loss is an economic as well as an environmental and safety problem. Even so, many homeowners and other stakeholders may not be aware of sediment issues unless their homes and neighborhood streets are damaged by mudslides or stream bank or lakeshore erosion.

A significant source of sediment is from urban run-off. Road construction and maintenance in or near streams can also be a source of sediment. Another transportation related source is Off-highway vehicle (OHV) use. OHV is a popular form of recreation in many places and local government agencies and private entities provide recreational areas for this purpose. These OHV recreation areas need to implement a range of storm water 'best management' practices to protect water quality. Additionally, unauthorized and unmanaged OHV areas can become erosion problems and discharge polluted storm water. With limited resources, maintaining and policing these areas can be a challenge.

Sedimentation can be a problem in the construction and operation of many mines. Increased potentials for erosion and sedimentation at mines are related to mine construction and facility location. Tailings dams, waste rock and spent ore storage piles, leach facilities, or other earthen structures are all potential sources of sedimentation to streams. Road construction, logging, and clearing of areas for buildings, mills, and process facilities can expose soils and increase the amount of surface runoff that reaches streams and other surface water bodies. These activities increase the potential for rill and inter-rill erosion and can increase peak stream flows, increasing the potential for channel erosion. Unusually high peak flows can erode stream banks, widen primary flow channels, erode bed materials, deepen and straighten stream channels, and alter channel grade (slope) as discussed above.

Some local governments (city and county) have also begun to support Low Impact Development (LID), including it as part of their planning and development ordinances. LID features design elements, including hydro-modification, that address sedimentation at the source. Resources, including model regulations, are available to help municipalities interested in incorporating sediment source management into their planning portfolios. From a human perspective, sediment has a dual nature—desirable in some locations and unwanted in others. Sediment can be used for many beneficial uses such as to create or restore beaches and other coastal habitats and to renew wetlands. Sediment is also needed to renew stream habitat. Spawning gravels for inland fisheries need replenishment, and fine sediment (often nutrient rich) is also needed to maintain, enhance, or restore good quality native riparian vegetation. Sediments can also be used for land reclamation and construction material. Such activities are referred to as beneficial uses. Excessive sediment, above natural loads, can cloud water, degrade wildlife habitat, form barriers to navigation, and reduce storage capacity in reservoirs for flood and water conservation. Contaminated sediment can contaminate the food chain for marine plants, animals, and humans.

Surface water with high sedimentation may negatively affect beneficial uses by increasing turbidity. Sediment affects sight-feeding predators in their ability to capture prey, clogs gills and filters of fish and aquatic invertebrates, covers and impairs fish spawning substrates, reduces survival of juvenile fish, reduces fishing success, and smothers bottom dwelling plants and animals. It may also physically alter streambed and lakebed habitat. Non point source (NPS) pollution is basically polluted runoff, which is diverse and each discharge may contribute only a small quantity of pollutants. Much of NPS pollution is sediment or pollutants carried by sediment. NPS pollution can be divided into the following six categories: (1) agriculture; (2) forestry; (3) urban areas; (4) marinas and recreational boating; (5) hydro-modification activities; and (6) wetlands, riparian areas, and vegetated treatment systems.

Nutrients (such as phosphorus) and toxic pollutants (contaminants such as trace metals and pesticides) are often associated with fine-grained sediment. In some cases suspended sediment particles increase growth of bacteria which can concentrate these nutrients from the water column. Toxic pollutants from storm water may also be absorbed onto sediments. Concentrated pollutants can greatly impair water quality if they are remobilized back into the environment.

Deposited sediment can reduce the hydraulic capacity of stream channels, causing an increase in flood crests and flood damage. It can fill drainage channels, especially along roads, plug culverts and storm drainage systems, and increase the frequency and cost of maintenance. Sedimentation can decrease the useful lifetime of a reservoir by reducing storage capacity. This loss in storage capacity affects the volume of stored water available for municipal supplies and the volume available for floodwater storage. Sedimentation of harbors and drainage systems results in higher maintenance costs and potential problems associated with disposal of removed material. The accumulation of sediment in recreational lakes affects boating activity in the shore zone, and can lead to demands for dredging to deepen marinas and channels. Whether sediment is desirable or not, its location and movement can have large economic and ecological consequences. For example, excess sediment in shipping channels may cost ports millions of dollars in delayed or limited ship access, while in other locations insufficient sediment deposits could result in the loss of valuable coastal wetlands.

Due to numerous factors, including geology, climate and population, sediment management issues vary significantly from one site to another. For that reason sediment management is best done on a watershed-wide scale. A major goal in this management is to try to make the watershed as stable as possible for sediment production (meaning to try to mimic natural sediment production, not to eliminate it). Watershed stability is determined by performing geomorphic assessments of the waterways within that watershed. Then, for the sediment that is produced, make efforts to use this sediment most beneficially throughout the watershed using a concept termed—regional sediment management (RSM) [Ffolliott and Brooks, this volume]. A stream that has excessive erosion and sedimentation may be determined by the Water Boards (or similar) to be unable to support its designated beneficial uses as it may exceed the Total Maximum Daily Load. In such cases there may be need to devise a remediation plan because excessive sediment in streams is the number one water quality problem in many countries. Many local agencies, for example, Water Quality Control Boards are working to reduce excessive sediment within streams when it occurs within their regions.

Some goals and objectives of sediment management are:

- Maintain stable watersheds
- Maintain infrastructure capacity
- Achieve stormwater permit requirements

Methods include supporting:

- Low Impact Development projects
- Clean Sediment (devise and enforce total maximum sediment load standards)
- Use of land use planning ordinances
- Devise and implement Regional Sediment Management Plans

7.1.2 Sediment Transport management

Sediment Transport management is the process of introducing or leveraging natural functions that create optimum sediment transport. This involves managing the speed and flow of the sediment conveyance and the natural or built structures to achieve a properly distributed balance of sediment types in the habitat. Properly managed transport of sediments will result in the best deposition of sediments. For example, sand bypass structures in flood control channels are starting to see some use. Such structures placed into flood channels allow the coarse-grained sediments to be diverted to a settling pond where they can be excavated and used for construction, while the fine-grained sediments are diverted to (say) a wetland where they add to the size of the wetland.

8 Summing Up and Conclusions

The aquatic and riverine environments of many river basins have developed over long period of time, during which they were subject to droughts and floods in recurring sequences. The drastic regulation of many river systems as now practiced, has reduced the frequency, magnitude and coverage of these natural events to the extent that many of the riverine and floodplain ecosystems have ceased to function, or at least have been radically altered. By developing the land we have changed its ability to absorb rainfall while at the same time allowed hundreds of thousands of people to live, work and play in the vicinity of rivers. How will we choose to accommodate the increasing population of the urban conurbations that crowd the floodplains of many major cities? This question lies at the heart of the choice we have to cover the soil with pavement or preserve its capacity to absorb the heavy rainfall that must come periodically and which can wreak havoc with transport and communication and damage people and their property.

In the developed world, slowly but surely, the combined weight of newly-educated public opinion, unavoidable evidence of degradation and world-wide awareness of environmental problems will combine to pressurize landholders, basin managers and water users into more environmentally sensitive methods of resource management and lifestyles. The future in those economies in transition to a market-oriented approach is more problematical. These economies face rapidly rising populations, over-commitment of existing water resources (surface and groundwater) and an economic imperative to generate GDP, which may lead to short-term planning horizons (Du et al., this volume; Xu et al., this volume). Bear in mind the impact that land-use practices can have on the hydrology, management and sustainability of natural resources and the longer-term economic status of various river basis as so well-illustrated by the case studies presented elsewhere in this book.

References and Further Reading

Arnalds, A. 2005. Approaches to landcare—A century of social conservation in Iceland. Land Degradation and Development 16(2): 113–125.

Cash, David.W, W. Neil Adger, Fikret Berkes, Po Garden, Louis Lebel\, Per Olsson, Lowell Pritchard and Oran Young. 2006. Scale and Cross-Scale Dynamics: Governance and Information in a Multilevel World. Ecology and Society 11(2): 8.

Costanza, R. 2012. Ecosystem functions and services. Ekonomia I Srodowisko (Economics and Environment) 2: 8–17.

Costanza, R. and M. Ruth. 1998. Using dynamic modeling to scope environmental problems and build consensus. Environmental Management 22: 183–195.

Cramb, R. 2007. Participation in community landcare groups in the Philippines: a social capital perspective. Australasian Journal of Environmental Management 14(2): 93–102.

Curtis, A. and M. Lockwood. 2000. Landcare and catchment management in Australia: lessons from state-sponsored community participation. Society and Natural Resources 13(1): 61–73.

Daniell, K.A. 2011. Enhancing collaborative management in the Basin. pp. 413–438. *In*: D. Connell and R.Q. Grafton (eds.). Basin Futures: Water Reform in the Murray-Darling Basin. ANU E-Press, Canberra, http://epress.anu.edu.au/basin_futures/pdf/ch26.pdf.

Du, Q., Z. Chao and S. Spooner. 2014. Sustainable Management of Water Resources in the Yellow River Basin: Main Issues and Legal Approaches. pp. 243–274. *In*: V.R. Squires, H.M. Milner and K.A. Daniell (eds.). River Basin Management in the Twenty-first Century: Understanding People and Place. CRC Press (this book).

Ffolliott, P. and K. Brooks. 2014. Integrated Watershed Management of Water and Other Natural Resources in River Basins of the United States. pp. 60–81. *In*: V.R. Squires, H.M. Milner and K.A. Daniell (eds.). River Basin Management in the Twenty-first Century: Understanding People and Place. CRC Press (this book).

Kingsford, R., J. Costelloe and F. Sheldon. 2014. Lake Eyre Basin—challenges for managing the world's most variable river system. pp. 346–370. *In*: V.R. Squires, H.M. Milner and K.A. Daniell (eds.). River Basin Management in the Twenty-first Century: Understanding People and Place. CRC Press (this book).

Krutov, A., S. Rahimov and A. Kamolidinov. 2014. Republic of Tajikistan: its role in the management of water resources in the Aral Sea Basin. pp. 325–345. *In*: V.R. Squires, H.M. Milner and K.A. Daniell (eds.). River Basin Management in the Twenty-first Century: Understanding People and Place. CRC Press (this book).

Lauber, T.B., D.J. Decker and B.A. Knuth. 2008. Social networks and community-based natural resource management. Environmental Management 42(4): 677–687.

Li, Y. and V. Squires. 2014. Integrated water resources management of inland river basins in the Hexi Corridor, Gansu, China with special reference to the Shule River Basin. pp. 82–100. *In*: V.R. Squires, H.M. Milner and K.A. Daniell (eds.). River Basin Management in the Twenty-first Century: Understanding People and Place. CRC Press (this book).

Loch, A., S. Wheeler and D. Adamson. 2014. People versus place in Australia's Murray-Darling Basin: balancing economic, social ecosystem and community outcomes. pp. 275–303. *In*: V.R. Squires, H.M. Milner and K.A. Daniell (eds.). River Basin Management in the Twenty-first Century: Understanding People and Place. CRC Press (this book).

Ludwig, J.A. and G. Bastin. 2008. Rangeland condition: its meaning and use: A Discussion Paper prepared for the Australian Collaborative Rangelands Information System (ACRIS)

Management Committee www.environment.gov.au/land/publications/acris/pubs/rangeland.
Mitchell, B. 2014. Addressing implementation deficits related to IWRM in Canada. pp. 42–59. *In*: V.R. Squires, H.M. Milner and K.A. Daniell (eds.). River Basin Management in the Twenty-first Century: Understanding People and Place. CRC Press (this book).
Soebels, J., A. Curtis and S. Lockie. 2001. The role of Landcare group networks in rural Australia: exploring the contribution of social capital. Journal of Rural Studies 17: 265–276.
Sullivan, C.A. 2014. Challenges for management of the Orange/Senqu River basin. pp. 189–215. *In*: V.R. Squires, H.M. Milner and K.A. Daniell (eds.). River Basin Management in the Twenty-first Century: Understanding People and Place. CRC Press (this book).
Thuo, S. and P. Riddell. 2014. Political economy versus comparative advantage in the Nile Basin: short term advantages or long term gain? pp. 216–240. *In*: V.R. Squires, H.M. Milner and K.A. Daniell (eds.). River Basin Management in the Twenty-first Century: Understanding People and Place. CRC Press (this book).
UNEP. 2006. Ecosystems and their services millennium ecosystem report http://www.unep.org/maweb/documents/document.300.aspx.pdf).
van den Belt, M. 2004. Mediated Modeling: A Systems Dynamics Approach to Environmental Consensus Building. Island Press, Washington, DC.
Voinov, A. and F. Bousquet. 2010. Modelling with stakeholders. Environmental Modelling & Software 25: 1268–1281.
Wenger, C. 2014. Sink or swim: alternative approaches to flood disaster reconstruction and mitigation pp. 418–445. *In*: V.R. Squires, H.M. Milner and K.A. Daniell (eds.). River Basin Management in the Twenty-first Century: Understanding People and Place. CRC Press (this book).
Xu, H., L. Sui, D. Zhang and Y. Li. 2014. The role of water users associations in integrated water resource management of Zhangye City in Heihe River Basin, China. pp. 304–324. *In*: V.R. Squires, H.M. Milner and K.A. Daniell (eds.). River Basin Management in the Twenty-first Century: Understanding People and Place. CRC Press (this book).

17

From Policy to Practice: Institutional Arrangements for Integrated River Basin Management in Nepal

Neera Shrestha Pradhan,[1,*] Suman Sijapati,[2] Umesh Parajuli[3] and Bhawani Shankar Dongol[4]

SYNOPSIS

Nepal is rich in water resources with more than 6,000 rivers and rivulets. This chapter will emphasise the shift from sectoral to integrated management of water resources in Nepal and the development of institutional arrangements to enable this process. We will also present the policies and governance practices proposed to realize an integrated approach to water management. The lessons learnt are outlined.

Keywords: China, climate change, drivers of change, glaciers, governance, Himalaya, holistic, India, legislation, NGO, operationalization, public participation, sectoral management, stakeholders

[1] GPO Box 3226, Kathmandu, Nepal.
Email: nspradhan@icimod.org
[2] P.O. Box 8975 EPC 1115, Kathmandu, Nepal.
Email: sijapati@wlink.com.np
[3] 132 Ichchu Nadi Marga, Baluwatar, Kathmandu, Nepal.
Email: unparajuli@gmail.com
[4] P.O. Box 7660, Kathmandu, Nepal.
Email: bhawani.dongol@wwfnepal.org
* Corresponding author

1 Introduction

Integrated Water Resources Management (IWRM) is a basin-wide approach of water resources management in a sustainable and balanced way considering social equity, environmental sustainability and economic efficiency (CDRI 2008; GWP and INBO 2009; Pollard and Toit 2011). It is increasingly seen as an important tool for sustainable use of water resources for development (CDRI 2008) acknowledging linkages in scale and across disciplines and adoption of iterative and adaptive approaches (Pollard and Toit 2008). For the successful implementation of IWRM, and sustainable water resources development and management, Global Water Partnership (GWP) has proposed "effective water governance" to be placed at different levels of society (Rogers and Hall 2003; K'Akumu 2007). It identifies accountability, participation, predictability and transparency as key elements in good governance. The Global Water Partnership defines water governance broadly as "the range of political, social, economic and administrative systems that are in place to regulate the development and management of water resources and provision of water services at different levels of society" (K'Akumu 2007). Asian Development Bank (ADB) describes governance as "promoting sound development management in which power is exercised in the management of a country's social and economic resources for development" (ADB 1995). Good governance is also considered as to comprise the rule of law, effective state institutions, transparency and accountability in the management of public affairs, respect for human rights, and the participation of all citizens in the decisions that affect their lives (CDRI 2008). According to Rogers and Hall (2003), creating an institutional and administrative framework is one of the key elements of governance within which various actors with different interests can discuss and agree to co-operate and coordinate their actions. According to Kaufmann et al. (2000), the most important and least studied aspects of institutional change is the context of governance in which water resource organizations operate. In this backdrop, this chapter looks into the role of institutions and governance structure for river basin management options, which are being implemented in Nepal.

Nepal is rich in water resources with more than 6,000 rivers and rivulets. The surface water available in the country is estimated to be about 225 bcm per annum, equivalent to an average flow of 7,125 m^3/s (WECS 2002). The total drainage area of these rivers is around 194,471 km^2, out of which 76% lies in Nepal. Around 79% of the average flow in the country is available in the four major basins (Mahakali, Karnali, Narayani and the Koshi), 8% in the medium basins (Babai, West Rapti, Bagmati, Kamala and Kankai) and 13% in the numerous small southern rivers of the Terai (Fig. 17.1).

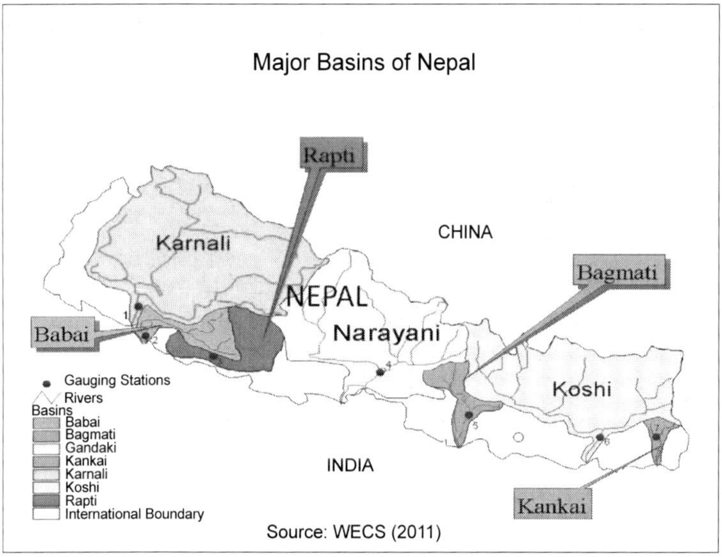

Figure 17.1. River System of Nepal.

Color image of this figure appears in the color plate section at the end of the book.

Out of the total population of 26.5 million (THT 2012) in Nepal, 42% of the population resides in the major basins, 17% in the medium basins and 41% in southern rivers of the Terai region. The basin-wise distribution of population and water availability has resulted in gross water surplus in some basins and acute deficit in others (Fig. 17.2) (WECS 2011).

In terms of time and space, the variation of water availability does not match the water supply due to its demand for domestic, industrial, irrigation, hydropower generation, environmental requirements, etc. They are interlinked with each other, but the traditional way of addressing the problem in a fragmented way is not efficient. While the sectoral approach has its own advantages, it has become necessary to look into broader and integrated approaches of resource management. These problems can be effectively dealt with by collaborative effort of stakeholders and interdisciplinary technical inputs (Abdullah and Christensen 2004), which calls for the integrated approach of water management (Jønch-Clausen and Fugl 2001). The management of water resources in Nepal is guided by several sectoral policies. Thus, it has often led to governing bodies with conflicting interests. Considering the integrated management of water resources, the fundamental question is, how should integrated management of water resources be implemented and which arrangements can be made to bring policy into practice? (Jaspers 2003). Therefore, an urgent call is foreseen to depart from the sectoral management of water

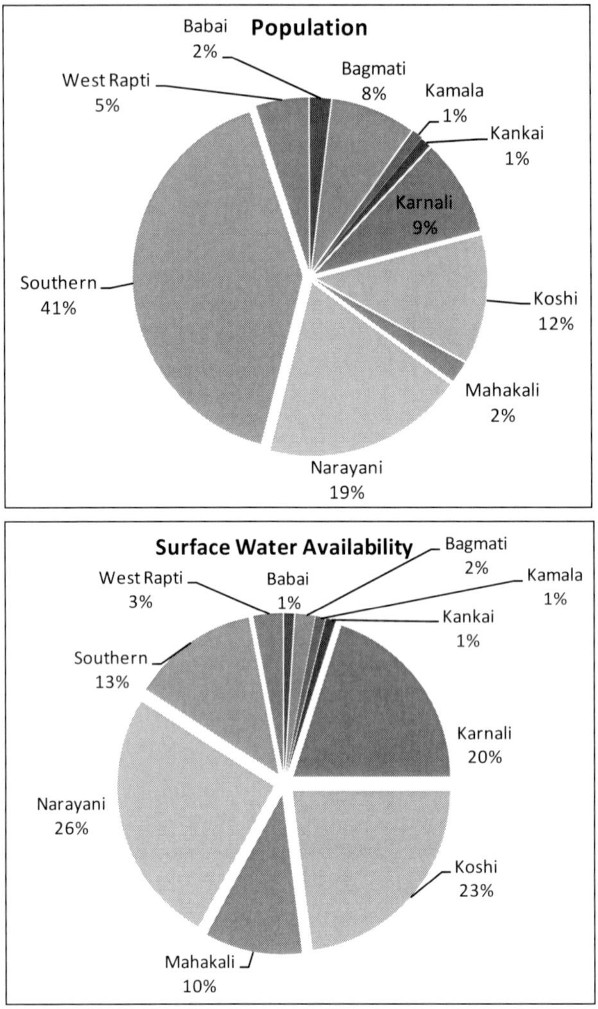

Figure 17.2. Distribution of water availability and population in the various basins of Nepal (WECS 2011).

resources and adopt integrated river basin management. This requires the institutional arrangements to enable communities to implement higher level of integration in water resources management (Jaspers 2003).

This chapter will emphasise the shift from sectoral to integrated management of water resources in Nepal and the development of institutional arrangements to enable this process. The chapter will also present the policies and governance practices to realize departing from water management in isolation to integrated approach and lessons learnt.

2 Terms and Definitions

In this chapter, a number of terms will be used, that have a variety of meanings in discussions, proceedings, publications and other communications. In order to provide streamlined discussions and ensure the same level of understanding, a number of the most important and frequently used terms are described below.

Institutions: Humanly created formal and informal mechanisms that shape social and individual expectations, interactions, and behavior (Ostrom 1990 as cited in (Agrawal 2008)). For this chapter, we will briefly mention about the policies and laws but focus on administration, which includes organizations at policy level for resources management and organizations at implementation level for delivery management (Bandaragoda 2002).

Integrated river basin management: Management of all surface and subsurface water resources of the river basin in its entirety with due attention to water quality, water quantity and environmental integrity. A participatory approach is followed, focusing on the integration of natural limitations with all social, economic and environmental interests (Jaspers 2003).

Integrated water resources management approach: Participatory approach to manage and develop water resources in a sustainable and balanced way, taking account of social, economic and environmental interests (GWP and INBO 2009).

Co-management: Sharing of power and responsibility between the government and local resource users, whereby arrangements are made to look beyond government, and towards public–private–civil society partnerships, as a way of dealing with the shortcomings (Berkes 2009).

3 Legal Framework

Since ancient times, natural resources (including water) in Nepal were mainly being managed by the local communities. It was only from the 1950s, that the government of Nepal focused on planning for the utilization of natural resources. Nepal's water resources management is guided by a number of sectoral policies focused on different use of water. Some such policies are: National Agriculture Policy 2004; Irrigation Policy 2003; Hydropower Development Policy 2001; Master Plan for the Forestry Sector 1989 and National Water Supply and Sector Policy 1998. These policies are being implemented by respective sub-sector government agencies, which are still tailored to sectoral water development rather than integration, coordination, and decentralization. In many cases, the sectoral agencies have multiple roles and they perform in parallel as planners, implementers, monitors and regulators. These multiple roles for agencies are contrary to current best practice which aims for separation of roles for transparency and efficiency. Other acts that also control management of water are: Electricity Act—1992, Environmental Protection Act—1997, Local Self Governance Act—1999, Aquatic Animal Protection Act—1960, Natural Disaster Relief Act—1982, and Country Codes (Muluki Ain).

The Water Resource Act (WRA) (1992) was issued by the Government of Nepal for the management of Water Resources that focused on creating ownership and the utilization of all the water resources of Nepal. WRA regulates water through a system of permits. District Water Resources Committee (DWRC) is empowered for regulating water allocation and distribution. But DWRC is under resourced and operates without useful assessments of water availability and on-going commitments. The WRA and related regulations also provide mechanisms for water users associations.

The Water Resources Strategy (WRS) (2002), stipulates that the development and management of the water resources shall be in a holistic and systematic manner relying on IWRM. It also embodies the policy principles of social equity, environmental sustainability and economy efficiency in any ongoing water resources development project and has laid out several principles, of which the following two principles are the key for the integrated management of water resources:

- Development and management of water resources shall be undertaken in a holistic and systematic manner, relying on integrated water resources management.
- Water utilization shall be sustainable to ensure conservation of the resources and protection of the environment. Each river basin shall be managed holistically.

At present, the global debate of water management is driven by various drivers of change including climate change. Change represents a new dimension of stress on water resources and has significant environmental, social and economic effects. The risks associated with climate change are real but highly uncertain, and the situation worsens to those parts of societies dependent on resources that are sensitive to changes in climate. Agriculture, fisheries and other options of rural livelihood populations in developing countries (including Nepal) tend to be more vulnerable to these changes (Mlote et al. 2002; Adger et al. 2003). Impacts of climate change are already measurable, with annual mean temperatures rising by an average of 0.06°C per year, and by even more in the highlands, where the glaciers are retreating rapidly. Uncertainty of future impacts creates pressure as the warming trends accelerate leading to increasing flood risks and ultimately water scarcity (Shrestha and Aryal 2011).

In order to address the climate change impacts, the Government of Nepal has developed the Climate Change Policy in 2011, which reflects the country's vision in dealing with and adapting to, the adverse impacts brought about by climate change across the different sectors and goes on to suggest the establishment of a Climate Change Centre. The National Adaptation Plan of Action (NAPA) document, released in 2010 is based on a rigorous consultation process with multiple stakeholders and expert groups, is considered as the first comprehensive and climate change dedicated government document. In order to mainstream climate change activities under NAPA and the risk management and development activities at national, sectoral and local planning process, it was realized that institutional mechanisms need to be developed.

Therefore, for the implementation of NAPA, a Local Adaptation Plan of Action (LAPA) framework was envisioned aiming as an interface of development process and adaptation activity at local level. The LAPA Framework (Fig. 17.3) supports the operationalization of the policy objectives in NAPA, the National Climate Change Policy and Climate Resilience Planning by facilitating the integration of climate change resilience into local-to-national development planning processes and outcomes (GoN 2011). In 2010, LAPA was designed and piloted in 10 districts across Nepal (Achham, Ilam, Udaypur, Dhadhing, Kaski, Kapilvastu, Pyuthan, NawalparasiRukum and Kalikot) by the Climate Adaptation Design and Piloting-Nepal Project (CADP-N). Currently the LAPA framework is being tested in selected districts in Nepal to enhance cross-sectoral coordination to build climate resilience.

For integrating climate change resilience into local-to-national development planning processes and outcomes, the VDC and the Municipality have been identified as the most appropriate unit. These Committees are responsible for planning, integration of the LAPA into sectoral and development planning processes, coordination, and service delivery and monitoring and evaluation. To ensure the delivery of adaptation services in a timely and effective manner, capacity building of these local bodies would be necessary.

4 Nepal National Water Plan 2005 and Clear Vision

The Government of Nepal has developed the National Water Plan (NWP) in 2005 (WECS 2005) through a series of stakeholders' consultations for the implementation of WRS. It adopted the following four doctrines as guiding principles.

- Integration (sectoral integration on water uses, integration on water policies and institutions)

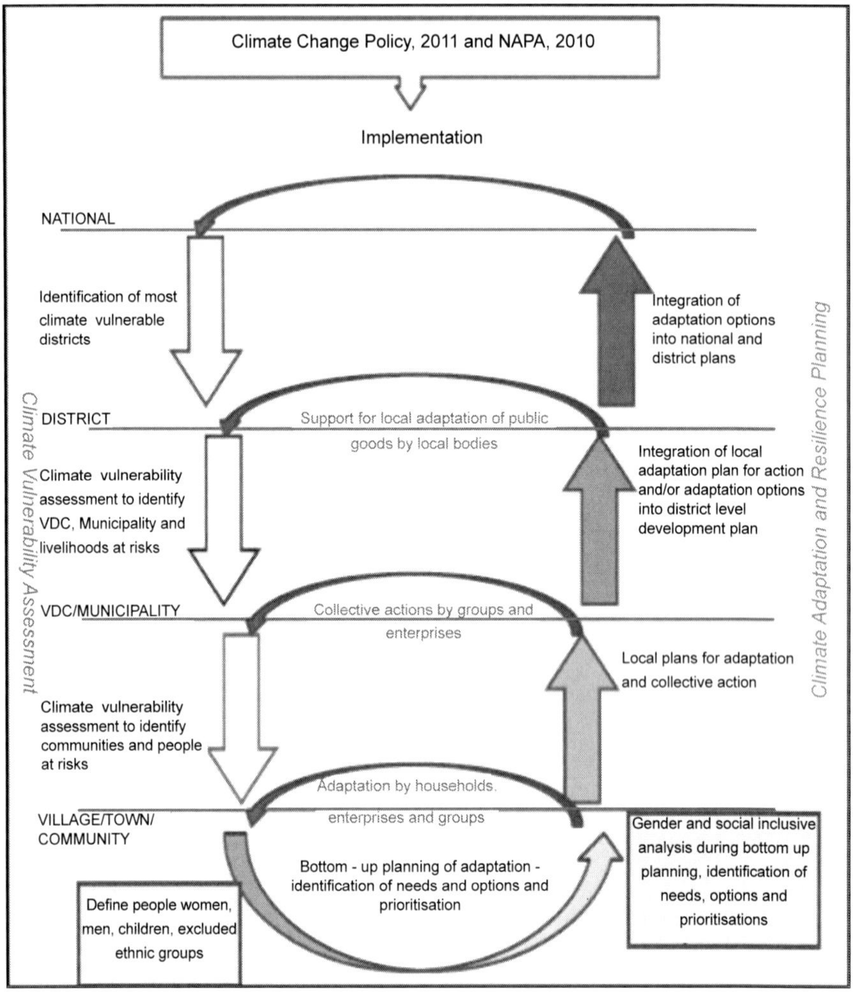

Figure 17.3. LAPA Planning and Implementation Framework (GoN 2011).

- Coordination (coordination between several Stakeholders)
- Decentralization (empowering users)
- Participation (stakeholders in decision making)

Preparation of NWP was challenging as the development issues of the country were getting increasingly complex due to the shift in their paradigms towards human, societal and environmental aspects from the traditional economic growth models. Also, governance issues were becoming more pronounced as demands for popular participation grew. The NWP recognizes these broader development needs; thus included them as critical factors in the policy framework for achieving targeted national goals.

IWRM was adopted as one of the principal themes of the NWP. IWRM principles profess that water must be viewed from a holistic perspective, both in its natural state and in balancing the competing demands on it, e.g., domestic, agriculture, hydropower, industrial, cultural and environmental. Hence, NWP emphasised that management of water resources services needs to reflect the interactions between different demands and so, must be coordinated within and across the sectors. It stresses that more equitable, efficient and sustainable regime will emerge if the crosscutting requisites are met along with horizontal and vertical integration within the management framework of the water resources and services.

The NWP is guided by the following policy principles:

- Development and management of water resources shall be undertaken in a holistic and systematic manner, relying on the principles of IWRM;
- Water utilization shall be sustainable to ensure the conservation of the resource and protection of the environment. Each river basin system shall be managed holistically;
- Delivery of water services shall be decentralized in a manner that involves autonomous and accountable agencies (e.g., public, private, community and user-based agencies);
- Economic efficiency and social equity shall guide water resources development and management;
- Participation of, and consultation with, all stakeholders shall constitute the basis of water sector development and management;
- Sharing of water resources benefits among the riparian countries shall be on an equitable basis for mutual benefit;
- Institutional and legal frameworks for coordination and transparency shall be an essential feature of water sector management; and
- Wider adoption of the best existing technologies and practices, and rapid innovation and adaptation of both institutional arrangements and new technologies shall be ensured.

With the national water sector goal of 'significantly improving the living conditions of the Nepali people in a sustainable manner' by the end of the plan, NWP has spelled out the activities for short-term (by 2007), medium-term (by 2017) and long-term (by 2027) stressing the efficient but sustainable use of the resource. Hence, the NWP is still effective and, it terms of time, the nation is currently about mid-way through the planned period.

Even though the NWP is fairly comprehensive document and the activities are well laid out, its progress in terms of implementation has not been very satisfactory. An assessment carried out for the irrigation sub-sector after the short-term timeframe (2007) has clearly revealed significant gaps exists between the targets set, proposed activities, actually implemented activities and the achievements (Sijapati 2008).

5 Institutional Arrangements

As climate change impacts continue to shift the availability of water resources across the Asian continent, the need for new and innovative water governance structures and mechanisms becomes more pressing (Godden et al. 2011). Jønch-Clausen and Fugl (2001) emphasized that most of the water crises are mainly due to management and governance crises. Hence, there is a need to find integrated approaches to coordinate policy-making, planning and implementation at across sectoral, institutional, professional and basin boundaries.

River basin management emphasises the use of hydrological boundaries, using basin/sub-basin/catchment as the basic water management unit (Jønch-Clausen and Fugl 2001; Jaspers 2003). Likewise, the NWP also considers the river basin as a planning unit and a sub-basin (or a watershed) as the lowest planning unit. Geographically, the boundary of sub-basin does not coincide with the boundary of districts in Nepal. A sub-basin may have one or more districts and likewise, a district may have two or more sub-basins within its boundary considering the natural flow of water in the watershed. The governance structure for water resources management is envisioned as top-down and bottom up as per the NWP. It envisages formation of river basin organization at the national level and sub-basin committees at local level for management of water resources. The sub-basin committee would have representation from all sectoral water users associations, womens' groups, other resource based user groups of the area, and the Village Development Committee (VDC). The roles of the sub-basin committee are to assess water demand in time and space for all users in the sub-basin, prepare and implement integrated water resources management program, maintain a database of the sub-basin, and assist sectoral water users in managing their share of water.

At the district level, the NWP envisages re-structuring the existing District Water Resources Committee (DWRC) for facilitating water resources management therein. The new DWRC would have representation from all sub-basin committee, and the chairman of the District Development Committee (DDC) would lead the DWRC. The DWRC is mandated with regulatory and facilitating roles. If a basin is constituted of two or more districts, NWP envisages formation of an inter district water resource committee (Fig. 17.4).

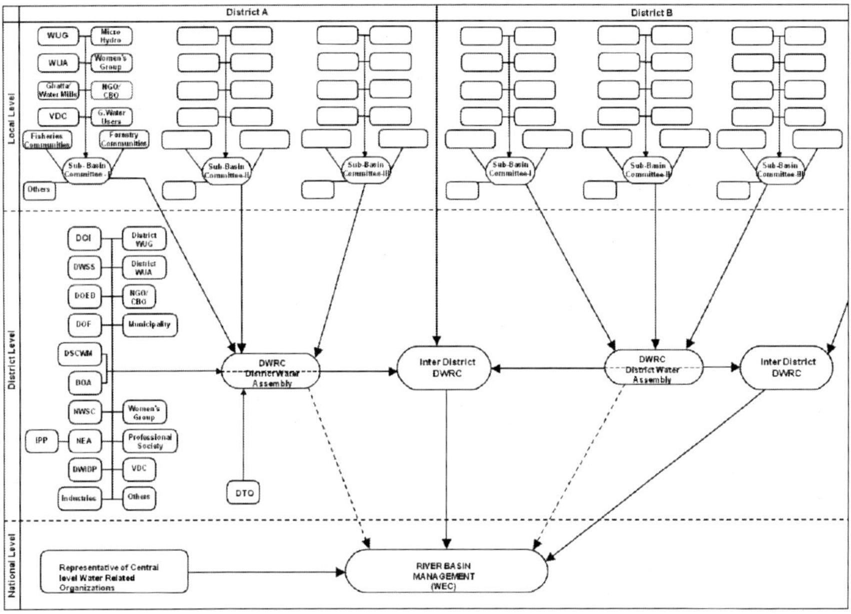

Figure 17.4. Institutional arrangement for River Basin Management as per NWP (WECS 2005).

6 Management Practices

The National Water Plan (2005) highlighted the need for IWRM and IRBM with a well defined framework for its operationalization at different levels of administration. The history of conducting studies on integrated management of water resources within hydrological units (river basins/sub-basins) in Nepal is observed to have started from the late eighties. The first of such endeavours can be mentioned as the collaborative action research carried out by Water and Energy Commission Secretariat (WECS) and IWMI Nepal from 1986–89 at Indrawati sub-basin (Yoder 1991). The study was funded by Ford Foundation and was mainly aimed at designing effective interventions for the development of water resources in the sub-basin.

The next study was on East Rapti sub-basin and was also carried out by IWMI from 1996 to 1999. The study was also mainly focused on carrying out water accounting of the basin and in assessing existing resource base and institutional arrangements (IWMI 2000). A similar study of Manusmara sub-basin, one of the sub-basins of Bagmati Basin, was also carried out by the Department of Irrigation (DOI) from 1999 to 2001 (Sijapati 2001). As part of the Comprehensive Assessment, another study was carried out by IWMI in collaboration with Stockholm Environmental Institute (SEI) from 2001 to 2003 in Begnas Sub-basin (Pant 2003). All these studies have highlighted the need of integrated planning of available water resources in the respective basin for optimizing the benefits.

Institutionally organized efforts in establishing river basin based management were commenced in Nepal only after the formulation of the National Water Plan (NWP) in 2005. WECS has been observed as the main agency working in the direction. Even though what has been achieved is far behind what has been stipulated in the NWP, activities are currently being carrying out in two sub-basins of Koshi Basin (namely, Dudh Koshi and Indrawati) and two medium size river basins in the western part of Nepal, viz. Babai and West Rapti.

The IRBM approach as described below, tries to address the co-management approach, which involves public participation and problem solving at the lowest level of organization (Berkes 2009). As per Ostrom (2005), the cases also discuss identification of appropriate local institutions and building on their strengths, or establishing new ones where existing ones do not work or are not appropriate. However, the sustainability of the institutions depends on the support provided by the Government organizations, Non Governmental Organization (NGOs) and other organizations to establish and implement the activities at local level together with local organization. Several pilot projects implemented with local water users and institutions as key stakeholders are analyzed for three major river basin management systems.

6.1 Babai and West Rapti Basin Management

Babai and West Rapti river systems are the two key medium size river basins located in mid-western region of Nepal. They originate from Mahabharat range and drain south towards India. Babai has a catchment area of 3,400 sq. km that falls in three districts: Salyan, Dang and Bardiya while West Rapti drains 6,300 sq. km that touches Rolpa, Pyuthan, Arghakhachi, Dand and Banke districts. About 65% of the catchment of these two basins is covered by natural vegetation and about 25% is cultivated land. As per the 2011 census, the population in Babai and West Rapti basins are 665,739 and 800,393 respectively and the annual population growth rate is about 1.5%.

In these river basins, WECS had conducted an action research together with Consolidated Management Services (CMS) Nepal on stakeholders' role in management of available resources. The bio-physical, geographical, natural, socio-economic and institutional aspects of resource availability were studied. Delineation of the sub-basins and their catchments were done to develop a better understanding of the details of the basin. Through rigorous stakeholders' consultation, the basin and sub-basin committees were established and Water Allocation and Access Mechanism (WAAM) were envisioned.

The stakeholders were divided into internal and external stakeholders. The internal stakeholders were the users within the basin who contribute to the labor required for construction of facilities and management of water and other natural resources. The external stakeholders were those who support the users by providing funds for infrastructure development and facilitate in resource mobilization through users' participation. After the formation of sub-basin committees, the basin committee was formed through consultation with the executive members of the sub-basin committees in Babai and West Rapti rivers.

Similarly, GIS analysis was carried out to study water availability, its use and the existing water allocation and access scenario at the basin/sub-basin level. Moreover, case studies focusing on resources, environment, socio-economic, institutional and technical aspects were carried out to prepare a framework for WAAM of the respective basins.

Although the institutional arrangements for the Babai and West Rapti River had been established with thorough consultation at local level and scientific analysis, the sustainability of the institutes are in question due to the lack of funds and fund raising mechanisms. The basin and sub-basin committees are yet to be formalized to function as a legal body to execute development works along with government agencies.

6.2 Koshi River Basin Management

Koshi River Basin is one of the three major river basins of Nepal with seven major sub-basins. Lying in the Himalayan region, it is a transnational river and flows across the international boundaries of China, Nepal and India. Koshi basin is very significant as it is a source of water and livelihoods of the millions of people in the basin and in the downstream. For the integrated management of Koshi river basin in Nepal, WECS with support from WWF Nepal initiated Koshi River Basin Management (KRBM) program. It is tasked with making wise use of Koshi Basin's water and related resources to promote socio-economic development for the benefit of people in the basin, while maintaining the ecological balance. A KRBM cell has been established at WECS to strategically manage the basin/sub-basin/catchment level activities.

To operationalize the IRBM approach, the institutional arrangement for KRBM was formulated based on the governance structure proposed by NWP and modified considering the local situation. In order to finalize the governance structure for KRBM, a series of stakeholder's consultations were held at local, regional and national level, where the institutional mechanism was presented and widely discussed. The stakeholders' consultation came out with the structure and implementation modality of institutions from catchment to River basin level (Fig. 17.5). This included institutions at four levels: catchment, district, sub-basin and river basin.

At catchment levels, Integrated Resource Management Committees (IRMC) were formed. District Water Resource Committee (DWRC) existed in each district under the umbrella of District Development Committee (DDC), a local development body of the Government. Depending on the number of catchment and number of districts on one sub-basin, there might be more than one IRMC or DWRC in one sub-basin. At the sub-basin and river basin level, Sub-Basin Management Committee (SBMC) and River Basin Organization (RBO) were established respectively. As guided by the National Water Plan, it was envisioned that these institutions deal at their own levels for IWRM/IRBM planning and implementation.

The major role of IRMC is to design, plan, coordinate, implement and monitor the IWRM approach at the catchment level. IRMCs are responsible for collating and disseminating catchment level data and information and passing to the DWRCs and other stakeholders as necessary. It also acts as the coordinating agency between local beneficiaries and the sub-basin committee, and also the local stakeholders. It will facilitate coordination with the local government authorities and line agencies such as DDCs. It

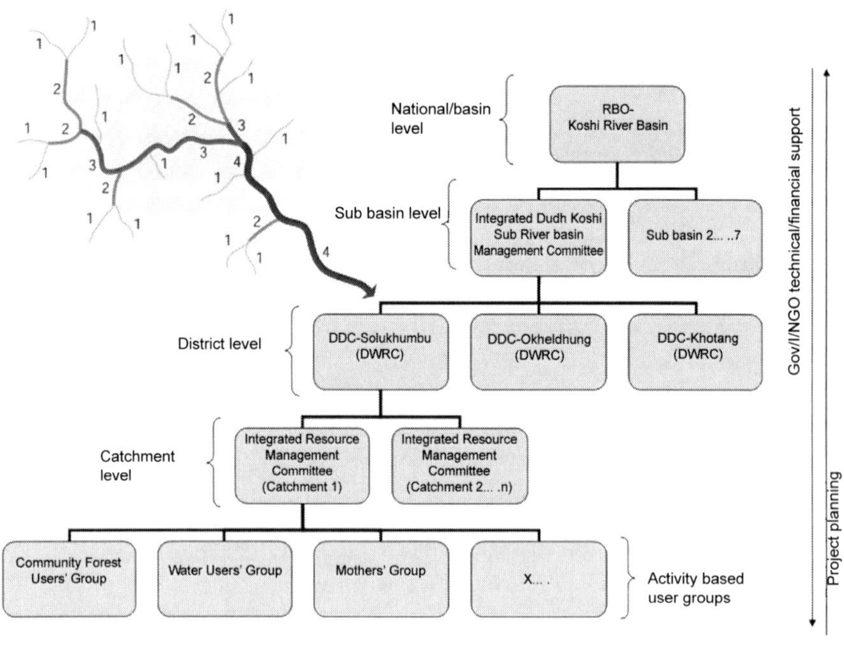

Figure 17.5. Institutional Mechanism of Koshi River Basin Management (WWF 2009).

will also be involved in awareness raising and capacity building activities at the catchment level.

SBMC exists at the sub-basins to look after overall coordination between different IRMCs, district line agencies such as DDCs and other relevant organizations to facilitate IWRM implementation. It functions as an interface between IRMCs, DWRCs and River basin office including information sharing. It helps IRMCs to coordinate and leverage funds from possible sources. It also acts as an agency to implement sub-basin level IWRM project activities and review the programs and progresses regularly. In this way, SBMC establishes vertical linkage with RBO and the IRMCs, and also establishes horizontal linkage with DWRCs.

RBO at the basin level are responsible for overall coordination, management of information, generation of resources and technical backstopping including capacity building to SBMCs and IRMCs. It formulates policies and strategies as necessary. RBO facilitates coordination between different stakeholders and agencies at river basin level and helps mainstream water issues/plans into sectoral issues/plans and vice versa. It facilitates sharing of information between different stakeholders and coordinates with other projects at basin level. RBO also acts as an agency to implement river basin level projects.

Based on the outcomes of the consultations, pilots of the river basin approach were initiated in two catchments in Dudhkoshi sub-basin, and were up scaled in nine catchments of Indrawati sub-basin, two sub-basins of Koshi basin. IRMCs were formed and established in 11 catchments: two in Dudhkoshi and nine in Indrawati sub-basins, and they are functional. The IRMC was composed of representatives from the activity level user groups such as: water user groups, irrigation user groups, farmer's group, community forest user groups, womens' groups and other appropriate groups from different VDCs and Wards from within a particular catchment. Ward is the smallest unit of VDC and VDCs will add up to the District. There is Ward level and VDC level sub committees from the representatives of different user groups, which will nominate the executive members of IRMC in each of the selected catchments. In addition representation from geographical coverage (upper, middle and downstream areas) was also duly considered while forming IRMCs. IRMCs once formed they were registered in respective District Administration Offices (DAO). They are provided orientation on IWRM, institutional supports to function as an institution and other capacity building trainings as necessary.

With the ongoing pilot project led by WECS and supported by WWF, a participatory planning process was initiated by IRMC for the management of water and related natural resources in each of the catchments. Once planning was done, the IRMC members and Project conducted a joint feasibility survey. Overall, the project works in close cooperation with the

IRMC, which acts as a guide in identifying needs, selection of locations and as clearing houses in allocating/disbursing funds to the individual villages. Interacting closely with the IRMC members, the Project's field-based social mobilizers work with local communities for the interventions, monitor and regularly update on the progress, and report back to field office. The Project also conducts regular monitoring and supervision in the Project areas. In this way, there is a set structure through which information and funds can flow. The project provides technical and financial support to local users and IRMCs for the planning and implementation of the IWRM interventions in the selected catchments. It also helps IRMCs coordinate with the district government line agencies to leverage funds and technical expertise.

The Dudh Koshi sub-basin project was successful in building the capacity of IRMCs in the Siku and Panku Khola (two intervention catchments). IRMCs (Integrated Resource Management Committees) at catchment level are implementing IWRM. The Siku and Panku Khola IRMC prepared their plans and approached respective DDCs (District Development Committees) for securing budget for its implementation (scaling-up of activities). They were able to get funds from DDCs and some other line agencies too. Influenced by the success of the IRMCs in Siku and Pankhu catchments, the communities of Solukhola in Dudh Koshi sub-basin formed IRMC and is under the process of registration to function as a legal institution. The Indrawati sub-basin project supported 9 IRMCs and built the capacity of these committees to prepare and implement their IWRM plans. With a seed fund support to the catchments, the IRMCs were successful in planning and implementation of activities (scaling-out with project support). IRMCs in Indrawati were strengthened and capacitated to leverage funds from local government line agencies for the implementation of IWRM plan.

6.3 Begnas Basin Management

The Begnas Sub-basin which is the micro-basin of the Seti River, lies in the country's middle mountain region and has an area of 34.06 sq. km. Physiographically, this sub-basin is divided into two distinct areas namely the upper watershed and the lower valley floor with the Begnas Lake at their interface. The upper watershed is mountainous with 28.5 per cent of its area covered by community forests. Several small streams flow down the hill through its gullies, many of which merge to form a single stream that ultimately drains into the Begnas Lake. Waters from these streams are used for irrigation and domestic consumption through series of open channels and buried pipes. Within this sub-basin, it is primarily the upper watershed which has considerable sources of waters (springs, streams and water catchment) that are collected in the Begnas Lake for the use by people residing in the valley floor.

An action research project led by IWMI on "Resource Management for Sustainable Livelihoods" was conducted between 2005 and 2008 in the upper catchment in Nepal (Begnas Sub-basin) and India with an objective to examine expanded mandates for Community Based Integrated Natural Resources Management (CBINRM) institutions by strengthening users' roles and linkages with external resources leading to integrated water resources management and sub-basin level planning. In this context, an integrated resources management platform was visualized and formed where resource users, managers and other stakeholders were brought together to discuss issues and opportunities of resources management leading to integrated management of natural resources.

Several organizations (formal and informal) were active in managing waters and other natural resources of the Begnas Sub-basin. These include irrigation water users groups (traditional and formal canals), boaters' association, fishermen's group, mothers' groups, youth clubs, several community forestry users groups, Begnas municipality, and village/district development committees. However, no formal relations existed among these organizations and all of them worked independently resulting in conflict in managing water resources. For example, during long dry spells; irrigation groups wanted to release more water from the lake to irrigate their dry lands, whereas, the fishermen's associations objected as it lowers the water level in the reservoir threatening the fishing activity. In order to address these issues, IRMC was formed at the level of Begnas Sub-basin. Firstly, the available resources of the Begnas Sub-basin and livelihood strategy of the people were assessed to understand linkages between resources and their use pattern. This assessment suggested that there are opportunities for integration of resources management activities for their optimum utilization. Secondly, consensus building among resources users and managers for implementation of IWRM were conducted. In this process, mechanism of forming a general assembly of proposed IRMC was also discussed. It included Water Users' Association (WUA) of the Begnas Irrigation System (BIS), Community Forest Groups, Water Users Group of Farmer Managed Irrigation System (FMISs), Fishermen's Associations, Boat Owners' Association, Fishery Development Board, Government line agencies (Irrigation, Forestry, Agriculture, Livestock, Fisheries, etc.), Local NGOs like Youth Clubs, Mothers' Groups, Eco-tourism Committee, World Vision, IDE, SORUP, Local governments like Begnas Municipality, and Kaski DDC, and other key stakeholders of the basin area (upper and lower watershed) representing different categories (well being, ethnicity, gender, etc.). Thirdly, a General Assembly (GA) for IRMC was formed with representative from all the above organizations depending on the extent of their association with the Begnas Sub-basin, especially from the perspective of livelihood and management responsibility with higher representation from irrigation water users' community, fisheries community, boaters' community, etc. Lastly, a participatory action plan for Begnas Sub-basin management was developed in two stages—prepared by the ad hoc executive committee, and presented to the GA where it was discussed and approved.

The action research found that the IRMC became involved only during the planning phase, but could not implement the plan successfully. Though the IRMC was registered at District level local government, it could not get administrative mandate to implement the management plan due to lack of legal tools. It was envisaged that with the seed support from the project, IRMC would be enabled to raise funds and function sustainably on its own resources. However this was not a success and the IRMC is virtually dysfunctional.

7 Challenges and Lessons Learnt

IRBM is a new approach for water resources management in Nepal. After the endorsement of the NWP 2005, there have been efforts to implement the IRBM in various river basins. The three cases presented in this chapter analyzed the co-management approach, where the government, non-government organizations and local community organizations work together for IRBM in various scenarios of institutional and financial support. One of the commonalities was the participatory approach taken to establish river basin/sub-basin committees. Based on the clear institutional structure given by the NWP for the implementation of IRBM, all the initiatives considered the local situation analysis and voice of stakeholders to modify the institutional arrangements to suit the local context.

From the policy perspective, though the Water Resources Strategy Nepal and the National Water Plan have been approved by the government in 2002 and 2005 respectively, they are not yet fully implemented as the existing sectoral water policies, legal instruments and institutional arrangements are yet to be harmonized with the NWP provisions. So, WRSN and NWP are regarded as a guiding document for approaching towards IWRM.

The three cases also pointed out the challenges regarding the administrative/political boundaries. Until now, the local plans were prepared by the district, VDCs and wards, except for some pilot basins where integrated plans were developed. As a river basin/sub-basin includes one or more districts, this has challenges in terms of coordination between different administrative boundary based organizations with river basin/sub-basin based management. Unless the legal identity/mandate is clear for IRMC, SBMC and/or RBOs, it will be difficult to streamline IRBM approach in the development planning.

One of the major challenges faced by the river basin management initiatives was the issue of sustainability after the completion of the project and/or support for the implementation primarily due to lack of financial support and ownership by the local communities and also lack of legal support. Babai and West Rapti river basin management initiated by WECS established the governance structure on the ground but due to lack of financial support to implement the activities, the IRBC was defunct. Similarly, the Begnas Management committee established by IWMI was also dysfunctional due to lack of funding after the completion of the project. By contrast the Koshi river basin management initiative led by WECS and supported by WWF Nepal has projects in Dudh Koshi and Indrawati sub-basins. The project support for the coming few years is sufficient for functional IRMCs and enabled the IRMC members to leverage funds from DDC. The IRMCs have showcased the implementation modality (institutional mechanism) at the catchment level.

8 Conclusion

Integrated river basin management approach is yet a new approach in Nepal and could be an example for the management of water and related natural resources that may lead to sustainable and economic development. Until now, the sectoral development activities have bearing on the livelihood of rural community to address the issues on low agricultural productivity, high population growth rates, livelihood enhancement of marginalized section of the community and an increasing rate of urbanization. To achieve more effective IRBM, the key is to adopt the co-management approach, where the public sector with support from private or civic sectors enables existing and/or new local level institutes to prepare and implement the management plan in a sustainable manner. This includes involving key stakeholders from the planning phase; creating an enabling environment which facilitates efficient implementation of management plans; and providing coherent legal framework with a strong governance mechanism and clear roles and responsibilities of related institutions.

In case of Nepal, the Climate Change related policies are new and some of them have just initiated as a legislative framework. The implementation mechanism as field tested in few basins/sub-basins needs to be analyzed by the government and considered for the review of NWP. The major thrust needs to be given to key questions like "What would be the legal status of the river basin/sub-basin committees and what should be the mechanism for forming of these committees? What role will they have in formulation and implementation of District Development Plan? How to enhance collaboration between different user groups for improved livelihood?" Against the backdrop of NWP and the large number of sectoral policies and legislations, an appropriate institutional and legal structure for coordinated and integrated water resources management needs to be practiced.

9 Acknowledgement

The authors would like to thank Victor Squires and Hugh Milner for providing the opportunity to contribute in writing the book chapter and timely feedback at various stages during the writing. We are thankful to Dr Rucha Ghate, Senior Governance Specialist at ICIMOD for reviewing the book chapter and providing valuable inputs. The authors would also like to acknowledge the respective organizations where they were/are involved in their work assignment whose experiences were drawn in this writing.

References

Abdullah, D.I.H.K. and B. Christensen. 2004. Integrated River Basin Management. Buletinningenieur 21–23.

ADB. 1995. Governance: Sound Development Management.

Adger, W.N., S. Huq, et al. 2003. Adaptation to climate change in the developing world. Progress in Development Studies 3(3): 179–195.

Agrawal, A. 2008. The Role of Local Institutions in Adaptation to Climate Change, Social Dimensions of Climate Change, Social Development Department, The World Bank, Washington DC.

Bandaragoda, D.J. 2002. A Framework for Institutional Analysis for Water-Resources Management in a River Basin Context. Regional Workshop on Integrated Water-Resources Management in a River-Basin Context: Institutional Strategies for Improving the Productivity of Agricultural Water Management. Bryan Bruns, D.J. Bandaragoda and M. Samad (eds.). Malang, Indonesia, January 15–19, 2001, IWMI.

Berkes, F. 2009. Evolution of co-management: role of knowledge generation, bridging organizations and social learning. J. Environ. Manage 90(5): 1692–1702.

CDRI. 2008. Framing Research on Water Resources Management and Governance in Cambodia: A Literature Review.

Godden, L., R.L. Ison, et al. 2011. Water governance in a climate change world: appraising systemic and adaptive effectiveness. Water Resources Management 25(15): 3971–3976.

GoN. 2011. National Framework on Local Adaptation Plans for Action. Government of Nepal, Ministry of Science Technology and Environment, Kathmandu, Nepal.

GWP and INBO. 2009. A Handbook for Integrated Water Resources Management in Basins.

Jaspers, F.G.W. 2003. Institutional arrangements for integrated river basin management. Water Policy 5: 77–90.

Jønch-Clausen, T. and J. Fugl. 2001. Firming up the conceptual basis of integrated water resources management. International Journal of Water Resources Development 17(4): 501–510.

K'Akumu, O.A. 2007. Toward effective governance of water services in Kenya. Water Policy 9(5): 529.

Kaufmann, D., A. Kraay, et al. 2000. Governance matters: from measurement to action. Finance & Development 10–13.

Mlote, S.D.M., C. Sullivan, et al. 2002. Water poverty index: a tool for integrated water management. 3rd WaterNet/Warfsa Symposium Water Demand Management for Sustainable Development, Dar es Salaam, 30–31 October 2002.

Ostrom, E. 2005. Understanding Institutional Diversity. Princeton University Press.

Pant, D., S. Thapa, A. Singh, M. Bhattarai and D. Molden. 2003. Integrated Management of Water, Forest and Land Resources in Nepal: Opportunities for Improved Livelihood. IWMI Comprehensive Assessment Report.

Pollard, S. and D.d. Toit. 2008. Integrated water resource management in complex systems: how the catchment management strategies seek to achieve sustainability and equity in water resources in South Africa. Water SA 34(6): 671–679.

Pollard, S. and D. Toit. 2011. Towards adaptive integrated water resources management in Southern Africa: the role of self-organisation and multi-scale feedbacks for learning and responsiveness in the Letaba and Crocodile catchments. Water Resources Management 25(15): 4019–4035.

Rogers, P. and A.W. Hall. 2003. Effective Water Governance, Global Water Partnership.

Shrestha, A.B. and R. Aryal. 2011. Climate change in Nepal and its impact on Himalayan glaciers. Regional Environmental Change 11(1): 865–877.

Sijapati, S. 2001. Water Accounting and Water Institutions' Study of Manusmara River Basin. 2001 Water Management Conference on Transbasin Water Transfers. USCID, Denver, Colorado, USA.

Sijapati, S. and S. Dhungel. 2008. Adoption of National Water Plan in Irrigation Sector. 2008 National Irrigation Seminar. Merabel Resort, Dhulikhel, Kabrepalanchok, Nepal.

THT. 2012. Nepal's population rises to 26,494,505. Retrieved 2 August 2013, from http://www.thehimalayantimes.com/fullNews.php?headline=Nepal's+population+rises+to+26%E2%80%9A494%E2%80%9A505+&NewsID=355699.

WECS. 2002. Water Resources Strategy Nepal. W.A.E.C.S. His Majesty's Government of Nepal, Singha Durbar, Kathmandu, Nepal, January 2002.

WECS. 2005. National Water Plan—Nepal. S.D. Water and Energy Commission Secretariat, Kathmandu, Nepal.

WECS. 2011. Water Resources of Nepal in the Context of Climate Change. S.D. Water and Energy Commission Secretariat, Kathmandu, Nepal.

WWF. 2009. National Launching of Koshi River Basin Management Program, Kathmandu, Nepal.

Yoder, R. 1991. Assistance to Farmer Managed Irrigation Systems: Experiences from WECS/IIMI/FORD Action Research Project in Indrawati Watershed Basin. Kathmandu: Ministry of Water Resources, Department of Irrigation, Planning Design and Research Division.

18

Sink or Swim: Alternative Approaches to Flood Disaster Reconstruction and Mitigation

Caroline Wenger

SYNOPSIS

Climate change and population growth are expected to worsen flooding globally, leading to escalating recovery costs that countries can ill-afford. Improving disaster resilience as part of post-disaster recovery is crucial to minimising these losses.

The first section, *Building Back Better*, looks at disaster costs following recent Australian floods. It examines post-disaster reconstruction policies in three different countries and the legislative and funding provisions that support them. Relocation is identified as a significant strategy in some countries.

Ecosystem Approaches to Flooding highlights an innovative resilience strategy used in many parts of the world that not only reduces flood risk but is also expected to buffer the effects of climate change. It generally involves relocation or changing land use. Discovering what makes a successful relocation scheme is thus important to implementing this approach. In contrast to some of the other countries studied in this chapter, achieving relocation in Australia is a challenge. Recent Australian examples of relocation are studied in the final section, *Relocation Policies in Australia*, and timing, funding, and social factors are all found to be significant for success.

Australian National University, Canberra, Australia.
Email: caroline.wenger@anu.edu.au

In the *Discussion*, lessons are drawn from all sections and countries on how to achieve improved flood resilience, particularly for countries such as Australia, where barriers such as cost impede the incorporation of betterment into post-disaster reconstruction.

Keywords: Australia, China, Dongting lake, ecosystem approaches, flood hazard mitigation, floodplains, land swap, legislation, levees, Murray-Darling Basin, New South Wales, polders, post-disaster reconstruction, Queensland, Regge River, risk assessment, 'room for river' measures, sea level rise, The Netherlands, USA, Victoria, wetland, Yangtze River

1 Introduction

Legally, rivers are defined by their banks, but in ecological terms, rivers also consist of the areas they occasionally swell into: their floodplains, ephemeral tributaries and anabranches (Taylor and Stokes 2005; Wenger et al. 2013). Perhaps because humans and rivers exist within different timescales, people think of such areas as being land and rarely recognize them as being part of the river. Floodplains attract settlement because of their alluvial fertility, access to drinking water, river transport, and resources such as fish. However, these bounties come at a cost and periodic inundations destroy assets, lives and livelihoods. Over time, towns and cities may form around vulnerably-located pioneer dwellings, land use changes and exposure to flood grows (Smith 1998; Squires, this volume).

Floods are a consequence of a complex interplay between climate, hydrological cycles, catchment topography and human land management. Factors that can worsen flooding of human settlements include:

- **narrow catchments or bottlenecks:** water has less room to spread and can rise rapidly and to great depth even in modest floods
- **dry catchments, vegetation clearance and urban development:** increase the amount and speed of run-off, and increase erosion and sediment movement
- **subsidence:** can be caused by extracting groundwater or draining wetlands
- **hotter climates:** higher sea surface and air temperatures result in greater evaporation and an increased water holding capacity of the air. Thus greater quantities of rain fall in intense bursts. Continental ice melt raises sea levels. Seasonal snow melt accelerates, flooding river valleys
- **structural measures to prevent flooding such as flood levees and flood gates:** cut off rivers from their floodplains and wetlands, reducing the overall flood storage capacity of the landscape

- **inappropriate development:** where inadequate controls are in place, land use and development form may be incompatible with the flood prone nature of the land, resulting in losses

People living on floodplains have traditionally chosen localized measures, such as raising houses on stilts or mounds, or building flood barriers and drainage channels to reduce damages. However, such measures are only effective locally and some of them exacerbate flooding within the catchment as a whole. Modern flood management has broadened adaptation options through the use of non-structural measures, such as development planning, insurance, management, flood warnings, preparation and awareness (White 1945).

Flood management is commonly divided into different phases, broadly 'before the flood', 'during' and 'after'. Thus flood information, risk assessment, flood hazard mitigation/prevention, human awareness and preparation all come before a flood, response during, and relief and recovery after. When prevention is incorporated into recovery it becomes a less linear model. The linking of these two phases is often neglected and this chapter will explore some options for achieving this outcome.

Flood risk is managed by governments in various ways. At the more coercive end of the scale, legislation may be used to regulate development and to protect catchments, vegetation, soil and water resources. In some countries legislation is accompanied by penalties such as prohibiting government investment or disaster relief in non-compliant jurisdictions. Other tools are more cooperative and exhortative, and can include whole of government strategies, partnership agreements and formalised arrangements that articulate responsibilities and funding by different levels of government and other stakeholders. Policies, procedures, guidelines and standards assist on-the-ground managers to determine which measures should be used, how and under which circumstances. The generation and provision of flood risk information is another significant government role. The government level at which these functions are performed and the mix of tools used varies from country to country. Market mechanisms can also be used to manage flood risk, and various private sector industries have a role in influencing policy and on the ground management (Handmer and Dovers 2007; Wenger 2013).

In recent years the capacity of current flood management techniques to cope with future threats has come into question as damage costs continue to rise. Population growth, urbanisation and climate change are increasing flood risks globally and increased flood frequency, variability and severity is expected for the majority of major global river basins. The Murray-Darling Basin in Australia could be one of the worst affected, with a recent multi-model study indicating that a twentieth century 1:100 year flood may occur as often as every ten years by the late 21st century (Bates et al. 2008; Hirabayashi et al. 2013; Jha et al. 2012; Munich RE 2013). In 2010–11, eastern Australia experienced some of the most damaging floods on record, with 80% of Queensland declared a disaster zone (QFCI 2011). However, comparison of Australia's flood prevention and recovery policies with those overseas indicate an over-reliance on out-dated approaches that will do little to minimize future flood risk.

Following the 2010–11 Australian floods a number of reviews were generated that explored the factors contributing to losses, providing an opportunity to investigate how Australian policy compares with those of flood prone countries elsewhere. In this study, Australian flood reviews, government policy and grants documents were analysed. These were compared with similar documents and case studies from the Netherlands, China and the United States of America to gain insights into how Australia could enhance its resilience to future flooding (Wenger et al. 2013).

Analysis of overseas flood policy and reviews showed that all have a strong traditional reliance on structural means of flood mitigation such as levees and flood gates. However the limitations of using these approaches in the face of future risks are now apparent. While levees can reduce flood losses for smaller floods, if they breach or overtop during large floods, damages are likely to be worse than if there had been no levees. Levees also adversely affect water quality, catchment-wide flood risks and ecosystem health. These problems have led to numerous reforms in flood management overseas, such as more rigorous land planning, incorporation of resilience into rebuilding and ecosystem approaches that recognise accommodation of floods as a legitimate land use (Wenger et al. 2013). Attempts have been made to achieve some of these approaches in Australia, but most are hampered by lack of coordination between jurisdictions, conflicting policy objectives and an unwillingness to invest in flood prevention (Wenger 2013). This chapter will explore some of the measures that countries are using to improve their flood resilience, with a particular focus on relocation and disaster recovery funding.

2 Building Back Better or Reinvesting in Disaster?

Countries vary widely in their approach to recovering from major floods. In Australia, recovery is funded primarily through the Natural Disaster Relief and Recovery Arrangements (NDRRA) grants process, with the proportion of assistance funded by the federal government dependant on the cost of damages. For large scale disasters, the cost share arrangement between governments is 75% federal: 25% state. This primarily funds public infrastructure replacement but not private losses, which are expected to be recouped through private insurance. Government and insurance expenditure following the 2010–11 floods was over A$10 billion (Table 18.1). This figure does not include uninsured losses, but it does include some indirect costs such as emergency accommodation. A breakdown of damage and loss in different sectors indicates substantially greater losses, close to $20 billion for Queensland alone (Table 18.2). For a country with a population of 22.5 million, costs have been enormous.

Recovery funds following the 2010–11 floods only restored assets to their previous condition which exposes Australia to the risk of incurring repeat damages. Reviews prepared following the floods reveal that at the time they were written, 'betterment', or rebuilding to a more disaster resilient standard, was theoretically eligible through NDRRA, but no betterment project had ever been approved (Comrie 2011). A once-off betterment fund for Queensland infrastructure has since been announced (Gillard 2013a),

Table 18.1. Recovery expenditure following the Australian floods of 2010–11.

Recovery Funding Source	**Expenditure A$ ('000,000,000)**	**Source**
Natural Disaster Relief and Recovery Arrangements (QLD)	6.8*	(QRA 2011b)
Natural Disaster Relief and Recovery Arrangements (VIC)	0.90	(VAGO 2013)
State of Victoria (additional to NDRRA cost share amount)	0.071	(VAGO 2013)
Flood insurance (QLD)	2.388	(ICA 2013)
Flood insurance (VIC)	0.1265	(ICA 2013)
Charitable donations (QLD)	0.266*	(QRA 2011b)
Total	**10.5515**	

* These figures do not separate flooding and Cyclone Yasi

Table 18.2. Estimated damage and loss by sector following 2010–11 floods and Cyclone Yasi (QLD).

Sector	**Damage & Loss A$ ('000,000,000)**
Mining	5.7*
Infrastructure	5
Housing	4
Commercial properties	2.6
Agriculture	1.6
Tourism	0.6
Total	**19.5**

* Loss at the end of the financial year 30 June 2011. The coal sector had only recovered to 75% as at May 2011, so this is likely to be an underestimate of total losses (QFCI 2012). The remaining figures were provided in a joint report by the World Bank and the Queensland Reconstruction Authority (The World Bank 2011).

but this is not standard policy. In June 2013, the Victorian State Government reported it had not received funding for betterment projects it submitted in 2011[1] (VAGO 2013). If the return period for large floods decreases due to climate change, the lack of federal investment in resilient reconstruction will make Australia increasingly vulnerable to future damages.

By contrast, Chinese disaster recovery aims to improve its future capacity to withstand floods. A series of floods along the Yangtze in the 1990s culminated in catastrophic floods in 1998 that killed over 4,000 people. Cities along the Yangtze, such as Wuhan, sustained enormous damage. Following the floods, relocation emerged as a significant policy to reduce the cost of future flood damages. Farming communities were relocated out of wetlands to increase the capacity of the landscape to store floodwaters (see 'ecosystem approaches' section) and relocation was incorporated into a package of measures to reduce future susceptibility to flooding, involving substantial funding commitments. US$3.2 billion was allocated for floodplain management and $30 billion for improved land management in upper catchments. While legislation made relocation mandatory, compensation was available and subsidies, such as the "Grain for Green" scheme, provided farmers with immediate livelihood benefits that enabled them to transition to flood compatible land uses (Wenger et al. 2013).

In the USA, the cost of repetitive damage provides the federal government with a strong incentive to invest in buyback as it is liable for insurance claims to private properties in flood prone areas through its National Flood Insurance Program (NFIP). This program represents the second highest potential demand on the federal treasury behind social security (Wright 2000). There is less financial incentive for Australian governments to fund voluntary purchase, as in Australia it is private individuals and insurers that are responsible for costs of rebuilding private assets. Insurance payouts only cover the cost of the actual damage, and are therefore not sufficient to fund an improved standard of repair, or assist residents to relocate.

Research from the USA into repetitive damage is compelling. In one study, cumulative damage costs were worth up to seven times the value of the original property (NWF 1998). In 1993, floods in the upper Mississippi caused a major shift in disaster relief and there was a "consensus that rebuilding or restoring to pre-flood conditions was not an acceptable policy position" (Wright 2000). During the 1990s, recovery and mitigation became increasingly integrated in the United States and for some disasters they completely merged. With the *Hazard Mitigation and Relocation Assistance*

[1] 23 betterment projects for public infrastructure were submitted in September 2011, the betterment component of which totalled A$13.3 million, or 28% of total project costs.

Act (1993), mitigation funding increased, and 15% of all federal disaster assistance funds became available to be spent on relocation, elevating structures and land acquisition. In some cases this meant whole communities could be relocated (Wright 2000:69, 78–9).

The conditions for voluntary purchase in the USA include the complete removal of structures and a requirement that land purchased revert permanently to open space uses compatible with fulfilling 'natural and beneficial floodplain functions', such as recreation and flood mitigation (FEMA 2010). The *Hazard Mitigation and Relocation Assistance Act (1993)* prohibits any future federal expenditure on disaster relief or rebuilding on land purchased through the program. Other legislation relating to relocation is found in Section 555 of the *National Flood Insurance Reform Act (1994)*. This requires that communities participating in the National Flood Insurance Program (NFIP) adopt laws requiring that 'substantially damaged' properties (where cost of reconstruction is equal to or greater than fifty per cent of the pre-flood market value of the property) be elevated to the 1 in 100 year flood level or removed from the floodplain (FEMA 2010; NWF 1998). Through its supplementary mitigation insurance provisions, the NFIP assists compliance with legislation by providing supplementary payouts. This enables rebuilding activities to comply with current building code standards. Eligible expenditure includes the cost of elevating, demolishing, flood proofing or relocating substantially damaged buildings (IFMRC 1994; Wright 2000). According to Wright, writing in 2000, the program had bought and removed an estimated 20,000 structures since its inception (Wright 2000).

Cost benefit analysis following the 1993 floods found that voluntary buyout was most feasible for repetitively damaged properties with a history of high insurance payouts (NWF 1998). Analyses of avoided flood damages indicate that US investment in preventative measures such as relocation have saved considerable amounts of money. In the Upper Mississippi, the 1993 flood caused US$20 billion in damages. Following this, $150 million was spent on relocation as part of the recovery effort. In 2008, a similar-sized flood occurred in the same area but the later flood had a much lower damages bill of $2 billion partly because the vacated land was available for flood storage (Freitag et al. 2009:5–6).

The failure of hundreds of flood levees during the 1993 Upper Mississippi floods was a significant factor that triggered the shift in policy towards relocation; while federal USACE flood levees generally performed to design standards, the consequence of over-relying on levees became evident (Wright 2000). While levees can reduce flood frequency, US levee reviews demonstrate that they also encourage additional development (and potential damages) in the areas 'protected'. Inadequate building standards, excessively high levees and lack of preparedness mean damages can be catastrophic if they breach or overtop (ASFPM 2007; Freitag et al. 2009; Galloway et al. 2007; ILPRC 2006; NCLS 2009; Wenger et al. 2013).

Since the 2010–11 floods, the Australian insurance industry has successfully lobbied government to increase investment in structural mitigation such as levees to reduce the frequency of damages (Gillard 2013; Milliard 2012; Suncorp 2012). While levees may be appropriate in some instances, overseas experience suggests that Australia would do well to consider options such as relocation that are more effective, reliable and permanent.

3 Ecosystem Approaches to Flooding

Ecosystem approaches to flooding are increasingly used in countries such as China, the Netherlands, the USA and the UK. These approaches often rely on the strategic relocation of incompatible development. They are characterised by vegetation management, 'room for river' measures that allow water more room to spread, and integrated, catchment-scale planning and management.

'Room for the river' is an approach that was developed in the Netherlands following a series of major floods in the 1990s. While dykes did not overtop or breach, they *almost* did, and the floods caused widespread evacuations and property damage. This reinforced the high consequence of failure and the limitations of relying too heavily on dykes (Deltacommissie 2008; Dutch Government 2000; 2006; Dutch Government, nd[a]).

In the room for the river program, floodwaters are allowed to spread over a wider area, reducing flood depth. This not only improves safety but means that dykes do not have to be raised higher. The Dutch achieve this by a number of means. Dykes are set back further from the river bank, the meanders of artificially straightened river channels are restored, secondary water channels are excavated and river beds, groynes and floodplains are lowered. For some polders (areas enclosed by dykes), partial lowering of dykes enables water to flow through polders when water levels are high. Development in these poldered areas is still permitted but is relocated to mounds outside the flow path. This approach has been implemented nationally along Dutch rivers, including on the Rhine, the Meuse, the Waal and the Regge (De Boer and Bressers 2011; De Hartog 2012; Dutch Government, nd[a]; Dutch Government, nd[b]). Similar national room for river policies have been adopted in other countries, including in Britain and China (DEFRA 2005; Pittock and Xu 2011).

In the USA, influential proponents of ecosystems approaches include the Association of State Floodplain Managers and the Federal Emergency Management Agency (ASFPM 2007; 2008; FEMA 2011; Freitag et al. 2009).

Funding, with the dual aims of increasing community safety and restoring flood storage, is available through pre-and post- disaster voluntary acquisition and relocation programs (FEMA 2010). The United States Army Corps of Engineers (USACE) is primarily involved in structural mitigation, but it also recognises the considerable benefits of incorporating greater floodable area into its levee systems. The Mississippi River and Tributaries project, for example, incorporates approximately two million acres of floodways and backwaters into its management options. When activated, these divert excess flows and relieve pressure on levees. The floodways are part of a highly engineered system but they provide much greater flexibility than the 'levees only' policy that was in place up until 1927. When the extent of the 1927 Mississippi flood was compared with the similar-sized 2011 flood, it was found that the use of floodways had reduced flood extent by 62% (Mississippi River Commission 2011).

Releasing floodwaters from the strict confines of mainstem levees has other benefits. When Mississippi floodways were opened in 2011 they deposited much needed sediment on degraded coastal marshes. The marshes form a natural defence against storm surge for New Orleans but confinement of sediment between narrow levees has prevented deposition, contributing to the loss of 3,700 km^2 of coastline (Aldous and Jabr 2011; Solomon 2011). Other USA examples of ecosystem approaches to flood management often involve the purchase of land or easements (e.g., covenants on land use) and offer significant side benefits for water quality, fisheries, wildlife and tourism. In one example, marginal agricultural returns made it more economically feasible for farmers to dissolve a levee district and sell the land for wetland reserve than to continue to maintain the levee system (Freitag et al. 2009; Kousky et al. 2011).

In China, land reclamation was a key strategy for grain production, particularly from 1950 to the late 1970s. As a result lakes and wetlands were cut off from the Yangtze River and polders were constructed in and around lakes. At least 1100 lakes in the middle and lower reaches of the Yangtze River have disappeared, while Dongting Lake, a major flood retention area in the middle reaches of the Yangtze, has reduced to a third of its original size, from 6,300 km^2 in 1825 to less than 2,000 km^2 in 2000. Nearby Jianhan Plain has lost 80% of its wetlands to agriculture since 1840. Compounding this problem, extensive deforestation in the upper catchment has increased sediment loads. Silt deposition raises river beds and fills lakes and dam reservoirs, contributing to about 30% of flood storage losses. Overall, it is estimated that land reclamation and siltation have reduced the landscape's water holding capacity by 75 per cent (CCICED 2004; Guangchun Lei, pers. comm.; An et al. 2007; Yin and Li 2001; Yu et al. 2009; Zhang 2004).

Devastating floods in 1998 triggered widespread support for ecosystem solutions, and existing laws and policies were revised and more rigorously enforced.[2] These supported revegetation and improved soil management of upper catchments, while in the middle and lower reaches polders were removed and sluice gates were seasonally opened to reconnect floodplain wetlands to the river (Pittock and Xu 2011; Te Boekhorst et al. 2010; Yu et al. 2009; Zhang 2004). This has involved large scale relocation of floodplain populations. During the 1998 floods, 2000 polders breached. During post-flood reconstruction between 1998 and 2003, 1,461 polders were demolished and 2.42 million people were relocated to higher ground. These measures resulted in 2,900 km^2 of land being returned to lakes and rivers, increasing flood storage capacity by 13 billion m^3 (Cheng 2004).

China and the Netherlands have both found that for relocation to be successful, it needs to benefit the people being relocated. Regardless of whether acquisition is voluntary or compulsory, this reduces resistance and legal opposition and hastens purchase processes. Demonstration projects in China provided those relocated with alternative livelihoods, increased incomes, improved water quality, health, housing, and clean energy from biogas, as well as reduced flood risk. These outcomes led to broad support from provincial and local governments and uptake elsewhere (Pittock and Xu 2011; Te Boekhorst et al. 2010; Yu et al. 2009). In the Netherlands, similar 'win-win' outcomes are sought, and a multifunction approach to land use has also resulted in stimulation of local economies and amenity. This cross-sectoral approach is supported by integrated water resource management and basin scale planning (CCICED 2004; De Boer and Bressers 2011; Dutch Government, nd[b]; EU 2000; Nijssen 2011).

Ecosystem approaches have been introduced not only in response to recent large floods, but specifically to address future climate change (Dutch Government 2000; Government of China 2007). Climate change is expected not only to increase severe flooding but also drought, while higher temperatures may exacerbate water quality problems. Ecosystem approaches to flooding help to address these issues. They increase flexibility in the system, soaking up water when it is in oversupply, filtering and releasing it slowly. Retaining water for longer increases groundwater infiltration, and improves water availability during drought. The replenishment of groundwater also helps form a barrier to prevent saltwater intrusion in coastal areas affected by sea level rise. Meanwhile, vegetation restoration not only reduces erosion and sediment load but can moderate water

[2] Key laws include the Water Law (1988) which was revised in 2002 to incorporate integrated water resource management; the Law of Soil and Water Conservation (1991), and the Law of Flood Control (1997). The latter already prohibited reclamation of land in wetlands and watercourses and supported planned resettlement. Following the floods these laws were reinforced and supplemented by the 1998 '32 Character Policy', and at the same time a National Logging Ban was imposed.

temperature and algal growth by providing shade. Rather than putting additional pressure on threatened riparian habitats, ecosystem approaches help ensure their survival. This is a flexible approach that incorporates greater redundancy into the system and improves water security.

Examples of ecosystem approaches in Australia are isolated and localized. They include wetland restoration at Leeton, NSW (Wenger et al. 2013) and an award-winning wetland reconnection program near Grafton, NSW (Clarence Valley Council 2013). These projects yielded multiple economic, social and wildlife benefits. Other innovative proposals have been less successful. A Moreton Bay payment for ecological services proposal to reduce sediment loads through vegetation management has so far failed to gain political support, despite being self-funding (QCC 2012), and a proposal to set back levees in the Lower Goulburn floodplain in Victoria failed as a result of opposition to land acquisition and withdrawal of federal funding (Pittock, pers. comm.; Water Technology Pty Ltd 2005). Nevertheless there has been recent interest in such measures (Parliament of Victoria 2012; Queensland Government 2012a; Rutherfurd et al. 2007).

4 Relocation policies in Australia

Relocation can either involve rebuilding elsewhere or moving an existing structure to higher ground.[3] As illustrated above, relocation can effectively eliminate risks for those exposed to frequent flooding and enable land to be used for flood storage. Thus relocation and modified land use feature highly in overseas 'room for the river' flood policies. However, relocation has many drawbacks, the most significant for government being the high cost of buying back land. Another problem is that despite repeated flooding and purchase offers, relocation is often resisted by residents. Notwithstanding emotional attachment to place, flood free land can be too expensive compared with the buyback amounts offered for the flood-prone property, or unattractive if the relocation site is too remote (Wenger et al. 2013).

Relocation has been used numerous times in Australia's history and the towns of Bega,[4] Nowra, Gundagai, Clermont and Smithfield (now a suburb of Cairns) all owe their present locations to catastrophic flooding (Coates 2011).

Figure 18.1 is a map of eastern Australia showing the locations of Australian places referred to in this section.

---

[3] Note that in the USA there is generally a distinction made between 'voluntary acquisition' and 'relocation', the latter referring to lifting an existing structure off its foundations and moving it to a new site. However, in this paper relocation is a more inclusive term whereby assistance is given to help people resettle out of floodplains using various strategies.

[4] For Australian places and associated waterways referred to in this section, GPS coordinates are provided in Table 18.4.

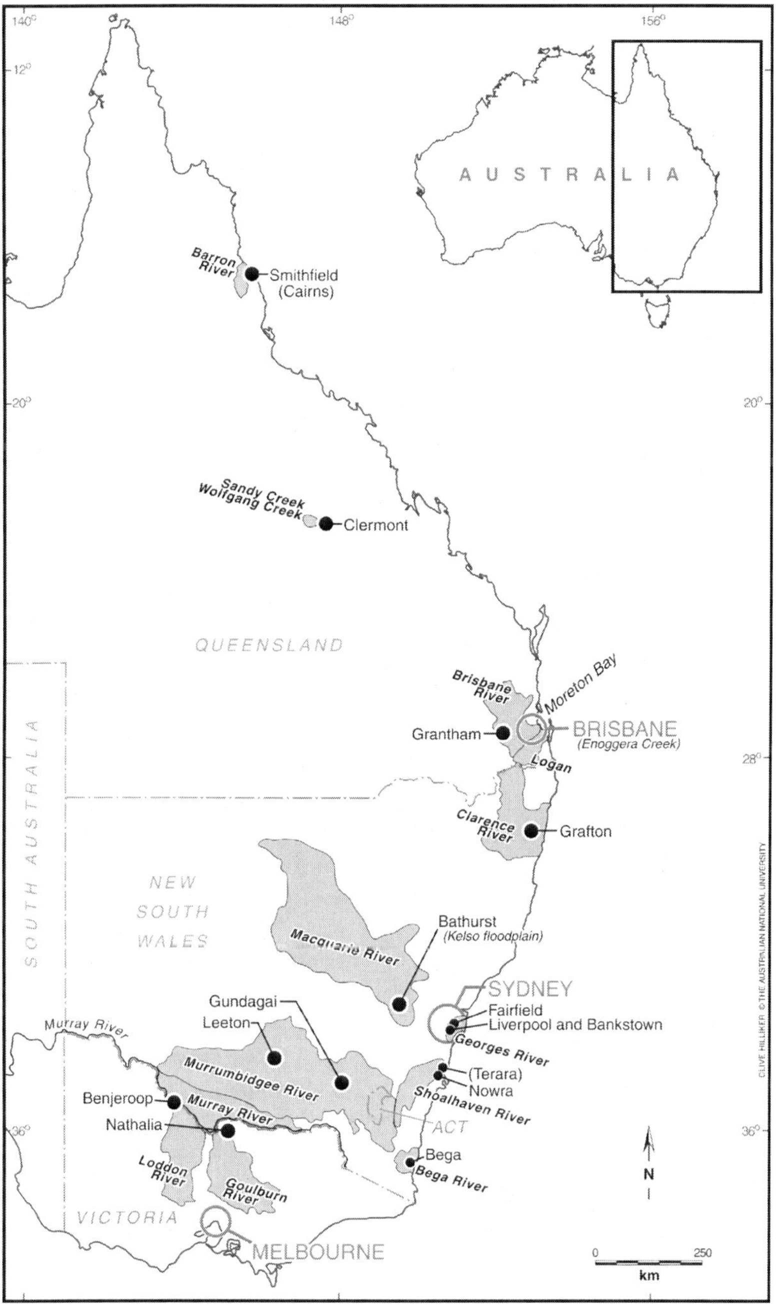

Figure 18.1. Map of eastern Australia showing the Location of Australian places referred to in the text and their river basins.

Recent reviews following Australia's 2010–11 floods offer several examples of relocation, including the town of Grantham in Queensland and the Lower Loddon (an agricultural area in Victoria). Queensland's capital city of Brisbane also has a voluntary purchase scheme that was noted in flood reviews. In all these cases, relocation was implemented by voluntary means, either through buyback, land swap or through compensation for rezoning or establishing covenants on land.

4.1 Brisbane

Brisbane is one of Australia's most flood prone cities and was badly affected by the 2010–11 floods, with 22,696 residential properties partially or wholly inundated, primarily in the older part of the city. Since 2006 Brisbane has had a Voluntary House Purchase Scheme targeting properties that are most frequently flooded (2 year average recurrence interval) and 525 properties have been identified in this category. However Queensland does not have a state-wide policy for purchasing or relocating flood prone properties and the program is wholly funded by local government (QFCI 2012). According to a review undertaken by the Brisbane City Council following the 2010–11 floods, the uptake of the scheme by Brisbane residents was only modest, though the reasons for this were not examined (Arnison et al. 2011). The terms of the scheme offer pre-flood property values. However even where a fair market value is offered this may be insufficient to purchase similar property elsewhere, particularly as flood prone land is often less valuable and residents poorer. Moreover, Brisbane home owners are expected to pay all costs associated with the sale such as conveyancing, transfer costs and stamp duty. Interest in the Brisbane scheme rose following the 2010–11 floods and in the following year the Council doubled the program budget to A$10 million per year to meet demand; 46 properties were purchased between 2006 and May 2011, this figure growing to 73 by September 2012 (Brisbane City Council 2011; Feeney 2012). This represents close to 14% of targeted properties over 6 years.

4.2 Grantham

Relocation of entire communities is less common but can be feasible for small towns, such as Grantham. In January 2011, floods in the Lockyer Valley claimed 19 lives, including 12 at Grantham. Throughout Lockyer Valley, 10 properties were completely destroyed, 19 were beyond repair and 119 properties were severely damaged. To help the community recover and to prevent recurrence the town was relocated to higher ground through voluntary land swap. The project was planned and organised by the local council. The Queensland Reconstruction Authority[5] amended planning provisions to ensure the project could be fast-tracked (QRA 2011a; Queensland Government 2012b).

Land swap is not a measure that will be universally applicable as it depends on the availability of nearby undeveloped and unconstrained land and the local government's financial resources (QFCI 2012). However, land swap has some significant advantages over buyout. It can be cheaper for councils to afford and it means that people remain in the community. It is also equitable as it doesn't discriminate on the basis of land values (Simmonds, pers. comm.).

Project costs included A$7.5 million paid by the local government for acquiring nearby flood free land and $18 million contributed by the state and federal governments for infrastructure and service development. A further $30–40 million is expected to be invested in community infrastructure by the council over a number of years (Gillard 2011; Lockyer Valley Regional Council 2011; van den Honert 2013).

Land swap participants were provided with ample assistance. Stamp duty was waived and legal firms provided conveyancing and other legal costs *pro bono*. Those owning investment properties were granted capital gains tax roll-over relief by the Australian Tax Office[6] (Simmonds, pers. comm.). Participants were responsible for the cost of demolishing damaged structures on their old sites and construction on their new sites. However, $2.835 million was made available through a state-wide donations appeal to help people with these costs (Queensland Government 2012b). Within two years, approximately 85% of Grantham residents had moved into their new homes (Australian Government 2011; Lockyer Valley Regional Council 2011; van den Honert 2013). Of the 120 new residential sites, all were fully subscribed two and a half years after the disaster, though some transactions were still being finalised (Simmonds, pers. comm.). Vacated flood zone land becomes the property of the Lockyer Valley Regional Council, to be used for parklands, community market gardens and farming (QRA 2011a).

Following the relocation of Grantham, a second large flood occurred in February 2013. In the Grantham area, this caused only $20,000 worth of damages to 3 properties. An estimated $30 million was saved in avoided damages (Lockyer Valley Regional Council 2013).

[5] An agency established for a limited 2-year term in February 2011 (since extended) to coordinate and manage Queensland's recovery from the 2010–11 floods. Under the *Queensland Reconstruction Authority Act 2011* it was given strong development powers including the right to compulsorily acquire land.

[6] Land value for a standard block dropped from approximately $100,000 to $20,000 post-flood. Investors wanting to swap their property with a new block worth $120,000 would therefore have incurred significant capital gains liabilities.

The scale of the disaster and financial assistance available contributed to the success of the Grantham scheme. However it was also due to several other factors, one of the most important being speed. This required strong local leadership as the project had to proceed without funding certainty. The disaster took place in January; the new site had been purchased and infrastructure construction begun by April; the first land ballot took place in August; and the first family moved in by November. This meant people were able to invest their insurance and assistance money into relocation instead of rebuilding or repairing damaged property. Council lobbying eventually led to funding support. Another important factor was that the scheme was voluntary which avoided opposition. The only restriction that applied to rebuilding on the same site was that where a house was destroyed, floor height needed to be raised. Existing planning scheme provisions that regulated building height were broadened to take account of the area inundated and water depth (Simmonds, pers. comm.).

4.3 Lower Loddon

Following the 2010–11 floods in Victoria, the state government implemented a buyback scheme, the Lower Loddon Irrigators Recovery Package, to irrigators in the Lower Loddon. The primary aim of the scheme was to convert the land from irrigated uses (broad acre and dairy) to more flood resistant rain-fed farming systems. The land most at risk of flooding was first determined by identifying the boundaries of the floodplain. Participating farmers were offered two options. They could either sell their irrigated land at pre-flood values (to be subsequently resold as a rain-fed farm to recoup some of the program costs). Alternatively they could remain on their land and receive compensation to place a covenant on the title of their land, restricting its use to rain-fed practices. Compensation was calculated as a percentage of the pre-flood land value. A$12.3 million was allocated to the package. Flood affected dwellings with minimal residual value were to be demolished, some relocated and some protected individually by ring levees (DSE 2011; Rural Finance 2012).

One of the most notable elements of the Lower Loddon program was its funding arrangements, as it opportunistically piggybacked onto a modernisation of irrigation infrastructure initiative. This scheme aimed to reduce over-allocated water in the Murray Darling Basin as part of the Murray Darling Basin Plan. Contraction of the area of irrigated land and its conversion from vulnerable irrigated to flood-resistant rain-fed uses were thus mutually supportive. This provided an additional source of funding to help people to relocate. The program was successful in being able to purchase all but two properties in the area identified as having the greatest risk (2,751 ha), partly attributable to above market value purchase offers (McBeath 2011; Rural Finance 2012; DSE 2011).

4.4 New South Wales

In recent decades, NSW is the state that has most actively pursued voluntary purchase, and state funding is currently available to local governments for high hazard areas on a 2:1 basis at market value. While voluntary purchase is initiated by individual councils, they have financial and policy support from the State government and schemes are designed with reference to the NSW Government Floodplain Development Manual. Risk to life is central to hazard rating and numerous factors are considered, including catchment characteristics, ease of evacuation, potential flood velocity, depth and duration (NSW Government 2005; NSW Government 2013b). Between 1999 and 2007 federal funds contributed to the NSW scheme, with voluntary purchase or house raising accounting for close to 25% of allocations from the Federal Government's Regional Flood Mitigation Program (Australian Government, nd). Comparable figures for Queensland and Victoria were 5% and 2.3% respectively, suggesting significantly lower investment in relocation in these states.[7]

NSW funding for voluntary purchase has declined markedly since 2009. Recent funding mechanisms include the NSW Ministry of Police and Emergency Services natural disaster resilience grants (2009–2011), where 17% of flood-related grants were for voluntary purchase, and the Office of Environment and Heritage floodplain management grants (2011–2013) (NSW Government 2011; NSW Government 2013a). In the latter scheme, the only NSW government voluntary purchase related grant in the past two years has been a feasibility study in Harden Shire.

Some noteworthy examples of past NSW voluntary purchase schemes include the Kelso floodplain at Bathurst and the Sydney suburbs of Fairfield, Liverpool and Bankstown (Table 18.3). Voluntary purchase was offered in areas of high risk and house raising[8] for areas of lower risk. In the case of

[7] Some caution is needed in interpreting these figures as an indication of state government support for relocation. Funding amounts provided by the Commonwealth were appropriations, not actual expenditure. Neither did they include relocation projects that state or local governments may have funded independently. Where the purpose of property purchase was to enable structural mitigation works, these figures were omitted from calculations. Some figures incorporated house raising. The Australian Government's Natural Disaster Mitigation Program also ran during this period (2003–2009) but it covered all natural hazards and percentages are harder to determine. However, divisions between states for investment in relocation reflect the same trend.

[8] House raising refers to the elevation of structures so that the floor height is at the 'defined flood level' (usually that of a 1:100 year flood event), plus freeboard, which is commonly between 300–500 mm.

Table 18.3. Uptake of voluntary purchase schemes in Australia.

Location	Number of Properties Targeted	Period	Properties Purchased	Source
Kelso floodplain, Bathurst, NSW	102	1986–2004 (18 years)	72 (70.6%)	(Australian Government, nd) (BTRE 2002) (Bathurst City Council 1995)
Fairfield, Sydney, NSW	96	1990–2010 (20 years)	76 (79.2%)	(Bewsher Consulting Pty Ltd 2010)
Liverpool & Bankstown, Sydney, NSW	202	1984–2004 (20 years)	170 (84.2%)	(Bewsher Consulting Pty Ltd 2004)
Brisbane, QLD	525	2006–2012 (6 years)	73 (13.9%)	(Arnison et al. 2011; Feeney 2012)
Grantham, QLD	120*	2011–2013 (2–3 years)	120 (100%)	(Simmonds, pers. comm.)
Lower Loddon, VIC	25	2011–2012 (1 year)	23 (92%)	(Neil McBeath, pers. comm.; Rural Finance 2012)

* 120 new lots were developed for the land swap, available to all eligible property owners in the Lockyer Valley (both residential and vacant blocks). The number of eligible properties is unavailable as this was assessed on application.

Fairfield, the local council commissioned a floodplain management report in 1990 that identified properties in high risk areas for voluntary purchase. A review in 2010 found that 76 out of 96 properties had been bought and returned to flood resilient open space. In Bathurst, 102 properties located in a floodway were targeted for voluntary purchase following floods in 1986. There was low uptake of the scheme until a second flood in 1990. Of these, around 30 remained to be purchased in 2004. In Bankstown and Liverpool 170 properties were purchased between 1984 and 2004 and 32 remained to be bought or moved. The cost of purchasing the remaining properties was high (A$30 million in 2004 figures), a problem compounded by the withdrawal of federal funding and dramatic property value increases. This led to private sector funding sources being explored and land already purchased was proposed for flood compatible uses such as golf courses and sand mining.

Australian relocation examples show that voluntary purchase is expensive and can take decades to achieve. However, people are more amenable to it immediately following a major flood particularly if amounts offered are sufficient to acquire equivalent property elsewhere, as at Grantham and Loddon. Thus timing and adequate funding are both factors in successful relocation. It is noteworthy that many of the suburban examples of flood prone properties are inhabited by people on lower incomes as these properties are more affordable. The Brisbane suburb of Rocklea is semi-industrial, while Fairfield and the Kelso floodplain also have a high proportion of low income earners (BTRE 2002; Sellers and Mooney 2012). It can be a substantial barrier to expect people to pay property transaction costs, demolition and moving costs, as well as having to move to a more expensive property that is less central or lower quality, albeit in

Table 18.4. Location of Australian places referred to in the text and their waterways.

Place	GPS coordinates	Significant Waterways
Smithfield, Cairns, Queensland (QLD)	16°49′14.01″S; 145°41′31.06″E	Barron River
Clermont, QLD	22°49′29.20″S; 147°38′25.15″E	Wolfgang Creek, Sandy Creek
Brisbane, QLD	27°28′15.36″S; 153°1′24.61″E	Brisbane River, Enoggera Creek
Grantham, QLD	27°34′45.58″S; 152°11′44.63″E	Lockyer Creek, Rocky Creek, Monkey Waterholes Creek
Moreton Bay, QLD	27°12′21.81″S; 153°15′1.28″E	Logan, Bremer, Lockyer river catchments lead to the bay
Grafton, New South Wales (NSW)	29°41′28.04″S; 152°55′59.52″E	Clarence River
Fairfield, Sydney, NSW	33°52′14.83″S; 150°57′19.76″E	Prospect Creek (tributary of Georges River)
Liverpool & Bankstown, Sydney, NSW	33°55′12.07″S; 150°55′26.86″E	Georges River
Kelso floodplain, Bathurst, NSW	33°25′3.55″S; 149°34′51.71″E	Macquarie River and tributaries
Leeton, NSW	34°33′5.01″S; 146°24′23.95″E	Main Canal (Yanco Irrigation Area), Murrumbidgee River, Fivebough and Tuckerbil Wetlands
Nowra (relocation site); Terara (original site), NSW	34°52′30.62″S; 150°35′44.81″E	Shoalhaven River
Gundagai, NSW	35°3′55.30″S; 148°6′25.90″E	Murrumbidgee River, Morley's Creek
Bega, NSW	36°40′27.04″S; 149°50′34.50″E	Bega River
Lower Goulburn (near Nathalia), Victoria (VIC)	36°3′23.45″S; 145°12′25.41″E	Goulburn River, Broken Creek, Murray River
Lower Loddon (near Benjeroop), VIC	35°28′35.62″S; 143°48′59.10″E	Loddon River, Murray River, Little Murray River, Murrabit River

a safer location. Simply agreeing to pay market value (even at pre-flood values) may be insufficient to achieve relocation.

5 Discussion

Susceptibility to flood is to some degree unavoidable in Australia as the oldest farming settlements were located on fertile floodplains close to water. Badly located development is often found in the oldest part of town, a legacy of the distant past when there was less information on flooding and development controls were less sophisticated or non-existent. However, reviews following the Australian floods show that irresponsible development decisions are still being made (QFCI 2012; Wenger et al. 2013). This reflects a nation-wide failure to adequately address disaster prevention[9] (Wenger 2013).

It is easier to prevent than to reverse bad land use decisions. Where relocation is unavoidable, the great dilemma for governments is cost. However, examples in this chapter demonstrate that this has not deterred all flood prone countries from investing in relocation. A number of observations can be made about the examples studied:

1. Countries relocate development for different reasons. In Australia, the primary object is to reduce threat to life of those located in dangerous areas. A secondary reason is to enable the construction of structural forms of flood control. In the Netherlands and China, the aim is to create more flood storage on a catchment scale, ultimately lessening massive actual or potential losses. In the USA, relocation is given impetus by a huge federal liability for the private losses of those participating in the National Flood Insurance Program, as well as disaster relief costs.
2. Relocation spikes immediately after a flood. This partly reflects greater willingness of residents to move following floods. However, it is also the most logical point in time for buyout to occur. It makes no economic sense to wait for money to be invested in rebuilding a property before deciding to buy and demolish it. This suggests that efficient relocation requires processes and funding to be in place as part of disaster recovery.
3. Voluntary purchase is most affordable where property is of low value, is frequently flooded, and has a history of high-cost repetitive damage.

---

[9] Federal funding for flood prevention (not to be confused with post-disaster betterment) is through the National Partnership Agreement on Natural Disaster Resilience which, combined with the smaller National Emergency Management Projects grants scheme, provides approximately A$28.7 million per year. This amount is divided between all natural disaster types in all states and territories.

4. In the USA, eligibility criteria for voluntary purchase are based on calculations of cumulative past damage and damage sustained in the most recent event. This makes it easy to incorporate relocation into post-disaster recovery following damage assessment processes. This contrasts with NSW's multi-faceted risk assessment approach which is arguably more tailored to pre-disaster mitigation.
5. Examples in this chapter indicate that for relocation to be successful, it has to benefit those being relocated. This results in less resistance, rapid uptake and positive social and economic outcomes. This is particularly important because flood prone areas are often inhabited by disadvantaged people who are least able to afford moving costs. Purchase offers need to be sufficient to cover all the costs of relocation, e.g., through over-market price offers, waiver of property transfer costs and taxes, charitable appeal funds or supplementary insurance.
6. Many different strategies are used to remove people from harm and to ensure land use is compatible with flood mitigation functions. Not all of them involve purchase, which can be viewed negatively if it prevents compatible development ('land sterilisation') or entails ongoing land management obligations by governments. Some strategies identified in the examples covered in this chapter include: compensation for land use covenants on rural land (e.g., Lower Loddon, Regge River); cost share by multiple agencies where mutual benefits are identified (e.g., Lower Loddon, Regge River); legislated protection of lakes and wetlands (e.g., China); reservation of first preference rights to purchase land in strategic locations (e.g., Netherlands); land swap to avoid unaffordable buyback costs (e.g., Grantham); land swap to prevent opportunistic purchase price elevation (e.g., Regge River); supplementary mitigation insurance (e.g., USA); 'payment for ecological services' or farming subsidies for compatible land management (e.g., Netherlands, Australia, China); flexible land use plans that facilitate the reversion of land to natural uses (e.g., Netherlands); relocation of existing structures to higher ground (e.g., USA, Australia); voluntary purchase (e.g., Australia, USA, Netherlands); financial penalties for communities that permit rebuilds to pre-flood standards (e.g., USA); and compulsory acquisition, particularly where it is in the national interest (e.g., China, Netherlands).
7. Countries with significant relocation programs have strong national government support (e.g., policies, coordination mechanisms, legislated funding ratios to support betterment, purchase conditions, incentives and penalties).
8. In areas where relocation has been carried out, large cost savings have been calculated following subsequent floods.

9. In Australia, flood mitigation is very much a local concern, which results in a patchwork of different measures and safety standards. This favours locally-implemented structural approaches. Countries with a more nationalised view of flood management find it easier to apply ecosystem approaches (and relocation), which benefit from catchment-scale planning and implementation, coordinated across jurisdictions and sectors.

Flood prevention policies studied in this chapter demonstrate that it is feasible to incorporate flood prevention into post-disaster recovery, and this is the approach taken by many countries around the world. However, countries such as Australia that only rebuild to pre-existing standards will be exposed to repeated large damage costs. These costs are likely to escalate in the future due to climate change, population growth and urbanization.

Incorporating flood resilience into post-disaster reconstruction can involve improved rebuild standards, house raising and flood proofing, or relocating development to a new site. Of these, relocation can have significant social, economic and ecological co-benefits. It can also buffer the anticipated effects of more severe floods (and droughts) as a consequence of climate change. Use of this strategy is applicable to river basins around the world that are likely be affected by climate change.

The examples studied in this chapter suggest that relocation is easier to achieve in the immediate aftermath of a disaster. Thus policies and funding mechanisms need to be in place to ensure opportunities to improve resilience are not missed.

In the case of Australia, the nine observations above suggest that Australia would benefit from greater federal support for relocation programs, particularly during post-disaster recovery. This could entail adoption of the USA requirement that 15% of disaster relief funds be available for spending on relocation and improved rebuild standards. Similarly, eligibility requirements and financial incentives and penalties, could be used to prevent the perpetuation of bad land planning decisions.

Intervention during disaster recovery has many advantages, and there is potential for governments to collaborate with the insurance industry to reduce costs of voluntary purchase. For example, if insurers paid for the actual damage sustained by the property, governments would only need to pay the balance to enable relocation. The advantage for insurers is the removal of properties that cost the most in claims, and an increase in insurance affordability and availability.

For such a scheme to be effective, assessment needs to be rapid following a disaster and property-level information is needed to determine eligibility for buyout. Eligibility criteria currently used in Australia require detailed flood studies which could prevent rapid assessment. As risk is related to damage, another possibility would be to follow the USA method where damage (including cumulative damage from previous floods) needs to be equal to or greater than fifty per cent of the pre-flood market value of the property. Insurers collect flood information on a property level to set premiums so such information may already exist.

Supplementary mitigation insurance products could be developed to cover the cost of house raising or flood proofing if owners elect to stay, or relocation expenses (demolition, moving costs) if they move. This strategy will not be suitable for all post-disaster buyouts, as for many flood insurance is unavailable or unaffordable. However, extending the role of the insurance industry in this way would benefit insurers, governments and owners alike. As out of pocket expenses are a significant barrier to relocation, state and municipal governments could also consider waiving stamp duty or other property transaction fees. These could be offset by lower flood risk and emergency service liabilities and alternative uses for the vacated space.

Investment in relocation is also needed prior to disasters, particularly for at-risk properties and in strategic areas that would result in significant flood mitigation gains, such as the Lower Goulburn levee setback proposal. In Australia, greater efforts could be made to identify synergies with multiple agency portfolios that could lead to creative ways of funding relocation. Some relocation programs, such as Grantham and the Regge River, Netherlands, demonstrate that land swap strategies can also improve the affordability of relocation, though this will not be suitable in all cases.

6 Conclusion

Floodplains are the interface between land and river and present a challenge for human occupation. Where development has been built in the wrong place, it is tempting to band-aid it with structural mitigation, and recent Australian floods have seen successful lobbying attempts to increase government expenditure on levees. Political pressure to build levees as a result of more frequent or severe flooding is likely to be a common response that will not be restricted to Australia. However the shortcomings of levees, and the failure of cost-benefit analyses to factor in external costs are well documented.

Many countries around the world now recognize that flood threats are growing due to factors such as increasing populations and climate change, and that existing approaches to managing flood risk will not be sufficient to address future risks. Alternative solutions such as improved rebuild standards, relocation and associated ecosystem approaches are now being implemented in flood prone countries around the world. These methods offer a more permanent and effective way of achieving community resilience to flooding and have added benefits in that they can buffer against future threats to water security, and protect the natural resources on which livelihoods depend. Various means are used to achieve these alternatives, including legislation, funding incentives and disincentives, development and land management controls and market mechanisms. In Australia, measures used tend to be almost exclusively cooperative and exhortative, while overseas more coercive means are included in the policy mix. Another difference is that integrated water resource management is often central to overseas flood management. By contrast, Australian flood management is largely conducted on a local scale and catchment-wide flood management would require significant institutional reform in some Australian states. Australia and other flood prone countries can learn much from past examples of relocation and progressive overseas strategies to ensure their management of flood is adaptive to anticipated changes in future flood threats.

7 Acknowledgement

This chapter builds on research carried out for two National Climate Change Adaptation Research Facility projects, which paid for preliminary work. The author gratefully acknowledges the work of Clive Hilliker from the Australian National University for producing a map of places referred to in the text. Thanks are also due to Jamie Pittock and Karen Hussey for their support for this work.

References

Aldous, P. and F. Jabr. 2011. Reshaping the Mississippi: after the recent floods, how much worse will it get with climate change, and could floodwaters be put to better use? New Scientist 210(2814): 6–7.

An, S., H. Li, B. Guan, C. Zhou, Z. Wang, Z. Deng, Y. Zhi, Y. Liu, C. Xu, S. Fang, J. Jiang and H. Li. 2007. China's Natural Wetlands: Past Problems, Current Status, and Future Challenges, Ambio 36(4): 335–344.

Arnison, P., R. Gotterson and C. Apelt. 2011. Brisbane Flood January 2011: Independent Review of Brisbane City Council's Response 9–22 January 2011, Brisbane City Council.

ASFPM. 2007. National Flood Policy Challenges. Levees: the double-edged sword, Association of State Floodplain Managers. Available: http://www.floods.org/PDF/ASFPM_Levee_Policy_Challenges_White_Paper.pdf [Accessed 18 July 2012].

ASFPM. 2008. Natural and Beneficial Floodplain Functions: Floodplain Management—More than Flood Loss Reduction. Position Paper adopted by the ASFPM Board 16 September 2008, Association of State Floodplain Managers. Available: http://www.floods.org/PDF/WhitePaper/ASFPM_NBF%20White_Paper_%200908.pdf [Accessed 19 July 2013].

Australian Government. 2011. Queensland communities boosted by continued reconstruction investment: media release 17 October 2011. Available: http://www.pm.gov.au/press-office/queensland-communities-boosted-continued-reconstruction-investment [Accessed 24 June 2013].

Australian Government. (nd). Regional Flood Mitigation Program Approved Projects 1999/2000–2006/2007, Emergency Management Australia. Available: http://www.em.gov.au/Documents/RFMP+Approved+Projects.pdf [Accessed October 2012].

Bates, B., Z.W. Kundzewicz, S. Wu and J. Palutikof (eds.). 2008. Climate Change and Water. Technical Paper of the Intergovernmental Panel on Climate Change. IPCC Secretariat, Geneva.

Bathurst City Council. 1995. Bathurst City Council Flood Management 1995. Available: http://www.bathurst.nsw.gov.au/images/stories/outline_summary_flood_plain_management_plan.pdf [accessed 15 June 2013].

Bewsher Consulting Pty Ltd. 2004. Georges River Floodplain Risk Management Study and Plan. Available: http://www.bankstown.nsw.gov.au/index.aspx?NID=501 [Accessed 19 June 2013].

Bewsher Consulting Pty Ltd. 2010. Prospect Creek Floodplain Management Plan Review.

Brisbane City Council. 2011. Voluntary Home Purchase Scheme: information for flood affected residents. Brisbane City Council. Available: http://www.brisbane.qld.gov.au/prdc/groups/corpwebcontent/documents/documents/045171.pdf [Accessed 21 June 2013].

BTRE. 2002. Benefits of Flood Mitigation in Australia. Bureau of Transport and Regional Economics, Department of Transport and Regional Services.

CCICED. 2004. Promoting Integrated River Basin Management and Restoring China's Living Rivers. CCICED Task Force on Integrated River Basin Management, Beijing.

Cheng, X. 2004. Asian Regional Workshop on Flood Mitigation Initiative: Final Report 2nd Asia Pacific Association of Hydrology and Water Resources Conference. Public Works Research Institute Side Event, Singapore.

Clarence Valley Council. 2013. Clarence Floodplain Project [Online]. Available: http://www.clarence.nsw.gov.au/cp_themes/metro/page.asp?p=DOC-LLW-02-05-64 [Accessed 6 May 2013].

Coates, L. 2011. Floods and land-use planning—a brief Australian history of town relocations. Risk Frontiers 11(2): 3–4.

Comrie, N. 2011. Review of the 2010–11 flood warnings and response: Final Report, State Government of Victoria.

De Boer, C. and H. Bressers. 2011. Complex and Dynamic Implementation Processes: the renaturalization of the Dutch Regge River. University of Twente, in collaboration with the Dutch Water Governance Centre.

De Hartog, H. 2012. C-Change Expert Joint Planning Groups Report: Wealthy Waal Case Study, Province of Gelderland, Netherlands, Province of Gelderland, Netherlands. Available: http://www.cchangeproject.org/jsp/uploaded_files/documents/ROOT/Case%20Study%20Wealthy%20Waal%20-%20Province%20of%20Gelderland.pdf [Accessed 6 August 2012].

DEFRA. 2005. Making space for water: taking forward a new Government strategy for flood and coastal erosion risk management in England. First Government response to the autumn 2004. Making space for water consultation exercise March 2005, Department for Environment, Food and Rural Affairs. Available: http://archive.defra.gov.uk/environment/flooding/documents/policy/strategy/strategy-response1.pdf [Accessed 12 July 2013].

Deltacommissie. 2008. Working together with water: a living land builds for its future. Findings of the Deltacommissie 2008, Deltacommissie 2008.

DSE. 2011. Lower Loddon Irrigators Recovery Package: voluntary buyback and incentives. The Victorian Government Department of Sustainability and Environment, Melbourne. http://www.water.vic.gov.au/__data/assets/pdf_fi le/0009/118584/Lower-Loddon-Irrigators-Recovery-Package-FACT-SHEET.pdf [Accessed 14 February 2012].

Dutch Government. 2000. A Different Approach to Water: Water Management Policy in the 21st Century. Ministry of Transport Public Works and Water Management.

Dutch Government. 2006. Spatial Planning Key Decision: Room for the River. Approved decision 19 December 2006. Dutch Cabinet.

Dutch Government. (nd[a]). Room for the River Brochure. Room for the River Programme Directorate. Available: http://www.ruimtevoorderivier.nl/meta-navigatie/english/publications/.

Dutch Government. (nd[b]). Tomorrow's Meuse. Ministry of Transport Public Works and Water Management. Available: http://www.rijkswaterstaat.nl/images/Kennismakingsbrochure%20Tomorrows%20Meuse_tcm174-287295.pdf [Accessed 24 May 2012].

EU. 2000. Directive 2000/60/EC of the European Parliament and of the Council of 23 October 2000 establishing a framework for Community action in the field of water policy. European Union. Available: http://eur-lex.europa.eu/LexUriServ/LexUriServ.do?uri=CONSLEG:2000L0060:20090625:EN:PDF [Accessed 20 July 2013].

Feeney, K. 2012. Flood buyback will transform 6 ha of properties into parks. Brisbane Times, 21 September 2012.

FEMA. 2010. Substantial Improvement/Substantial Damage Desk Reference. Federal Emergency Management Agency.

FEMA. 2011. Floodplain Management Course (Graduate Level) [Online]. Federal Emergency Management Agency Emergency Management Institute. Available: https://training.fema.gov/emiweb/edu/fmgl.asp [Accessed 19 July 2013].

Freitag, B., S. Bolton, F. Westerlund and J. Clark. 2009. Floodplain Management: A New Approach for a New Era. Island Press, Washington.

Galloway, G.E., J.J. Boland, R.J. Burby, C.B. Groves, S.L. Longville, L.E. Link, J.F. Mount, J. Opperman, R.B. Seed, G.L. Sills, J.J. Smyth, R. Stork and E.A. Thomas. 2007. A California Challenge—Flooding in the Central Valley: A Report from an independent review panel to the Department of Water Resources, State of California. Department of Water Resources, State of California.

Gillard, J. 2011. Queensland communities boosted by continued reconstruction investment. Media release 17 October 2011, Prime Minister of Australia. Available: http://www.pm.gov.au/press-office/queensland-communities-boosted-continued-reconstruction-investment [Accessed 24 June 2013].

Gillard, J. 2013. Government to increase affordability of insurance: Media release 28 February 2013, Prime Minister of Australia. Available: http://www.pm.gov.au/press-office/government-increase-affordability-insurance [Accessed 1 March 2013].

Gillard, J. 2013a. $85 million for Floods Support Package and Flood-Proofing Projects in Queensland: Media Release 28 February 2013, Prime Minister of Australia. Available: http://www.pm.gov.au/press-office/85-million-floods-support-package-and-flood-proofing-projects-queensland [Accessed 1 March 2012].

Government of China. 2007. China's national climate change programme (English version). National Reform & Development Commission, Government of China. Beijing.

Guangchun Lei. (pers. comm.). Professor, Beijing Forestry University.

Handmer, J. and S. Dovers. 2007. Handbook of Disaster & Emergency Policies & Institutions. Earthscan, London.

Hirabayashi, Y., R. Mahendran, S. Koirala, L. Konoshima, D. Yamazaki, S. Watanabe, H. Kim and S. Kanae. 2013. Global Flood Risk Under Climate Change. Nature Climate Change: Letters.

ICA. 2013. Historical Disaster Statistics [Online]. Insurance Council of Australia. Available: http://www.insurancecouncil.com.au/industry-statistics-data/disaster-statistics/historical-disaster-statistics [Accessed 7 July 2013].

IFMRC. 1994. Sharing the Challenge: Floodplain Management into the 21st Century. Interagency Floodplain Management Review Committee.

ILPRC. 2006. The National Levee Challenge: levees and the FEMA Flood Map Modernization Initiative. Interagency Levee Policy Review Committee, Washington DC.

Jha, A.K., R. Bloch and J. Lamond. 2012. Cities and Flooding: A Guide to Integrated Urban Flood Risk Management for the 21st Century. The World Bank, Washington DC.

Kousky, C., S. Olmstead, M. Walls, A. Stern and M. Macauley. 2011. The Role of Land Use in Adaptation to Increased Precipitation and Flooding: A Case Study in Wisconsin's Lower Fox River Basin. Resources for the Future.

Lockyer Valley Regional Council. 2011. Grantham Township—Rebuilding for the Future: Grantham Master Plan Factsheet. Available: http://www.lockyervalley.qld.gov.au/images/PDF/grantham_factsheet_final.pdf [Accessed 24 June 2013].

Lockyer Valley Regional Council. 2013. Media Release: Strengthening Grantham project saves the community millions. Lockyer Valley Regional Council. Available: http://www.lockyervalley.qld.gov.au/news-events/news/1575-strengthening-grantham-project-saves-community-millions [Accessed 24 June 2013].

McBeath, N. 2011. Transcript. Environment and Natural Resources Committee Inquiry into flood mitigation infrastructure in Victoria. Rochester—15 November 2011. Available: http://www.parliament.vic.gov.au/images/stories/FLOOD/Transcripts/Rochester/Lower_Lodden_Irrigators_Recovery_Taskforce.pdf [Accessed 25 July 2013].

Milliard, M. 2012. Roma flood mitigation measures: letter to Councillor Robert Loughnan, Mayor of Roma. Available: http://extras.theglobalmail.org/uploads/flood-maps/Mitigation_Ltr_MM-Roma-04042012.pdf [Accessed 20 August 2012].

Mississippi River Commission. 2011. Mississippi River Commission 2011 MR&T Flood Report, Mississippi River Commission.

Munich Re. 2013. Natural catastrophes worldwide 1980–2012. Münchener Rückversicherungs-Gesellschaft, Geo Risks Research, NatCatSERVICE. Available: http://www.munichre.com/app_pages/www/@res/pdf/media_relations/press_releases/2013/natural-catastrophes-2012-wold-map_en.pdf [Accessed 21 May 2013].

NCLS. 2009. Recommendations for a National Levee Safety Program: a report to Congress from the National Committee on Levee Safety. National Committee on Levee Safety.

Neil McBeath. (pers. comm.). Chair, Lower Loddon Irrigators Recovery Taskforce.

Nijssen, P. 2011. Integrated approach for river safety & urban development [Online]. Available: http://keynote.conference-services.net/resources/444/2653/pdf/IWAWCE2012_0774.pdf [Accessed 24 May 2012].

NSW Government. 2005. Floodplain Development Manual: the management of flood liable land. Appendix L: hydraulic and hazard categorisation. Office of Environment and Heritage.

NSW Government. 2011. Natural Disaster Resilience Program 2009/10-2010-11: NSW Natural Disaster Resilience Grants Scheme Approved Projects, Ministry for Police and Emergency Services. Available: http://www.emergency.nsw.gov.au/ndrgs [Accessed 5th November 2012].

NSW Government. 2013a. Floodplain Management Grants: funded floodplain management projects [Online]. NSW Office of Environment and Heritage. Available: http://www.environment.nsw.gov.au/coasts/Floodgrants.htm [Accessed 1st July 2013].

NSW Government. 2013b. Floodplain management program guidelines for voluntary purchase schemes. Office of Environment and Heritage. Available: http://www.environment.nsw.gov.au/resources/water/coasts/20130055fmpvolpurchase.pdf [Accessed 22 June 2013].

NWF. 1998. Higher Ground: A Report on Voluntary Property Buyouts in the Nation's Floodplains: a common ground solution serving people at risk, taxpayers and the environment. National Wildlife Federation, Washington.

Parliament of Victoria. 2012. Environment and Natural Resources Committee Inquiry into Flood Mitigation Infrastructure in Victoria. Parliament of Victoria.

Pittock, J. (pers. comm.). Senior Lecturer, The Australian National University. Correspondence 2013.

Pittock, J. and M. Xu. 2011. World Resources Report Case Study. Controlling Yangtze River Floods: A New Approach. Washington DC. Available: http://www.worldresourcesreport.org/files/wrr/wrr_case_study_controlling_yangtze_river_floods.pdf [Accessed 8 July 2012].

QCC. 2012. Saving Moreton Bay: investing in our future. Queensland Conservation Council Inc. Available: http://www.savemoretonbay.com.au/assets/untitled%20folder/SMB%20-Sept12-FINAL3%5B1%5D%20copy.pdf [Accessed 24 October 2012].
QFCI. 2011. Queensland Floods Commission of Inquiry: Interim Report. Brisbane.
QFCI. 2012. Queensland Floods Commission of Inquiry: Final Report. Brisbane.
QRA. 2011a. Rebuilding Grantham Together: development scheme Grantham reconstruction area. Available: http://qldreconstruction.org.au/u/lib/cms2/rebuilding-grantham-full.pdf [Accessed 24 June 2013].
QRA. 2011b. Rebuilding Queensland after the natural disasters of the summer of 2010–11, Queensland Reconstruction Authority. Available: http://www.budget.qld.gov.au/budget-papers/2011-12/rebuilding-queensland.pdf [Accessed 15 July 2013].
Queensland Government. 2012a. Natural assets for flood and cyclone: review of planning mechanisms used worldwide for mitigation of natural disasters using a natural assets approach. Department of Environment and Heritage Protection.
Queensland Government. 2012b. Reconstruction journey will continue as thousands rebuild their lives. Media release 9 January 2012. Queensland Premier's Office. Available: http://www.cabinet.qld.gov.au/MMS/StatementDisplaySingle.aspx?id=78384 [Accessed 1 March 2012].
Rural Finance. 2012. Rural Finance Annual Report 2012, Rural Finance Corporation of Australia.
Rutherfurd, I., B. Anderson and A. Ladson. 2007. Managing the effects of riparian vegetation on flooding. pp 63–84. *In*: S. Lovett and P. Price (eds.). Principles for Riparian Lands Management. Land and Water Australia, Canberra.
Sellers, E. and A. Mooney. 2012. So just how exceptional is Fairfield? Contesting the 100 year ARI plus 0.5 metre flood planning level. 2012 Floodplain Management Association National Conference. Batemans Bay, Australia.
Simmonds, J. (pers. comm.). Project Manager, Strengthening Grantham Project. Lockyer Valley Regional Council. Conversation 15 August 2013.
Smith, D.I. 1998. Water in Australia: resources and management. Oxford University Press.
Solomon, S. 2011. The Mississippi, No Longer Muddy, Still a Threat. Available: http://www.bloomberg.com/news/2011-06-21/the-mississippi-no-longer-muddy-still-a-threat-steven-solomon.html [Accessed 24 July 2012].
Squires, V. 2014. People and place as determinants of success or failure to meet management objectives in river basins. pp. xx. *In*: V.R. Squires, H.M. Milner and K.A. Daniell (eds.). River Basin Management in the Twenty-first Century: Understanding People and Place. CRC Press (this book).
Suncorp. 2012. Risky Business: Insurance and Natural Disaster Risk Management. Suncorp Personal Insurance Public Policy. SuncorpGroup Limited.
Taylor, M.P. and R. Stokes. 2005. When is a river not a river? Consideration of the legal definition of a river for geomorphologists practising in New South Wales, Australia. Australian Geographer 36(2): 183–200.
Te Boekhorst, D.G., T.J. Smits, X. Yu, L. Li, G. Lei and C. Zhang. 2010. Implementing integrated river basin management in China. Ecology and Society 15(2).
The World Bank. 2011. Queensland Recovery and Reconstruction in the Aftermath of the 2010/2011. Flood Events and Cyclone Yasi: a report prepared by the World Bank in collaboration with the Queensland Reconstruction Authority Available: http://www.qldreconstruction.org.au/u/lib/cms2/world-bank-report-1.pdf [Accessed 15 July 2013].
VAGO. 2013. Flood Relief and Recovery. Victorian Auditor-General's Office. Available: http://www.parliament.vic.gov.au/file_uploads/20132606-Flood-relief_2BgxQ4Y0.pdf [Accessed 15 July 2013].
van den Honert, R. 2013. Nomination for the Sasakawa Award 2013. Risk Frontiers Natural Hazard Research Centre. Available: http://www.preventionweb.net/files/workspace/30411_nominationforsasakawaaward2013lvrc.pdf [Accessed 24 June 2013].

Water Technology Pty Ltd. 2005. Lower Goulburn floodplain rehabilitation scheme hydraulic modelling Report.
Wenger, C. 2013. Climate Change Adaptation and Floods: Australia's institutional arrangements. National Climate Change Adaptation Research Facility. Gold Coast.
Wenger, C., K. Hussey and J. Pittock. 2013. Living with Floods: Key Lessons from Australia and Abroad. National Climate Change Adaptation Research Facility. Gold Coast.
White, G.F. 1945. Human Adjustment to Floods: A Geographical Approach to the Flood Problem in the United States. PhD, University of Chicago.
Wright, J.M. 2000. The Nation's Responses to Flood Disasters: a historical account. Association of State Floodplain Managers.
Yin, H. and C. Li. 2001. Human impact on floods and flood disasters on the Yangtze River. Geomorphology 41: 105–109.
Yu, X., L. Jiang, L. Li, J. Wang, L. Wang, G. Lei and J. Pittock. 2009. Case study: Freshwater management and climate change adaptation: Experiences from the central Yangtze in China. Climate and Development 1(3): 241–248.
Zhang, H.-L. 2004. China: Flood Management. Case study for WMO/GWP Associated Programme on Flood Management. World Meteorological Organisation & Global Water Partnership.

19

The Role of Territorial Intelligence: The Case of the Thau Territory, Southern France

Roel Plant,[1,*] **Pierre Maurel,**[2] **Olivier Barreteau**[3] **and Yann Bertacchini**[4]

SYNOPSIS

This chapter aims to illustrate through a real-world case study, the Thau Territory, situated on the French Mediterranean coast, how IWRM and territorial development can be combined to better incorporate people and place in IWRM approaches. Developing a territory with an explicit water dimension requires knowledge brokers between top-down technocratic public policies (including water policies) and bottom up development projects—supported by local population and politicians. In the Thau Territory, this role is played since 2005 by a multidisciplinary engineering structure, the SMBT (Syndicat Mixte du Bassin de Thau). Specific STICA (socio-technical information and communication arrangements) can be imagined by these knowledge brokers to bridge the gap between engineers, experts, local people and elected officials. An example is presented which shows how local knowledge of fishermen has been used to make an inventory of the Thau lagoon's seagrass.

[1] Irstea, UHR TETIS Lab, Montpellier, France/ISF, and University of Technology, Sydney, Australia.
Email: Roelof.Plant@uts.edu.au

[2] Irstea, UMR TETIS lab Montpellier, France.

[3] Irstea, UMR G-Eau Lab Montpellier, France.

[4] I3M Lab, University of the South Toulon-Var, France.

* Corresponding author

Keywords: algal bloom, demographic pressures, eutrophication, facilitation, France, free divers, groundwater, information and communication, IWRM, knowledge brokers, seagrass, shellfish, spatial representation, territorial intelligence, Thau, tourism, urban sprawl

1 Introduction

Elsewhere (Maurel et al., this volume) we have addressed how IWRM implementation could be addressed through an approach centered on the territorial dimension of development at the local scale which includes the water component. The term "territorial" is used here to represent its interpretation in the French geographical tradition (Debarbieux 1999). This interpretation of "territory" is more akin to the Anglo-Saxon concept of "place" than to the concept of "territory". The key question that we addressed in the previous chapter was: what concepts and lessons of territorial development could be integrated with IWRM to better incorporate people and place in river basin management? The current chapter aims to illustrate through a real-world case study, the Thau Territory, situated on the French Mediterranean coast, how IWRM and territorial development can be combined.

The chapter is organized as follows. Section two briefly presents the Thau Territory to contextualize the implementation of a combined IWRM and territorial intelligence approach. This section describes the progressive implementation of the integrated approach, based on various innovations such as new institutions, new planning instruments, new governance systems, and new socio-technical information and communication arrangements (STICAs). Section three focuses on an engineering structure, the SMBT (*Syndicat Mixte du Bassin de Thau*), a major innovation established in 2005, which greatly contributed to promoting and supporting an integrated approach. Section three also details a specific STICA that was developed during the planning phase of the decision making process in order to illustrate how local knowledge can be used to strengthen the self-observation capacities of a territory. Section three concludes with a summary of some key innovative elements of the Thau territorial project and their bearings on IWRM. In the final section, we highlight lessons learnt from this work and how these could be transposed to other contexts.

2 The Thau Territory

The Thau Territory is located on the French Mediterranean coast in the department of Hérault, about 20 km west of Montpellier, the regional capital.

Sète, Mèze and Frontignan are the major cities in this area between land and sea. With 117,000 inhabitants in 2010, the Thau basin is the third most populated area of the Department of Hérault.

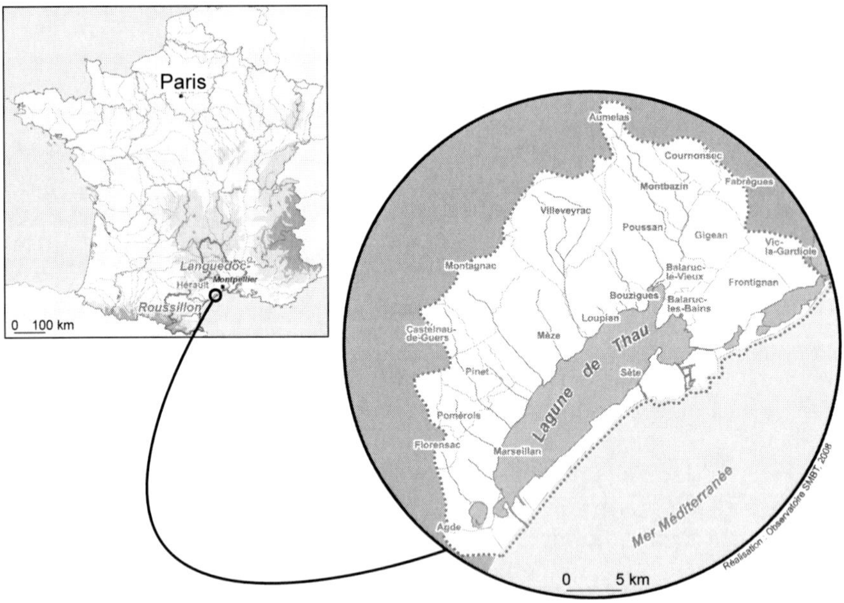

Figure 19.1. Location of the Thau Territory.

Twenty-two municipalities are located in the catchment (water territory). Since the early 2000s the 14 municipalities closest to the lagoon have been joined into two public bodies for inter-communal cooperation (political territories): Thau Agglo, formerly denominated CABT (*Communauté d'Agglomération du Bassin de Thau*) in the south and the CCNBT (*Communauté de Communes du Nord Bassin de Thau*) in the north.

The Thau Territory is dominated by water: 30 km of coastline, several lagoons, including the Thau lagoon (7,500 ha) which represents the unifying element of the Territory (or indeed its great controversy, depending on a stakeholder's perspective), a catchment of 44,300 ha of which 7,500 ha are brackish waters of the Thau lagoon, 27% of the area being covered by water (the highest rate among French coastal territories). Thau is also located upon three groundwater bodies: (i) the alluvial plain of the Hérault River (North-West of the lagoon), (ii) the Jurassic calcareous areas of West Montpellier (North and North-East of the lagoon) and, (iii) the Astien sands (west of the lagoon). Most of the drinking water resources are provided by the Hérault River groundwater table. Neighbouring groundwater resources are already affected by over-pumping in summer due to the tourist industry's demand for drinking water (Orb plain and North Montpellier). The Astien sands are under threat from salt intrusion

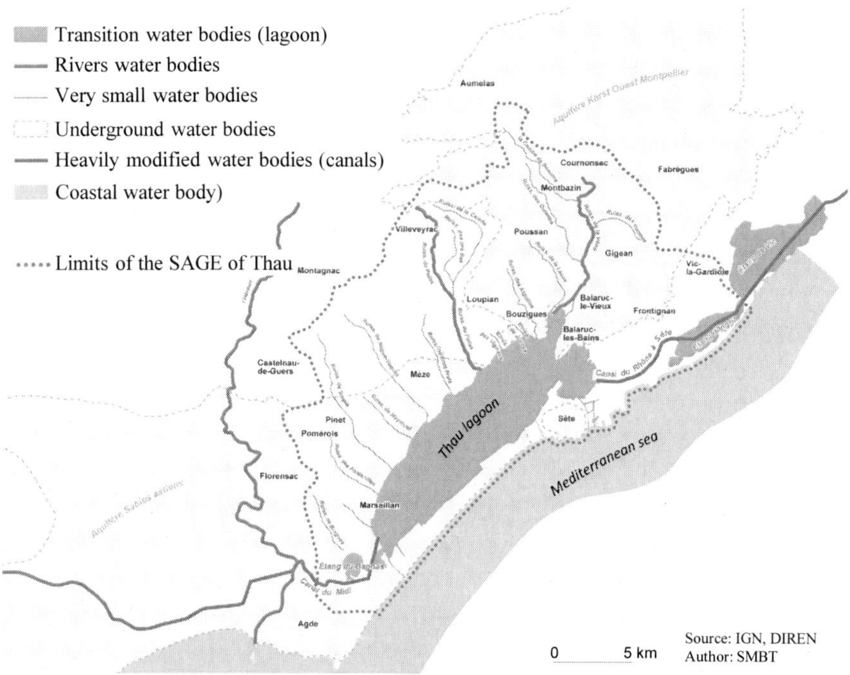

Figure 19.2. Main water bodies of the Thau Territory and surrounding areas.

The Thau Territory is extremely rich in terms of biodiversity and landscapes, between land and sea, coastal plain and wooded reliefs, among lagoons, wetlands and agricultural plains. It covers the catchment area of the Thau lagoon, the main French Mediterranean lagoon due to its size (18,500 acres) and depth (5 m in average). The lagoon sustains a wide number and diversity of marine activities and recreational uses. Major economic activities include Sète's harbor industries, shellfish farming and fishing, wine growing, thermal baths, tourism and leisure activities. Since the 1970s, shellfish farming (oysters and mussels) became the key economic activity. It covers about 20% of the whole lagoon area and produces about

15,000 tons of shellfish annually providing work for approximately 2,000 persons. This activity also contributes significantly to the symbolic image and cultural heritage of the Thau Territory.

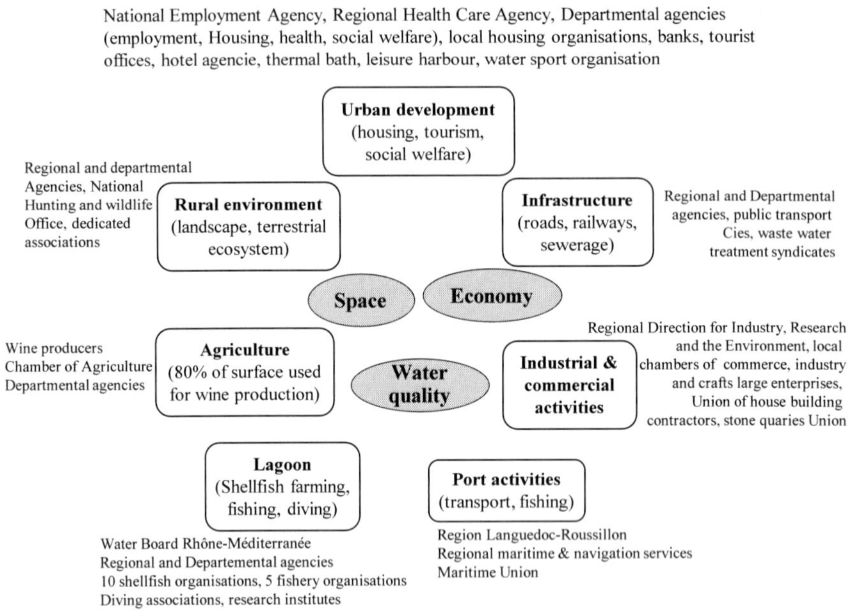

Figure 19.3. Activities and stakeholders in the Thau Territory.

However, due to its location on the Mediterranean coast near Montpellier and Béziers, its intersection by the A9 and A75 motorways and its accessibility by train, this highly attractive area is subject to strong demographic pressures and urban sprawl, especially in summer during the peak tourist season. Projections for 2030 predict Thau Territory to have one of the fastest growing populations in France. These dynamics threaten ecosystem stability, generate tensions between traditional economic activities and the residential economy as well as social inequity due to steep real estate price increases. Industrial and wine production are in crisis although some local production has remained viable. Broader-scale changes, including climate change, also affect the Thau Territory with impacts such as coastline erosion and sea flooding risk.

The lagoon's production activities depend to a large extent on nutrient inputs into the ecosystem, supplied mainly from fresh water. Its catchment area is drained by numerous small streams with intermittent flows. Anthropogenic pressure on the catchment area is due to agriculture (mainly vineyards), agro-food and fertilizer industries, and domestic sewage. Due to these pressures and to low water exchange, in the past decades the Thau lagoon has experienced acute eutrophication problems with anoxic crises (*malaïgue*), resulting in a progressive improvement of certain facilities such as networks and treatment plant infrastructures. In 1989, just before the Christmas and New Year celebrations a historical health crisis occurred. This major crisis had a disastrous effect on the shellfish farming economy. It revealed the dependency and vulnerability of lagoon activities in relation to water quality and the need to better regulate the pressures exerted on the catchment. To face this issue, the French government implemented a Sea Exploitation Scheme (*Schéma de Mise en Valeur de la Mer—SMVM*) in 1995. This implementation was the first in France, although the State-controlled regulatory instrument was introduced in the legislation in 1986. The SMVM of the Thau Territory has mainly been used as a zoning instrument to regulate land use allocation in urban planning documents of all the municipalities located on the shores of the lagoon. The main objective of the SMVM was to ensure the maintenance of shellfish and fishing activities in the lagoon because the State considered these as the best guarantee for preserving the quality of lagoon water. The SMVM was combined with two bay contracts (1990–1995, 1996–2001) that brought significant budgets to upgrade water treatment facilities. The first bay contract was designed and implemented directly by the State engineers due to the urgency of the situation. The State and the Water agency then tried to convince the local authorities to create an engineering structure to manage the second bay contract. In 1999, the authorities finally negotiated the creation of a simple association, *Apogée*, rather than a more ambitious engineering structure.

Despite these various initiatives and other sectorial measures, aquatic health crises and conflicts between activities still occurred. In August 1997, nearly one third of the annual oyster production was lost. This happened again in 1998, due to harmful algal blooms (*Alexandrium*) with direct impact on shellfish production and commercialization. Because of limited human and financial resources related to its legal status, as well as a lack of technical expertise, it appeared that the *Apogée* association was unable to impose its legitimacy and bring effective responses to deal with health crises of the lagoon.

Since the 1980s, the Thau Territory's symbolic image—closely related to the produce of the lagoon—has been deteriorating sharply due to repeated crises, all widely publicized in local and national media. The decline of industrial and port activities, with the emergence of brownfields as a result, has accentuated this phenomenon. A survey conducted in 2005 revealed that scientists and local stakeholders mainly attributed the negative image of the lagoon to the social disorganization of the professional sector (Rey-Valette et al. 2005). The survey also highlighted a general lack of data to support and justify policy decisions. Yet all stakeholders were convinced of the need to restore a positive image of the lagoon which all considered to be an exceptional ecosystem. This point appeared as an explicit objective in the third Lagoon Contract.

If we represent the Thau Territory as a complex and dynamic system based on the territorial model that we presented elsewhere (see Maurel et al., this volume), the Territory appears in steep decline (negative drifts, various hazards), both at the physical level (e.g., lagoon crises, social and economic crises) and symbolically at the identity level (e.g., negative publicity in the media and lack of shared Thau identity among the population), despite innovations at the logical level (e.g., establishment of SMVM and *Apogée* association) and an action program (second Lagoon Contract) embedded in the physical level.

To address this continuing deterioration of the Thau lagoon, two major new innovations at the logical level were then negotiated between the State and Thau's local authority in the early 2000s. The first innovation was the creation of a multidisciplinary engineering structure, the *Syndicat Mixte du Bassin de Thau* (SMBT), in 2005. This structure itself carried (and carries) additional innovations, including new governance mechanisms and new ways for information sharing and communication. The second innovation encompassed the simultaneous development of several additional planning tools, in addition to the implementation of a third Lagoon Contract. This integrated resource planning, which ran from 2005 to 2012, finally led to a regional project which featured new regulatory mechanisms at the physical level as well as the emergence of a new vision for the Thau Territory at the existential level. These changes can be ascribed to the establishment of a territorial intelligence approach presented elsewhere in this volume (see Maurel et al., this volume) and allows the gradual re-emergence and re-empowerment of the Territory.

3 Two Innovations as an Illustration of a Territorial Intelligence Approach

To illustrate how a territorial intelligence approach can contribute to IWRM, we will now present two of the innovations that have been implemented in the Thau Territory. The first innovation is the creation of a new institution, the *Syndicat Mixte du Bassin de Thau* (SMBT). The second innovation describes a socio-technical information and communication arrangement (STICA) specifically designed to promote the incorporation of local ecological knowledge in territorial decision-making.

3.1 Innovation 1: The role of the SMBT and its "animateurs"[1] as knowledge brokers

The Thau Territory comprises two inter-communal authorities (Thau Agglo and the CCNBT) which have been cooperating since 2005 through the SMBT as a joint planning-engineering structure. The purpose of this structure is to deal more effectively with issues related to the dynamics of deterioration described above. The establishment of the SMBT was imposed jointly by the State and the Water Agency *Rhône Méditerranée* under the funding structure of the third Lagoon Contract (2005–2012). With a small team of six engineers at its core, the SMBT became the technical partner that the State and the Water Agency had been seeking for several years.

The State exercised the Solidarity and Urban Regeneration Act (*Loi SRU*) of 2000 and the Water Framework Directive (WFD) requirements to provide the newly established SMBT with jurisdiction over several planning instruments: (i) the third Lagoon Action Plan; (ii) a strategic Land Master Plan (*SCoT—Schéma de Cohérence Territoriale*) covering the two inter-communal authorities; (iii) a maritime extension of the SCoT to better integrate coastal issues; (iv) a Water Development and Management Plan (*SAGE—Schéma d'Aménagement et de Gestion des Eaux*) covering the lagoon catchment; and (v) a Natura 2000 conservation plan for the entire Thau lagoon and its surrounding wetlands. The perimeters of the SCoT and the SAGE are nearly the same—a rare exception in France—and thus facilitate integrated land and water management. The integrated development of these various planning instruments took place between 2006 and 2012.

Immediately after its creation, SMBT successfully responded to a national call for Integrated Coastal Zone Management (ICZM) pilot projects. This pilot gave the Thau Territory national visibility as well as an obligation to develop an exemplary integrated approach. Encouraged by their engineers, the elected officials of SMBT took the decision to carry out an ambitious participatory process which engaged stakeholders from various sectors and thus supported cross-sectoral visioning.

The ambition of the State and the SMBT to apply the ICZM principles to the Thau Territory has, however, been challenged with the fact that each sectoral public policy imposed its own instruments and governance system. This has resulted in particularly complex and burdensome overall governance arrangements when it comes to the joint development of various planning documents (Fig. 19.4).

[1] In our case study, an "animateur" or facilitator is a person who steers a specific planning process (SCoT, SAGA, Natura 2000) and fosters communication among stakeholders. He or she also bridges the gap between technical knowledge, state procedures and local stakeholders.

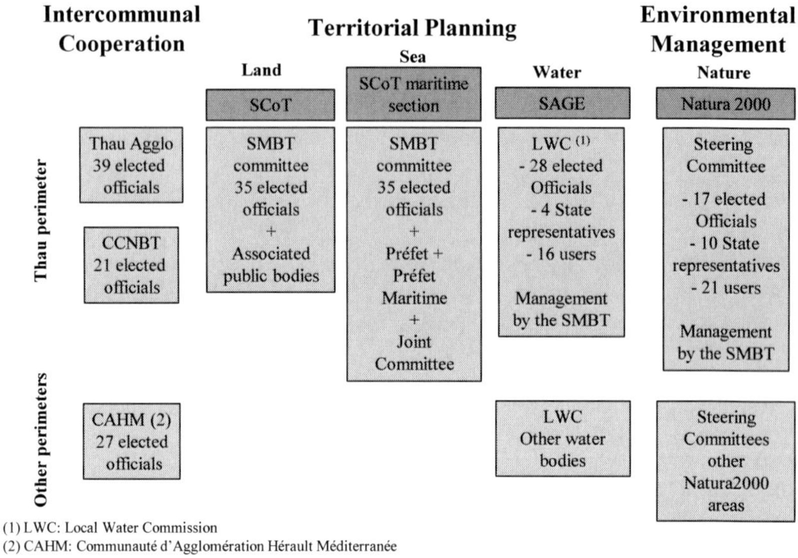

Figure 19.4. Governance system for the integrated planning of the Thau Territory (2006–2012).

This handicap has been partially offset by the work of integration and mediation across sectors and scales carried out by SMBT engineers, both within their own team and with the participants of the different governance systems. From the very beginning, in order to facilitate collective learning and a holistic understanding across sectors, SMBT tried as much as possible to enrol the same elected officials and the same participants in the various steering committees and technical groups. The initial integrated management ambition has thus been transformed into an integration effort, first between different sectoral public policies, and subsequently between these policies and local territorial development projects (Fig. 19.5).

SMBT also succeeded in providing a collaborative environment where the *animateurs* of SCoT, SAGE, Natura 2000 and the Lagoon Contract were hired by the same legal entity and placed in one building, in offices close to each other. They perform similar duties such as collecting and analysing information and preparing syntheses for different audiences (politicians, State engineers, stakeholders, local population). They are bound to exchange information and share views on-site. They share the organization of events such as a public exhibition of the SCoT or the public debate on WFD matters. They play the symmetric role of intermediate actors between the water sector and the Thau Territory, bridging the gap between territorial representatives and water or land-use specialists. Thanks to their mediation,

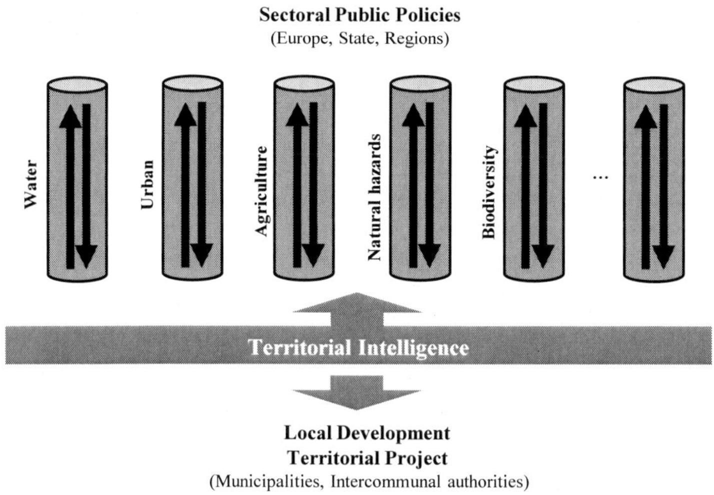

Figure 19.5. Territorial Intelligence as a broker between top-down sectoral public policies and bottom-up local territorial project.

generalists from the Territory may participate in water-related events whilst water specialists can be heard speaking about land-use planning.

Animateurs in Thau also deal with coordination across scales since the Thau Territory encompasses source waters flowing into the lagoon and the sea which in turn have connections with other sub-catchments. There are some interdependent groundwater aquifers which are used by inter-municipal bodies outside the SMBT structure (among them *Montpellier Agglomeration*). *Animateurs* from different sub-catchments interact in an inter-SAGE process which involves meetings and exchange of information to harmonize water plans. Large projects of significance for the whole district like the Rhone aqueduct (*Aqua Domitia*) are also discussed locally by *animateurs*. During thematic commissions and informal meetings, *animateurs* frequently interact with technicians from the *Conseil Général*, the local delegation of the Water Agency, and the State Environmental Department.

The legitimacy of *animateurs* comes from the networks they build around water professionals, scientists, private consultancies, state offices, local authorities and their services, elected officials, fishermen and oyster producers, NGOs, schools, journalists and the larger public. Moreover, the SMBT *animateurs* do not rely just on their own expertise. Rather, they seek to benefit from other skills and knowledge. To this end, they foster and support other experts' participation in and commitment to their networks. Keen on promoting environmental outreach, they also seek partnerships with NGOs, for example to develop activities with schools and the public at large. Proximity to multiple research hubs in Montpellier also creates opportunities for Thau's *animateurs* to improve expertise on their river basin and Territory (Mazouni et al. 2006).

At the logical level of the territorial meta-model, the SMBT's activities considerably strengthened the volume and quality of representations of the Territory (maps, data sets, brochures, artist impressions, etc.). These facilitated the drawing up of the different strategic planning documents (SCoT, SAGE and Natura 2000) that are currently being finalised. From a territorial intelligence perspective, all this contributed to increase the formal capital of the Territory (Bertacchini 2004). The most emblematic measure resulting from this process has arguably been the adjustment (in the SCoT) of population growth and urbanization rate to the carrying capacity of the Territory. Future growth rates will have to be lower than in recent years. The SCoT stipulates that future urban development will be concentrated in the south of the catchment at a location suitable for a high-performing public transportation service linked to the SNCF railway which connects Sète to Montpellier (SMBT 2013). This will help to reduce negative water quality impacts from runoff on impervious areas. A similar policy choice was made to preserve the environmental quality of natural areas and agricultural livelihoods in the north of the Territory.

A new six-year Action Plan (2012–2017), the Integrated Management Contract for the Thau Territory, has been drawn up to implement the land planning and water management orientations contained in the SCoT and the SAGE documents in the Territory at the physical level (SMBT 2012). A new governance system has been designed to support this process (Fig. 19.6). It is based on a permanent strategic committee which represents contracting authorities (municipalities and inter-communal bodies, professional unions), funding partners, and prime contractors. The SMBT is the executive structure of this fourth Contract.

At the identity level of the territorial meta-model, SMBT engineers also assist their elected officials in translating expert knowledge, technical information and statements from actors with different interests and representations into a common understanding and vision for the Territory (see Maurel et al., this volume). For different reasons, including political rivalry, local political leaders have translated and appropriated these elements in a variety of political projects of territory rather than in a single unified project.

Despite this apparent discord in the political stewardship of a unified project for the Thau Territory, a detailed analysis of the content of the various projects and narratives developed by their respective political leaders shows that ultimately they all share the same fundamental and meaningful orientation at the identity level (Maurel 2012): a model of development closely tied to environmental quality, ecological integrity

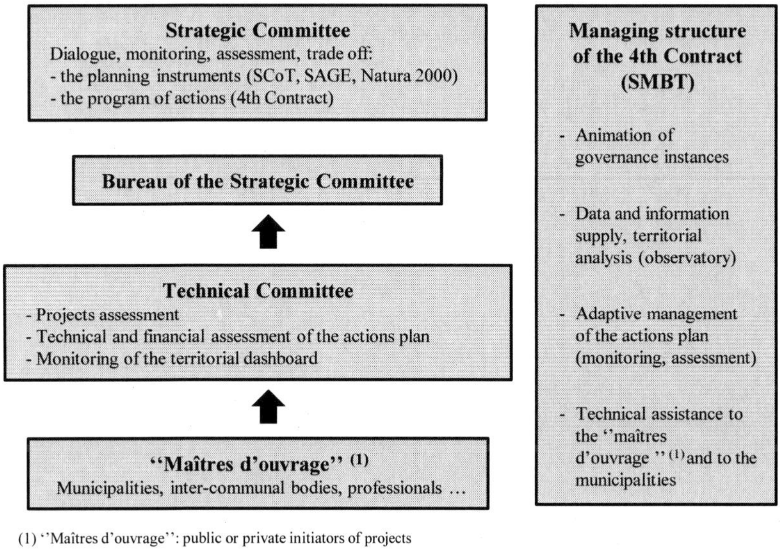

Figure 19.6. Permanent governance system for the adaptive implementation and monitoring of the 4th program of actions (2012–2017).

and traditional economic activities in the lagoon. This vision conveyed by politicians, engineers and relayed in various public spaces, is taking root and is gradually transforming the image of the Thau Territory's identity, shifting from a territory in crisis towards a model of "eco-territory". The transition path and the new model have been a source of inspiration for other territories (Barone et al. 2009).

3.2 Innovation 2: Connecting decision making processes with information and communication dimensions

In the field of spatial planning and IWRM, the decision-making process is in practice still directly related to the model of bounded and procedural rationality (Simon 1982; Jeffrey and Gearey 2006). This model represents a cyclical process, organized in several iterative and intertwined phases, in response to a problem, either identified locally or imposed externally (e.g., a SCoT imposed by the State to coastal municipalities): (i) the organization of actors affected by the problem; (ii) a phase of intelligence (inventory, diagnosis, prioritized issues); (iii) a prospective phase to imagine the future to address these issues; (iv) a modeling and choice phase (comparison of scenarios); (v) a phase of development and implementation of an action plan; and (vi) a phase of monitoring and evaluation to measure the effects of actions and possibly redirect any remaining actions.

Public policies are also accompanied by specific "policies of representation" (in the cognitive, informational and communicational senses of the term), that is to say arrangements of formal representations of reality, as highlighted by the sociology of quantification (Desrosières 2000) and illustrated for instance in the field of flood risk management (Le Bourhis 2007). These representations can take the form of measurements, numbers, statistics, databases with their associated conceptual models and semantic descriptions, maps and their legends, indicators, and model results. Representations play a fundamental role for the public policy custodian. Indeed, they can impose analytical frames of reference (Goffman 1974; Minsky 1974), especially through mapping, for the interpretation of a complex reality to develop shared representations, both of the object to manage and of the stakeholder community. They also reinforce the authority of the policy maker, legitimate public actions and regulate the practices of local actors.

This approach requires a change of attitude from the local stakeholders in terms of their treatment of the semiosphere and information (Lotman 1990). They need to abandon their communication routines to adopt a new logic of a shared territorial project. Attitudinal change also implies the adoption of a collective territorial intelligence stance to anticipate the risks of breakdown (Herbaux 2009). Below we will discuss several kinds of STICA that can foster and support these changes.

Event-based/temporary and long-term/permanent STICAs

Within multi-actor decision-making processes, we distinguish between temporary STICAs, used to support specific events (workshops, public meetings, etc.), and STICAs that have greater permanence in time (online newspapers, websites, steering committees, municipal councils, etc.). We highlight two permanent STICAs, both used to manage the formal capital, that we will call "*Heritage-type STICA*" and "*Observatory-type STICA*" (Fig. 19.7).

The "*Heritage-type STICA*" aims to fuel various ephemeral STICAs with cartographic supports (maps and aerial or satellite imagery used as background layers) and information content (data, reports, thematic maps, etc.) from the beginning of the decision cycle up until the choice of a strategic development project. The "*Heritage-type STICA*" provides a horizon of meaning, legitimized by the local political authorities and an action plan. In return, it receives, classifies (for instance through metadata cataloguing), archives and makes accessible the media products produced by temporary STICAs and its partners.

The "*Observatory-type STICA*" supplements the "*Heritage-type STICA*" from the moment the territory has decided on a strategic development

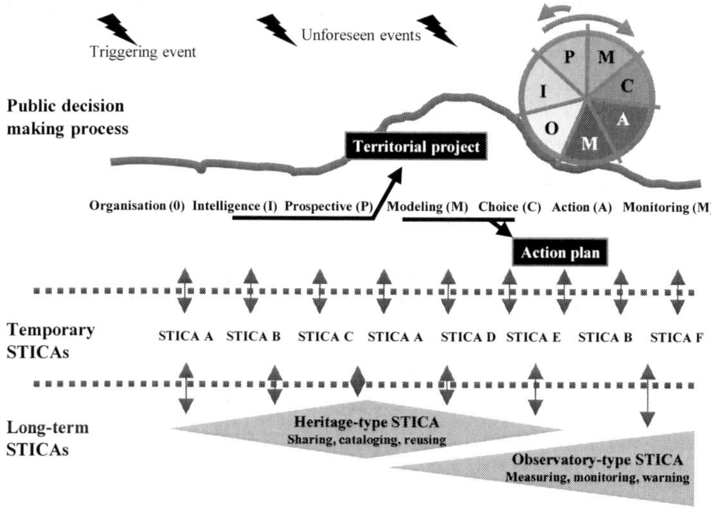

Figure 19.7. Relation between territorial decision making process and STICAs.

project, to be implemented by those involved in the action plans. Based on monitoring and warning indicators which make sense to local actors, this STICA allows measurement of changes at the physical level of the territorial meta-model. Such an arrangement requires a preliminary analysis (scoping) and a design phase to identify relevant media products with all stakeholders. In the "area of social cooperation for reception", sharing and interpreting this information and their associated signs allows a regular assessment of the progress of the territory towards the desired horizon in relation to the actions which have been implemented. Corrective actions can then be decided depending on the results of the previous analysis. Such an approach is consistent with adaptive land and water management principles, acknowledging that decisions about future impacts are inherently based on partial knowledge and emphasize learning and adjustment as principles of best practice (Walters 1986; Timmerman and Pahl-Wostl 2008).

Potential functionalities of spatial representations and gaps in current territorial engineering practice

Among the media products within a STICA, spatial representations are of special interest (Maurel 2001). They includes both mental and material representations (maps, aerial photographs or field pictures, satellite images, 3D models, etc.) of land features and territorial phenomena. The issue of territorial development and IWRM concerns in essence, spatial phenomena with both individual and collective dimensions, for example identity, territoriality, attachment to place, familiar spaces and emotions (Bailly 1991; Raffestin 2012). Communications on space and territory can, of course, be verbal only. However, there are also phenomena that cannot easily be spoken about. Such phenomena can be brought to the fore using spatial representations and allowing actors to discuss them. The spatial dimension of territorial reality offers real potential for mediation between heterogeneous social groups and creation of collective meaning. In this case, spatial representations can play the role of "boundary object" (Star and Griesemer 1989).

Defining and designing a STICA based on spatial representations calls for questioning the roles that such design artifacts can play to establish specific communication between participants in a development project, whether in the "area of social cooperation for production" or in the "area of social cooperation for reception". The scientific literature highlights six major categories of functionality that can be performed by various spatial representations, each within a particular register: (i) analytical (understanding); (ii) creative (imagining); (iii) cognitive (learning); (iv) relationship—between individuals, social groups, institutions—(connecting); (v) decision (changing), and (vi) operational (acting). A detailed list of these functionalities is given in Bertacchini et al. (2013). Furthermore, analysis of 13 guideline documents used by territorial engineering practitioners shows that the model of expertise and technical rationality, dominated by the traditional coupling of elected officials and technicians, continues to drive the content of current methodological handbooks dealing with public participation and communication strategy. Cognitive and relational functionalities of spatial representations are rarely taken into account in territorial mediation practices, thus confirming observations already made by territorial development specialists (Debarbieux and Lardon 2003).

A STICA as mediator: an example to promote local knowledge of seagrass ecology

Several STICAs have been designed and used during the integrated planning process of the Thau Territory between 2006 and 2012, such as: a web site as a heritage-type STICA, 3D physical models for mediation purposes (Maurel and Bertacchini 2008), sustainable development indicators workshops (Rey-Valette et al. 2007) and a warning information system to prevent microbiological pollution (Brocard et al. 2010). An observatory-type STICA (see above) is currently under development to evaluate the new action plan (2012–2017) and, once implemented, regularly monitor and assess the impact of the actions on the Thau Territory (Lemoisson et al. 2012). This STICA will foster an adaptive management approach.

The STICA described below demonstrates how local knowledge can be used to better integrate people and place within IWRM and territorial intelligence approaches. This STICA deals with the achievement of an inventory of zostera seagrass in the Thau lagoon within the framework of a Natura 2000 project. Zostera seagrass is considered a meaningful proxy indicator of the degree of conservation of the Lagoon. Seagrass is a habitat provider for important biodiversity and contributes to water aeration. Seagrass also mitigates lagoon bank erosion.

To achieve the seagrass inventory, the SMBT set up a hybrid working group consisting of local lagoon professionals (fishermen), ecologists and geo-information scientists, and diving and marine ecology associations. As there was no set national methodology for seagrass mapping, three complementary methods were used based on the expertise and skills of the group members: (i) direct observation by freedivers towed by a boat equipped with a GPS; (ii) observation by remote sensing techniques; (iii) mapping based on the knowledge of professional fishermen operating on the lagoon. Here we present the third approach.

We designed a STICA in partnership with the SMBT in order to produce a zoning map of the lagoon seagrass by transforming tacit knowledge of

Figure 19.8. A STICA to set up an inventory of zostera seagrass of the Thau lagoon based on local knowledge.

Color image of this figure appears in the color plate section at the end of the book.

local fishermen representatives into objective data. The first design activity was to produce a cartographic background to help fishermen to 'outsource' their local knowledge by spatializing it as accurately as possible. Preliminary exchanges with the fishermen and other stakeholders of the working group showed that fishermen get their bearings using observable landmarks on the lagoon (bathymetry, shallows named "*toc*" in French which run parallel to the coastline) and ashore (permanent fishing stations equipped with nets, mobile fishing stations located by fishermen within two meters based on landmarks such as shellfish tables, water towers, groves of trees, steeples or other tall buildings).

The background map was developed on the basis of pragmatic ergonomic criteria. The aim was to provide a geo-referenced support with meaningful features for the fishermen. On the other hand, this background map had to be free of any information that would bias the cognitive activity of the fishermen based on their mental representations. This cartographic document should also support an exercise of co-design mapping done by various people working simultaneously and interactively. This document eventually took the form of a 1: 10,000 scale base map, two meters long and 80 cm wide, based on the terrestrial part of the reference IGN (French National Geographic Institute) map and for the lagoon area and a bathymetry grayscale map representing depth ranges. The above-mentioned landmarks were added and highlighted: lagoon shellfish tables, number of each fishing station, "*toc*" shallows (as mentioned below), lighthouses and towers.

The actual drawing exercise took place over less than half a day. It was recorded and filmed for subsequent detailed analysis. During the workshop, the first discussion focused on the alignment of the above-described backdrop landmarks on those mentally used by fishermen to exercise their profession. This alignment was done through a series of individual activities and several exchanges between fishermen and other participants, leading, for example, to modify the "*toc*" shallow location, which was incorrectly positioned. Delineation of seagrass was then performed rapidly, within the time span of half an hour. During this exercise, we observed the fishermen suspending their drawing on several occasions, looking for landmarks with their eyes and through gestures (e.g., counting the rows or columns of shellfish tables or tracing gestures of imaginary lines with landmarks) and then resuming their drawing. They also told several stories about their practical experiences. They expressed sometimes extremely detailed knowledge of the lagoon, providing interpretations of its ecological functioning, related to perceived signs, and of seagrass changes over time. They also managed to draw and date former areas of seagrass.

At the end of the drawing exercise, which lasted less than half an hour, the map produced by the fishermen was used as a visual support for informal discussions, extended and renewed by overlaying the map with another one made by a scientist in 1994. The hand-drawn map produced by fishermen was digitized and given to the SMBT's GIS engineer to derive various maps. Maps were subsequently compared with those obtained through two other methods (visual surveys by freedivers towed by a boat and remote sensing mapping) during a workshop with all the participants and the scientific reporter of the Thau Natura 2000 project. The few discrepancies identified between the drawings were discussed one by one, each bringing some explanations. The record of the meeting shows that these differences were mainly due to differences in the geo-positioning techniques and the categorization systems of reality. If the maps provided by freedivers towed by a boat equipped with GPS are the most detailed, then hybridization and complementarily between the three methods have increased the reliability and the recognition of the final maps produced by this working group.

Feedback from participants, including selected scientists, highlighted that the main benefits of this participatory experience were both mutual recognition and creation of shared representations of reality from scattered knowledge. Although several points for improvement were suggested, the complementary of the methods and their cross-validation had reassured all participants after the initial uncertainties due to the novelty of the approach compared to the as-usual professional practices to implement environmental public policies. Scanning the hand-drawn map into the GIS also helped to give it technical credibility among the State administration and the domain experts.

The results were formally presented to the steering committee of the Natura 2000 project, a formal body of elected representatives, governmental services, local authorities and professional users of the lagoon. Validation of the results by all participants explicitly meant the recognition of local stakeholders' knowledge and skills, thereby reinforcing their commitment to the project. It also demonstrated the ability of SMBT to develop innovative approaches. The maps produced by the freedivers and the fishermen have been included in the regulatory documents of the initial seagrass inventory. This test of external legitimacy of the approach and the data produced has been successfully extended through presentations of the project by various participants who acted as spokesmen of this approach within their respective professional networks.

The above example of mapping based on complementary approaches and the mediation of local ecological knowledge demonstrates that it is possible to rely on local resources to increase the formal capital of the territory as well as to strengthen local actor networks and involvement. This result fulfils the conjunction of the first two of the three factors required for the development of a territorial intelligence process: (i) agents share information, and (ii) they give credit to the received information because it provides them with a benefit. The established communication process is a precursor for the third factor: actors developing appropriate networks to transfer their expertise for the development of a policy.

4 Discussion

To advance practical IWRM implementation we proposed (Maurel et al., this volume) a complementary approach centered on the territorial dimension of development at the local scale which includes the water component. Below we highlight the key concepts and lessons of this approach that are, in our view, most amenable to integration into IWRM so as to better incorporate people and place in river basin management.

The case study that we used to illustrate the territorial approach—the Thau Territory in Southern France—had favorable initial conditions such as: the near-perfect alignment of the catchment and the perimeters of political territories; the willingness of the state to defend the priority given to traditional activities on the lagoon (oyster production and fishing) because they guarantee the quality of water resources; and the historical importance of water in the bio-physical space, economic activities and local culture.

The main innovations observed in the Thau Territory that promote the development of an IWRM approach in synergy with territorial development and which could be transposed elsewhere are: (i) the creation of a single engineering structure (SMBT) as coordinator of several sectoral policies; (ii) the integration of expertise within such a structure composed of "*animateurs*" with complementary mandates and skills (technical pluri-sectoral knowledge, skills in human relationships); (iii) the construction of a shared political vision of the territory; (iv) the attention paid to actions of mediation between the technocratic sphere, local people and elected officials; (v) the design of innovative governance arrangements and information and communication systems (STICAs) to support this goal of mediation; and (vi) to allow the promotion of lay expertise (see Maurel et al., this volume for a fuller discussion of these matters).

Several general limitations should be highlighted. The absence of full electoral support for inter-municipal institutions weakens the engagement of the elected officials and those of citizens in building a shared and legitimate vision for the future at the scale of inter-municipal territories where the implementation IWRM is relevant. Furthermore, the technical instruments of planning and management employed under the innovative institutional arrangements often remain sectoral. Their concurrent use with stakeholders can lead to governance systems with sometimes prohibitive transaction costs. The first tier of impact concerns the engineers who often have to cope with participation fatigue. They must also find a balanced relationship with their elected officials to avoid their disengagement and to help them understand this inadvertent additional territorial complexity. Thus, it seems that integrated management through a territorial approach by coordinating various scales and sectoral policies with their own instruments has some limits. It is likely that new paradigms and associated tools—fewer, simpler but intrinsically more integrative tools—will need to be developed in due course.

References

Bailly, A. (ed.). 1991. Les concepts de la géographie humaine. Armand Colin, Paris.

Barone, S., C. Gramaglia and G. Bouleau. 2009. Bridges between lagoons: the political construction and uses of a success-story in Thau and Berre (South France). 5th International Conference on Interpretative Policy Analysis. Grenoble.

Bertacchini, Y. 2004. Le territoire, une entreprise d'intelligence collective à organiser vers la formation du capital formel. Communication & Organisation 25: 35–51.

Bertacchini, Y., P. Maurel, P. Deprez and R. Plant. 2013. Spatial information & communication arrangements: a contribution to territorial intelligence. International Journal of Computer and Information Technology 4: 19–28.

Brocard, G., V. Derolez, O. Serais, A. Fiandrino, C. Lequette, C. Lescoulier, M. Benedetti, P. Couton and D. Marty. 2010. OMEGA Thau: outil de management environnemental et de gestion de l'avertissement des pollutions microbiologiques du Bassin de Thau. Novatech 2010—7ème conférence internationale—Techniques et stratégies durables pour la gestion des eaux urbaines par temps de pluie. Lyon.

Debarbieux, B. 1999. Le territoire: Histoires en deux langues (a Bilingual (his-)story of territory).

Debarbieux, B. and S. Lardon. 2003. Les figures du projet territorial, La tour d'Aigues, Editions de l'Aube.

Desrosières, A. 2000. La politique des grands nombres. Histoire de la raison statistique, La Découverte Poche.

Goffman, E. 1974. Frame Analysis. Harper and Row, New York.

Herbaux, P. 2009. Anticipation and Territory: Does collecting information need citizens? International Conference of Territorial Intelligence, Besançon 2008. Papers on Tools and methods of Territorial Intelligence. Besançon: MSHE, Besançon 2009.

Jeffrey, P. and M. Gearey. 2006. Integrated water resources management: lost on the road from ambition to realization? Water Science & Technology 53: 1–8.

Le Bourhis, J.-P. 2007. Du savoir cartographique au pouvoir bureaucratique. La politique des risques en France 1970–2000. Genèses 68: 75–96.

Lemoisson, P., J.-P. Tonneau and P. Maurel. 2012. Penser et piloter l'action au cœur des territoires, le projet d'observatoire du Bassin de Thau. Symposium PSDR Les chemins du développement territorial. Clermont-Ferrand.

Lotman, Y.M. 1990. Universe of the Mind: A Semiotic Theory of Culture. I.B. Tauris & Co. Ltd., London & New York.

Maurel, P. 2001. Les représentations spatiales: concepts de base et éléments de typologie. *In*: S. Lardon, P. Maurel and V. Piveteau (eds.). Représentations spatiales et développement territorial. Hermès, Paris.

Maurel, P. 2012. Signes, Données, Représentations Spatiales: des éléments de sens dans l'élaboration d'un projet de territoire intercommunal. Application au territoire de Thau. Thèse de doctorat en Sciences de l'Information et de la Communication (sous la dir. Yann Bertacchini), Université du Sud Toulon/Var.

Maurel, P. and Y. Bertacchini. 2008. Conception, representation and mediation in participatory land planning projects: 3D physical models artefacts. *In*: Caenti (ed.). Territorial Intelligence and Governance Conference Besançon.
Maurel, P., R. Plant, O. Barriteua and Y. Bertacchini. 2014. Beyond IWRM: developing territorial intelligence at the local level. pp. 22–41. *In*: V.R. Squires, H.M. Milner and K.A. Daniell (eds.). River Basin Management in the Twenty-first Century: Understanding People and Place. CRC Press (this book).
Mazouni, N., L. Loubersac, H. Rey-Valette, T. Libourel, P. Maurel and J.-C. Desconnets. 2006. SYSCOLAG: a transdiciplinary and multi-stakeholder approach towards integrated coastal area management. An experiment in Languedoc-Roussillon (France). Vie et Milieu—Life and Environment 56: 265–274.
Minsky, M. 1974. A framework for representing knowledge. *In*: P. Winston (ed.). The Psychology of Computer Vision. McGraw-Hill, New York.
Raffestin, C. 2012. Space, territory, and territoriality. Environment and Planning D: Society and Space 30: 121–141.
Rey-Valette, H., Carrié and F. Valette. 2005. Analyse de la perception de l'étang de Thau: résultats de l'enquête auprès des populations de Sète, Mèze et Montpellier. Rapport projet DITTY. Montpellier.
Rey-Valette, H., S. Damart and S. Roussel. 2007. A multicriteria participation-based methodology for selecting sustainable development indicators: an incentive tool for concerted decision making beyond the diagnosis framework. International Journal of Sustainable Development 10: 122–138.
Simon, H.E. 1982. From substantive to procedural rationality. Models of Bounded Rationality 2: 424–443.
SMBT. 2012. Contrat de Gestion Intégrée 2012–2017. 325.
SMBT. 2013. SCOT du Bassin de Thau. Projet d'Aménagementet de Développement Durables du SCoT et de son chapitre individualisé valant SMVM. Février 2013.
Star, S.L. and J. Griesemer. 1989. Institutionnal ecology, 'Translations' and boundary objects: amateurs and professionals on Berkeley's museum of vertebrate zoology. Social Studies of Science 19: 387–420.
Timmerman, J. and C. Pahl-Wostl. 2008. The adaptiveness of IWRM. Analysing European IWRM research, London, Moltgen, J (eds.) IWA Publishing.
Walters, C.J. 1986. Adaptive Management of Renewable Resources. McGraw-Hill, New York.

20

Uniting Perspectives: People and Place in River Basin Management

Katherine A. Daniell,[1,*] Hugh M. Milner[2] and Victor R. Squires[3]

SYNOPSIS

This chapter brings together some of the main issues related to people and place in river basin management that need to be acknowledged and are presented in this book. It first examines the power of people and their relationships in defining the scope and focus of water management, including how river basins are socially constructed and how power can shape IWRM in practice. It next focuses on the importance of place and contextualized knowledge for developing improved river basin management practice. It finally considers the complexity of water governance systems, governance mechanisms and the need for continuous adaption of brokering mechanisms for integration and innovation.

Keywords: Integrated Water Resources Management, River Basin Management, water governance, power, relationships, inequality, knowledge, stakeholder engagement, participatory process, brokering, complexity, integration, ecosystem services, economic instruments, market mechanisms, boundary spanners, facilitation

[1] Fellow, HC Coombs Policy Forum, ANIPP, Crawford School of Public Policy & Centre for Policy Innovation, Research School of Social Sciences, The Australian National University, Canberra ACT 0200, Australia.
Email: katherine.daniell@anu.edu.au

[2] Freelance International Water Resources Management Consultant, Sydney Australia.
Email: hugh.milner@gmail.com

[3] Visiting Professor, Gansu Agricultural University, Lanzhou, China, International Consultant, Adelaide, Australia.
Email: dryland1812@internode.on.net

* Corresponding author

1 Introduction

Rivers, and the waters within and connected to them, are the lifeblood of the places and people where they are found. This fundamental life-giving or life-removing capacity of water gives its management special importance in people's lives and their aspirations for the future. As populations have grown and settled in greater concentrations, their reliance on common sources of water—rivers, lakes, springs, aquifers—in their proximate landscape has led to a variety of challenges (Squires, Daniell and Milner, this volume) requiring increased attention to be given to the management of water. Over human history, those people and places that have succeeded in managing their water for the benefit of a majority of people and ecosystems have typically prospered and those who have struggled have either witnessed decline, death or important societal and ecological reconfigurations. More recently, technological developments and the power of the state have resulted in populations local to rivers and their special places losing control over the changes in these landscapes. Developments in water management have sought to achieve a balance between local, commercial, state and international interests and plans.

In this final chapter, we summarize some of the main threads of thinking and issues for river basin management practice that have emerged through the book and seek to reflect on them in relation to other global academic literature and practice examples.

2 Acknowledging the Power of People and Their Relationships in Defining the Scope and Focus of Water Management

Today we understand that the processes of water management are rarely simple and note their typical complex, uncertain and conflict-ridden nature. How water is managed is inextricably linked to power, knowledge and the social structures that are developed, maintained and adapted over time around it. As highlighted in the 2006 United Nations Development Program's Human Development Report, "*The scarcity at the heart of the global water crisis is rooted in power, poverty and inequality, not in physical availability*" (UNDP 2006). Although, of course, there are still important technical challenges to be overcome in developing access to adequate quantities and qualities of water for the people and places that require and seek it, it needs to be acknowledged that addressing some of the social and political challenges of water management is likely to more quickly lead to improvements in the lives of many, including the almost 780 billion people who do not have access to sufficiently clean drinking water and 2.5 billion who do not have access to adequate sanitation (UNICEF/WHO 2012). Even though most water managers, and people and places with a stake in its management, tend not to acknowledge power and the role that power plays in shaping water management systems and societal structure, developing such an understanding, and of its mechanisms and enactments, and then changing current water management institutional processes based on this understanding, is necessary in order to develop more sustainable water and river basin management practice around the world (see, for example, Swyngedouw 2004; Mollinga et al. 2007; Molle et al. 2009).

2.1 The social construction of river basins

Extending this understanding, we emphasize that river basin management is a social process, not just a technical one, and that even the process of defining what a specific river basin encapsulates is a question that can only be socially, not technically, defined and is contingent on the time and place where this process of social construction occurs (see also Bourblanc and Blanchon 2014, for a recent discussion on this issue).

As demonstrated by the discussion by Du et al. (this volume) about the Yellow River, we can see that, based on a technical hydrological definition of a river basin linked to topography (see Ffolliott and Brooks in this volume for these definitions), as a part of the river is now raised above the surrounding land it has no contributing "basin" on either side of it. However, if we were to consider the people living around this section of the river, it is likely that they would strongly identify with the Yellow River and the potential of being flooded by it if its levees were breeched. In other words, they would include themselves in the "basin" of the river's influence. Riddell and Thuo (this volume) also discuss the importance of the Nile's physical basin and the contributors and users of its water, as well as its influence over the wider region. They explain how this influence needs to be understood within its historical and current political context, in order to propose processes within which knowledge of alternative water allocations can be considered.

River basins are commonly administratively defined for management purposes, although this is not always along hydrological boundaries but sometimes along jurisdictional ones for a range of reasons, including ensuring political power is maintained over certain water resources or for reasons of government cost linked to economies of scale. The other obvious example occurs when hydrologically distinct basins are linked via infrastructure such as pipes that enable inter-basin transfers. In this case, the people and place with common water to manage is expanded and a new socially and physically connected basin could be defined. Put more simply, river basin management does not just occur around a so-called "watershed", but needs to be considered within the scope of a socially defined "problemshed" (Allan 1998; Loucks 1998). All this means that exactly *who* is involved in river basin management, *when* and *how* is of utmost importance to the eventual management focus, outcomes and who will most benefit or be marginalized or disadvantaged by the decisions taken and eventual actions taken.

The answers to who, when and how are culturally and socially influenced. This book gives a range of examples of institutional processes that have been developed, along with discussion of their effectiveness.

2.2 Power and IWRM in practice

Globally, integrated water resources management (IWRM), also known as "integrated watershed management" has emerged as the most commonly referred to and promoted "standard" for improving management of water and river basins. Both Mukhtarov and Cherp (this volume), and Maurel et al. (this volume) detail the diversity of definitions of IWRM, its historical rise as the dominant water management paradigm or concept on the world stage and how its implementation is notoriously difficult and increasingly criticized. IWRM has conceptual appeal to many (like sustainability) as it is seen as a positive step in the right direction, as "integration" of a range of parts of water and related systems is seen as important to overcome challenges caused by: 1) a lack of communication between individuals, groups, sectors and levels of governance (*social and institutional integration*), such as policy makers and administrative managers, community members, technical experts, and NGOs or local, regional, national and international levels of administration and governance; and 2) consideration of complementary or connected domains or issues (*functional integration*) such as land and water, conservation and development, economic sectors like industry and agriculture, surface and ground water, socio-economic and environmental issues, blue and green water, upstream and downstream concerns, coastal and inland waters, water quality and quantity, water supply and demand management, flood and drought considerations, and water and wastewater (GWP 2000; Pahl-Wostl et al. 2005; Claassen 2013; Lubell and Edelenbos 2013; Verkerk and van Buuren 2013; Mitchell, this volume). Yet, as the studies on integration and fragmentation in the governance literature note, not everything and everyone can be successfully integrated into any specific management or policy process, so choices about what is to be left aside must be made (e.g., Peters 1998; Bache and Flinders 2005).

It is this important issue of choice around the scope of what particular people see as important to implement in any particular manifestation of IWRM around the world that leads to its perceived usefulness or uselessness as a concept by different practitioners and commentators. For example, having the power to decide who will be included in a "participation process" or what the "integrated plan" will encompass will have significant ramifications on the version of IWRM that is implemented. Stakeholder legitimization as well as the provision of a broader and more relevant knowledge and resource base on which to build decisions and actions are two important reasons for the importance of the developing participatory planning approaches to IRWM (see Ffolliott and Brooks, this volume, for further reasons and discussion) that will ideally lead to co-management practices for its on the ground implementation (Berkes 2009; Pradhan et al., this volume). However, for these participatory approaches to be effective, attention must be given to the organizational challenges and power struggles associated with their development, design, implementation and evaluation (Daniell 2012). Maurel et al. (this volume) and Plant et al. (this volume) describe the importance of facilitation in resolving such conflicts of power and influence, and give a practical example of how specific recognition and appointment of individuals to this facilitation role leads to successful outcomes.

Kingsford et al. (this volume) provide an example of how power to focus water management more on conservation or development in the Lake Eyre Basin was affected by the Queensland Government's choices on who was to be involved in the official advisory processes, with the exclusion of one state and environmental interests by the new Government leading to much more "pro-development" conclusions than a similar process under a previous Government.

Through analysis of water management issues in the Orange River Basin in southern Africa, Sullivan (this volume) also alludes to issues of power, cautioning that not dealing effectively with stakeholder representation in water management processes and decisions can lead to the disenfranchisement of weaker social groups and strengthen existing social inequities. This can be tricky as those with the power to decide on who is included in such processes may prefer to retain power and their vision of key values in water management, rather than have to negotiate it with others. This occurred in the first phase of development of the Murray-Darling Basin Plan in Australia where a top-down planning process disempowered and angered stakeholders who were used to having a say in how the basin was managed (Daniell 2011), and eventually lead to major changes in the management structure and more genuine stakeholder engagement being reintroduced thus altering the chosen balance in the plan between environmental and socio-economic interests (Loch et al., this volume). Despite not all stakeholders being satisfied with the outcome, the more collaborative process taken in this second phase of planning has allowed more in-depth exchanges of views and local knowledge to be considered, building legitimacy that was not previously forthcoming for water management decisions. This has allowed the Murray-Darling Basin

Plan to eventually be passed into law in 2012. Across the basin there has to be a focus on the development of water resource plans for sub-basins, and work with local communities to understand their specific needs, aspirations and capacities to adapt to the required water-constrained futures under the "Sustainable Diversion Limits" that will be imposed from 2019 (MDBA 2013).

Unless communities and impacted stakeholders are empowered to fully participate in this process, these political reforms and the health of the basin's ecosystems and communities will be further jeopardized. The appropriate and acceptable processes for participation must vary depending on the cultural and social setting. What works in Australia or USA is unlikely to work without careful adaptation in a different cultural context.

3 Acknowledging the Importance of Place and Contextualized Knowledge for Developing Improved River Basin Management Practice

Big visions for river basin management are important, but without considering the specificities of place, including different hydrologies, climates, population structures, histories, cultures, socio-economic situations and population capacities, may lead to inappropriate or unimplementable water management actions at the local scale (see also Mitchell, this volume, for further discussion). Understanding how people in different places have previously managed water and conceptualized their water systems, territories and roles and meanings within them is vital for developing appropriate IRWM for different contexts. This means having an understanding of their underlying mental models and their preferences towards decentralized/centralized, public/private, and ecosystem vs. human system-orientated management models, which is likely to influence how they appropriate particular versions of IWRM in practice. Such practice was used, for example, as part of the IWRM and associated territorial planning processes in the Thau basin, France, where sea-grass mapping in the Thau lagoon was carried out through participatory processes where the knowledge of locals, including shell-fish farmers, was used to produce maps and discuss differences with other maps created by visual surveys by freedivers towed by a boat and by remote sensing mapping. In this case, the knowledge was formalized in GIS maps to enhance its legitimacy and worth in the eyes of other territorial decision-makers (Plant et al., this volume).

3.1 History and development of local and contextually appropriate knowledge for river basin management

Different places around the world have varying histories related to the underlying values of water that communities and managers accord with and how these manifest in the local water management practices, systems and governance structures. For example, some areas have typically relied upon largely decentralized systems with individual householders, farmers and businesses managing their own water (e.g., through individual groundwater pumps/sources, rainwater collection systems and wastewater disposal and treatment systems such as septic tanks or wetlands), whereas others have very long histories of more centralized management structures owing to large infrastructural systems such as irrigation and drainage systems, dams, dykes, desalination and large-scale stormwater, wastewater treatment and sewerage systems, which are typical in areas of higher population density. Over time, practices are adapted to manage and reinforce these systems of functioning, creating place-specific water management cultures built upon both tacit and explicit knowledge of the inhabitants and key water managers. Yet, such systems, although often strongly supported at the local or higher levels, may also have created problems in the local area or further afield in the basin, for example pollution, environmental degradation such as biodiversity loss, land salinization or acidification, incapacity to support socio-economic development including food production, and human health problems (e.g., through lack of access to clean water and sanitation, and exposure to pesticides and other organic or inorganic contaminants). Thus, in many basins there is a challenge in trying to both overcome water management issues associated with historical use and management practices, and learn about what worked, has not worked so well and could work better in the future in a range of contexts (Squires, this volume).

In India, for example, in the Uttah Pradesh region where there are agricultural productivity issues linked to the management of groundwater table heights and their effects on water-logging and salinity/sodicity levels, it has been found that neither purely centralized systems of canal management nor decentralized systems of using groundwater pumped from tube-wells is likely to manage the issue alone. It has thus been proposed that a mixed system based on IWRM principles, with centralized management at the canal district level and decentralized management of tube-well irrigation, could help. This underlines the importance of farmers' roles in maintaining the groundwater at optimal levels for agricultural productivity and avoiding over-use of canal water that can lead to water-logging and reduced crop yields (Marr and Raut, this volume).

Likewise, in China water managers are attempting to find a balance between centralized infrastructure development and management and decentralized management systems to overcome their current challenges. The traditionally centralized management systems have thus been complemented with local Water User Associations (WUAs) in order to improve water management use efficiency and maintenance of infrastructure (see Aarnoudse et al. 2012, on the Shiyang, Li and Squires, this volume, on the Shule Basin and Xu et al., this volume, on the Heihe Basin). Although there is promising development, there are still a number of issues to resolve to enhance the WUAs to effectively fulfill their role (Xu et al., this volume). Finding the balance between central (national) and provincial development and management is another challenge in China. This is discussed in the case of the Yellow River in by Du et al. (this volume).

Learning from the history of transboundary disputes in the Tigris-Euphrates, Kibaroglu and Ahmetova (this volume) highlight some of the common impacts of water privatization from cases around the world, and what this could mean for Turkey and its consideration of a new privatization agenda for parts of the trans-boundary Tigris-Euphrates basin. Specifically, they caution that with privatization, water can be treated as a basic requirement and commercial good by private investors, which could lead to the public sector being open to large compensation claims if this water is not made available (see also Barlow 2001). As well as being economically costly, such a development could also enflame political tensions with other states in the Basin. Kibaroglu and Ahmetova (this volume) thus conclude that if such an agenda is to be pursued, Turkey needs to look very carefully at its national legal frameworks and the international frameworks to which it must adhere (which can override national laws), and be careful to develop any water policy responses in an integrated way with its foreign policy and security ministries. Such analyzes would also be worthwhile for other countries considering or currently rolling out privatization initiatives, as will be further discussed in Section 4.

3.2 Place-based implementations of IWRM: lessons for the future from current practice

River basins can vary greatly in their geography, including topology, climate, geology, hydrology and land-use patterns. These geographical aspects of place should drive the type of IWRM that is enacted as not one form of implementation and principles may function effectively in all areas. In particular, current practice shows that the kinds of intensification of agricultural and industrial systems that may function relatively well in temperate environments with good soils and constant high water availability are unlikely to be sustainable in arid environments where soil nutrient levels are poor and water availability is highly variable and on average low (even if it is economically and physically feasible to increase this water supply), as environmental degradation leading to the destruction of the underlying ecosystem functions can occur (see MEA 2005; Squires, this volume, for more on ecosystem service provision). Such issues are clearly visible, for example, in the arid inland (endoheic) river basins in China (see Li and Squires, this volume) and Australia (see Kingsford et al., this volume) where careful management needs to develop or take into account hydrological and ecological studies of the current environments in order to avoid over-exploitation of often scarce water and soil resources and irreversible damage to their ecosystems, as well as participatory IWRM work with local indigenous groups who can contribute local and traditional knowledge, and with farmers and other community members who can act as land and water stewards.

This knowledge of place is just as valuable when attempting to deal with an excess, rather than a scarcity, of water. As Wenger (this volume) points out from studies across the Netherlands, China, the USA and Australia, there is a need for authorities in different countries to consider ecosystem approaches to managing floods (e.g., leaving space for water and slowing down water velocities with vegetation and by recreating meanders in straight channels) and learning from current experiences of flood disasters to build back better and consider relocating people to places out of likely harm. Such approaches aim to enhance social-ecological resilience and are applicable in other areas of IWRM practice such as wetland management through environmental watering (Pittock and Finlayson 2013).

Current practice in IWRM implementation also shows that the focus and values associated with water management change over time, with movement between conservation and development (e.g., Loch et al., Kingsford et al., Squires, this volume) particularly common. However, as the environmental and social results of an important wave of economic development in the Aral Sea Basin in the last century showed (see Pearce 2004; Krutov et al., this volume), only some economic development decisions may be wise in the long-run while others, especially when large-scale, cannot be undone and may impart irreparable damage to the system over the long term, requiring significant new cooperation to search for more mutually beneficial management decisions and improve the situation for those in different parts of the basin (upstream/downstream) who experience varied impacts.

Looking at differentiated impacts across basins, in the Orange River basin in Southern Africa, for example, being the upstream water provider does not mean that Lesotho has less water issues than downstream countries. In fact, Lesotho appears to be more "water poor" (see Sullivan 2002) due to having less access to water and capacity to manage what they have, which in turn leads to poorer economic and social outcomes for its populations. Issues of serious inequity and how water is managed into the future have the potential to alter this structure for better or worse, although all future IWRM efforts in the basin should aim to have benefit-sharing and prosperity for all as its key objectives (Sullivan, this volume).

4 Acknowledging the Complexity of Water Governance Systems and the Need for Continuous Adaption of Brokering Mechanisms for Integration and Innovation

The upstream/downstream, public/private/community and conservation/development issues mentioned in the last section are only three examples of the complexity of water governance systems. Other integral issues for effective water governance include the need for cross-sectoral integration (Pradhan et al., this volume), such as the water/foreign policy/security issues pointed out by Kibaroglu and Ahmetova (this volume) associated with transboundary management (see also Hassenforder and Noury, this volume), and with sediment management (Squires, this volume), land management (Ffolliott and Brooks, this volume), energy development and management (Hussey et al. 2013; Krutov et al., this volume) and agricultural production (Marr and Raut, this volume), climate variability, change and management of extreme events (Daniell et al. 2011; Daniell 2013; Pradhan et al., this volume; Wenger, this volume) and population growth and urbanization of basins (see Squires, this volume, on low impact development, and Wenger, this volume, on flood risk). Such complexity has led to the multiplication of mechanisms used to manage these different issues under the auspices, or connected to, IWRM, as well as the need for specific people and organizations to act as brokers between communities focusing on these different facets linked to water management (see Section 4.2).

4.1 Multiplication of water governance mechanisms

Management of water and associated aspects of river basins is carried out through often complex multi-level governance systems, where a variety of social, legal, economic, technical and other mechanisms are employed to influence behaviors of actors in the system. A number of mechanisms, such as the social mechanisms of Water User Associations in China (e.g., Xu et al., Li and Squires, this volume), and legal instruments such as Basin Plans (e.g., Loch et al., this volume) or international investment and trade laws (Kibaroglu and Ahmetova, this volume) have already been discussed, but others, such as the growing number of economic instruments and market mechanisms being promoted for water governance, have not yet received much attention.

As an example, in a number of countries around the world, moves are under way to develop systems of water trading. Australia, and particularly the systems in the Murray-Darling Basin, are perhaps the furthest advanced (see Loch et al., this volume), although other countries, such as China (Xu et al., this volume) are developing reforms that would allow water right transfers to take place within their allocation systems (see also Du et al., this volume, for discussion on wet, normal and dry annual scenarios as a basis of water allocation in the Yellow River basin and moves toward markets with water right transfers). Challenges of transfers of water rights and trading in international free-market systems have already been alluded to (Kibaroglu and Ahmetova, this volume), although these are not the only kinds of economic instruments that can be employed.

Other important economic instruments include: insurance schemes, which are specifically used to enhance governance around extreme water related events (e.g., floods, droughts, cyclones, storms, bushfires) that can be developed as either part of public or private systems (see Handmer and Dovers 2007; Wenger, this volume); and taxes, levies and tariffs, which can be designed for a variety of objectives, including recovering infrastructure costs and enhancing government or private sector revenue for general or specific-purposes, such as environmental works and measures. Another increasingly common and perhaps less well known economic mechanism that can form part of a water governance system is eco-compensation schemes.

Eco-compensation schemes set up agreement and payment for ecosystem services between two or more land and water users (e.g., Li et al. 2011). The agreement may be, for example, over the quality of water to be provided to the downstream user. Agreement is reached between the parties on the water quality standard to be met and the payments to be made both in the event of meeting the standard and failing to meet the standard. In a pilot program in central China, a fund has been set up to protect the quality of water in the Xin'an River. Agreement has been reached that the total contribution to this fund is to be 500 million yuan ($82 million) a year. The river rises in Anhui Province and flows to Zhejiang Province. Anhui and Zhejiang provinces each contribute 100 million yuan a year to the program. The central government has agreed to contribute 300 million yuan each year. However, the payments made by the two provinces are dependent on the achievement of the water quality standard. The funds are being used by the upstream Anhui Province to compensate land and water users who are required to cease activities which pollute water in the Xin'an River. If the water reaches the quality standard by 2015, then downstream Zhejiang Province will pay Anhui 100 million yuan. If not, Anhui will pay Zhejiang 100 million yuan. Regardless of water quality, Anhui will receive 300 million yuan each year from the national government in compensation.

The pilot program is one of many which are being established to define ecological "red lines" across China (see also Xu et al., this volume). "The red line is to limit economic development of environmentally vulnerable regions, such as river sources."[1] These red lines establish ecological objectives from which eco-compensation schemes can be developed. One of the red lines for Anhui and Zhejiang is the water quality standard in the Xin'an River. Another red line has been established in Jiangsu province in eastern China, where the conservation area takes up more than 20 percent of provincial land. Approval for development requires accompanying agreement to set aside an equivalent conservation area, for which landholders are compensated.

Another early example of an eco-compensation scheme is the Salinity and Drainage Strategy in the Murray-Darling Basin, Australia. Here, the states agreed on a "red line" for river water salinity and the Commonwealth government agreed to contribute to a fund to construct groundwater interception works to reduce the flow of saline water entering the river. The Strategy was developed in response to a study that showed that irrigation development was leading to increased river salinity and affecting water supply to the city of Adelaide in South Australia (see Loch et al., this volume).

4.2 The need for brokers and boundary spanners

The complex nature of interlinked components with water and combination of governance mechanisms leads to increasing information needs to support effective water management. Although some advocate increasing information via a "scientific approach", especially around water supply scarcity and allocation issues (e.g., ADB 2010), this is typically insufficient and means that engaging with a wider range of stakeholders and knowledges is required for effective water governance and IWRM (e.g., Thomas 2004; Daniell 2012). Bridging different information sources and stakeholder groups to develop integrated responses to river basin management requires a range of specific people and content-management skills, in particular a capacity to coordinate others, and translate and synthesize information from one language (used by a scientific discipline or stakeholder group) to another; for example, from policy and legislation to what this means for engineers designing infrastructure and water operations and farmers' practices.

People or organizations acting in this role are typically termed brokers, boundary spanners, or bridging organizations, and are vital for

---

[1] Xia Guang in Global Times—Xinhua 15 Dec 2013, China weighs eco-compensation. http://www.globaltimes.cn/content/832237.shtml#.Uq8NwtJDuE4.

effective river basin management to occur, in particular for overcoming implementation issues by merging IWRM principles and practice with that of territorial development and ensuring the brokering arrangements are effective in facilitating multi-stakeholder engagement and decision-making processes (Maurel et al., this volume). This theme is picked up in a number of ways by other chapters in the book, including that of Xu et al. who write about the importance of "water brokers" for establishing trading between buyers and sellers in China, and Plant et al., this volume, who highlight the importance of "animateurs" or facilitators who broker knowledge between top-down and bottom-up management processes in the French Thau basin, or simply "champions" who bring people together and drive IWRM processes forward (Mitchell, this volume).

Riddell and Thuo (this volume) in their analyses of the Nile transboundary basin point to the importance of the Nile Basin Initiative in terms of its capacity to initiate work in synthesis of knowledge and brokering across the large and complex basin. In a similar way, Hassenforder and Noury (this volume) investigate the need for bridging organizations to support trans-boundary basin projects and to outline a range of characteristics that is necessary for them to have in order to be viable and adaptive to coping with rapid changes and uncertainty, including being creative, multi-scalar (including stakeholders from various levels/sectors), having individuals with strong leadership skills and a capacity to learn in an ongoing manner. They also highlight the importance of monitoring and evaluation in order to understand and learn what is occurring in river basins and their complex governance systems, concluding that "such monitoring should be integrated, adaptive, trans-disciplinary, multi-scale and participatory" (Hassenforder and Noury, this volume). Their work and views support the growing body of work about moving beyond linear forms of command-and-control management to adaptive management; views that are also strongly promoted by Mitchell (this volume) based on analyses in the Canadian watershed management context.

In a slightly more unusual way, Plant et al. (this volume) examine the importance of "socio-technical information and communication arrangements" as mediators for observing what is going on in a territory (e.g., monitoring) and triggering action and adaptive management as required, as well as sharing and cataloguing knowledge (e.g., developing meaning). Here, it is not just humans or groups of people (e.g., organizations, institutions) who can act as brokers, but also non-human actors (e.g., Callon et al. 1986; Latour 2005), such as models, that, by building relations to other actors, can support more participatory forms of IWRM (see also Barreteau et al. 2010).

5 Conclusions

One-size-fits-all integrated water resources management is not possible and would not do justice to the diversity of people and places around the world for which river basin management is an important issue. In each individual river basin, and sub-basin or catchment, however they end up being defined, it will be vital to take a long-term perspective of needed reforms, management action impacts and potential necessary behavioral changes. A focus on visions of the future and the values underlying these, as well as an appreciation of historical policies and practices, should underlie these long-term perspectives, as short term plans are unlikely to treat many of the underlying issues affecting river basins, which include environmental and societal pressures and changes, and power struggles over resources and their management. As highlighted succinctly by Mitchell (this volume): "Several elements provide legitimacy or credibility for IWRM: a legal foundation; political commitment; appropriate policies, administrative support, and human resources; sufficient funding; and suitable governance arrangements, including partnerships. The more of these present, the more probable will be successful implementation." Although IWRM has developed an impressive global following, as we have seen in this book, it can be interpreted in multiple ways that may seem legitimate and credible, yet still does not live up to the expectations that it will improve river basin management and the people and places reliant upon it (e.g., Mukhtarov and Cherp, this volume). Therefore, our collective challenge in the future will be to not just give preference to the more expert-based conceptions of how IWRM should work and the associated government responses, but to fully engage with people and place in river basins, including all responsible public, private and community stakeholders, to manage water more equitably and beneficially for all into the future.

6 References

Aarnoudse, E., B. Bluemling, P. Wester and W. Qu. 2012. How WUAs facilitate direct groundwater regulation. A case study of Minqin County, China. Water Policy Research Highlight 23: 1–5.

ADB. 2010. People's Republic of China: River Basin Water Resources Allocation and Management Policy, Project Number: 40687 September 2010, Report prepared by GHD Pty Ltd Australia for the Ministry of Water Resources, People's Republic of China, Asian Development Bank.

Allan, A.J. 1998. Watersheds and problemsheds: explaining the absence of armed conflicts in the Middle East. Middle East Review of International Affairs 2(1): 49–51.

Bache, I. and M. Flinders (eds.). 2005. Multi-Level Governance. Oxford University Press, Oxford, UK.

Barlow, M. 2001. Blue Gold: The Global Water Crisis and the Commodification of the World's Water Supply, http://www.ratical.org/co-globalize/BlueGold.pdf.

Barreteau, O., P.W.G. Bots and K.A. Daniell. 2010. A framework for clarifying "participation" in participatory research to prevent its rejection for the wrong reasons. Ecology and Society 15(2): 1, http://www.ecologyandsociety.org/vol15/iss2/art1/.

Berkes, F. 2009. Evolution of co-management: role of knowledge generation, bridging organizations and social learning. Journal of Environmental Management 90(5): 1692–1702.

Bourblanc, M. and D. Blanchon. 2014. The challenges of rescaling South African water resource management: Catchment Management Agencies and Interbasin Transfers, Journal of Hydrology (In Press).

Callon, M., J. Law and A. Rip. 1986. Mapping the Dynamics of Science and Technology: Sociology of Science in the Real World. Macmillan, London.

Claassen, M. 2013. Integrated water resource management in South Africa. International Journal of Water Governance 1: 323–338.

Daniell, K.A. 2011. Enhancing collaborative management in the Basin. pp. 413–438. *In*: D. Connell and R.Q. Grafton (eds.). Basin Futures: Water Reform in the Murray-Darling Basin. ANU E-Press, Canberra, http://epress.anu.edu.au/wp-content/uploads/2011/08/ch26.pdf.

Daniell, K.A. 2012. Co-engineering and Participatory Water Management: Organisational Challenges for Water Governance. Cambridge University Press, Cambridge, UK, 346 p.

Daniell, K.A. 2013. Practical responses to water and climate policy implementation challenges. Australian Journal of Water Resources 17(2): 111–125, http://dx.doi.org/10.7158/W13-WC01.2013.17.2.

Daniell, K.A., M.A. Manez Costa, N. Ferrand, A.B. Kingsborough, P. Coad and I.S. Ribarova. 2011. Aiding multi-level decision-making processes for climate change mitigation and adaptation. Regional Environmental Change 11(2): 243–258.

Du, Q., Z. Chao and S. Spooner. 2014. Sustainable management of water resources in the Yellow River Basin: mainissues and legal approaches. pp. 243–274. *In*: V.R. Squires, H.M. Milner and K.A. Daniell (eds.). River Basin Management in the Twenty-first Century: Understanding People and Place. CRC Press (this book).

Ffolliott, P. and K. Brooks. 2014. Integrated watershed management of water and other natural resources in River Basins of the United States. pp. 60–81. *In*: V.R. Squires, H.M. Milner and K.A. Daniell (eds.). River Basin Management in the Twenty-first Century: Understanding People and Place. CRC Press (this book).

GWP. 2000. Integrated water resources management, TAC Background Paper no. 4. Global Water Partnership, Stockholm. http://www.unep.org/civil_society/GCSF8/pdfs/IWRM_water_efficiency_eng.pdf.

Handmer, J. and S. Dovers. 2007. Handbook of Disaster & Emergency Policies & Institutions. Earthscan, London.

Hassenforder, E. and B. Noury. 2014. Managing transboundary river basins. pp. 129–161. *In*: V.R. Squires, H.M. Milner and K.A. Daniell (eds.). River Basin Management in the Twenty-first Century: Understanding People and Place. CRC Press (this book).

Hussey, K., N. Carter and W. Reinhardt. 2013. Energy sector transformation: implications for water governance. Australian Journal of Water Resources 17(2): 170–179, http://dx.doi.org/10.7158/W13-025.2013.17.2.

Kibaroglu, A. and R. Ahmetova. 2014. The privatization initiatives in the Turkish water sector: reflections on the transboundary water politics in the Euphrates-Tigris River Basin. pp. 162–188. *In*: V.R. Squires, H.M. Milner and K.A. Daniell (eds.). River Basin Management in the Twenty-first Century: Understanding People and Place. CRC Press (this book).

Kingsford, R., J. Costelloe and F. Sheldon. 2014. Lake Eyre Basin—challenges for managing the world's most variable river system. pp. 346–369. *In*: V.R. Squires, H.M. Milner and K.A. Daniell (eds.). River Basin Management in the Twenty-first Century: Understanding People and Place. CRC Press (this book).

Krutov, A., S. Rahimov and A. Kamolidinov. 2014. Republic of Tajikistan: its role in the management of water resources in the Aral Sea Basin. pp. 325–345. *In*: V.R. Squires, H.M. Milner and K.A. Daniell (eds.). River Basin Management in the Twenty-first Century: Understanding People and Place. CRC Press (this book).

Latour, B. 2005. Reassembling the Social: An Introduction to Actor–Network–Theory. Oxford University Press, Oxford, UK.

Li, J., M.W. Feldman, S. Lia and G.C. Daily. 2011. Rural household income and inequality under the Sloping Land Conversion Program in western China. Proceedings of the National Academy of Sciences, www.pnas.org/cgi/doi/10.1073/pnas.1101018108.

Li, Y. and V. Squires. 2014. Integrated water resources management of inland river basins in the Hexi Corridor, Gansu, China with special reference to the Shule River Basin. pp. 82–100. *In*: V.R. Squires, H.M. Milner and K.A. Daniell (eds.). River Basin Management in the Twenty-first Century: Understanding People and Place. CRC Press (this book).

Loch, A., S. Wheeler and D. Adamson. 2014. People versus place in Australia's Murray-Darling Basin: balancing economic, social ecosystem and community outcomes. pp. 275–303. *In*: V.R. Squires, H.M. Milner and K.A. Daniell (eds.). River Basin Management in the Twenty-first Century: Understanding People and Place. CRC Press (this book).

Loucks, D.P. 1998. Watershed planning: changing issues, processes and expectations, water resources update, Universities' Council on Water Resources, Spring Carbondale, IL 111: 38–45.

Lubell, M. and J. Edelenbos. 2013. Integrated water resources management: A comparative laboratory for water governance. International Journal of Water Governance 1: 177–196.

Marr, A. and A. Raut. 2014. The critical role of local farmers in IWRM in Ghaghra-Gomti Basin, India. pp. xx. *In*: V.R. Squires, H.M. Milner and K.A. Daniell (eds.). River Basin Management in the Twenty-first Century: Understanding People and Place. CRC Press (this book).

Maurel, P., R. Plant, O. Barreteau and Y. Bertacchini. 2014. Beyond IWRM: developing territorial intelligence at the local scales. pp. 22–41. *In*: V.R. Squires, H.M. Milner and K.A. Daniell (eds.). River Basin Management in the Twenty-first Century: Understanding People and Place. CRC Press (this book).

MDBA. 2013. Basin Plan Implementation, Murray-Darling Basin Authority, Australian Government, Canberra, http://www.mdba.gov.au/what-we-do/basin-plan/basin-plan-implementation.

MEA—Millennium Ecosystem Assessment. 2005. Ecosystems and Human Well-being: The Assessment Series. Island Press, Washington, DC.

Mitchell, B. 2014. Addressing implementation deficits related to IWRM in Canada. pp. 42–59. *In*: V.R. Squires, H.M. Milner and K.A. Daniell (eds.). River Basin Management in the Twenty-first Century: Understanding People and Place. CRC Press (this book).

Molle, F., P. Mollinga and P. Wester. 2009. Hydraulic bureaucracies and the hydraulic mission: flows of water, flows of power. Water Alternatives 2(3): 328–349.

Mollinga, P., R. Meinzen-Dick and D. Merrey. 2007. Politics, plurality and problemsheds: a strategic approach for reform of agricultural. Water Resources Management, Development Policy Review 25(6): 699–719.

Mukhtarov, F. and A. Cherp. 2014. The hegemony of integrated water resources management as a global water discourse. pp. 3–21. *In*: V.R. Squires, H.M. Milner and K.A. Daniell (eds.). River Basin Management in the Twenty-first Century: Understanding People and Place. CRC Press (this book).

Pahl-Wostl, C., T. Downing, P. Kabat, P. Magnuszewski, J. Meigh, M. Schlueter, J. Sendzimir and S. Werners. 2005. Transition to adaptive water management: the NeWater project, University of Osnabrück, Osnabrück, Germany.

Pearce, F. 2004. Keepers of the Spring: Reclaiming Our Water in an Age of Globalisation. Island Press, Washington DC.

Peters, B.G. 1998. Managing horizontal government: the politics of coordination. Public Administration 76(2): 295–311.
Pittock, J. and C.M. Finlayson. 2013. Climate change adaptation in the Murray-Darling Basin: reducing resilience of wetlands with engineering. Australian Journal of Water Resources 17(2): 161–169, http://dx.doi.org/10.7158/W13-021.2013.17.2.
Plant, R., P. Maurel, O. Barreteau and Y. Bertacchini. 2014. Development of territorial intelligence in the Thau, France. pp. 446–466. *In*: V.R. Squires, H.M. Milner and K.A. Daniell (eds.). River Basin Management in the Twenty-first Century: Understanding People and Place. CRC Press (this book).
Pradhan, N.S., S. Sijapati, U. Parajuli and B.S. Dongol. 2014. From policy to practice: alternatives for governance and institutions for river basin management in Nepal. pp. 397–417. *In*: V.R. Squires, H.M. Milner and K.A. Daniell (eds.). River Basin Management in the Twenty-first Century: Understanding People and Place. CRC Press (this book).
Squires, V. 2014. People and place as determinants of success or failure to meet management objectives in river basins. pp. 371–396. *In*: V.R. Squires, H.M. Milner and K.A. Daniell (eds.). River Basin Management in the Twenty-first Century: Understanding People and Place. CRC Press (this book).
Squires, V., K.A. Daniell and H. Milner. 2014. Preamble: an introduction to people and place in river basin management. pp. xxvii–xxxii. *In*: V.R. Squires, H.M. Milner and K.A. Daniell (eds.). River Basin Management in the Twenty-first Century: Understanding People and Place. CRC Press (this book).
Sullivan, C.A. 2002. Calculating a water poverty index. World Development 30: 1195–1210.
Sullivan, C.A. 2014. Challenges for management of the Orange/Senqu River Basin. pp. 189–215. *In*: V.R. Squires, H.M. Milner and K.A. Daniell (eds.). River Basin Management in the Twenty-first Century: Understanding People and Place. CRC Press (this book).
Swyngedouw, E. 2004. Social Power and the Urbanization of Water—Flows of Power. Oxford University Press, Oxford, 209 p.
Thomas, R.L. 2004. Management of freshwater systems: the interactive roles of science, politics and management, and the public. Lakes & Reservoirs: Research and Management 9(1): 65–73.
Thuo, S. and P. Riddell. 2014. Political economy versus comparative advantage in the Nile Basin: short term advantages or long term gain? pp. 315–340. *In*: V.R. Squires, H.M. Milner and K.A. Daniell (eds.). River Basin Management in the Twenty-first Century: Understanding People and Place. CRC Press (this book).
UNDP. 2006. Beyond scarcity: Power, poverty and the global water crisis, Human Development Report 2006. New York: United Nations Development Programme.
UNICEF/WHO. 2012. Progress on Drinking Water and Sanitation: 2012 update, WHO/UNICEF Joint Monitoring Programme for Water Supply and Sanitation, UNICEF and World Health Organisation, New York, http://www.unicef.org/media/files/JMPreport2012.pdf.
Verkerk, J. and A. van Buuren. 2013. Integrated water resources management in the Netherlands. Historical trends and current practices in the governance of integration. International Journal of Water Governance 1: 427–451.
Wenger, C. 2014. Sink or swim: alternative approaches to flood disaster reconstruction and mitigation. pp. 418–445. *In*: V.R. Squires, H.M. Milner and K.A. Daniell (eds.). River Basin Management in the Twenty-first Century: Understanding People and Place. CRC Press (this book).
Xu, H., L. Sui, L. Yuhong and D. Zhang. 2014. The role of water users' associations in IWRM in Heihe River Basin, China. pp. 304–324. *In*: V.R. Squires, H.M. Milner and K.A. Daniell (eds.). River Basin Management in the Twenty-first Century: Understanding People and Place. CRC Press (this book).

Glossary of Terms

Anastomosing channels, a stream consisting of a series of channels that wander, branch, and reconnect, creates a braided pattern, known as anastamosis; often near a river mouth or near terminal lakes

Animateurs, communicate to build a common ground and promote a basin public interest vis-à-vis more local claims

Chemin faisant, (in French) is "inevitably" or "inexorably", or "as a result".

Conveyancing, in law, the transfer of legal title of property from one person to another, or the granting of an encumbrance such as a mortgage or a lien

Endoheic, a closed drainage basin that retains water and allows no outflow to other external bodies of water, but converges instead into lakes or swamps, permanent or seasonal, that equilibrate through evaporation

Free diving, a form of underwater **diving** that relies on a diver's ability to hold his or her breath until resurfacing rather than on the use of an air tank

Gigaliter, a unit of volume equivalent to 10^9 litres (one billion [US billion] liters). Symbol: Gl

Gobi, a large barren expenses of gravel plains and rocky outcrops

Hermeneutics, the branch of philosophy that is concerned with the study and interpretation of human behaviour, structures of society, and how people function within these structures

Holon, is a component or unit that is simultaneously a whole and a part

Jal Nigam, in India, a body responsible for supply of drinking water for urban areas

Juridical of or relating to law, to the administration of justice

Kankar, in India, calcium carbonate nodules

Kharif, In India, Monsoon season

Maîtres d'ouvrages", i.e., local authorities in France such as the municipalities competent in water projects involving private land owners

mcm, million cubic meters

megaliter, Ml—is measure of water volume and is the same as 1000 m^3

Mentha, a plant that is source of peppermint oil

Mombio, woodlands in South Africa, arid or semi-arid area characterized by scrubland

Mu, a Chinese land measure 15 mu = 1 hectare (ha)

Muluki Ain, in Nepal, is a country code or law.

Neo-Gramscian, applying the ideas of Gramsci to current international politics. Gramsci, one of the most important Marxist thinkers in the 20th century, is known for his theory of cultural hegemony, which describes how states use cultural institutions to maintain power in capitalist societies. Neo-Gramscians argue that, because of globalization, a neoliberal transnational historic bloc exists or is coming into existence

Nyaya Panchayats, in India, is a group of people appointed to resolve disputes that may occur between villagers. The group has responsibility for an area which extends to include a few natural villages.

Oblast, an administrative division in Slavic countries, including some countries of the former Soviet Union corresponding to an autonomous province

Panchayats, in ***Uttar*** Pradesh, India the third tier of government

Persian wheel, is a mechanical water lifting device operated usually by draught animals like bullocks, buffaloes or camels. It is used to lift water from water sources, typically open wells, widely used in India

Piezometer, is a device used to measure static liquid pressure in a system by measuring the height to which a column of the liquid rises against gravity. Used in soil water determination.

qtl/ha, a common way to express grain yield in India and elsewhere

Quintal, qtl, equivalent to 100 kg

Rabi, In India, Winter season

RMB, The unit of currency in China. In 2013 $1 = 6.5 RMB

Semiosphere, a signifying system of signs from which the actor draws the informational resources for action

Toc, in French bathymetry, shallows which run parallel to the coastline

Zaid, In ***Uttar*** Pradesh, India Pre-monsoon season

Subject and Author Index

D

E

N

O

P

Q

R

S

About the Editors

Dr Victor R. Squires is an Australian who as young man studied Botany/Ecology. He has a PhD in Rangeland Science from Utah State University. He is a former Dean of the Faculty of Natural Resources at the University of Adelaide and was the Foundation Director of the National Key Center for Dryland Agriculture and Land Use Systems.

Since retirement from the University, Dr Squires was a Visiting Fellow at the East West Center in Hawaii and is currently an Adjunct Professor in the University of Arizona, Tucson and at Gansu Agricultural University, Lanzhou, China. He has been a consultant to World Bank, various UN agencies and the Asian Development Bank. He was GEF Advisor on a World Bank pastoral development project in NW China in 2006-2007 and consultant to ADB/GEF in Tajikistan. He is author of over 100 papers in refereed journals and numerous invited chapters and author/editor of 6 books. He was awarded the 2008 *International Award and Gold Medal for International Science and Technology Cooperation* by the Government of China and in 2011 was awarded the *Friendship Award* by the government of China, The Gold Medal is the highest honour for a foreigner.

Mr Hugh M. Milner is an independent water resource consultant with 40 years experience working in Australia and in developing countries on:

- Policy and institutional development for Sustainable management of water resources
- River basin planning and irrigation and water supply management
- Surface water hydrology and data management for water resource systems
- Water allocation issues and property rights
- Environmental impact assessment

Hugh worked for the Department of Land and Water Conservation (and its predecessors) in Australia for 26 years. His final position in the Department was Senior Hydrologist, Water Resource Management Directorate. In this position he was responsible for a number of policy and technical aspects of inter-state and inter-governmental water resource planning and management, particularly the Murray Darling Basin Commission (total area about 500,000 km^2, affecting four State Governments and the Commonwealth of Australia) and the Snowy Mountains Scheme, jointly operated to produce electricity and divert water into the Murray-Darling Basin. Since the late 1990s he has worked as an international consultant and team leader in water resources management and water policy in several countries in Asia, including China, Afghanistan, Lao PDR, India and Cambodia for the World Bank, Asian Development Bank, AusAID and European Union. He is the author of more than 10 papers and of 2 chapters for the Encyclopaedia of Life Support Sciences on watershed management. In 2008 he was awarded the *International Award for Cooperation in Forestry* by the Government of China.

Dr Katherine A. Daniell is a Fellow in the Australian National University's Centre for Policy Innovation and the HC Coombs Policy Forum. Her current research focuses on participatory processes for developing coordinated policy and local action for sustainable development. In this field, she has recently worked in France, Australia and Bulgaria on projects related to water governance, risk management, sustainable urban development and climate change adaptation. Katherine is the author of the Cambridge University Press book "Co-engineering and participatory water management: organisational challenges for water governance", published in the UNESCO International Hydrology Series. She was also a guest editor for a special feature in the journal Ecology and Society on "Implementing participatory water management: recent advances in theory, practice and evaluation", a compilation of papers that brings together policy, management and research lessons from the European Union's AquaStress and NeWater Integrated Projects.

In a research management role, Katherine coordinates the ANU's involvement in the PACE-NET Plus EU project on strengthening bi-regional dialogue on science, technology and innovation between Europe and the Pacific. She also works in the HC Coombs Policy Forum at the Crawford School of Public Policy on a range of Australian Public Service—ANU cooperation projects and is a member of the National Committee on Water Engineering (Engineers Australia). Katherine has received many awards and honours for her work including a General Sir John Monash Award and being elected as a Fellow of the Peter Cullen Water and Environment Trust.

Color Plate Section

Chapter 6

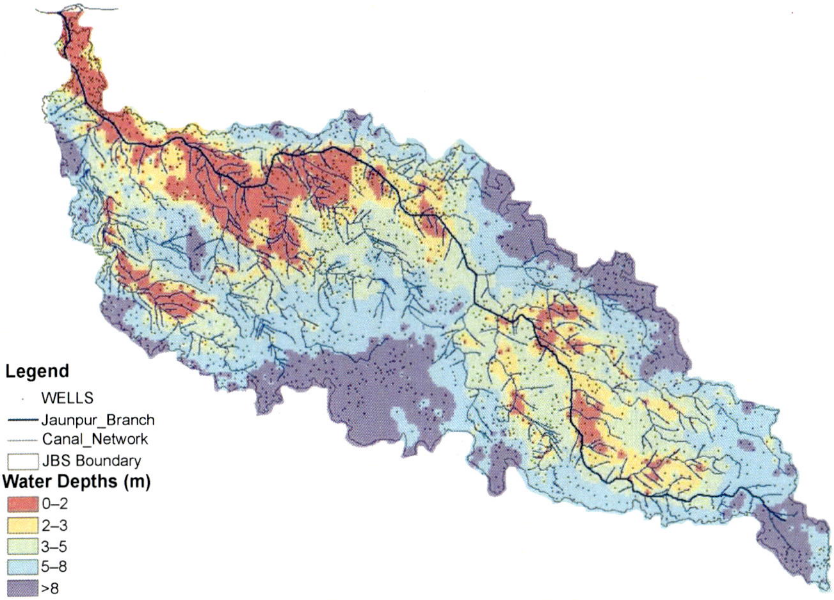

Figure 6.3. Distribution of canals and depth to the water table (Color chart).

Chapter 9

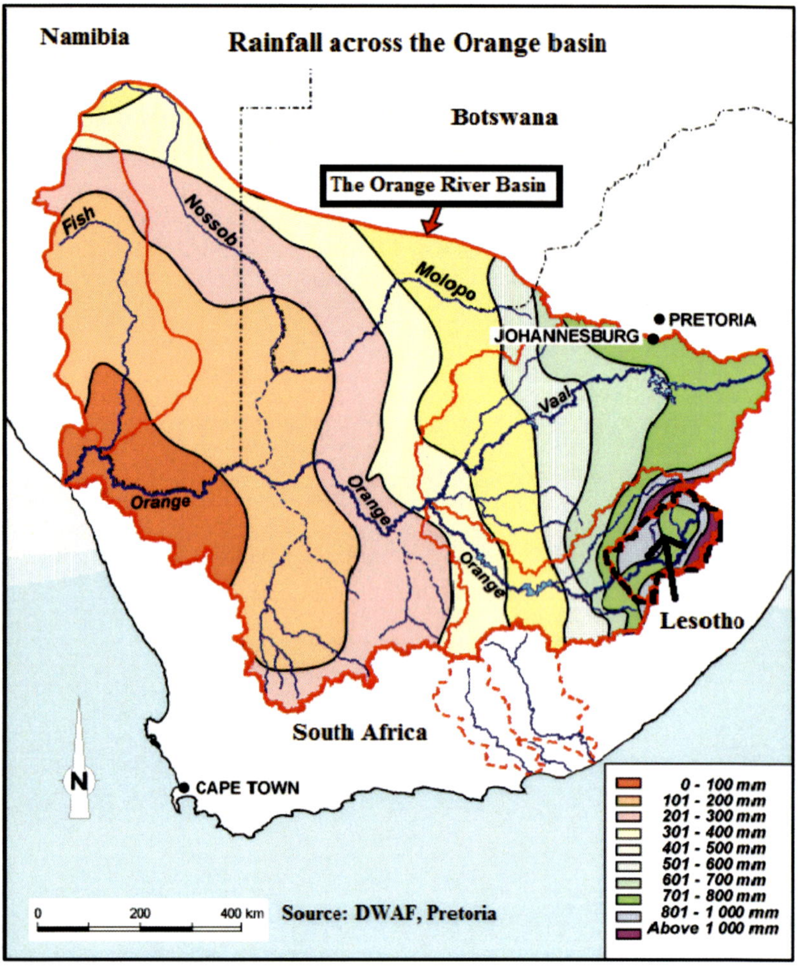

Figure 9.1. The Orange basin and its rainfall variability.

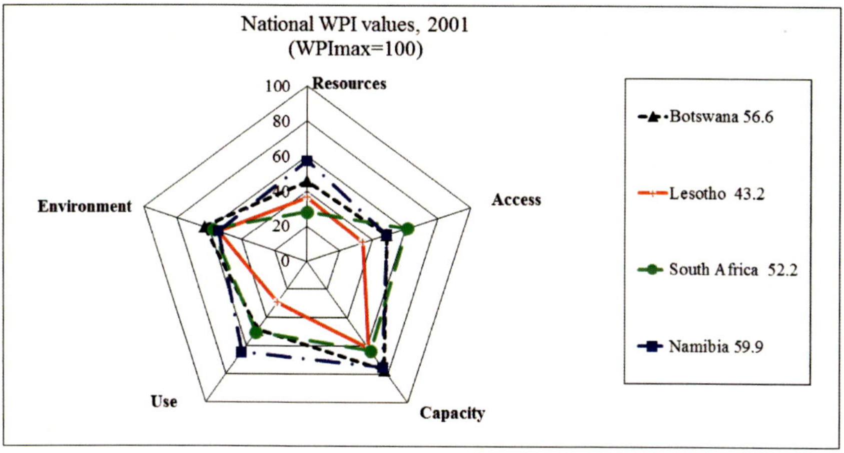

Figure 9.2. Comparing *Water Poverty* in the riparian states of the Orange basin.

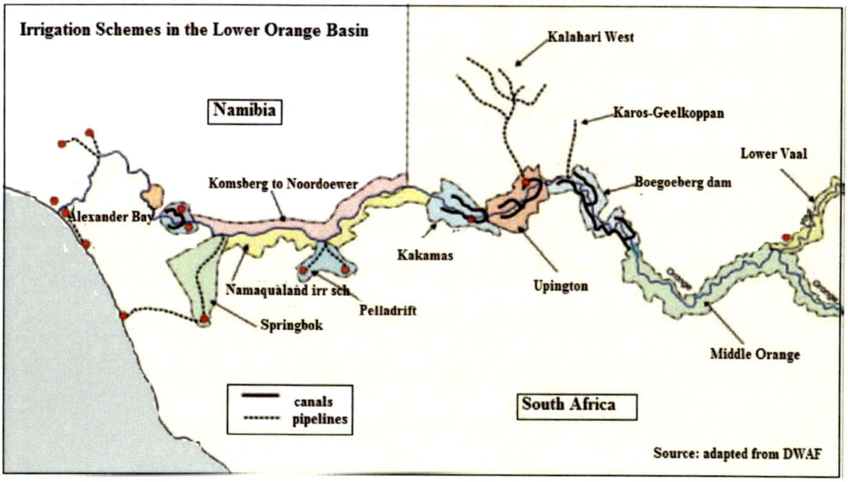

Figure 9.3. Irrigated agriculture in the South African part of the Orange basin.
Source: DWAF

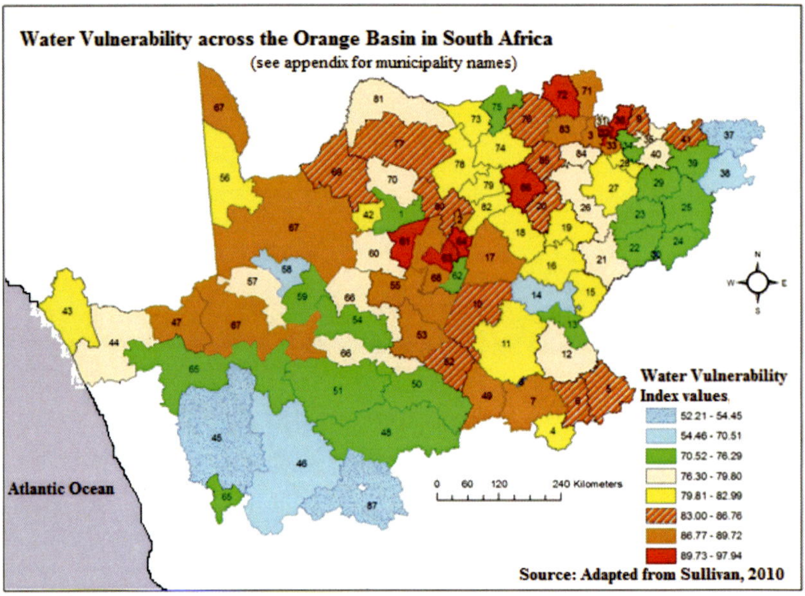

Figure 9.5. Water Vulnerability across the South African part of the Orange basin.
Source: Sullivan 2010

Chapter 13

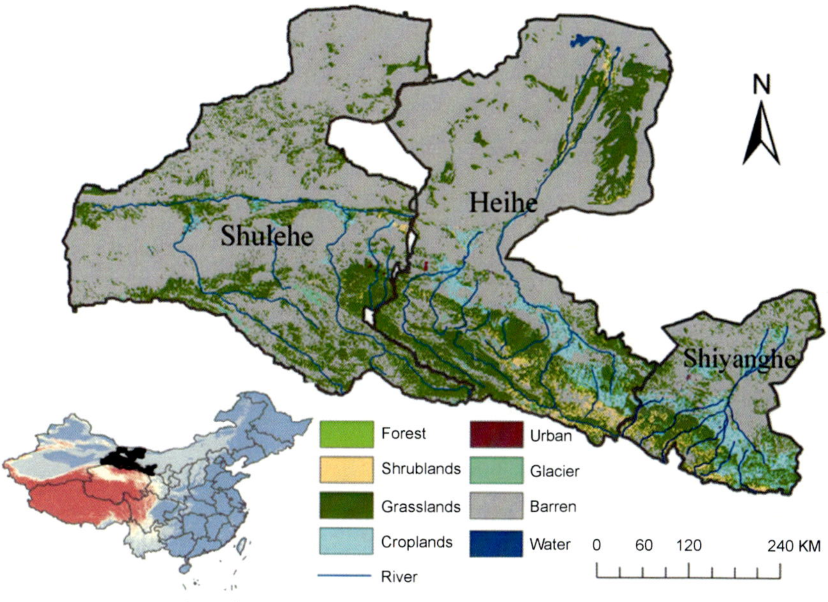

Figure 13.1. Map of three inland river basins of Hexi Corridor in Gansu Province.

Chapter 14

Figure 14.1. Major River Basins in The Republic of Tajikistan.

Chapter 15

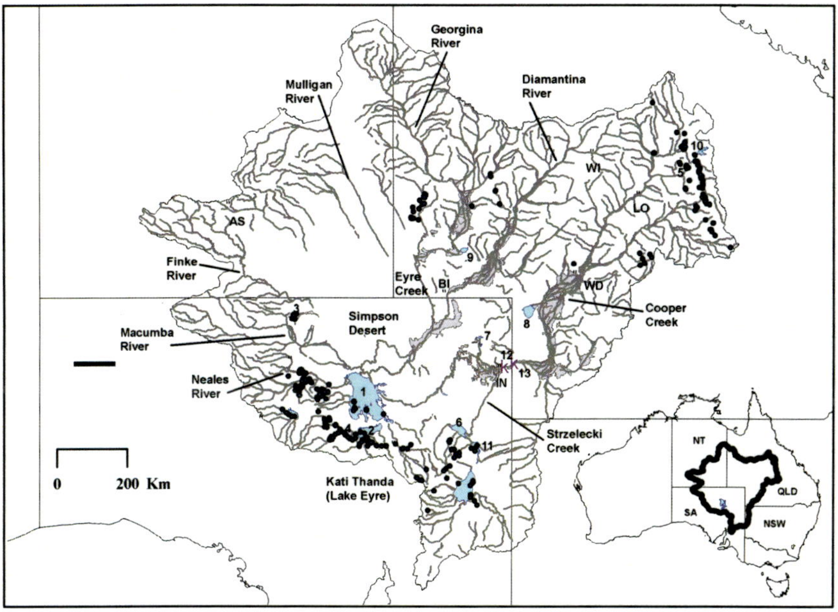

Figure 15.1. Location of the Lake Eyre Basin in central Australia (inset), including the three states (NSW-New South Wales, QLD-Queensland, SA-South Australia) and the Northern Territory (NT), showing its major rivers and wetlands in arid Australia. Numbers mark locations of the gauges (Cullyamurra (12), Nappa Merie (13)) used to show flow regime of Cooper Creek and major wetlands: Lake Eyre North-Kata Thanda (1); Lake Eyre South-Kata Thanda (2); Dalhousie Springs (3), Wabma Kadarbu springs (4); Aramac Springs (5); Lake Blanche (6), Coongie Lakes (7), Lake Yamma Yamma (8); Lake Machattie (9); Lake Galilee (10); Lake Callabonna (11). Major towns are shown (Alice Springs (AS), Winton (WI), Longreach (LO), Windorah (WD), Innamincka (IN) and Birdsville (BI)).

Chapter 16

Figure 16.2. In this aerial view of the Murray river near Renmark, South Australia (34.17S and 140.76E) we see the multiple land uses that the river creates. Urban, rural and conservation areas (riverine forest and wetlands) are all represented [Photo Department of Environment, Water and Natural Resources, South Australia, used with permission].

Chapter 17

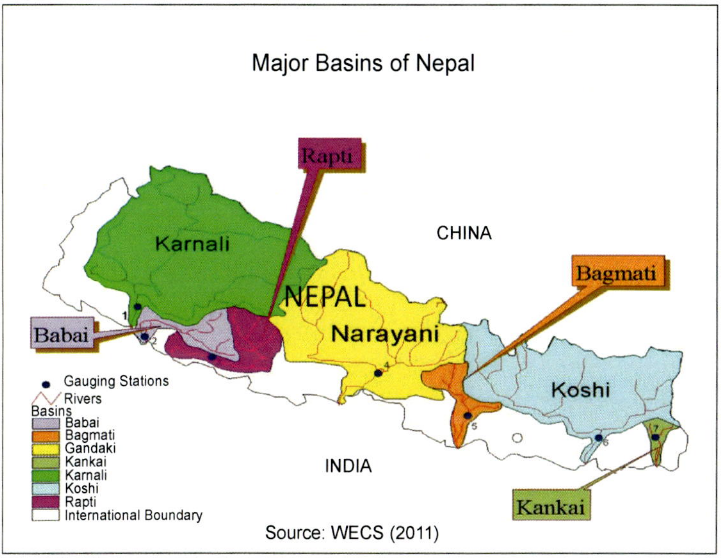

Figure 17.1. River System of Nepal.

Chapter 19

Figure 19.8. A STICA to set up an inventory of zostera seagrass of the Thau lagoon based on local knowledge.